Ultrasonic Vibration-Enhanced Nanolubricants Grinding Fiber-reinforced Composites:

From Mechanisms to Machinability

Teng Gao
Qingdao University of Technology, China

Changhe Li
https://orcid.org/0000-0003-4409-956X
Qingdao University of Technology, China

Xuesong Chang
Hisense Refrigerator Co., Ltd, China

IGI Global
Scientific Publishing
Publishing Tomorrow's Research Today

Published in the United States of America by
IGI Global Scientific Publishing
701 East Chocolate Avenue
Hershey, PA, 17033, USA
Tel: 717-533-8845 | Fax: 717-533-7115
Website: https://www.igi-global.com E-mail: cust@igi-global.com

Library of Congress Cataloging-in-Publication Data

Names: Gao, Teng, 1993- author
Title: Ultrasonic vibration-enhanced nanolubricants grinding fiber-reinforced composites : from mechanisms to machinability / authored by: Teng Gao.
Description: Hershey PA : IGI Global Scientific Publishing, [2027] | Includes bibliographical references. | Summary: "This book proposes a novel method of ultrasonic vibration-assisted nanofluid minimum quantity lubrication grinding for heterogeneous and anisotropic carbon fiber composites"-- Provided by publisher.
Identifiers: LCCN 2026005286 (print) | LCCN 2026005287 (ebook) | ISBN 9798337390130 hardcover | ISBN 9798337390147 paperback | ISBN 9798337390154 ebook
Subjects: LCSH: Lubrication and lubricants | Nanofluids | Fibrous composites
Classification: LCC TJ1077.5.N36 G36 2026 (print) | LCC TJ1077.5.N36 (ebook)
LC record available at https://lccn.loc.gov/2026005286
LC ebook record available at https://lccn.loc.gov/2026005287

British Cataloguing in Publication Data
A Cataloguing in Publication record for this book is available from the British Library.

Table of Contents

Detailed Table of Contents

FRCs have emerged as preferred materials for high specific strength and stiffness in advanced equipment. Precision grinding is an essential post-forming processing method for ensuring assembly accuracy of FRCs. However, it's considered difficult to machine due to their laminated structure, anisotropy, and heterogeneity. This chapter analyses the grinding damage mechanisms, including material removal behavior, thermomechanical damage responses and surface integrity. Subsequently, the geometric reconstruction of grinding wheels and material modeling are investigated. A comprehensive review is conducted on micro-scale analytical modeling methods for grinding forces under various grinding configurations, accompanied by comparative error analysis and identification of error sources. The influence of grinding parameters on grinding forces is systematically summarized. Furthermore, advanced grinding strategies are critically reviewed, including UVAG and MQL. Finally, the research significance of high-performance machining of FRCs and the logical framework of this monograph are presented.

This chapter analyzes the geometric kinematics of multiangle 2D UVAG grains and the workpiece, establishing a model for their relative motion trajectory. The cutting characteristics and the surface generation mechanism are revealed, including multiple buffing of tangential ultrasonic vibration,furrow widening of axial ultrasonic vibration,repeated cutting of single grains,differential cutting of follow-up grains. Excellent gain from UVAG can be obtained by adjusting grinding and ultrasonic vibration parameters. Furthermore, the permeation mechanism at the grinding interface and the hydrodynamics lubrication laws in ultrasonic-assisted nanofluid MQL grinding are investigated. A model for the lubricant film thickness was established. By formulating the hydrodynamic pressure equation, the variations in hydrodynamic pressure, lubricant flow rate, and flow velocity were clarified. The introduction of the ultrasonic vibration velocity increases the lubricant flow velocity and flow rate.

This chapter first investigates the dispersion performance of different surfactants, comparing the influence of their molecular structures and physicochemical properties on dispersion mechanism. The dispersion mechanisms of various surfactants for CNTs nanofluids were studied from the perspectives of space barriers, reversed micellization, hydrocarbon chain, london force, and synergism. Subsequently, the spreading and diffusion mechanisms of nanolubricant droplets on different FOA CFRP grinding surfaces and the influencing patterns on droplet contact angle were investigated. The mechanical behavior of the droplet spreading contact line and the anisotropic spreading mechanism are revealed. Finally, the influence of ultrasonic vibration on the spreading and diffusion was analyzed, and the mapping relationship between surface roughness and spreading rate was explored. This chapter provides a theoretical foundation and data support for the subsequent in-depth study of the tribological behavior and material removal mechanics in MQL UVAG of CFRP.

This chapter analyzes the tribological characteristics of transverse and longitudinal grinding of CFRP surfaces. Firstly, a contrast analysis on the coefficient of friction (COF) of different CNT nanofluids and the wearing loss of discs is performed to compare the dispersion mechanism. Secondly, the tribological characteristics in transverse grinding under different lubrications are analyzed, focusing on the COF and wheel blockage. Finally, the tribological behavior of the nanolubricant on longitudinal grinding surfaces of different FOA is studied, by evaluating the coefficient of friction, wear scar profiles, and wear volume. Compared to dry grinding, the COF of nanolubricant under four FOAs is reduced by 17.09%, 34.87%, 19.59%, and 27.13%, respectively. Notably, the lowest COF is observed at an FOA of 90°. Similarly, the anti-wear performance at 90° FOA is enhanced by 48.29% compared to the dry condition. This chapter provides a theoretical foundation and essential data to support further investigation into the material removal mechanics in MQL UVAG of CFRP.

Chapter 5

This chapter explores the CFRP grinding mechanism of single-grain of CNT nanolubricant MQL. A mechanical model is established based on the unique geometric interactions between the grain and fibers, which arise from random fiber arrangements and the grain topography under varying undeformed chip thicknesses. Firstly, four distinct contact and grinding models are developed, corresponding to different interaction states between the grain and fiber throughout the grinding arc length. Subsequently, the influence of different grinding parameters on the geometry of the cutting zone is elucidated. Then, a mechanical model for single-grain grinding is established, which incorporates the homogeneous randomness of fiber distribution across the entire grinding arc length. Finally, the model is numerically simulated and the trend of force along the entire grinding arc length is obtained. Experimental verifications demonstrate the approach for predicting the grinding force have acceptable accuracy and can successfully capture the mechanics of CFRPs.

Chapter 6

This chapter investigates the grinding performance of CFRP composites using NMQL, with a focus on the characterization and analysis of grinding surface integrity and machining-induced damage. Initially, a comparative analysis of surface roughness parameters was conducted. Subsequently, the fractal dimension was employed to quantitatively characterize the overall fractal features across different dimensions. Furthermore, multifractal spectrum with its width and difference was utilized to analyze the local topographic characteristics. Then, SEM was used for qualitative analysis of the surface micro-morphology, detailed machining defects, and damage under different conditions. Moreover, the energy distribution ratio of wavelet decomposition were applied to examine the high-frequency texture variations in both surface roughness signals and SEM images across different directions, enabling a quantitative characterization of grinding defects. Finally, the surface of grinding wheel under different conditions is characterized, and the diffusion state of lubricating oil film is analyzed.

Chapter 7

In the context of 2D multi-angle ultrasonic assisted nanofluid MQL grinding of CFRP, fundamental issues such as the material removal mechanics for different FOA remain unaddressed. This chapter elucidates the material removal mechanism in longitudinal grinding of CFRP with varying FOAs. It analyzes the fracture mechanics of fiber during negative-rake-angle cutting by abrasive grains, specifically investigating the buckling failure and elliptical contact mechanics of 0° FOA. The interlayer shear failure mechanism and the bending-deformation-fracture removal process of 45° FOA are clarified. Furthermore, the chapter reveals the bending deformation and fracture removal mechanics of 90° FOA, along with the matrix cracking caused by longitudinal compression and associated fiber shear fracture. The buckling failure and bulk removal process of fiber bundles in 135° FOA are also examined. The revelation and analysis of these mechanical behaviors establish boundary conditions and provide a theoretical foundation for developing a grinding force model applicable to different FOAs.

This chapter investigates the material removal mechanism of CFRP with different FOAs and establishes a force model of multiangle 2D ultrasonic vibration nanolubricant MQL longitudinal grinding. Firstly, a homogeneous random distribution model for fibers is developed, wherein the fiber diameter follows a normal distribution, to predict the mechanical properties of the interfacial phase. Then, a model of the diamond grinding wheel is developed, accounting for the normal distribution of both the protrusion height and the size of grains. Furthermore, based on the geometric kinematics of the grain trajectory and intermittent cutting state, the influence of 2D ultrasonic vibration angles on the dynamic stiffness of the grinding system was analyzed. Finally, grinding force models of entire domain of FOAs under different ultrasonic vibration modes and lubrications were interconnected. Experimental assessments demonstrated that the grinding force model has acceptable estimation error and effectively captures the mechanics of CFRPs with different FOAs.

In this chapter, Firstly, The instantaneous dynamic generation, conduction, accumulation of grinding heat and its thermal effects were revealed. Furthermore, a 3D microscale three phase finite element numerical model was established to determine the removal and damage mechanisms of different FOAs using single grain grinding. Finally, the grinding experiments were conducted under different ultrasonic and lubrication conditions. The results showed the damage formation and evolution mechanisms with different FOAs are various which complement the debris generation mechanisms. The damage is mainly distributed along the grinding direction, with the lowest damage at 0° FOA, and the highest at 45°. The

2D ultrasonic assisted nanolubricant MQL with 90° coupled vibration angle (θU) has an optimal damage reduction effect with a maximum of 27.56% compared to conventional grinding. However, 45° θU only showed excellent damage suppression on 135° FOA of 40.38%. This paper provides theoretical support for the damage suppression strategy of CFRP grinding.

The aeronautical industry constantly strives for efficient technologies to facilitate high surface integrity in CFRP structural components. This chapter conducts a multi-dimensional systematic evaluation of grinding surface integrity, ranging from global characteristics to local features in. Firstly, line and areal roughness are employed to characterize surface topography, functional performance, fluctuation intensity, and height distribution, with peak heights and valley depths specifically reflecting fiber pull-out damage. Secondly, 2D and 3D fractal dimensions, derived from line roughness profiles and surface microtopography, are used to characterize the overall morphological attributes. Subsequently, multifractal spectra including spectral width and difference, are applied to quantitatively analyze local morphological variations. Finally, SEM and laser confocal microscopy are utilized to observe 2D and 3D surface topography, enabling qualitative analysis of grinding-induced damage such as fiber (bundle) pull-out, fiber fragmentation, resin smearing, and fiber-matrix debonding.

This chapter summarizes the principal conclusions and innovations presented in this monograph. Primarily, it provides a focused analysis of the current research hotspots in the machining of FRCs and offers an outlook on future developmental trends. Research on the machining of FRCs, particularly CFRPs, is evolving toward a system characterized by clarified mechanisms, low-damage processes, intelligent control, and integrated systems. Key research hotspots reflect this evolution. First, understanding multi-scale material removal and damage formation is of paramount importance. Second, advanced tooling technologies and low-damage assisted processes are essential for improving machining quality and efficiency. Furthermore, challenging machining technologies for emerging materials and structures present new frontiers. Future breakthroughs will necessitate deep interdisciplinary integration to establish a complete, predictable, controllable, and optimizable precision machining technology system that links material microstructure to component service performance.

Preface

INTRODUCTION

Fiber-reinforced composites (FRC) has become the preferred materials for aerospace equipment due to its high specific strength and stiffness. Although it can be manufactured by near net forming, secondary processing is required for assembly connection surfaces to ensure accuracy and assembly requirements. Precision grinding is an irreplaceable machining method ensures the integrity and machining accuracy of the mating positioning surface. Carbon fiber-reinforced polymer (CFRP) composites represent the performance benchmark and technological forefront within the FRC system, boasting the broadest range of applications. They define the upper limit of mechanical performance for advanced composites and, through their highly mature industrial chain and cutting-edge applications, continuously drive technological innovation and industrial advancement across the entire field. However, its hygroscopic expansion characteristics limit the application of flood lubrication. Dry grinding can easily lead to damage affecting the load-bearing capacity and service safety. Minimum quantity lubrication (MQL) is an indispensable cooling lubrication can replace dry grinding. However, the gas barrier layer makes it difficult for lubricant to enter the grinding wedge area, causing nanolubricant cannot exert its anti-friction and heat transfer effects.

Drawing on the intermittent grinding and pumping effects of ultrasonic vibration is expected to solve the aforementioned technical bottlenecks. However, as a special composite material, CFRP exhibits significant differences between ultrasonic vibration and nanolubricant MQL coupled grinding and traditional grinding. Therefore, mechanics of ultrasonic nanolubricant grinding fiber composites study was carried out to address scientific issues. This book proposes a novel method of ultrasonic vibration-assisted nanofluid minimum quantity lubrication grinding for heterogeneous and anisotropic carbon fiber composites. It aims to: Reveal the mechanisms of ultrasonically enhanced infiltration of nano-lubricants into the micro-channel interfaces between wheel and workpiece. Elucidate the cross-scale material removal mechanics of CFRPs. Address the technical bottleneck of severe thermomechanical coupling damage in traditional MQL grinding of CFRPs. Achieve multi-parameter cooperative regulation and suppression of grinding damage in CFRPs. Provide technical support for precision manufacturing in high-end aerospace equipment.

ORGANIZATION OF THE BOOK

This book is organized into 11 chapters, with brief description of each chapter as follows:

Chapter 1: Fiber-reinforced composites (FRCs) have emerged as preferred materials for weight reduction and performance enhancement in high-end equipment such as aerospace systems, owing to their high specific strength, high specific stiffness, and capability for integrated design and manufacturing of materials, structures, and properties. Precision grinding is an essential post-forming processing method for ensuring assembly accuracy of FRCs. However, these materials are considered difficult to machine due to their laminated structure, anisotropy, and heterogeneity. This paper begins by analyzing the mechanisms underlying grinding-induced damage in FRCs and discusses fundamental grinding mechanisms, including material removal behavior, thermomechanical responses, surface integrity, and damage modes. Subsequently, the geometric reconstruction of grinding wheels and material modeling of FRCs are examined. A comprehensive review is conducted on micro-scale analytical modeling methods for grinding forces under various grinding configurations, accompanied by comparative error analysis of grinding force models and identification of error sources. The influence of grinding parameters and FOAs on grinding forces is systematically summarized. Furthermore, advanced grinding strategies, including ultrasonic vibration-assisted grinding and minimum quantity lubrication (MQL), are reviewed, with their technical advantages critically analyzed. Finally, the research significance of high-performance machining of FRCs and the logical framework of this monograph are presented.

Chapter 2: This chapter analyzes the geometric kinematics of multi-angle 2D UVAG abrasive grains and the workpiece, establishing a model for their relative motion trajectory, which reveals the cutting characteristics of the grains and the surface generation mechanism. It further investigates the influence of ultrasonic vibration on the permeation mechanism and hydrodynamics lubrication. To achieve superior gain effects of ultrasonic vibration, the grinding and ultrasonic vibration parameters were adjusted such that the ultrasonic vibration period is shorter than the time required for a grain to complete the cutting process within the grinding zone. This enables the abrasive grain to achieve a repeated cutting action. The repeated cutting action can be enhanced by adjusting the depth of cut and the grinding wheel speed, thereby improving the ground surface quality. Under the premise of ensuring grinding efficiency and accuracy requirements, the micro-cutting effect can be improved by appropriately increasing the ultrasonic frequency, decreasing the grinding wheel speed, and increasing the depth of cut, consequently enhancing overall grinding performance. Furthermore, this chapter elucidates the infiltration mechanism at the grinding interface and the hydrodynamics lubrication laws in ultrasonic-assisted nanofluid MQL grinding. A model for the lubricant film thickness was established. By formulating the hydrodynamic pressure equation, the variations in hydrodynamic pressure, lubricant flow rate, and flow velocity were clarified. The introduction of the ultrasonic vibration velocity increases the lubricant flow velocity and flow rate.

Chapter 3: This chapter first investigates the dispersion performance of different types of surfactants, comparing the influence of their molecular structures and physicochemical properties on the dispersion mechanism. The dispersion mechanisms of various surfactants for CNTs nanofluids were studied from the perspectives of steric hindrance, reverse micelle effect, hydrocarbon chain lipophilicity, dispersion forces, and synergistic effects. A dispersant optimization experiment was conducted, evaluating the dispersion stability and viscosity of CNTs nanofluids with different added surfactants. Subsequently, the spreading and diffusion mechanisms of nano-lubricant droplets on CFRP grinding surfaces with different FOA and the influencing patterns on droplet contact angle were investigated. This revealed the mechanical behavior of the droplet spreading contact line and analyzed the anisotropic spreading mechanism and capillary action on the 0° FOA CFRP ground surface. Finally, the influence of ultrasonic vibration on the spreading and diffusion of nano-lubricant droplets was analyzed, and the mapping relationship between surface roughness and spreading rate was explored. The application of 2D ultrasonic vibration reduced the stable aspect ratio of the nano-lubricant droplets. Compared to the non-ultrasonic condition, the droplet aspect ratio decreased by 28.34% on the FOA=0° CFRP surface and by 18.49% on the FOA=45° CFRP surface with coupled vibration angle $\theta_U = 90°$. The application of ultrasonic vibration accelerated droplet spreading, allowing the droplets to reach a stable state more rapidly. Furthermore, the 2D ultrasonic vibration at $\theta_U = 45°$ resulted in a slightly higher spreading rate compared to that at $\theta_U = 90°$. This chapter provides a theoretical foundation and data support for the subsequent in-depth study of the tribological behavior and material removal mechanics in ultrasonic-assisted nanofluid MQL grinding of CFRP.

Chapter 4: This chapter compares and analyzes the frictional characteristics and laws of transverse and longitudinal grinding of CFRP surfaces. Firstly, a contrast analysis on the friction coefficient of the lubrication of different CNT nanofluids and the wearing loss of discs was performed to compare and verify the dispersion mechanism. Secondly, this section presents a comparative analysis of the tribological characteristics of CFRP in transverse grinding under different lubrication conditions, focusing on the coefficient of friction (COF) and wheel surface loading. Finally, a comparative study was conducted on the tribological behavior of the nano-lubricant on longitudinal grinding surfaces of CFRP with different FOA, by evaluating the coefficient of friction, wear scar profiles, and wear volume. Compared to dry conditions, the nano-lubricant significantly reduced the coefficient of friction. The COF for the four fiber orientations was reduced by 17.09%, 34.87%, 19.59%, and 27.13%, respectively, when compared to dry grinding. Notably, the lowest COF was observed at an FOA of 90°. Similarly, the nano-lubricant substantially diminished the wear scar area and wear volume, with the most significant reduction of 32.05% also occurring at the 90° FOA. The anti-wear performance at this orientation was enhanced by 48.29% compared to the dry condition. This chapter provides a theoretical foundation and essential data to support further investigation into the material removal mechanics in ultrasonic-assisted nanofluid MQL grinding of CFRP.

Chapter 5: This chapter explores the grinding mechanism of single-grain engagement with CFRP involving CNT nanolubricant MQL. A mechanical model is established based on the unique geometric interactions between the abrasive grain and carbon fibers, which arise from random fiber arrangements and the grain's grinding -edge topography under varying undeformed chip thicknesses. Firstly, the material removal mechanism in CFRP during meso-scale single-grain grinding is revealed. Four distinct contact and grinding models are developed, corresponding to different interaction states between the grain and carbon fiber throughout the grinding arc length: 1) the Hertzian contact model between the grain's blunt conical tip and the carbon fiber, 2) the extrusion model of the grain against the carbon fiber end-face, 3) the elliptical localized contact model between the grain and carbon fiber, and 4) the single-fiber grinding model. Subsequently, the influence of different processing parameters on the geometry of the cutting zone is elucidated, specifically investigating how the geometric characteristics of the grinding zone evolve with variations in cutting parameters. The different contact states and material removal behaviors between the grain and carbon fiber at various stages along the grinding arc are revealed. Finally, a mechanical model for single-grain grinding of carbon fiber composites is established, which incorporates the homogeneous randomness of carbon fiber distribution across the entire grinding arc length. Finally, the model is numerically simulated and the trend of force along the entire grinding arc length is obtained. Experimental verifications demonstrate the approach for predicting the grinding force have acceptable accuracy and can successfully capture the mechanics of CFRPs.

Chapter 6: This chapter investigates the grinding performance of CFRP composites under NMQL conditions, with a focus on the characterization and analysis of grinding surface integrity and machining-induced damage. Initially, a comparative analysis of surface roughness parameters, including *R*a, *R*z, and *RS*m, was conducted. Subsequently, the fractal dimension was employed to quantitatively characterize the overall fractal features of the machined surfaces across different dimensions (2D and 3D). Furthermore, multifractal spectrum, along with its width and difference, was utilized to analyze the local topographic characteristics of the CFRP ground surfaces. Thereafter, SEM was used for qualitative analysis of the surface micromorphology, detailed machining defects, and damage under different working conditions. Moreover, wavelet analysis and the energy distribution ratio of wavelet decomposition were applied to examine the high-frequency texture variations in both surface roughness signals and SEM images across different directions, enabling a quantitative characterization of grinding defects. Finally, microscope imaging was used to capture the surface condition of grinding wheels after processing CFRP under different conditions. This allowed for the characterization and analysis of wheel loading and lubricant film spreading states across the various lubrication scenarios.

Chapter 7: In the context of 2D multi-angle ultrasonic-assisted nanofluid MQL grinding of CFRP, fundamental issues such as the material removal mechanics for different FOA remain unaddressed. This chapter elucidates the material removal mechanism in longitudinal grinding of CFRP with varying fiber orientations. It analyzes the fracture mechanics of fiber during negative-rake-angle cutting by abrasive grains, specifically investigating the buckling failure and elliptical contact mechanics of 0° FOA. The interlayer shear failure mechanism and the bending-deformation-fracture removal process of 45° FOA are clarified. Furthermore, the chapter reveals the bending deformation and fracture removal mechanics of 90° FOA, along with the matrix cracking caused by longitudinal compression and associated fiber shear fracture. The buckling failure and bulk removal process of fiber bundles in 135° FOA are also examined. The revelation and analysis of these mechanical behaviors establish boundary conditions and provide a theoretical foundation for developing a grinding force model applicable to different fiber orientations.

Chapter 8: This chapter investigates the material removal mechanism of CFRP with different fiber orientations and pioneers the establishment of a novel model for predicting grinding forces in surface grinding under multi-angle, two-dimensional ultrasonic-assisted nanofluid MQL conditions. Firstly, a homogeneous random distribution model for carbon fibers in CFRP was developed, wherein the fiber diameter follows a normal distribution, to predict the mechanical properties of the interfacial phase. Subsequently, a model of the diamond grinding wheel was established, accounting for the normal distribution of both the protrusion height and the grain size of the abrasive particles. Furthermore, based on the geometric kinematics of the abrasive grain trajectory in multi-angle 2D UVAG and the intermittent cutting state induced by different 2D ultrasonic vibration synthesis angles, the influence of 2D ultrasonic vibration on the dynamic stiffness of the grinding system was analyzed. Building upon this analysis, the mechanical behavior of material removal during negative-rake-angle cutting of CFRP with various fiber orientations was revealed. An interpolation method was employed to develop a grinding force model applicable across the entire domain of fiber orientations under different ultrasonic vibration modes and lubrication conditions. An experimental platform was designed and set up for validation. The maximum error between the predicted and experimental values was 12.82% for the normal grinding force and 16.97% for the tangential grinding force.

Chapter 9: A new CFRP grinding approach using multiangle 2D ultrasonic assisted nanolubricant MQL was proposed. However, the damage formation and evolution under inherent random properties of grains are still unclear. In this chapter, Firstly, The instantaneous dynamic generation, conduction, accumulation of grinding heat and its thermal effects were revealed. Furthermore, a 3D microscale three phase finite element numerical model was established to determine the removal and damage mechanisms of different FOAs using single grain grinding. Finally, the grinding experiments were conducted under different ultrasonic and lubrication conditions. The results showed the damage formation and evolution mechanisms with different FOAs are various which complement the debris generation mechanisms. The damage is mainly distributed along the grinding direction, with the lowest damage at 0° FOA, and the highest at 45°. The 2D ultrasonic assisted nanolubricant MQL with 90° θ_U has an optimal damage reduction effect with a maximum of 27.56% compared to conventional grinding. However, 45° θU only showed excellent damage suppression on 135° FOA of 40.38%. This paper provides theoretical support for the damage suppression strategy of CFRP grinding.

Chapter 10: This chapter conducts a multi-dimensional systematic evaluation of the grinding-induced surface integrity, ranging from global characteristics to local features. Firstly, line roughness and 3D surface roughness parameters were employed to characterize surface topography, functional performance, fluctuation intensity, and height distribution, with peak heights and valley depths specifically reflecting fiber pull-out damage on the ground CFRP surfaces. Secondly, 2D and 3D fractal dimensions, derived from line roughness profiles and surface micro-topography, were used to characterize the overall morphological attributes of the ground surfaces. Subsequently, multifractal spectra, including spectral width and difference, were applied to quantitatively analyze and discuss local morphological variations. Finally, SEM and laser confocal microscopy were utilized to observe 2D and 3D surface topography, respectively, enabling qualitative analysis of grinding-induced damage such as fiber (bundle) pull-out, fiber fragmentation, resin smearing, and fiber-matrix debonding.

Chapter 11: This chapter first summarizes the principal conclusions and innovations presented in this monograph. Subsequently and primarily, it provides a focused analysis of the current research hotspots in the machining of FRCs and offers an outlook on future developmental trends. Research on the machining of FRCs, particularly CFRPs, is evolving toward a system characterized by clarified mechanisms, low-damage processes, intelligent control, and integrated systems. Key research hotspots reflect this evolution. First, understanding multi-scale material removal and damage formation is of paramount importance. Investigations have now expanded from a sole focus on macroscopic cutting forces to encompass meso- and micro-scales, emphasizing competitive failure mechanisms such as fiber fracture, matrix cracking, and interfacial debonding, with delamination prediction constituting a critical subtopic. Second, advanced tooling technologies—encompassing innovations in macro-geometry, applications of super-abrasives, and intelligent wear monitoring—are essential for improving machining quality and efficiency. Third, low-damage assisted processes, such as ultrasonic vibration-assisted and cryogenic machining, represent frontier areas for proactively mitigating inherent damage by altering tool-workpiece contact states and suppressing thermal effects. Concurrently, process development is being transformed by multiscale simulation (FEM, DEM) and data-driven methods, such as machine learning, which enable rapid prediction and optimization. The drive toward intelligent manufacturing is fueling research into integrated, multi-sensor quality inspection and closed-loop adaptive control strategies. Furthermore, challenging machining technologies for emerging materials (e.g., ceramic matrix composites, thermoplastic CFRPs, multi-material stacks) and structures (e.g., metamaterials) present new frontiers. Sustainability imperatives are making the environmental and economic assessment of dry and minimum quantity lubrication (MQL) technologies, alongside the integration of digital twins and additive-subtractive hybrid manufacturing, significant trends. Future breakthroughs will necessitate deep interdisciplinary integration to establish a complete, predictable, controllable, and optimizable precision machining technology system that links material microstructure to component service performance.

Teng Gao
Qingdao University of Technology, China

Changhe Li
Qingdao University of Technology, China

Xuesong Chang
Hisense Refrigerator Co., Ltd, China

Acknowledgments

The whole book was compiled by PhD. Teng Gao. The book was reviewed and finalized by Professor Changhe Li and Xuesong Chang.

Otherwise, this book is financially co-supported by the following Foundation items: Postdoctoral Fellowship Program of CPSF (GZB20250337), Shandong Postdoctoral Science Foundation (SDBX2024013), National Key R&D Program of China (2020YFB2010500), National Nature Science Foundation of China (52375447, 52475469, 52305477), Innovation Capability Enhancement Project for Technology-based Small and Medium Enterprises of Shandong Province (2024TSGC0237), Shandong Provincial Natural Science Foundation of Youth Fund (ZR2022QE028), Shandong Taishan Scholar Funding Program.

This book has received substantial support and assistance from many experts and colleagues in the compilation process with a reference to related literatures by many professors and experts. Hereby sincere gratitude is extended to them.

Restricted to professional level and available time of the editor, there must be omissions and improper parts in this book, which are expected to be criticized and corrected by readers.

Chapter 1
Fiber-Reinforced Composites Machining:
Removal Mechanism, Mechanical Model, Advanced Processes, and Application

ABSTRACT

FRCs have emerged as preferred materials for high specific strength and stiffness in advanced equipment. Precision grinding is an essential post-forming processing method for ensuring assembly accuracy of FRCs. However, it's considered difficult to machine due to their laminated structure, anisotropy, and heterogeneity. This chapter analyses the grinding damage mechanisms, including material removal behavior, thermomechanical damage responses and surface integrity. Subsequently, the geometric reconstruction of grinding wheels and material modeling are investigated. A comprehensive review is conducted on micro-scale analytical modeling methods for grinding forces under various grinding configurations, accompanied by comparative error analysis and identification of error sources. The influence of grinding parameters on grinding forces is systematically summarized. Furthermore, advanced grinding strategies are critically reviewed, including UVAG and MQL. Finally, the research significance of high-performance machining of FRCs and the logical framework of this monograph are presented.

1.1 INTRODUCTION

In the global pursuit of high-end equipment manufacturing and sustainable development, high-performance composite materials have emerged as strategic frontier materials driving structural lightweighting and functional integration

DOI: 10.4018/979-8-3373-9013-0.ch001

(Zhang *et al.*, 2025). Fiber-reinforced composites (FRCs), owing to their exceptional specific strength, specific stiffness, and designability, are widely utilized in critical sectors such as aerospace, new energy vehicles, and advanced equipment (Wang *et al.*, 2025). However, FRCs are typical multiphase, anisotropic materials composed of reinforcing fibers and a matrix (Pereira *et al.*, 2025). Their heterogeneous structure and weak interlaminar interfaces render them highly susceptible to machining-induced damage, such as delamination, fiber pull-out, matrix cracking, and interfacial debonding, during conventional mechanical processing, severely constraining component service performance and reliability (Long *et al.*, 2025). This "difficult-to-machine" characteristic makes the machining of FRCs not merely a process implementation issue but a fundamental scientific challenge intersecting multiple disciplines including material mechanics, fracture mechanics, and tribology. Consequently, achieving high-quality, high-efficiency, and low-damage grinding of FRCs has become a major research focus for both academia and industry (Qu *et al.*, 2025).

From the perspective of manufacturing paradigm evolution, the global manufacturing sector is undergoing a profound transformation toward precision, intelligence, and sustainability, imposing stricter requirements on the control of both geometry and properties during machining. Particularly with the deepening integration of the concepts of high-performance manufacturing and green manufacturing, realizing high-precision, low-damage, and high-efficiency machining of FRCs has become a key technological link for overcoming performance bottlenecks in high-end equipment and enhancing industrial core competitiveness. Achieving this goal necessitates a deep understanding of material removal mechanisms during processing. It requires the establishment of accurate thermomechanical coupling models to reveal the intrinsic relationships between process parameters and machining-induced damage, thereby guiding process optimization and equipment development.

Current research has accumulated certain knowledge in areas such as tool development and parameter optimization. However, several fundamental challenges persist. First, the systematic understanding of material removal behavior across multiple scales (macro-, meso-, and micro-scale) remains incomplete, lacking a unified theoretical framework to describe the competitive mechanisms of fiber fracture, matrix shearing, and interface failure. Second, existing mechanical models are predominantly based on homogeneous assumptions or simplified conditions, making it difficult to accurately reflect the response of FRCs' anisotropic, multiphase, and discontinuous characteristics under actual dynamic cutting conditions. Third, traditional machining processes exhibit limitations in terms of quality consistency, surface integrity, and environmental friendliness, struggling to meet the stringent requirements of extreme operating conditions, such as those for aero-engine blades and spacecraft load-bearing components.

Therefore, facing challenges like difficulty in ensuring quality consistency, insufficient process adaptability, and a lack of dedicated equipment, conducting systematic research along the complete technological chain of "material removal mechanism, mechanical modeling, process innovation, equipment integration" is essential. This endeavor involves not only fundamental scientific questions and manufacturing process challenges stemming from the intersection of materials science, mechanics, and manufacturing technology, but also represents a necessary pathway toward achieving green and high-performance manufacturing for composite materials.

1.2 MATERIAL PERFORMANCE ADVANTAGES AND PARTICULARITIES

The superior performance of CFRPs fundamentally stems from their multiscale architecture and synergistic interactions among constituent materials. In terms of specific strength and specific modulus, CFRPs surpass most conventional structural materials. This characteristic primarily arises from the unique combination of high axial mechanical properties (tensile strength: 3–7 GPa, elastic modulus: 200–600 GPa) and low density (~1.8 g/cm^3) of carbon fiber reinforcements, coupled with efficient stress transfer through optimized fiber/matrix interfaces, as shown in Figure 1.

Notably, the fatigue behavior of CFRPs exhibits distinct damage mechanisms. Unlike the single crack propagation mode observed in metallic materials, CFRPs employ multiple energy dissipation mechanisms—including fiber/matrix interface debonding, delamination, and fiber bridging—that collectively retard damage accumulation. Studies demonstrate that CFRPs can retain 60–80% of their ultimate tensile strength after 10^6 loading cycles, a critical advantage for applications subjected to high-cycle fatigue loading, such as wind turbine blades.

Regarding thermophysical properties, fiber composites exhibit tailorable anisotropic characteristics. By optimizing stacking sequences and fiber orientations, their axial coefficient of thermal expansion (CTE) can be precisely regulated over a wide range, even approaching near-zero values. This unique attribute makes CFRPs ideal for support structures in high-precision aerospace optical systems, where they effectively compensate for dimensional variations induced by temperature fluctuations.

Furthermore, the damping capacity of CFRPs exceeds that of conventional metallic materials by approximately 2–3 orders of magnitude. Enhanced vibration energy dissipation is achieved through micro-frictional mechanisms at fiber/matrix interfaces and the viscoelastic behavior of the polymer matrix. This property holds significant engineering value for improving the stability of precision instrument platforms and reducing noise in moving components.

The challenging machinability of CFRPs, characterized by complex material removal mechanisms, poor processability, and severe machining-induced damage, primarily stems from their inherent anisotropic mechanical properties, heterogeneous microstructure, and the difficult-to-machine nature of their constituent materials (Ge *et al.*, 2025). As summarized in Figure 2, the difficult machinability of CFRPs can be attributed to the following factors:

The heterogeneous nature of CFRPs significantly influences cutting and grinding processes, mainly manifested through the alternating contact and interaction between the cutting tool and the dissimilar carbon fiber reinforcements and polymer matrix. Under the action of the same cutting edge, the material removal mechanisms for carbon fibers and the resin matrix differ substantially, leading to uncoordinated mechanical behavior during material removal. This mismatch readily induces severe machining damage such as fiber pull-out, interfacial debonding, and matrix cracking, thereby deteriorating the integrity of the machined surface (Shi *et al.*, 2025).

Figure 1. Advantages of material properties of fiber-reinforced composites

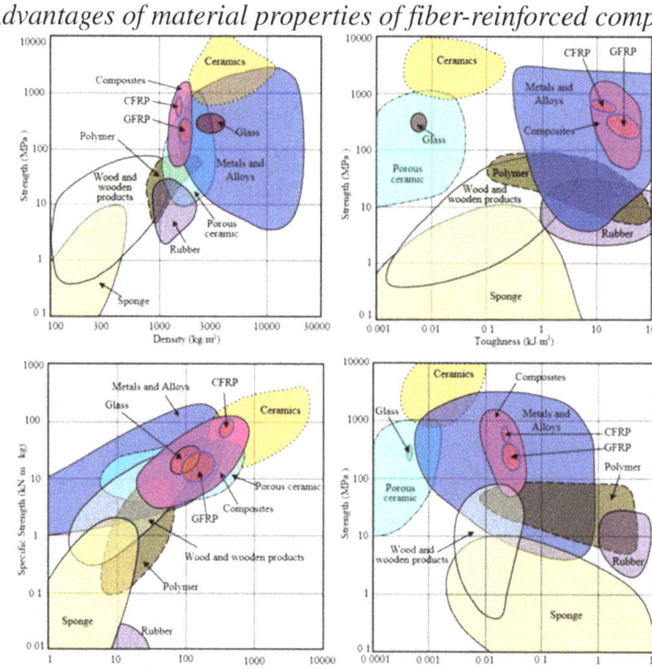

The anisotropy of CFRPs affects cutting and grinding processes mainly through the variation in fiber/resin removal mechanisms with the FOA, resulting in direction-dependent differences during machining. Furthermore, both the carbon fiber reinforcement and the polymer matrix in CFRPs are inherently difficult to machine.

Carbon fibers exhibit high strength, high hardness, and brittle characteristics, which often lead to rapid tool wear during cutting and grinding. In contrast, the matrix material behaves as a soft polymer, whose properties are susceptible to thermal softening and glass transition at elevated temperatures. This thermal susceptibility can cause the loss of necessary support and protection for the reinforcement, giving rise to defects such as delamination, burrs, and tearing. The overall machinability of CFRPs is principally determined by the physical and mechanical properties of the resin matrix and carbon fiber reinforcement, the fiber orientation, and the carbon fiber volume fraction.

Figure 2. Special properties and hard-machinability of CFRP

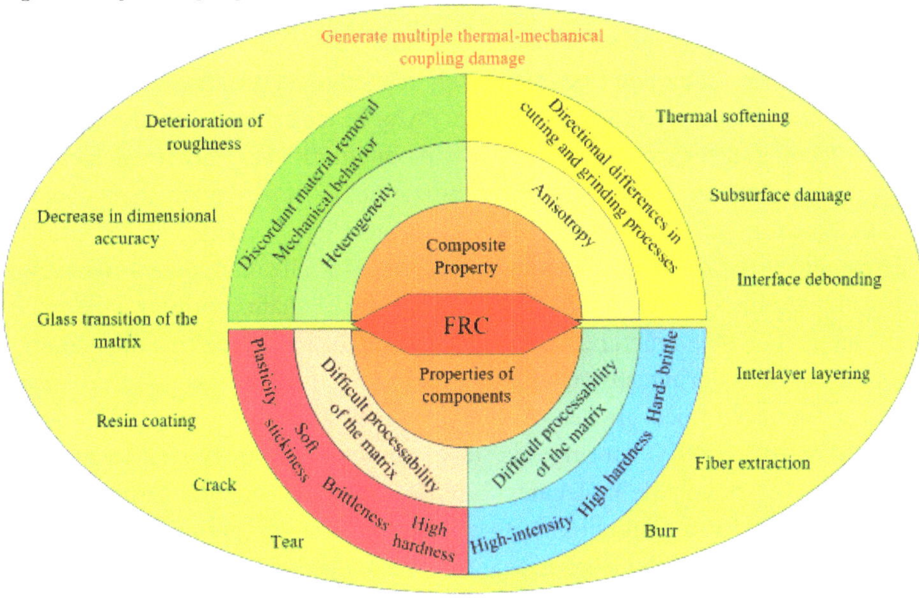

1.2.1 Carbon Fiber

Carbon fiber (CF) is a fibrous carbon material with a carbon content exceeding 90%, characterized by a microstructure composed of turbostratic graphitic carbon. It has become a cornerstone material in the field of high-performance composites, renowned for its exceptional specific strength and stiffness. As the primary reinforcement phase in polymer matrix composites (PMCs), carbon fiber enables the fabrication of lightweight structural components that are crucial for advancements in the aerospace, automotive, renewable energy, and sporting goods industries.

Carbon fiber, a non-metallic material, serves as an excellent composite reinforcement due to its high strength and specific modulus (tensile strength and modulus can reach 2–4 GPa and 400–700 GPa, respectively), chemical stability, superior thermal properties, and damping capacity for vibration attenuation and noise reduction. It differs significantly from both inorganic and organic fibers. With weak radial molecular bonding, its axial tensile strength is several times greater than its compressive strength. Compared to graphite, carbon fiber exhibits stronger axial intermolecular bonding, resulting in higher strength and modulus. It also demonstrates significant anisotropy, with axial thermal conductivity approximately 20 times higher than the radial value, while the radial coefficient of thermal expansion is an order of magnitude greater than the axial one.

The properties of carbon fiber are closely related to its precursor and manufacturing process. Currently, polyacrylonitrile (PAN)-based carbon fibers dominate the market (accounting for approximately 90% of production). The typical manufacturing process involves three core stages: (1) Stabilization: PAN precursor fibers undergo heat treatment in an oxygen-containing atmosphere (typically air) at 200–300°C. This complex exothermic oxidation process induces cyclization, dehydrogenation, and oxidative cross-linking of the linear PAN molecular chains, forming a thermally stable ladder polymer structure. This step is critical to prevent melting or decomposition of the fiber during subsequent high-temperature treatments. (2) Carbonization: The stabilized fibers are heat-treated in an inert atmosphere (e.g., high-purity nitrogen) at 700–1500°C. During this stage, non-carbon elements (such as H, O, N) are removed as volatile molecules, significantly increasing the carbon content to over 90% and forming preliminarily oriented turbostratic graphitic crystallites. The tensile strength of the fiber is primarily developed at this stage. (3) Graphitization: To produce high-modulus carbon fibers, graphitization is performed at even higher temperatures (typically >2000°C) in an inert atmosphere (e.g., argon). This process promotes further growth, rearrangement, and orientation of carbon crystallites, reducing lattice defects and thereby substantially enhancing the fiber's elastic modulus and thermal conductivity.

The properties of carbon fiber are intrinsically derived from its microstructure. It is an anisotropic material consisting of highly aligned graphitic crystallites embedded within an amorphous carbon matrix along the fiber axis. Based on mechanical performance, carbon fibers are generally categorized as follows: (1) Standard Modulus (SM): With a tensile modulus of about 200–250 GPa, these fibers exhibit high strength and good toughness, making them the most widely used general-grade fibers. (2) Intermediate Modulus (IM): Featuring a tensile modulus between 250–350 GPa, they offer an excellent balance of strength and stiffness and are currently the primary choice for aerospace structural components. (3) High Modulus (HM) and Ultra-High Modulus (UHM): These fibers, with a modulus exceeding 350 GPa and potentially reaching over 600 GPa, are produced through high-temperature graphitization. They possess a higher degree of graphitization, resulting in exceptional dimensional stability and thermal conductivity, though often at the expense of some tensile strength.

In addition to these core mechanical properties, carbon fibers also exhibit a low coefficient of thermal expansion, good corrosion resistance, and excellent fatigue resistance. However, as a brittle material, they demonstrate relatively low impact resistance and fracture strain. In practical applications, carbon fibers are primarily embedded in polymer (e.g., epoxy, thermoplastics), metal, or ceramic matrices to form composites. The key to their effectiveness lies in the fiber/matrix interface. Pristine carbon fiber surfaces are chemically inert and exhibit poor adhesion to the matrix. Therefore, two key techniques are commonly employed in industry: Surface Treatment: Methods such as electrochemical oxidation, gas-phase oxidation, or plasma treatment are used to introduce oxygen-containing functional groups (e.g., -COOH, -OH) and increase surface roughness, thereby enhancing chemical bonding and mechanical interlocking with the matrix. Sizing: A thin polymer coating (e.g., epoxy-based) is applied to the treated fiber surface. This sizing layer not only protects the fibers from processing damage but, more importantly, acts as a "bridge" that significantly improves fiber wettability and interfacial bond strength with the matrix resin, ultimately enhancing the composite's interlaminar shear strength (ILSS) and impact resistance.

Carbon fiber represents an outstanding example of modern materials science and engineering, with properties that can be precisely tailored through precursor selection and processing parameters. As the foundation of high-performance composites, it continues to drive the lightweight technology revolution. Future research will focus on developing low-cost, sustainable precursors (e.g., lignin), optimizing surface engineering to further enhance interfacial performance, and exploring new applications for carbon fibers in multifunctional composites (e.g., structural-power integration, self-monitoring smart structures). A deep understanding of its manufacturing science, structure-property relationships, and interfacial behavior remains the core driver for continued innovation in this field.

1.2.2 Resin Matrix

Based on the type of polymer matrix, carbon fiber-reinforced polymers (CFRPs) can be classified into thermosetting and thermoplastic composites. Historically, thermosetting resins were predominantly used in CFRP manufacturing due to their favorable strength, heat resistance, and chemical stability. However, these systems suffer from limitations such as non-reprocessability, slow curing cycles, and environmental concerns, which conflict with contemporary demands for green and sustainable development. Consequently, carbon fiber-reinforced thermoplastic composites (CFRTPs) have gained increasing attention in recent years.

Thermoplastic resins offer advantages including wide raw material availability and lower manufacturing costs, rendering them a promising class of matrix materials. Commonly used thermoplastics include polyethylene (PE), polycarbonate (PC), polyoxymethylene (POM), polypropylene (PP), polyamide (PA), polyphenylene ether (PPE), and polysulfone (PSU). Compared to conventional thermosetting composites, CFRTPs exhibit lower chemical toxicity, superior corrosion resistance, reformability, shorter processing cycles, reduced curing complexity, and suitability for large-scale integrated manufacturing. These characteristics contribute to enhanced production efficiency and significant cost reduction. Moreover, the linear molecular structure of thermoplastic resins imparts high fracture toughness, characterized by large fracture strain and high impact strength, along with excellent damage tolerance.

A distinctive advantage of CFRTPs is their ability to be remelted and reshaped, enabling material recycling and end-of-life reprocessing, thereby addressing critical concerns regarding the disposal of composite materials after service life. As a result, CFRTPs are increasingly applied in aerospace, defense, marine engineering, automotive lightweighting, shipbuilding, and wind turbine blades. High-quality machining technologies for these materials have thus become a key research focus in the field of composite applications.

Since the 1960s, developed regions such as Europe and the United States have prioritized the development and application of high-performance thermoplastic composites, achieving several breakthroughs. For instance, aerospace manufacturers including Boeing have incorporated CFRTPs into components such as aircraft skins, elevators, and radomes. Specific applications include polyphenylene sulfide (PPS) for wing leading edges, polyetherimide (PEI) for pressure bulkheads, and polyetheretherketone (PEEK) for access panels and doors in stealth unmanned aerial vehicles.

Owing to their linear molecular chains, which are more susceptible to scission under heat, thermoplastics melt more readily than thermosets. Thermoplastic resins also generally exhibit better ductility, whereas thermosets tend to be brittle. In addition, the thermal conductivity of thermoplastic resins is typically lower than that of thermosets, which may lead to inefficient heat dissipation during machining. Owing to these distinct thermomechanical properties, the cutting and grinding behavior of thermoplastic composites differs considerably from that of thermosets, often resulting in poorer machinability (Xu *et al.*, 2020).

Furthermore, compared to thermoset-based composites, the significantly enhanced toughness of the matrix in CFRTPs leads to pronounced plastic deformation under tool action, impeding efficient material removal and exacerbating machining-induced damage. Such defects can severely compromise the structural integrity of the workpiece, substantially reduce the load-bearing capacity of the component, shorten service life, and pose serious risks such as assembly mismatch or part rejection (Wang *et al.*, 2022).

1.2.3 Interface

In carbon fiber-reinforced polymer (CFRP) composites, the interphase—often referred to as the "interfacial region"—is not a simple two-dimensional plane, but rather a three-dimensional nanoscale domain between the carbon fiber reinforcement and the polymer matrix, characterized by gradients in chemical composition, molecular structure, and properties. The structural integrity and effectiveness of this interphase directly govern the efficiency of stress transfer, damage evolution mechanisms, and the ultimate macroscopic mechanical performance of the composite. A meticulously designed and controlled interphase is therefore a crucial prerequisite for achieving superior overall performance in composite materials.

The interphase fulfills multiple critical roles during composite service, which can be summarized as follows: (1) Stress Transfer Medium: When the composite is subjected to external loads, the interphase serves as the exclusive pathway for efficient stress transfer from the relatively compliant polymer matrix to the high-strength, high-modulus carbon fibers. A strong and tough interface ensures full utilization of the fiber reinforcement, enhancing the composite's strength and stiffness. (2) Energy Dissipation and Toughening Mechanism: Under impact or crack propagation, an optimally engineered interphase is not designed to be infinitely rigid. Instead, it enhances composite fracture toughness and impact damage tolerance through controlled debonding and fiber pull-out mechanisms, effectively absorbing and dissipating energy to prevent cracks from propagating directly through fibers or the matrix. (3) Physicochemical Transition Zone: Acting as a buffer layer, the interphase mitigates residual thermal stresses caused by the significant mismatch in coefficients of thermal expansion between the fiber and matrix, thereby suppressing microcrack formation during processing or under service conditions. (4) Environmental Protection Barrier: A dense, well-formed interphase can effectively hinder the penetration of moisture, chemicals, and other environmental agents to the fiber surface, delaying interfacial degradation and aging to ensure long-term durability of the composite.

As one of the three essential constituents of CFRPs, the interphase plays a decisive role in determining the composite's properties (Xu *et al.*, 2022). It is fundamental to the reinforcement function of carbon fibers and the stress transfer capability of the matrix. Macroscopically, loads borne by either the resin matrix or carbon fibers must be transmitted through the interphase. Thus, the interphase constitutes a distinct material entity with properties different from those of the fiber and matrix, and it significantly influences the material removal mechanisms and cutting/grinding behavior of composites (Wang *et al.*, 2019a).

Addressing the formation and control of the interface between carbon fiber and resin matrix represents the scientific essence of fabricating CFRP composites that meet practical requirements. Moreover, the specific physicochemical characteristics of both fiber and matrix contribute to the protection or functional performance of the composite, enabling its adaptation to diverse application environments. Bu et al. (2023) proposed a calculation method for the transverse tensile strength of UD-CFRP considering the interface cracking process. Based on a representative volume element model with crack interfaces, the influence of interface cracks on the stress field of composite materials was considered, and the propagation behavior of interface cracks was simulated using the Benzeggh Kenane criterion. The results indicate that interfaces can be classified into three types based on the strength ratio between the interface and the matrix, including the weakest interface, weak interface, and strong interface. When the interface belongs to the weakest or strongest interface, the transverse tensile strength is not affected by the interface strength and increases linearly with the increase of matrix strength. When the interface belongs to a weak interface, the transverse tensile strength increases linearly with the increase of interface strength and is not affected by the strength of the matrix. Hoang et al. (2025) simulated that under normal loading, chemical bonds are the only interaction at the interface, and the interface performance is proportional to the CF surface area. Under both horizontal and vertical loads, chemical bonding and mechanical interlocking coexist: chemical bonding dominates in the early stages of debonding, while mechanical interlocking becomes the main load transfer mechanism as debonding progresses. Research has found that the contribution of chemical bonds to interfacial response is minimal under the influence of surface roughness of carbon fibers.

1.3 ADVANCED EQUIPMENTS APPLICATION AND MACHINING DEMANDS

1.3.1 Aerospace Field

The prominent status of CFRP composites stems from a range of macroscopic performance advantages determined by their microstructure, which through careful material and structural design, translate into transformative engineering benefits in specific applications. Their most critical advantage lies in their exceptional specific strength and specific modulus. The high axial tensile properties and low density of carbon fibers, combined with efficient interfacial stress transfer mechanisms, enable CFRPs to achieve substantial structural lightweighting. This characteristic directly addresses the aerospace industry's persistent pursuit of weight reduction and performance enhancement as shown in Figure 3. For instance, the extensive replacement of conventional aluminum alloys with CFRPs in primary load-bearing structures (e.g., wings and fuselages) of commercial aircraft such as the Boeing 787 and Airbus A350 has led to significant improvements in fuel efficiency and reductions in emissions.

Owing to their lightweight and high-strength properties, carbon fiber composites are increasingly used in aerospace and space technology equipment (Jebaratnam and Hassan, 2025). However, manufacturing and machining precision requirements in high-end equipment applications are extremely stringent (Wang *et al.*, 2021). Although carbon fiber composites can be manufactured using near-net-shape techniques, subsequent machining processes such as milling, grinding, and drilling are indispensable to meet part accuracy and assembly requirements. For example, after the fuselage of the B787 is cured into an integrated structure, secondary machining is required to create openings for passenger doors and windows. Additionally, numerous aerospace components, including composite sealing adjustment pieces for aero-engines, composite outer bypass casings, leading edges of spacecraft wings, and heat-protection covers for missile sections, require such post-processing. In particular, precision components and mating positioning surfaces often necessitate precision grinding operations.

Due to the difficult-to-machine nature and inherent heterogeneity of CFRPs, cutting and grinding processes are highly prone to inducing various defects such as interfacial cracking, fiber fracture, fiber pull-out, and resin smearing. These defects degrade surface integrity and significantly compromise the load-bearing capacity and in-service safety of the components (Asmael *et al.*, 2021).

Figure 3. Aerospace applications and grinding requirements of FRC

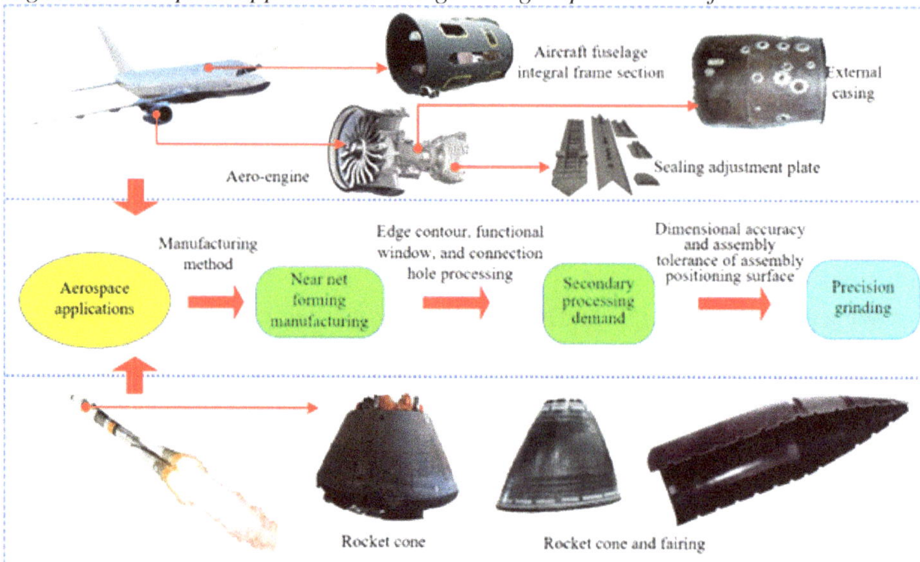

1.3.2 Marine Engineerings

In marine structural applications, carbon fiber combined with matrices such as resin forms lightweight and high-strength carbon fiber-reinforced composites. Compared to steel-based shipbuilding, these composites offer advantages including simplified manufacturing processes, reduced hull weight, and lower fuel consumption. Furthermore, CFRPs exhibit excellent chemical inertness and fatigue resistance, which effectively mitigate corrosion and marine biofouling (Zhang *et al*., 2024a).

The natural frequency of CFRP structures is relatively high, reducing the risk of resonance and thereby preventing sudden brittle fractures. The damping capacity of CFRPs is primarily attributed to the viscoelastic behavior of the polymer matrix. In addition, the interface between the reinforcing carbon fibers and the resin matrix can absorb and reflect vibrational energy, resulting in significant vibration damping. Under identical conditions, the damping performance of a CFRP beam can be several times greater than that of a lightweight alloy beam (El-Hofy & El-Hofy, 2019).

Consequently, in military vessel construction, CFRPs are employed to enhance stealth performance due to their superior vibration and noise attenuation characteristics. For instance, in nuclear submarines, where acoustic quieting is paramount, the reactor coolant pump represents a major noise source. A fundamental approach to improving submarine concealment is to reduce radiated noise transmitted into the surrounding water. Thus, CFRPs can be applied in fixed connection systems for condenser pumps within the propulsion system of nuclear submarines.

Moreover, the use of CFRPs in naval vessels contributes to weight reduction and provides radar wave absorption and electromagnetic shielding. In addition, to further reduce noise and vibration, carbon fiber composites are increasingly being adopted in propulsion systems, such as propellers and shaft systems, of cruise and reconnaissance ships (Hussien *et al.*, 2021).

1.3.3 Other Advanced Equipment Fields

Beyond fundamental specific strength and specific modulus, CFRPs demonstrate tailorable anisotropy and superior fatigue performance, which further broaden their application boundaries. Through precise control of stiffness and strength orientation via ply design, CFRPs enable optimal structural load paths—a critical feature in high-performance applications requiring specific stiffness distributions, such as satellite trusses, optical support platforms, and lightweight robotic arms.

Simultaneously, the inherent multiphase composite structure of CFRPs effectively deflects and bifurcates microcracks, endowing the material with exceptional resistance to fatigue damage. This capability is vital for components subjected to long-term cyclic loading—such as wind turbine blades and helicopter rotors—ensuring structural integrity and operational safety over extended service lives.

Furthermore, the intrinsic corrosion resistance and low coefficient of thermal expansion (CTE) of CFRPs make them uniquely valuable in harsh environments and precision equipment. The polymer matrix provides a natural barrier against chemical agents, allowing CFRPs to avoid electrochemical corrosion issues in marine engineering systems and chemical transport containers, thereby offering significantly superior long-term durability compared to metals.

Moreover, by tailoring the laminate sequence, the CTE of CFRPs can be engineered to approach near-zero values. This characteristic ensures exceptional dimensional stability under fluctuating thermal conditions, making CFRPs ideal for high-precision applications such as space telescopes and semiconductor manufacturing equipment.

In summary, from aerospace structures demanding extreme specific properties, to renewable energy and marine engineering applications requiring durability and reliability, and further to precision instruments and high-end consumer goods that rely on dynamic response and functional integration, the performance advantages of CFRPs do not exist in isolation. Instead, through systematic design and application, they form a logically coherent and deeply integrated solution that continuously advances modern engineering technology and drives innovation.

1.4 RESEARCH PROGRESS ON FIBER-REINFORCED COMPOSITES GRINDING

1.4.1 Material Removal Mechanism

The removal mechanism of fiber composites is complex because of the unique structural characteristics of heterogeneity, anisotropy, and multi-phase cross-scale composition. In addition, the inherent brittleness and interface complicates the study of the fundamental mechanism of removal process (Qu *et al.*, 2022). Even more, the randomness of grains in grinding leads to more complex contact state and cutting behavior between grains and workpiece. Figure 4 shows the material removal mechanism of fiber-reinforced composite grinding (Luna *et al.*, 2020; Hu and Zhang, 2003; Liu *et al.*, 2017a; Liu *et al.*, 2021). Hu and Zhang (2003) found that the fragments produced by multi-directional CFRP grinding showed a mixture of fine powder and broken fibers of different lengths (Figure 4(a)). This is different from unidirectional fiber-reinforced composite grinding. In UD-CFRP, the chip geometry mainly depends on the fiber orientation. The chips with 0° fiber orientation are mainly pulled out of broken fibers, and the chips with 90° fiber orientation are mainly fine powder. Qu et al. (2020a) showed that with the increase in grinding depth, a large number of damages was formed before the abrasive particles contact the fiber. Poor support conditions increased the debonding depth. The fracture surface of single fiber and fiber bundle becomes increasingly uneven and irregular. The grinding force and chip size increase gradually, and the grinding surface quality decreases gradually. According to Liu et al. (2017a) in the grinding of 2-dimensional (2D) C/SiC composites, the failure forms are mainly the combination of fiber fracture, matrix fracture, and interface debonding (Figure 4(b)). The grinding debris is composed of carbon fiber debris, carbon powder, and SiC matrix debris. The main removal mechanism of 2D C/SiC composites is brittle fracture. Zhang et al. (2016a) reached the same conclusion on the failure form of material removal in unidirectional C/SiC composite grinding. Yang et al. (2021a) found that most fibers and matrix were broken into fine debris during the grinding of unidirectional C/SiC. The average crack depth and the average length of debris increase gradually with the cutting depth. However, two completely different kinds of debris are formed in the grinding of 2.5D C/SiC. The appearance and size of warp wear debris are very similar to those of unidirectional C/SiC. The crack is more likely to propagate along the fiber direction in the weft, indicating the formation of long and large fiber debris. Further, Liu et al. (2018) analyzed the grinding mechanism of 2D C_f/C-SiC composites by single grain scratch test. The fiber bundle is damaged by the combined action of peeling and tearing, which depends on the FCA. The material removal mechanism is mainly brittle removal, that is, the combination of fiber layered brittle fracture and SiC matrix cracking.

The influence of carbon fiber on machining results is greater than that of matrix material. In each cutting mode, matrix cracking, matrix/carbon fiber bonding, carbon fiber fracture, and other phenomena occur, thereby resulting in the brittle spalling of the material. Different cutting methods have various effects on chip width, and the order of chip width is $\perp > \parallel > \odot$. Fibers scratched in the $\perp$ and $\parallel$ directions are often removed as blocks by peeling or pushing away (Liu et al., 2019a). Li et al. (2019a) reached a similar conclusion in the single grain scratch test and considered that the cutting force and direction would affect the failure mode and subsurface influence zone of the material. Yin et al. (2021) studied the effect of grinding speed on the material removal mechanism of SiC/SiC using single grain (Figure 4(c)). The results show that increasing the grinding speed can embrittle the material and strengthen the fracture of fiber. When grinding along the warp direction of the fiber, the fiber shows brittle fracture, and the matrix is torn during high-speed grinding. No fiber ploughing or substrate coating is observed in low-speed grinding. When grinding transversely along the fiber warp direction, increasing the grinding speed can completely remove the fiber, and a little cut fiber residue is found on the bottom surface of the groove, which improves the surface finish. Increasing the grinding speed can also improve the removal rate of SiC/SiC composites. Luna et al. (2020) studied the effects of grain shape, size, spacing, and fiber direction on the material removal mechanism of SiC/SiC composite grinding through a single grain scratch test (Figure 4(d)). The results show that the grain shape has a greater influence on the force than the fiber orientation because of the removal of brittle materials. Circular grain produces greater thrust than square and triangular grains because it has no stress concentration points. However, square grain shows excellent ability in reducing transverse damage of transverse fibers. The grain shape determines the location of cracks, and the fiber orientation determines the subsequent crack propagation, and preferentially along the interface direction. Wei et al. (2019a) conducted the frequency component analysis and damage pattern recognition of acoustic emission signals from 3D orthogonal SiO_2/SiO_2 composite grinding. The results show that the main sources of acoustic emission are fiber fracture, matrix crack, and debonding between fiber and matrix. The grinding study of 2.5D woven SiO_2/SiO_2 composite by Wang et al. (2019b) showed that crack propagation and material breakage can be found in the grinding area. The main removal mechanism is brittle fracture, which is similar to brittle homogeneous materials (Cao et al., 2014a; Anand et al., 2018). Inoue and Kawaguchi (1990) clarified the grinding mechanism of GFRP. Two internal failure modes are observed near the surface of the fiber bundle. One is the mode, in which deeper fiber debonding unfolds in the whole fiber bundle. The other is that the failure in the fiber bundle is relatively shallow and random in terms of depth and expansion, often in the form of needle cracks. There are two kinds of fiber bundle cutting profiles. One is the pit formed by the protruding fiber bundle under the action of digging up during grinding, and the other is the expansive bulge composed of uncut glass fibers. Chockalingam et al. (2013) showed that the protruding fibers are still obvious in GFRP dry grinding, although the surface is covered with broken matrix. The surface ground with synthetic coolant is clean, hard, and smooth. Chockalingam and Kuang (2015) compared the GFRP grinding performance of alumina grinding wheel and cubic boron nitride (CBN) grinding wheel. The results show that the grinding force ratio and machining efficiency of CBN grinding wheel is generally higher than that of alumina, and the grinding surface roughness is lower in most cases.

Figure 4. Material removal mechanism of fiber-reinforced composite grinding

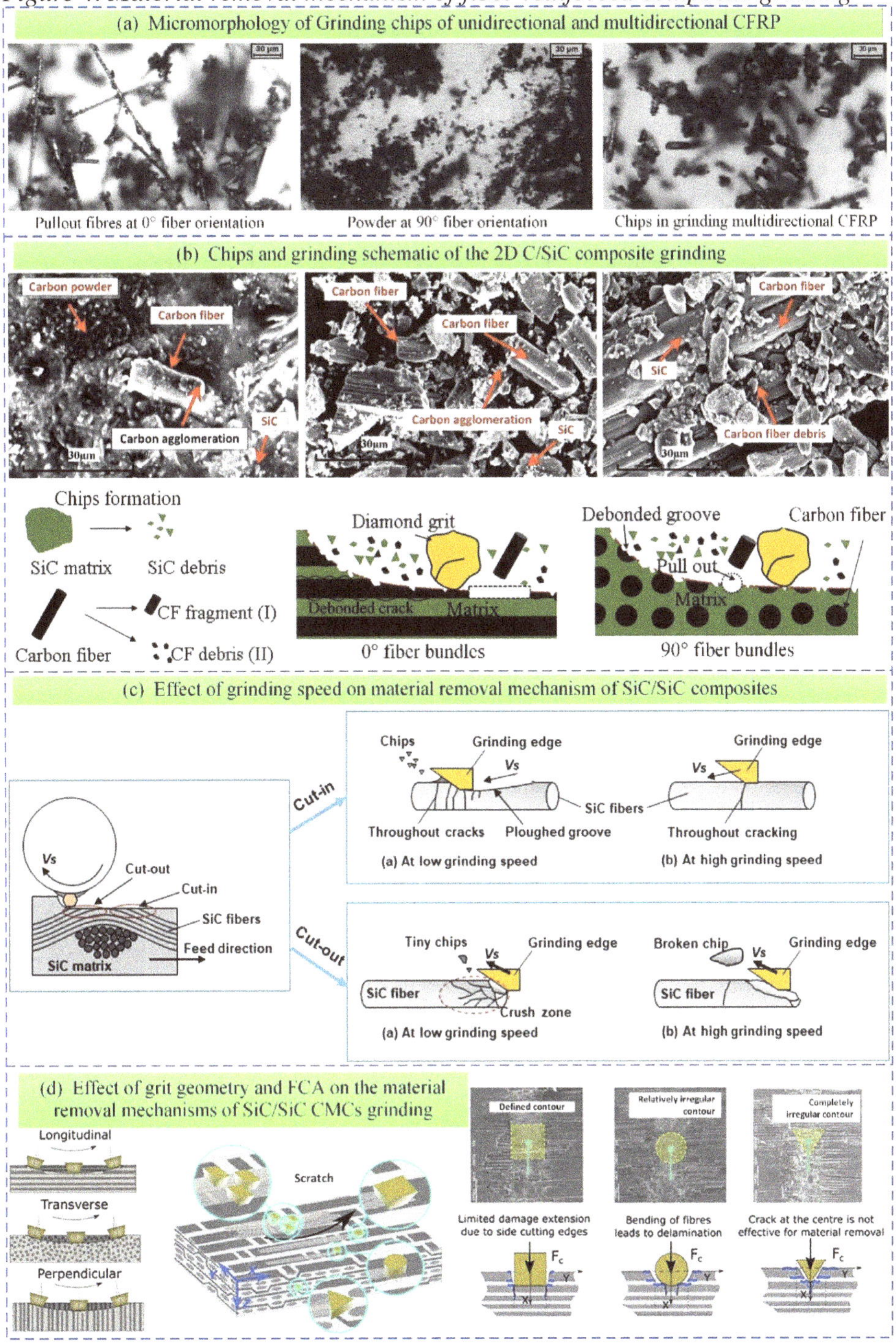

1.4.2 Thermal and Mechanical Behavior

Grinding force is an important index to investigate the influence of machining parameters on grinding performance (Wei et al., 2019b). It results from the elastic and plastic deformation of the workpiece and the interaction among grains, materials, and debris (Sambhav et al., 2011; Sun et al., 2016). It involves almost all factors of grinding and is the basis of researching grinding technology, designing appropriate cutting tools, and selecting appropriate machine tools. It is also an important index to evaluate the processing characteristics of materials. Figure 5 shows different grinding experimental equipment of single grain. Figure 6 shows the mechanical behavior of fiber-reinforced composite grinding. The grinding force of fiber composites is also highly dependent on the fiber orientation (Cao et al., 2015). Zhang et al. (2016a, 2019) investigated the grinding forces in three typical grinding directions of unidirectional C/SiC composites. The grinding force follows the order: normal > longitudinal > transverse (Figure 6(a)). Li et al. (2019a) designed a novel scratch test equipment (Figure 5(b)). The results show that the tangential force (F_t) is generally greater than the normal force (F_n) at the same cutting depth. In the longitudinal scratch direction, the carbon fiber is mainly pulled out and bonded with the fractured SiC matrix. In the transverse scratching, the carbon fiber is mainly cut off and slightly pulled away laterally with the fractured SiC matrix. The longitudinal cutting force is greater than the transverse cutting force because of the high axial Young's modulus, and the carbon fiber transmits its stiffness and resistance to the fiber direction. The grinding force of multi-directional CFRP (MD-CFRP) increases approximately linearly with the grinding depth and is generally greater than those of UD-CFRP composites. The reason may be that the layers with different fiber orientations have stronger mutual support. The grinding force increases with the feed speed and cutting depth and decreases with the increase in grinding wheel speed. Under the same experimental conditions, the force ratio (F_n/F_t) and specific grinding energy of 2D C/SiC composites are lower than those of traditional ceramics, such as Al_2O_3, SiN_4, ZrO_2, and SiC ceramics (Chen et al., 2010; Ding et al., 2017a). This result may be due to the introduction of brittle carbon fiber material, which causes the 2D weave C/SiC composite to have higher brittleness and easily damaged, so that its force ratio is lower than that of traditional ceramic materials. According to the weave and laminated structure of fiber bundles, Liu et al. (2019) selected three typical surfaces (e.g., SA, SB, and SC) for scratch test (Figure 5(a)) under different scratch speed and depth combinations. Under different scratch modes, the order of the maximum scratch force is normal > transverse > longitudinal, and the order of the average scratch force on the three surfaces is SB > SA > SC (Figure 6(b)). Under the same grinding parameters, the grinding force of 2.5D C/SiC is always greater than that of UD-C/SiC (Yang et al., 2021a). Considering that the weak coupling force between unidirectional fibers promotes the crack propagation further. In 2.5D C/SiC, the orientation of warp and weft yarns changes under the action of needle structure, which shows that the coupling between fibers is effective in obtaining higher strength. Wang et al. (2019b) found that the grinding of 2.5D woven SiO_2/SiO_2 composite can travel along the fibers. Therefore, for 2.5D woven SiO_2/SiO_2 composite, the influence range of grinding process is wider than that of other materials. As shown in Figure 6(d), only shear stress is generated on the fiber without tensile stress in the grinding direction of B2 and C2. Thus, the matrix is less damaged. Cao et al. (2015) found that the material removal of 2.5D woven SiO_2/SiO_2 composite was mainly caused by shear fracture and open crack when FCA = 0°. The force that separates the fiber/matrix interface is small, because the bonding strength between fiber and matrix is much lower than fiber strength. The material removal involves the shear fracture of fibers with the increase in FCA, and the grinding force increases due to the high shear strength of fiber. The vertical grinding force is the largest when FCA = 90°, because the fiber has the highest longitudinal strength and the resistance to the vertical compression of the grinding wheel increases. According to the weaving method (Figure 6(c)), the proportion of weft fiber is less than that of warp. Therefore, the normal grinding force of surface B is larger than A. The supporting conditions and debonding depth are the key factors that affect the grinding force. According to the interaction of forces, Qu et al. (2020) decomposed the grinding force model of C/SiC composites into the mechanical model of the fracture area of the SiC matrix and the formation area of the fiber grinding chips. Based on energy balance theory, Wei et al. (2019b) modified the specific grinding energy model, combined with the single abrasive experiment of 3D orthogonal SiO_2/SiO_2 composite (Figure 5(c)), a new semi-analytical force model of single grain grinding force is proposed. Liu et al. (2017b) found that the grinding force remained basically unchanged with the increase in grinding wheel particle size. The reason is that with the increase of particle size, the particle size of abrasive particles decreases while the number of active abrasive particles increases, resulting in slight changes in grinding force. Kodama et al. (2020) believed that carbon fiber composites produce two-stage heat affected layers during grinding. First, the heat affected layer of the glass transition of the resin matrix, and then the heat affected layer of the thermal replacement of the matrix resin is generated. The thermal deformation of matrix resin deteriorates the characteristics of grinding surface. From the point of view of machining quality and machining damage degree, small equivalent chip thickness trimming is the most suitable (Sheikh-Ahmad and Mohammed, 2014). Under fine machining conditions (e.g., minimum equivalent chip thickness, minimum cutting depth), no edge delamination and internal damage occur. Fan et al. (2019) established the simulation model of CFRP grinding heat distribution ratio and found that the grinding heat was preferentially conducted along the fiber because the thermal conductivity of fiber is obviously greater than that of resin. The resin surface temperature is higher because the thermal conductivity of the resin is low. Sheikh-Ahmad et al. (2019) proposed that inverse heat conduction method was an effective and efficient technique for determining the energy balance in machining CFRP. Qian et al. (2020) found that the heat generation mechanism at 90° was more complex compared with 0°, resulting in the highest cutting temperature when the FCA was 90°. The thermal damage depth is the minimum at 0° and the maximum at 90°.

Figure 5. Various experimental devices of single grain grinding

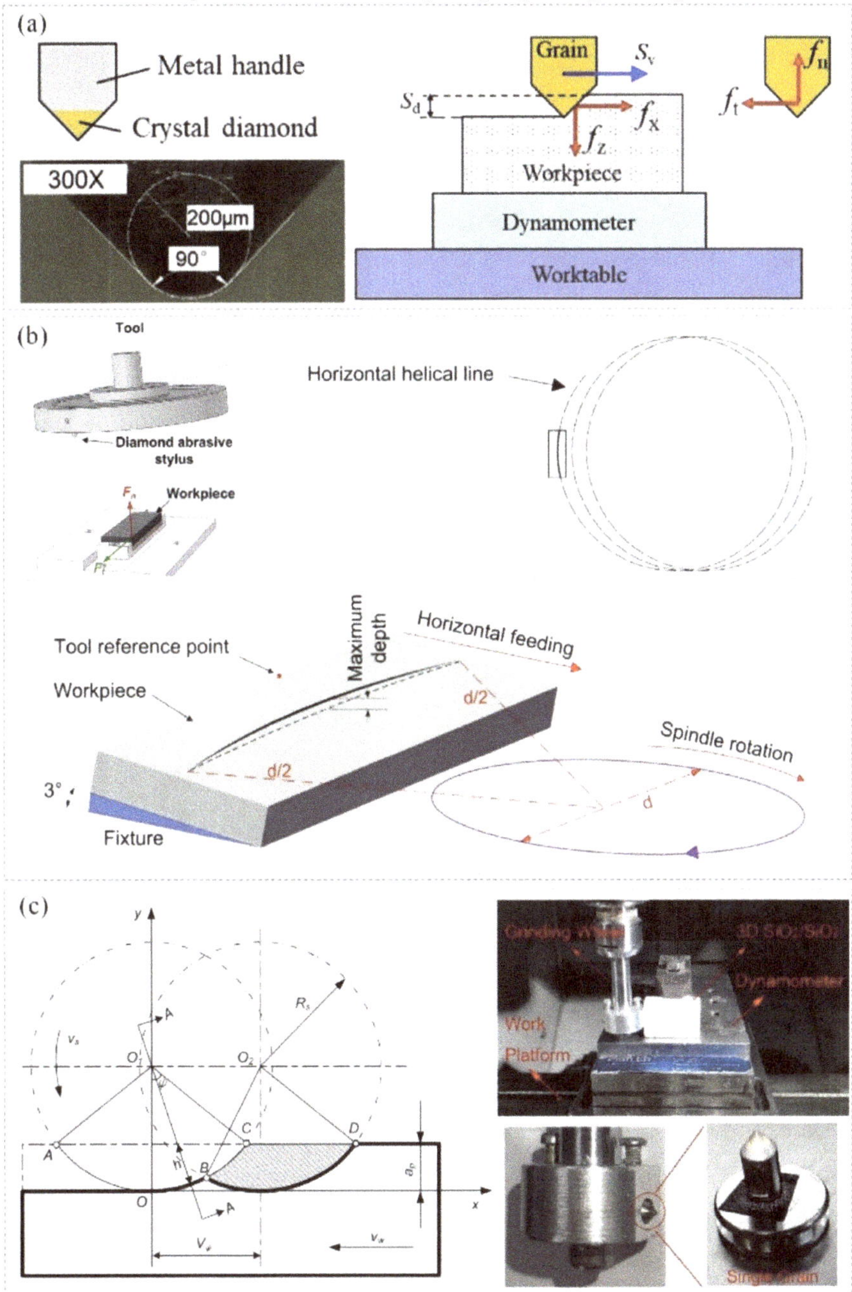

Figure 6. Mechanical behavior of fiber reinforced composites grinding

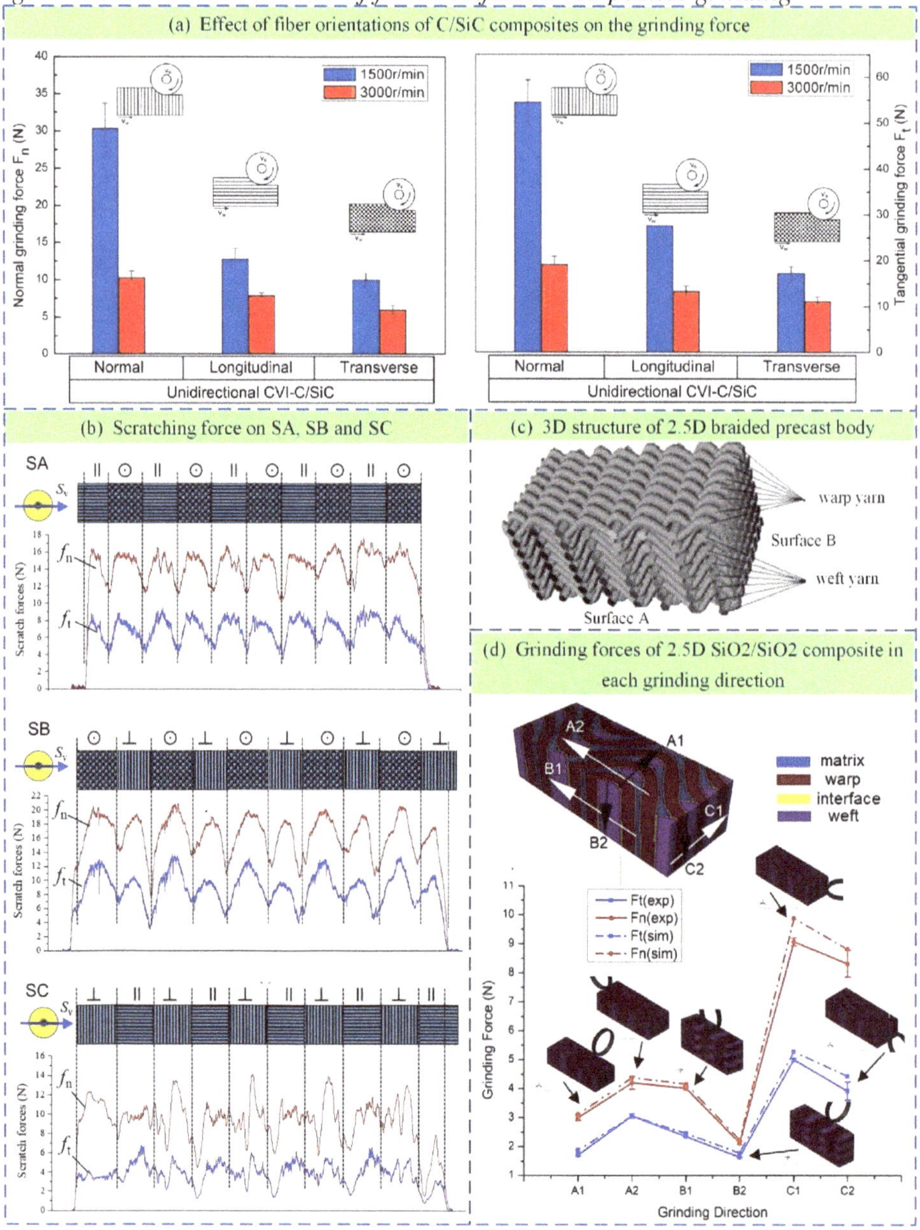

1.4.3 Damage and Surface Integrity

The damaged carbon fibers in the grinding of CFRPs include fiber pullout, voids, partially cut fibers, breakage and fracture of carbon fibers, and flowed matrix (Wang et al., 2018a; Choudhary et al., 2021). The surface roughness of MD-CFRP changes with the fiber's local orientation, and there are serious fiber pullout and other damage in 135° layer. Hanasaki et al. (2003) investigated the grinding of UD-CFRP and MD-CFRP and found an interesting phenomenon. As shown in Figure 7(a) (Wang et al., 2019b; Choudhary et al., 2021; Hanasaki et al., 2003), the workpiece surface roughness in the 90° direction is the smallest, and the surface roughness in the 135° direction is the largest during UD-CFRP grinding. However, the surface roughness of MD-CFRP composed of 45°/135° is less than that of UD-CFRP. The explanation for this phenomenon is detailed as follows. The unidirectional carbon fiber of 135° is bending and fracture, and the fracture point is unstable. However, the combination of 45° and 135° carbon fibers can stabilize the fracture point. Soo et al. (2012) found that tool wear, cutting force, and surface roughness were high using CBN abrasive for CFRP laminate edge grinding compared with diamond abrasive. The resulting damage includes slight edge fracture/chip and pores caused by fiber/matrix loss. The damage caused by both abrasives includes slight edge fracture/chip and pores caused by fiber/matrix loss. However, there is no obvious large-scale delamination in the two kinds of superbrasive coatings. From the surface micromorphology, matrix cracking, fiber pullout, and fiber outcrop are the main damage forms in the grinding of unidirectional C/SiC composites. In the grinding process, the debonding depth between matrix and carbon fiber depends on the sharpness and lubrication state of grains (Qu et al., 2020b). Using mechanical damage phenomenology, Cao et al. (2013) analyzed the formation mechanism of grinding surface ripple of CMCs composites. Fiber orientation plays a decisive role in waviness, which is different from the traditional theory of ripple formation caused by the vibration of machine tool system. Zhang et al. (2016a) found that the surface roughness of arithmetical mean deviation of the profile R_a and maximum height of the profile R_z of unidirectional C/SiC composites in grinding directions follow the order of longitudinal > normal > transverse. Fiber pullout and fiber outcrop are the basic damage modes of unidirectional C/SiC. Interface peeling, matrix cracking, fiber pullout and fiber outcrop are the main defect forms of 2.5D C/SiC. Compared with unidirectional C/SiC, the subsurface damage of 2.5D C/SiC is shallower and wider (Yang et al., 2021a). Wang et al. (2019b) considered that the surface damage of 2.5D SiO_2/SiO_2 composites mainly resulted from matrix crushing and fiber debonding. Matrix breakage is caused by two reasons: 1) The matrix is weak and damaged by grinding force; and 2) the matrix is squeezed by fibers (Figure 7(b)). The latter plays a greater role because it cannot only destroy the matrix in the grinding area but also destroy the matrix in the area far away from the grinding area. In the aspect of fiber debonding, there have been studies on the influence of fiber orientation on the cutting process and the damage of fiber debonding on the machined surface (Zhang et al., 2016a; Cao et al., 2015; Diaz and Axinte, 2017). However, these studies ignore the interaction among fiber bundles. Wang et al. (2019b) considered that the difference of warp and weft fiber bundle deformation capacity was the key factor that affected the overall deformation. The weft bundle is only inserted into the gap to form a weak connection. When the grinding force is in the direction shown in Figure 7, the weft fiber produces greater deformation than the warp fiber bundle. Then, the bonding connection between warp and weft is destroyed, thereby forming mechanical damage on the fiber bonding. Fiber debonding usually appears as a slight tear in the grinding area, and its damage to the surface quality is less than that of matrix breakage. Therefore, grinding along the weft axis is a better choice to reduce fiber debonding. Transversely oriented fibers are subjected to high normal force during grinding, thereby resulting in crack propagation along the fiber direction and delamination. Choudhary et al. (2021) proposed that the grinding scallop and the chip thickness of the cutting surface determine the characteristics and degree of fiber damage (Figure 7(c)). The crack deflection mechanism plays an important role in the orientation of fibers and limits the delamination of fiber matrix. High-speed grinding reduces the maximum undeformed chip thickness and grinding scallop and promotes the shear micro cutting of fiber and matrix rather than brittle fracture. High-speed grinding process is effective in machining defect-free C/SiC CMCs. Fiber orientation plays a key role in the grinding of surface micromorphology of SiO_2/SiO_2 composite (Cao et al., 2014). Grinding speed has the greatest influence on height and surface support performance, followed by grain size and depth of cutting. Grain size is the key factor that affects the change in surface micromorphology. Cao et al. (2015) also believed that the grinding speed and cutting depth had a great influence on the surface morphology. The surface topography height decreases with the increase in grinding wheel speed, and the deflection of topography distribution decreases with the increase in grinding depth.

Figure 7. Grinding damage mechanism and surface roughness trend

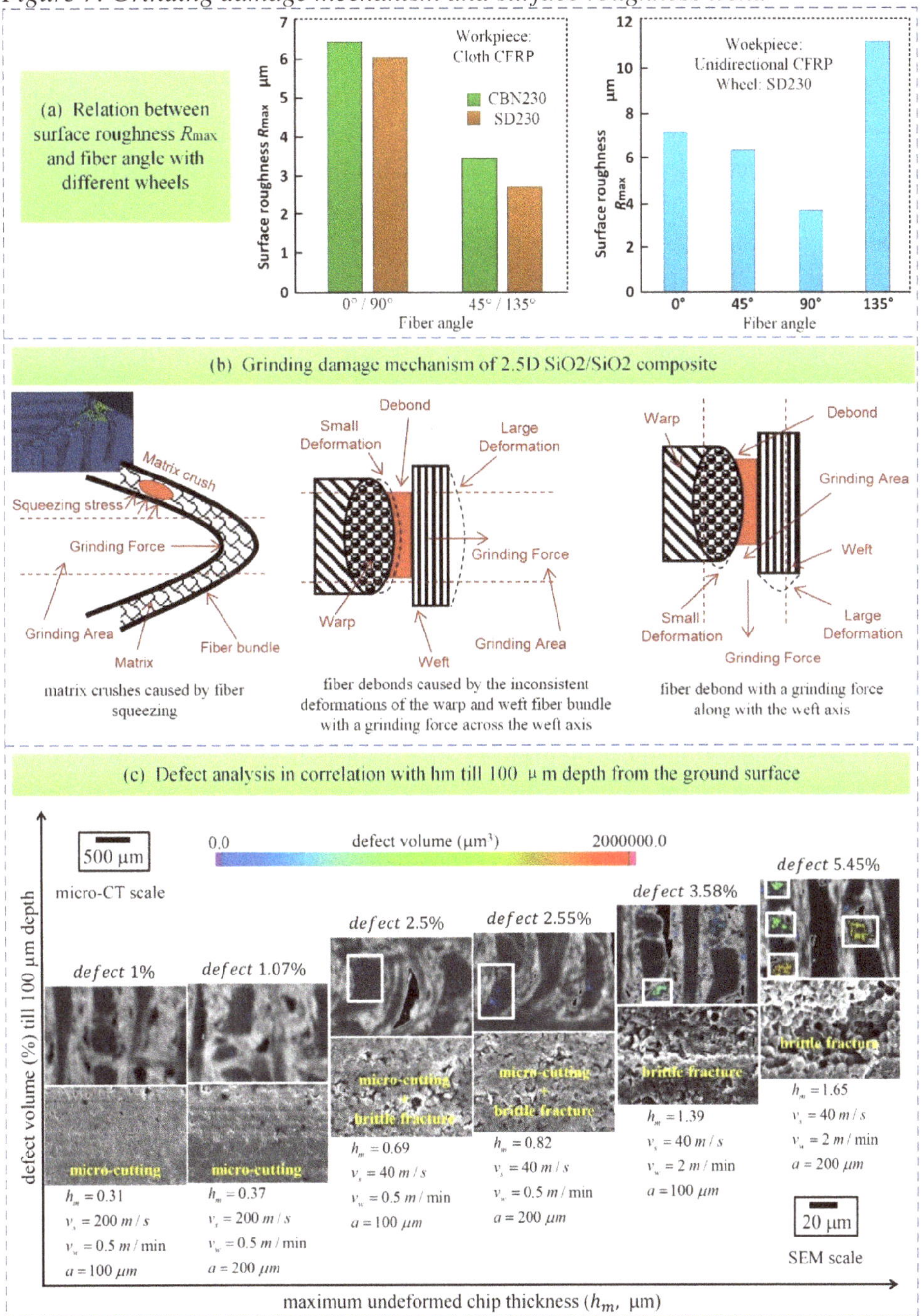

1.5 RESEARCH PROGRESS ON ANALYTICAL GRINDING FORCE MODELING

Grinding force is an important indicator of material removal process, which directly affects machining efficiency, surface quality, and tool life. The force model is the most direct and comprehensive mathematical expression for describing the mechanical behavior of material removal, and plays an important role in the application of grinding in various fields (Li et al., 2023). The mechanical modeling of grinding is also crucial for effectively controlling the occurrence of FRC machining damage. However, due to the non-uniformity, heterogeneity, anisotropy, and low heat dissipation characteristics of these FRCs, it is difficult to ensure high accuracy and low estimation errors of the force model.

1.5.1 Material Model

From the perspective of material model and structural characteristics, the fiber distribution, fiber volume fraction, weaving structure characteristics, interface performance parameters, and other material properties of FRC also have a significant impact on the grinding force model.

Among them, the constitutive model provides a mathematical framework for describing the mechanical behavior and deformation laws of materials by establishing the stress-strain relationship in the grinding zone, which can predict stress distribution and strain accumulation, and help analyze the mechanical response during material removal process (Zhang et al., 2022a). Reasonable selection and establishment of applicable constitutive models are beneficial for improving the accuracy and reliability of grinding process modeling. The workpiece grinding process usually undergoes large strain, high strain rate, and temperature rise. Therefore, the constitutive model of composite materials that can be effectively applied to grinding processing must be able to reasonably capture the effects of strain, strain rate, and temperature on the dynamic behavior of composite materials. It is still challenging to establish a universal or systematic constitutive model to describe FRC grinding. The constitutive models can be divided into two categories: phenomenological and materialistic. The commonly used models include the Bodner Partom model (BP model), Zerrli Armstrong model (ZA model), and Johnson Cook model (JC model), each with its own scope of adaptation and advantages. The phenomenological (Johnson Cook, JC) constitutive model and its improved forms are widely used in the field of cutting and are often used to describe the thermodynamic properties of materials. Currently, most constitutive models are based on single-phase homogeneous materials, which makes it difficult to directly apply them in FRC machining modeling. Gao et al. (2017) established a constitutive model for CFRP under quasi-static and dynamic loads, which integrates the contributions of the three-phase composition (fiber, matrix, and fiber/matrix interface) to capture the key micro mechanical effects of the interface on the macroscopic mechanical properties of composite materials. The author believes that the application of this model in glass/epoxy resin and carbon/epoxy resin composite materials is in good agreement with experimental measurement results, and can accurately characterize the static and dynamic mechanical properties of FRP composite materials, especially when the fiber volume fraction is high. This model is more effective.

Ran et al. (2023) analyzed the stress transfer characteristics between different components based on the theory of elastic deformation, and established a multi-scale model of the elastic mechanical properties of SiCf/SiC composite materials, including elastic modulus and Poisson's ratio. Based on the energy balance method, the occurrence and failure behavior of material damage were analyzed, and equivalent fracture toughness in different directions was obtained, providing data and theoretical support for the subsequent establishment of grinding force models. Liu et al. (2019b) used a dual perturbation algorithm to achieve periodic random arrangement of carbon fibers in the matrix. The dual perturbation algorithm consists of two parts: global perturbation and local perturbation. In addition, a combination of Weibull strength distribution theory and Monte Carlo method was proposed to achieve a random distribution of single carbon fiber strength. Mei et al. (2017) proposed a random method based on Gaussian distribution to simulate the random fiber positions in composite materials. Based on this, two random strategies were developed to randomly locate fibers corresponding to the brittle and ductile fracture of fibers during cutting process.

1.5.2 Grain and Wheel Models

Accurately considering the parameters of grinding wheel grains and material properties is an important prerequisite for the analytical modeling of FRP grinding forces (Cao et al., 2024). The shape of abrasive particles has a significant impact on the modeling of grinding forces, and different abrasive particle shapes can lead to different stress characteristics, resulting in differences in mechanical models (Liu et al., 2023). The closer the shape model of the abrasive particles is to the actual shape, the more accurate the force model will be. But it is extremely difficult and unnecessary to establish the precise shape of the abrasive. Most abrasive particles are irregularly shaped polyhedra. Some scholars simplify the abrasive particles as conical shapes for ease of modeling process analysis and calculation (Zhang et al., 2017), but in reality, the abrasive particles are not conical. As shown in Figure 8 (a), Dong et al. (2022) simplified the abrasive particles as frustum bodies for SiCf/SiC composite material grinding force modeling. The equivalent grinding wheel profile is the union formed by the superposition of the profiles of multiple abrasive grains during the grinding process. Yao et al. (2023a) simplified the projection cross-section of abrasive grains into triangles, therefore, the projection cross-section of abrasive grains can be determined by three-point coordinates. These three points are calculated based on the grinding position, protrusion height, and cone angle of the abrasive, as shown in Figure 8 (c). In addition, the distribution of abrasive particles also has a significant impact on grinding force. Zhang et al. (2022b) defined the random protrusion height characteristics of diamond abrasive particles using the Rayleigh distribution probability density function, as shown in Figure 8 (b), in order to analyze the interaction between the abrasive particles and the workpiece and establish a force model. Zhang et al. (2024b) used a random offset algorithm to simulate the surface morphology of the grinding wheel, as shown in Figure 8 (d), and revealed the discrimination mechanism of effective abrasive particles in the grinding area through discrete grinding wheel and abrasive particle motion trajectories. Based on the microstructure characteristics of the material, a C/SiC composite material grinding force model was established.

Figure 8. Geometric reconstruction and modeling of grinding wheel for FRC grinding force model

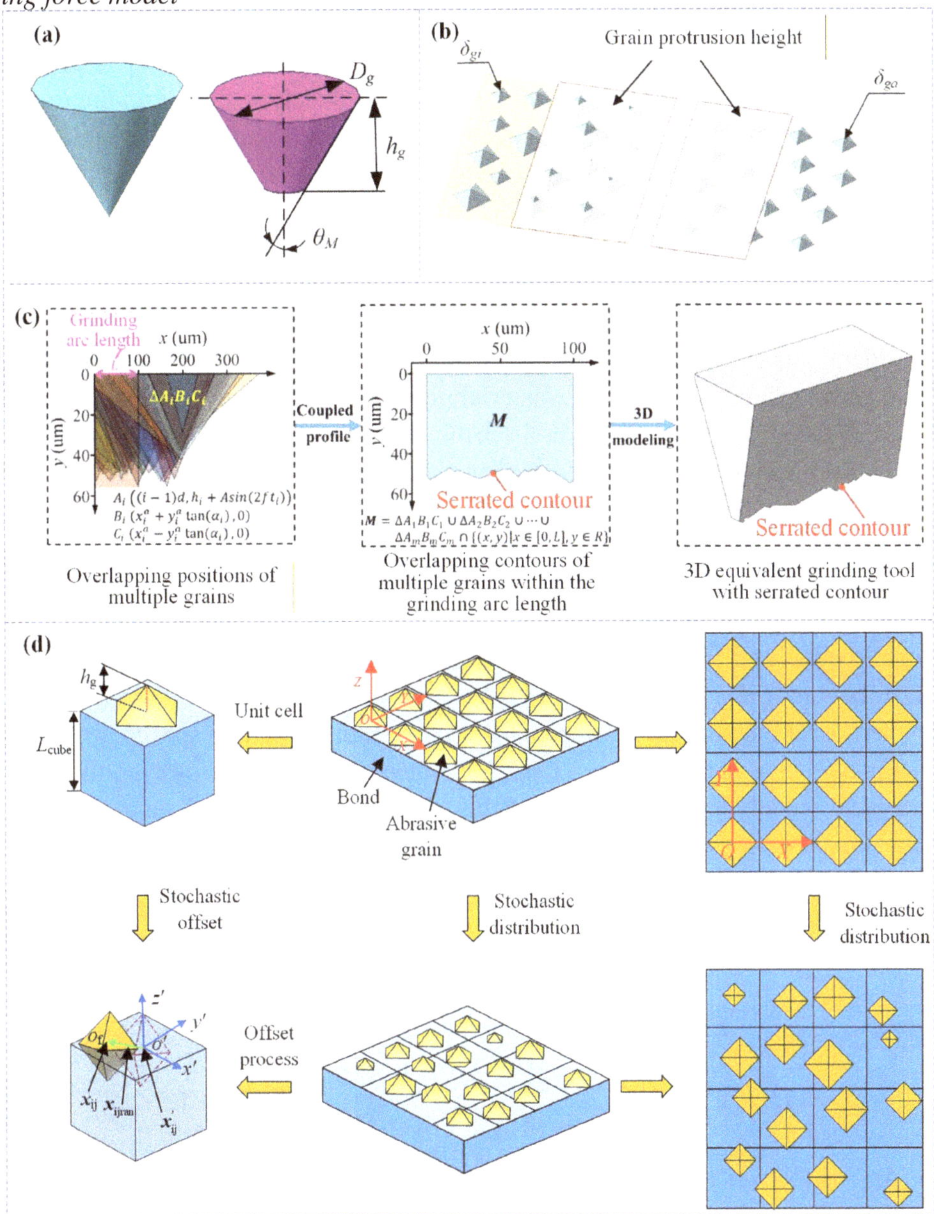

1.5.3 Analytical Force Model

Mechanical modeling is divided into three methods: empirical modeling, finite element modeling, and analytical modeling. Empirical modeling is based on a large amount of experimental data to establish a relatively reliable grinding force prediction model, but the model has poor universality and cannot reflect universal mechanical laws. Finite element modeling is the use of simulation platforms to analyze the relationship between abrasive particles and workpieces, which can simplify the modeling of grinding forces. However, it requires a large amount of computation and has a large prediction error between the simulation model and the actual experimental values. The analytical modeling method requires comprehensive consideration of other energy field factors such as wheel structure, abrasive particle parameters, grinding parameters, material properties, anisotropy, and cooling and lubrication conditions. It can also assist in the analysis and study of material removal mechanisms and surface damage formation mechanisms in grinding processes(Meng et al., 2023). So this section mainly elaborates on the analytical modeling method systematically.

Scholars have established different FRC grinding force models based on various theories such as material mechanics and fracture mechanics. Wei et al. (2019) established a semi analytical model for single abrasive grinding force considering 3D orthogonal SiO_2/SiO_2 composite FOA based on energy balance theory, using the mathematical relationship between tangential grinding force and specific grinding energy as shown in equation (1.1), as shown in equation (1.2). However, the semi analytical model lacks sufficient analysis and characterization of the mechanical behavior of material removal, and the prediction error exceeds 20%.

$$u' = \frac{E_w}{V} = \frac{F_t(v_s + v_w)}{2a_p^2 v_w \tan\theta} \cdot \frac{\arccos\left(1 - a_p/R_s\right) + \pi v_w/(v_s + v_w)}{\pi} \tag{1.1}$$

$$\begin{cases} F_t = \dfrac{2\pi R_s \tan\theta a_p^2 v_w u_f}{\sqrt{2a_p R_s (v_s + v_w) + \pi R_s v_w}} \\ F_n = (F_t \pm C_F)/\mu \end{cases} \tag{1.2}$$

Among them, u_f is the specific grinding energy related to the grinding direction and fiber orientation, μ is the friction coefficient between the abrasive particles and the material surface and debris, and C_F is a constant related to the grinding process. Amin et al. (2017, 2018a) established a grinding force model for CFRP axial and feed directions by characterizing and calculating the protrusion height of diamond abrasive particles on two types of grinding wheels, and introduced overlapping cutting quantification for the first time. However, the limitation of this model is that it assumes that all diamond abrasive grains have the same particle size and the material removal mode is rigid brittle fracture. Liu et al. (2017b) established the relationship between the depth of abrasive grain penetration and cutting force in CFRP as shown in equation (1.3), and used the removal volume and material removal rate of a single abrasive grain material to establish an analytical model for the axial force of CFRP in ultrasonic vibration assisted end face grinding as shown in equation (1.4). However, this model only considers the trajectory of abrasive particles and does not involve the shape and random arrangement of abrasive particles. Zhang et al. (2023) proposed a UVAG analytical force model considering the 2D-C_f/SiC ductile transition based on the shape and motion trajectory of abrasive particles and the effective number of abrasive particles. Among them, the transition critical cutting depth of removal mode h_c is used to distinguish between ductile and brittle fracture removal processes, as shown in equation (1.5). However, this model assumes that the abrasive particles are regular octahedra of the same size and evenly distributed, and does not accurately characterize the morphology characteristics of the grinding wheel. Shi et al. (2020) also considered the removal of brittleness and ductility in CFRP and proposed the ductility brittleness ratio K. They established a vertical feed direction force model for ultrasonic vibration assisted peripheral grinding, as shown in equation (1.6). However, this model has a certain drawback that the toughness brittleness ratio K needs to be obtained through experiments.

$$\delta = \sqrt[5/2]{\frac{3A(1-\nu^2)F}{(2d)^{1/2}/nE}} \tag{1.3}$$

$$F = n \cdot \sqrt[5]{K_C^3 H^2 \frac{2a_p R v_w}{2nfK_V \int_0^{\Delta t} \sqrt{\begin{pmatrix} \left(v_w - \frac{\pi}{30} n_s \sin\frac{\pi}{30} n_s t\right)^2 + \left(\frac{\pi}{30} n_s \bar{R} \cos\frac{\pi}{30} n_s t\right)^2 \\ + \left(2\pi f A_U \cos(2\pi f t + \varphi_0)\right)^2 dt \end{pmatrix}}^{4}}} \tag{1.4}$$

$$h_c = \cot\theta \sqrt{\frac{2\lambda}{\xi}} \left(\frac{K_{ID}}{H_V}\right)^2 \tag{1.5}$$

$$F_x = -C_3 \cdot C_2 \cdot \frac{C_a^{\frac{1}{3}}}{r} \cdot a_p \left(\frac{K}{1+K} \cdot F_{nd} + \frac{1}{1+K} \cdot F_{nb}\right) \tag{1.6}$$

Wang et al. (2019c) used CFRP brittle fracture removal to obtain the material removal volume through the kinematic law of a single abrasive particle and its interaction with the workpiece. Then, by considering the effective number of abrasive particles, cutting time, horizontal ultrasonic vibration of the workpiece (as shown in Figure 9(a)), and material removal rate, the impact force of abrasive particles and the grinding force in the feed direction of end face grinding were obtained using the relationship between impulse and force. Furthermore, Wang et al. (2020a) established a peripheral grinding force model for vertical ultrasonic vibration of grinding wheels (as shown in Figure 9(b)) based on ductile material removal and brittle fracture. Furthermore, as shown in Figure 9(c), Wang et al. (2020b) established a grinding force model for the end face grinding feed direction of ultrasonic vibration applied to the grinding wheel (vertical direction) and workpiece (horizontal direction), as shown in Equation (1.7).

$$\begin{cases} F = \left(\frac{C_a}{100}\frac{0.88\times 10^{-3}}{\rho(\pi d^3/6)}\right)^{\frac{2}{3}} \cdot \frac{DoC\cdot \pi R}{\pi f_x} \cdot \left(\frac{\pi}{2} - \arcsin\left(1 - \frac{\delta_x}{A_x}\right)\right) \cdot \frac{f_x \cdot E(8d)^{0.5}\delta_x^{1.5}}{3(1-\nu^2)} \\ F = \dfrac{(3\cdot DoC\cdot 2R\cdot f_r)^{\frac{4}{3}} \cdot \left(\frac{\pi}{2} - \arcsin\left(1 - \frac{\delta_x}{A_x}\right)\right) \cdot K_C}{n^{\frac{1}{3}} f_x^{\frac{4}{3}} \pi^{\frac{7}{3}} K_V^{\frac{4}{3}} (d\delta_x - \delta_x^2)^{\frac{2}{3}} \left(\int_0^{\Delta t} \sqrt{\left(\frac{\pi SR}{30}\right)^2 + \left(2\pi f_z A_z \sin(2\pi f_z t)\right)^2}\, dt\right)^{\frac{4}{3}}} \end{cases} \tag{1.7}$$

RAN et al. (2023) hypothesized that the plastic removal volume is included in the brittle removal volume and the material is completely removed through brittle fracture, proposing a multi-scale elastic mechanics model and equivalent fracture toughness for ultrasonic assisted grinding force modeling of SiC_f/SiC composites. This model considers the anisotropy and heterogeneity of composite materials, providing a new method for predicting grinding forces and insights into material removal mechanisms. Based on the interaction of forces, Qu et al. (2020a) decomposed the grinding force model of FRC into a matrix fracture zone mechanical model and a fiber debris formation mechanical model. The single fiber removal force model adopted the elastic foundation beam theory to solve the deformation control equation. However, the above FRC grinding force analysis models have not fully considered the influence of friction between the grinding wheel and the workpiece interface. Because friction plays a crucial role in the grinding process, the removal of grinding materials is considered a friction system with multiple input and output variables. However, these models only generated average grinding forces. Therefore, Zhang et al. (2024b) analyzed the removal mode of a single effective abrasive grain at different cutting depths on the cutting arc length based on the surface morphology of the grinding wheel, the microstructure characteristics of the material, and the transition of brittle removal mode. Based on the decomposition and summation of the effective abrasive grain force in the grinding area, a dynamic grinding force model was proposed. The experiment proves that the model can accurately predict the fluctuation amplitude, dynamic evolution, and distribution law of grinding force.

Figure 9. Grain trajectories and material removal volumes in end periphery grinding under different ultrasonic vibration modes

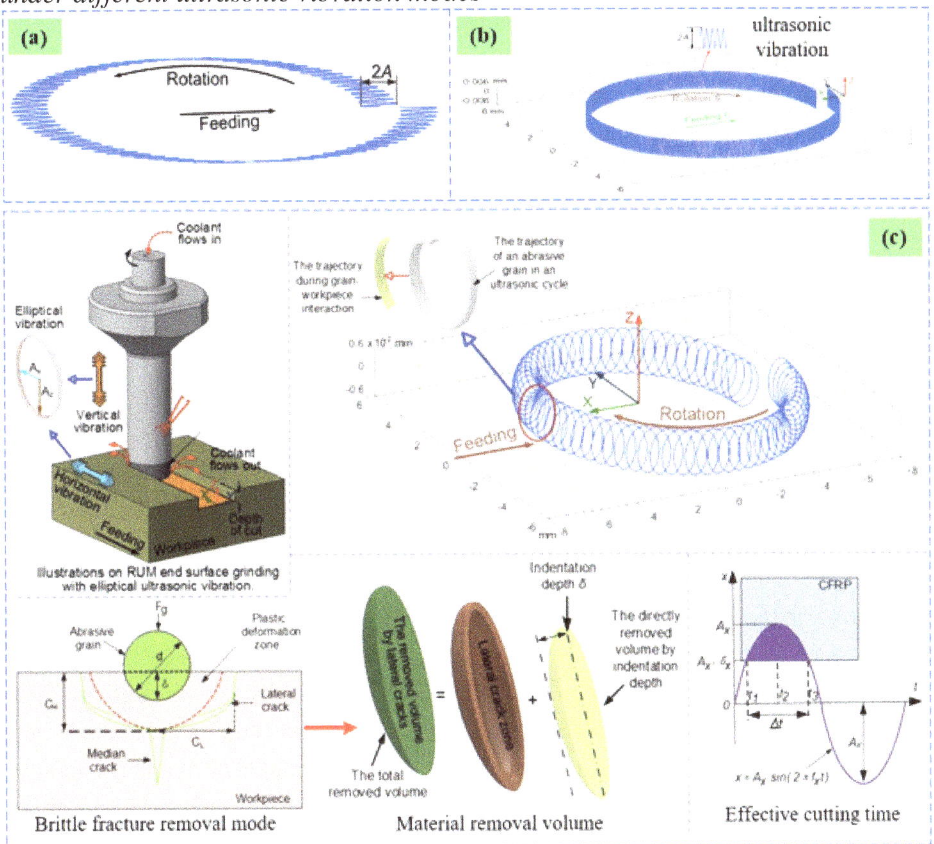

1.5.4 Model Error Assessment

The error between the grinding force model and the experimental results indicates the reliability of the predicted results. However, due to the influence of various factors during the preparation of FRC specimens and the measurement of grinding force, the errors of the models built by scholars are not the same. The comparison of the errors or correlations of mechanical models for different FRCs under different grinding methods is shown in Table 1. The errors in the CFRP grinding force model mainly come from the following sources: fiber arrangement is not completely straight and evenly distributed, equivalent homogeneous materials cannot completely replace the surrounding materials of fibers, idealized materials have no inherent defects, but there may be manufacturing defects such as voids in actual materials. During the grinding process of ceramic based FRC, the crack propagation system is significantly different from the crack propagation system in ceramics due to the presence of carbon fibers and interfacial phases. Based on the material removal mechanism, the transverse cracks in the normal FOA grinding area are confined within the grinding path. However, transverse cracks in traditional ceramics extend beyond the cutting path. Therefore, establishing a mechanical model for C/SiC composites using traditional ceramic indentation fracture mechanics may reduce prediction accuracy. RAN et al. (2023) believe that the main reason for the error is the uneven distribution of abrasive particles on the surface of the grinding wheel, which leads to inconsistent protrusion heights and affects the accuracy of the grinding force prediction model. In addition, the uneven distribution of fibers and the presence of preparation defects such as pores can also affect the accuracy of the grinding force prediction model. For the error of UVAG force model, Bie et al. (2023) believe that in addition to the non-uniformity of abrasive particles, the fluctuation of ultrasonic vibration frequency and amplitude with changes in process parameters may also be the cause of the error. FEM can be used as an auxiliary tool for analytical modeling. The main reason for FEM simulation errors is the neglect of the elasticity and damping of the grinding wheel, which can reduce the hard impact of abrasive on the workpiece. During the grinding process, the abrasive particles cut into the workpiece at high speed, generating significant impact force. In fact, the abrasive grains of the grinding wheel are connected to the substrate through bonding materials such as resin or electroplating layers, serving as a damping medium. When facing impact forces during grinding, the bonding agent can absorb energy and act as a buffer. However, due to FEM setting the grinding wheel as a rigid body, there is no buffering mechanism between the abrasive and the base ring, which can lead to grinding force errors (Wang et al., 2019b).

Table 1. Comparison of grinding force model errors

References	**Material**	**Grinding methods**	**Errors / Correlation coefficient R^2**
Ran et al. (2023)	UD SiC_f/SiC	Peripheral grinding	Maximum error F_n: 16.41%, F_t:10.79%; Average error: F_n: 9.29%, F_t: 6.16%
Qiao et al. (2022)	2.5D C_f/SiC	Face grinding	Maximum error: 18%
Zhang et al. (2019)	UD C_f/SiC	Surface grinding	$R^2 > 0.9$
Yao et al. (2023b)	2.5D SiO_{2f}/SiO_2	Face grinding	Axial grinding force error 16.02%, Transverse grinding force error 15.31%

continued on following page

Table 1. Continued

References	Material	Grinding methods	Errors / Correlation coefficient R^2
Zhang et al. (2023)	2D C_f/SiC	Peripheral grinding	Average error: F_n: 11.45% and F_t: 20.83%
Zhang et al. (2024b)	UD C_f/SiC	Peripheral-face grinding	Average error: 6.17%,
Ning et al. (2017)	MD CFRP	Peripheral grinding	Transverse grinding force error 22.80%
Zhou et al. (2023)	2.5D C_f/ SiC	Peripheral grinding	Average error: F_n: 8. 49%, F_t: 13.59%
Bie et al. (2023)	2D SiC_f/ SiC	Face grinding	Maximum error: F_n: 8.51%, F_t: 10.19%
Zhang et al. (2022b)	SiC_f/SiC	Peripheral grinding	Average error: 7.43%
Dong et al. (2022)	2D SiC_f/ SiC	Helical grinding	Maximum error of axial force 21.1%
Huang et al. (2023)	SiC_f/SiC	Helical grinding	Maximum error 17.34%
Wang et al. (2019b)	2.5D SiO_2/ SiO_2	Peripheral grinding	Maximum error of F_t and F_n: 13.09% and 8.91%
Gao et al. (2021a)	UD CFRP	Single grain	F_t and F_n: 4.76% and 16.14%
Gao et al. (2024)	UD CFRP	Surface grinding	Maximum error of F_n and F_t: 12.82% and 16.97%
Wei et al. (2019b)	3D SiO_2/ SiO_2	Single grain	Maximum error > 20%

1.5.5 Grinding Force Law and Influencing Factors

The grinding force of FRC is influenced by various factors such as material parameters, grinding wheel parameters, grinding amount, and working conditions, so the variation law of its grinding force is also extremely complex. Li et al. (2019a) conducted different single abrasive grain scratching and found that the longitudinal scratching force was greater than the transverse scratching force. Ran et al. (2024) also found a significant correlation between grinding force and fiber orientation through FEM simulation and experiments, as shown in Figure 10(a). The magnitude of the scratching force of a single abrasive grain in different directions is normal>transverse>longitudinal, while the degree of fluctuation follows transverse>longitudinal>normal. This is attributed to the influence of FOA on the crack propagation path; Meanwhile, the complex and diverse crack propagation paths cause fluctuations in grinding force and energy required for material removal to vary along different

FOAs, thereby affecting the magnitude of grinding force along different FOAs. Liu et al. (2019a) conducted experiments on three typical surfaces of 2D C_f/C-SiC composite materials with different combinations of scratching speeds and depths. The results showed that the maximum scratching force order for different cutting modes was normal>transverse>longitudinal, and the scratching force order for the three surfaces was SB>SA>SC, as shown in Figure 10(b). Under the same grinding parameters, the grinding force of MD FRC is always greater than that of UD. Because the weak coupling force between unidirectional fibers further promotes the propagation of cracks. In MD FRC, the orientation of warp and weft yarns changes under the action of needle punching structure, indicating that the coupling between fibers is effective, thereby obtaining higher strength. Cao et al. (2023) established a three-phase cutting model for 2.5D Cf/SiC composite materials considering the mechanical properties of the matrix, fibers, and interfaces. Combined with scratch experiments, the scratch forces in the three directions of MA (longitudinal and normal), MB (longitudinal and transverse), and MC (normal and transverse) were compared, as shown in Figure 10(c). The order of average scratch force magnitude is MA>MC>MB. Zan et al. (2022) found that the single abrasive scratch force of 3D SiC_f/C-SiC composite material follows SA>SC>SB on three surfaces as shown in Figure 10(d). However, Zhang et al. (2022c) found in their experiment that as the scratch depth increases, the scratch force and cross-sectional area gradually increase, but the specific scratch energy decreases. Guo et al. (2023) demonstrated through single abrasive grinding FEM that in high-speed grinding, the grinding force only slightly increases as the grinding speed increases.

Figure 10. Mechanical laws of single grain scratching along different FOAs

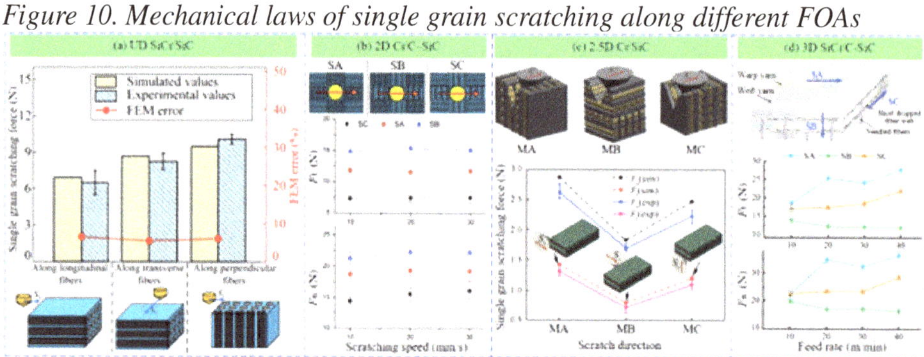

The mechanical laws of grinding wheel grinding vary with different FOAs. Zhang et al. (2016a, 2019) investigated the grinding forces in three typical directions of UD Cf/SiC composite materials, and the order of magnitude was: normal>longitudinal>transverse, as shown in Figure 11(a). Hu and Zhang (2003) found that the grinding force of UD CFRP was lower than that of MD CFRP, and believed that this was due to stronger interlayer interactions between different FOAs. Wang et al. (2019b) believed that the grinding force is transmitted along the fibers, and fiber deformation can be observed far away from the grinding zone. Wang et al. (2018b) found that the grinding force of CFRP end face periphery with FOA=0 ° is higher than that of 90°, mainly due to continuous cutting of carbon fibers in the 0° direction, which generates high friction force, while alternating cutting of carbon fibers and resin in the 90 ° direction. In addition, the grinding force decreases with the increase of tool speed, the decrease of feed rate, or the increase of ultrasonic power. LIU et al. (2017c, 2018) found that the grinding force follows the order of 45 °>30 °>0 ° under different FOAs. The order of grinding force in different grinding directions for grinding 2D Cf/C-SiC composite materials is: surface B>surface A>surface C, as shown in Figure 11(b). The grinding force increases with the increase of feed rate and grinding depth, and decreases with the increase of grinding wheel speed. This is because as the grinding wheel speed increases, the number of grains in the grinding zone increases, resulting in a decrease in the material removal volume per unit time of a single grain (Lin et al., 2021). Moreover, the grinding force ratio and specific grinding energy of 2D C/SiC are lower than those of traditional ceramics. In addition, Du et al. (2018) believe that due to the different fiber weaving structures, the tangential force generated during the peripheral grinding of 2D C/SiC composite materials may be greater than the normal force. Regarding the influence of grinding amount on grinding force, Yang et al. (2024) obtained the same pattern when grinding 3D Cf/SiC composite materials, but different conclusions were drawn when comparing the grinding force on different grinding surfaces, as shown in Figure 11(c). When grinding on surfaces A, B, and C of C_f/SiCs, the difference in grinding force was not significant. The reason is that 3D C_f/SiC composite materials have three types of tightly woven fibers on each surface, which give each surface similar mechanical properties and both can withstand significant external forces. On the contrary, the porosity of the stacked surface of 2D Cf/SiC composite material is higher than that of the 3D woven surface, resulting in a larger gap between the mechanical properties of the two surfaces. Lin et al. (2021) also found that the grinding force variation trends of single particle grinding and grinding wheel grinding of 3D SiO_{2f}/SiO_2 were significantly different, and believed that the main reason for the increase in grinding force was fiber fracture. Moreover, Wang et al. (2016a) found that the grinding force of CFRP decreases with increasing abrasive size and increases with increasing abrasive concentration. Yin et al. (2022) increased the grinding speed from 20m/s to 50m/s and 90m/s, reducing the width of the edge debris area by 30% and 60%, respectively. Increasing the grinding speed not only does not enlarge the chips, but also greatly improves the material removal rate, which is of great significance for the surface smoothness and damage control of SiCf/SiC. Xu et al. (2021)

found that under the same grinding parameters, thermoplastic resin based FRC has higher grinding force and temperature than thermosetting FRC, making the grinding wheel more prone to clogging and making the machining more difficult. However, the surface quality of thermoplastic resin matrix FRC with high ductility is much better than that of thermosetting resin matrix.

Figure 11. The influence of different FOA on grinding force

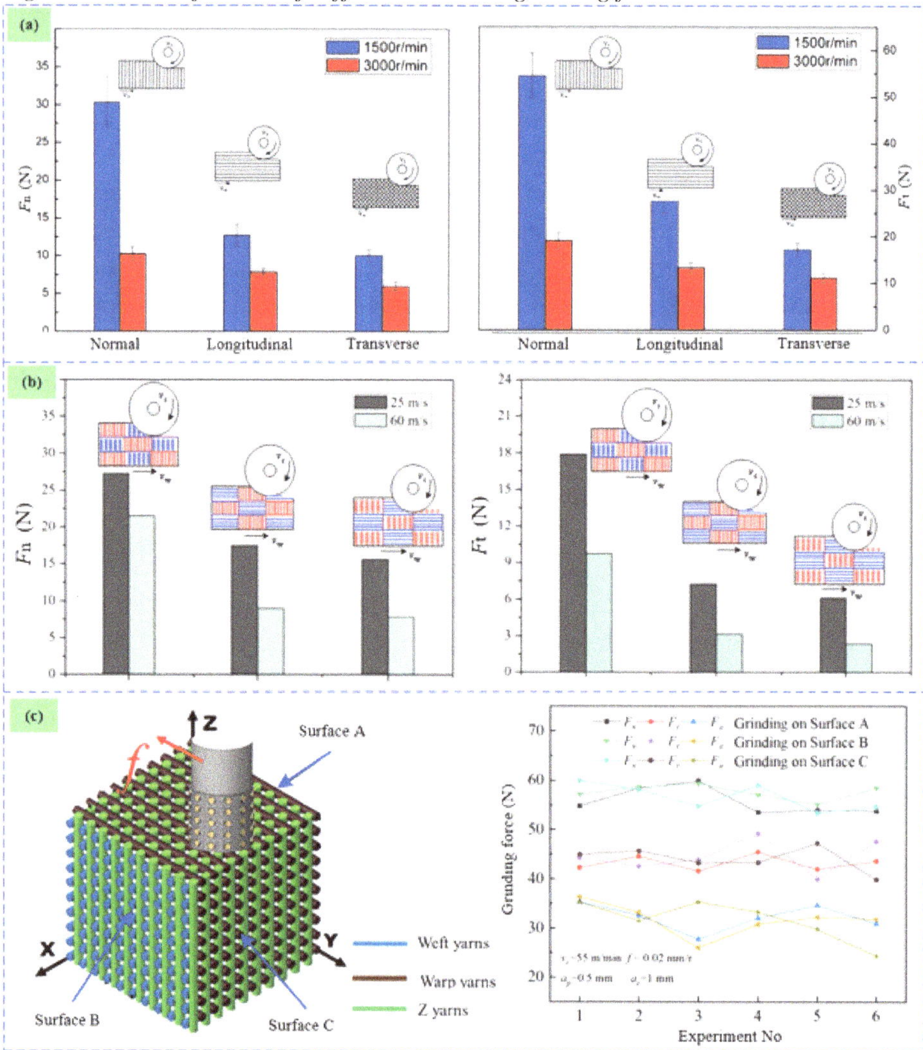

1.6 ULTRASONIC VIBRATION ASSISTED NANOLUBRICANT MINIMUM QUANTITY LUBRICATION

1.6.1 Ultrasonic Vibration Assisted Machining

Ultrasonic vibration-assisted machining (UVAM) is one of the newly developed machining technologies (Yuan et al., 2016; Xu et al., 2017b). It is a machining method that combines ultrasonic vibration with traditional machining (Babbar et al., 2019; Verma & Pandey, 2019). It adds high-frequency ultrasonic vibration to the tool or the workpiece, uses the energy of ultrasonic vibration to change the removal mechanism (Yuan et al., 2017a), improve the machining process, and obtain better machining performance (Gao et al., 2020a; Wang et al., 2016b; Ma et al., 2020). Figure 8 shows the damage suppression mechanism of ultrasonic vibration grinding (Chen et al., 2020a; Ding et al., 2017b; Ning et al., 2019; Xue et al., 2021a). UVAM adds a microscale amplitude displacement with ultrasonic frequency to the tool tip motion (Cong et al., 2011; Fernando et al., 2019) and changes the path of the tool tip so that the instantaneous cutting depth is much smaller than the fiber diameter, which can improve the surface integrity (Xu et al., 2014; Yang et al., 2020). UVAM with horizontal ultrasonic vibration generated fewer defects because of abrasive-grain trajectory overlapping. The smaller simultaneous depth of cut and smaller simultaneous cutting volume in UVAM contributed to the less severe fiber fracture, cracks, and damages on matrix in processing fiber composites (Wang et al., 2020; Wang et al., 2017). As shown in Figure 12, the literature review and analysis of UVAM are presented, with data sourced from the Web of Science (retrieved Dec. 2025).

UVAM has a significant effect on reducing cutting force, and the cutting force needs to be predicted and modeled for its controllability and minimization. The modeling of cutting force is also the key to effectively control the machining damage of fiber composites. However, due to the heterogeneity, anisotropy, and low heat dissipation characteristics of these composites, ensuring high accuracy and low estimation error of the force model is difficult. Amin et al. (2018a) established the axial and feed cutting force models of CFRP composites and characterized and calculated the exposed height of diamond grains in the modeling process. Furthermore, the force models of CFRP ultrasonic-assisted surface milling of two milling tools (cylindrical abrasive core tool and conical shaped core tool) are established (Amin et al., 2017). The expression of tool contact area is improved, and overlapping cutting quantification is introduced for the first time. The estimation error between the experimental and simulated values of cutting force is less than 10%. In recent years, ductile material removal mode and brittle mode can be applied in CFRP ultrasonic vibration-assisted grinding (UVAG). Wang et al. (2020a) established the force model of ultrasonic vibration-assisted edge surface grinding of CFRP composites based on ductile and brittle mode and used the critical indentation depth to distinguish plastic and brittle zones. The plastic zone increases, and the brittle zone decreases with the increase in tool speed or the decrease in feed rate. The brittle region increases with the cutting depth. Rotary ultrasonic machining using vertical ultrasonic vibration is an effective method to reduce the cutting force. However, the quality of the machined surface decreases due to the knocking effect of vertical vibration on the machined surface. Based on the removal mechanism of brittle fracture materials, Liu et al. (2017b) established a CFRP grinding force model under horizontal ultrasonic vibration. The results show that the grinding force decreases with the increase in ultrasonic amplitude. Ning et al. (2017) established a feed force model for CFRP surface grinding based on the material removal hypothesis of brittle fracture, by introducing the fracture volume factor of awesome material. The difference in the fracture behavior of carbon fiber leads to the difference in the material removal energy consumption. The difference of material removal mechanism directly leads to the difference of machined surface morphology. The material removal mechanism can be effectively characterized by machined surface roughness.

Figure 12. Literature review and analysis on UVAM

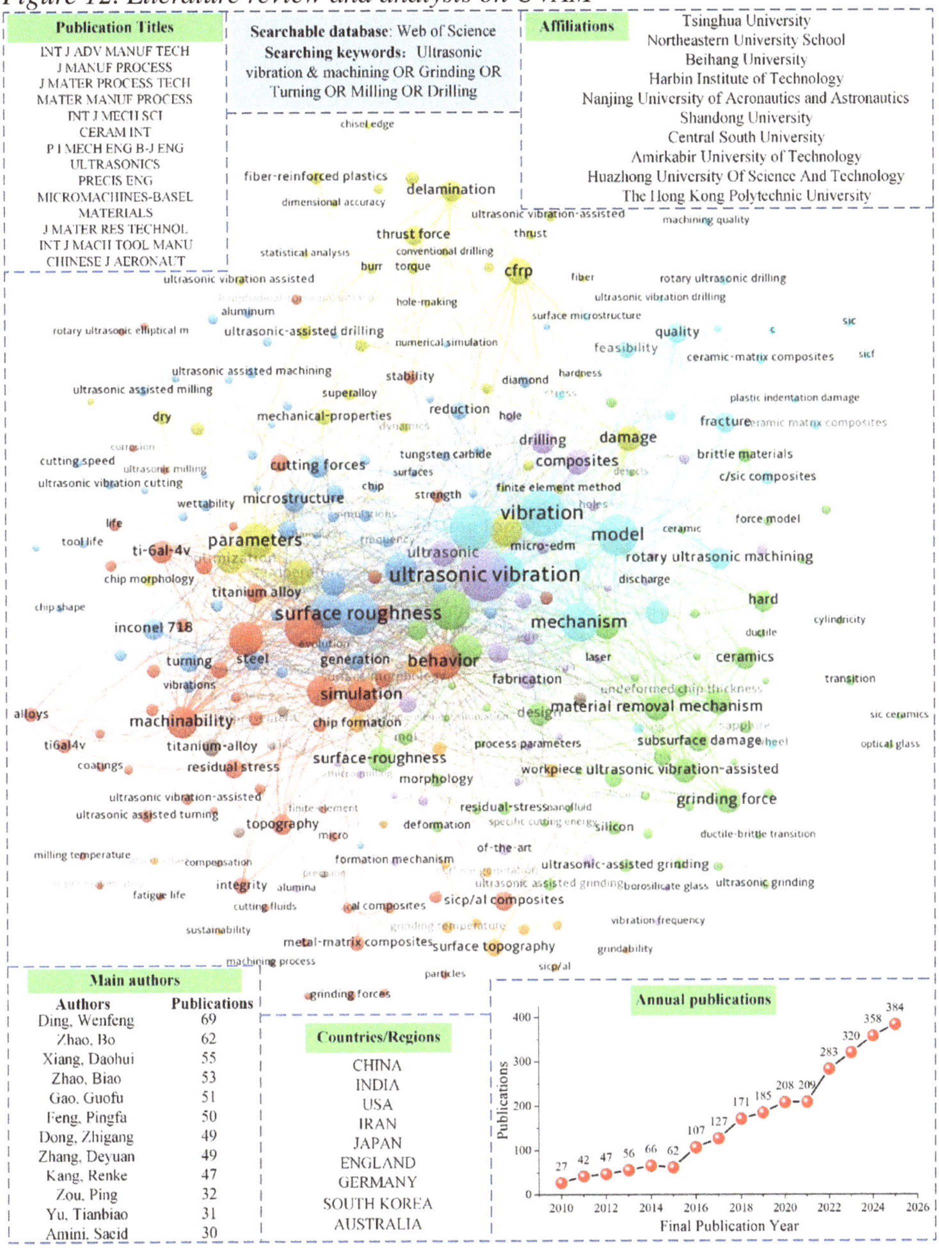

Chen et al. (2020a) considered that the transformation of material removal mechanism depends on the maximum undeformed chip thickness (h_{max}). Micro cracks are generated and propagated in the carbon fiber because of the extrusion of abrasive particles when h_{max} is small enough. SiC matrix breaks and carbon fiber debonds from the matrix when h_{max} is large enough. The main fracture mechanisms are bending fracture, compression fracture, and shear fracture. As shown in Figure 13(a), UVAG promotes the removal of carbon fiber in nano brittle fracture by reducing the maximum undeformed chip thickness to improve the machined surface quality. The improvement effect increases with the amplitude. Ding et al. (2017b) considered that UVAG reduced the layered brittle fracture and pit group caused by fiber fracture and pull-out to varying degrees compared with conventional grinding (CG) because UVAG could reduce the grinding force that determined fiber fracture. Therefore, due to less fiber breakage and reduction of fracture size (Figure 13(b)), the surface roughness of arithmetical mean deviation of regional morphology S_a obtained by UVAG is lower than CG, with a maximum decrease of 12%. Wang et al. (2018a) reported the influence of machining variables (e.g., ultrasonic power, grinding speed, feed rate and depth of cutting) on output variables (e.g., cutting force, torque, and surface roughness). The results show that increasing the ultrasonic power and grinding speed and decreasing the feed rate can reduce the force and the depth of damage. The surface roughness can be lessened by reducing ultrasonic power, feed rate, and cutting depth and increasing grinding speed. The up-grinding force is evidently greater than that of down grinding, and down grinding produces low surface roughness. Li et al. (2019b) discussed the reasons for the differences in cutting force and surface integrity. The surface morphology shows that brittle fracture is the main material removal mode of CFRP composite grinding. The chip size of the resin, the fracture size of carbon fiber, and the material removal size are small using down grinding. Therefore, ultrasonic vibration-assisted down-grinding CFRP is the optimal scheme to obtain high surface quality and the required dimensional accuracy. Wang et al. (2018b) found that ultrasonic vibration-assisted milling along 90° orientation can obtain less cutting force and torque. The main reason is that when the machining orientation is 0°, the tool continuously cuts carbon fiber layers, accelerate the tool passivation, and increase the friction between the tool and the workpiece. The tool can alternately cut carbon fiber and epoxy resin in the FCA of 90°. Smaller surface roughness can be obtained in 0° because the fiber is not successfully cut in 90°, thereby resulting in fiber pulling out and burr. The heavier the fracture of the carbon fiber is, the more chips and cracks are produced. Ning et al. (2019) found that CFRP was brittle removed, resulting in matrix damage and serious fiber pullout and macro cracks in the conventional scratch test. However, ultrasonic vibration-assisted scratching (UVAS) of CFRP has a large ductile removal area before continuous brittle fracture and crack. Debonding and pulling of fiber matrix are also significantly reduced, and only matrix buckling and fiber fracture occur in the groove (Figure 13(c)). Xue et al. (2021a) showed that the surface residual compressive stress of C/SiC composites is approximately 2 GPa because of the continuous impact of high-frequency and low-amplitude vibration (Figure 13(d)).

Residual compressive stress restrains most of the interface cracks and hinders the propagation of fiber cracks to improve the toughening effect and machined surface quality of fiber. The surface quality of fiber composites is closely related to the fracture mechanism (Xue et al., 2021b). High-frequency and low-amplitude vibration changes the cutting angle of the fiber and increases the shear stress of the fiber through the axial to radial deflection friction of the fiber. It greatly promotes the transformation from shear fracture mode to dominant fracture mechanism. Ultrasonic vibration increases the proportion of FCA under shear fracture mode by 30%. Table 2 shows the vibration parameters of ultrasonic vibration and their comparison with the results in traditional machining.

Figure 13. Damage suppression mechanism of UVAG

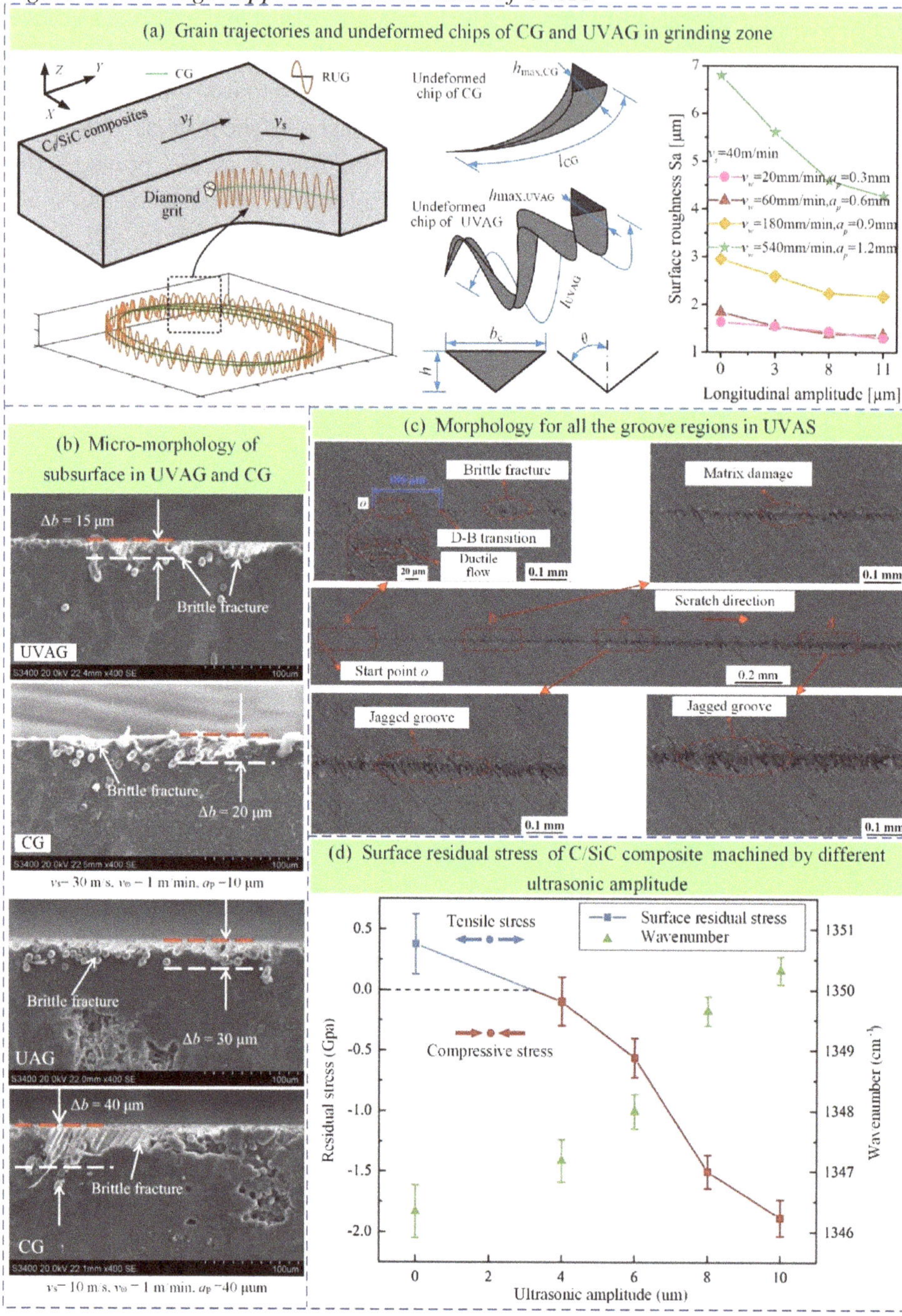

Table 2. Ultrasonic vibration parameters and output parameters

Reference	Vibration mode	Amplitude/μm	Frequency/kHz	Composites	Output parameters (compared with CM)
Wang et al. (2019c)	1D horizontal	2; 4; 5	28	UD-CFRP	–
Liu et al. (2017b)	1D longitudinal	10	20	MD-CFRP	–
Wang et al. (2018a)	1D longitudinal	–	20	UD-CFRP	–
Xu et al. (2021)	Longitudinal–torsional	6; 3	22.5	AFRP	Cutting force: ↓ 15%–48%
Chen et al. (2020a)	Longitudinal-torsional	Longitudinal amplitude (LA)/ torsional amplitude (TA) = 14/5; LA = 3; 8; 11	19.3	2D C/SiC	Specific grinding energy: ≈ ↓ max 50%; S_a: ≈ ↓ max 38%
Ding et al. (2017b)	1D longitudinal	3.85	21.5	2D C/SiC	F_n: ↓ 9%–21%; F_t: ↓ 9.7%–19.4%; S_a: ↓ 6.6%–12%; Fracture size: ↓ 25%
Ning et al. (2019)	Horizontal–vertical	3	20	MD-CFRP	–
Xue et al. (2021a)	1D longitudinal	4; 6; 8; 10	20	3D C/SiC	Standard deviation of height of each point in the area S_q: ↓ 37.9%; fatigue damage rate: ↓ 31%–80%
Wang et al. (2020c)	Horizontal; horizontal–vertical	Horizontal 4; vertical 6;	28; 20;	CFRP	Cutting force: ≈ ↓ 21%–36%; surface roughness: ≈ ↓ 50%–62%
Amin et al. (2017, 2018a, 2018b)	1D longitudinal	10	16; 20.5	MD-CFRP	–
Wang et al. (2020a)	1D longitudinal	4; 6; 7; 8	20		–
Ning et al. (2017)	1D longitudinal	4; 5; 6; 7; 8	20	MD-CFRP	–
Li et al. (2019b)	1D longitudinal	6	20	MD-CFRP	–
Wang et al. (2018b)	1D longitudinal	–	20	MD-CFRP	–
Xue et al. (2021b)	1D longitudinal	10	20	3D C/SiC	Distance between the highest and lowest points of the contour R_t: ↓ 37.5%; R_a: ↓ 32.4%; R_z: ↓ 32.8%

continued on following page

Table 2. Continued

Reference	Vibration mode	Amplitude/μm	Frequency/ kHz	Composites	Output parameters (compared with CM)
Yuan et al. (2017b)	1D longitudinal	10; 15	--	MD-CFRP	–
Liu et al. (2021)	1D longitudinal	2; 4; 6; 8; 10	21.19	C/SiC	F_x: ↓ 43.7%; F_y: ↓ 9.16%; F_z: ↓ 68.09%; R_a: ≈ ↓ 12%–16.7%
Chen et al. (2020b)	Longitudinal-torsional	LA/TA=14/5; LA =2; 4; 6; 7	19.32	2D C/SiC	Temperature: ↓ 30.4%; specific milling energy: ≈ ↓ 33.3%; S_a: ≈↓ 28.6%
Xie et al. (2021)	1D longitudinal	2; 4; 6; 8; 10	23.98	2.5D C/SiC	F_x: ↓ 27%; F_y: ↓ 49.5%; F_z: ↓ 28.6%; S_a: ↓ 53%
Geng et al. (2015)	Double bending	9.6	20.73	CFRP	F_x: ↓ 8%–27%; F_y: ↓ 12%–43%; F_z: ↓ 2%–40%; surface roughness: ↓ 54%
Shu et al. (2021)	1D longitudinal	10	30	C/C	Fiber pull-out length: ↓ 10%–50%
Zhang et al. (2016b)	1D longitudinal	15	21.5	C/SiC	–
Islam et al. (2020)	1D longitudinal	–	17	C/SiC	–
Liang et al. (2019)	1D longitudinal	5	20-30	MD-CFRP	–
Li et al. (2016)	1D longitudinal	–	26.5	2.5D SiO_2/ SiO_2	Grinding forces: ↓ 30%–35%; S_q: ↓ 12.5%; maximum height of 2D profile S_z: ↓ 12.3%
Wang et al. (2020d)	1D longitudinal	4	21.2	C/SiC	Grinding forces: ≈ ↓ max 60%
Wang et al. (2020e)	Vertical; vertical–horizontal	4; 6	–	CFRP	Vertical vibration: F_x: ↓ 11%–20%; F_z: ↓ 12%–40%; S_a: ↓ 12%–21%; vertical–horizontal vibration: F_x: ↓ 21%–57%; F_z: ↓ 30%–52%; S_a: ↓ 50%–60%

continued on following page

Table 2. Continued

Reference	Vibration mode	Amplitude/μm	Frequency/ kHz	Composites	Output parameters (compared with CM)
Azarhoushang and Tawakoli (2011)	1D tangential	8	20	C/SiC	Grinding forces: ≈ ↓ max 20%; surface roughness R_a and R_z: ≈ ↓ max 30%; G-ratio: ≈ ↑ 30%–40%; radial wear: ≈ ↓ max 28%–45%
Chen et al. (2019)	1D longitudinal	14	29.7	CFRP	-
Liang et al. (2021)	1D longitudinal	4	26	UD-CFRP	F_t: ↓ max 19% at 135°; F_n: ↓ max 7.2% at 135°; maximum fiber chip length: ≈ ↓ 66%–78%; S_a: ↓max 17.7%

1.6.2 Minimum Quantity Lubrication

The hygroscopicity of resin matrix will lead to the water absorption and expansion of CFRP composites, seriously affects its mechanical properties which and limits the application of casting cooling lubrication in the grinding of CFRP composites (Larbi et al., 2015; Scida et al., 2013; Wu et al., 2021). Therefore, most of them currently adopt dry processing. However, the dry machining of CFRP composites usually leads to surface quality deterioration, dust pollution (Haddad et al., 2014), wheel blockage, and various damages (e.g., fiber pull-out, fiber fracture, resin coating, interface cracking, fiber matrix debonding, and delamination). In dry grinding, the grinding temperature in the contact area of the grinding wheel is higher than the thermal substitution temperature of the matrix resin, and the thermal deformation of the matrix resin destroys the grinding surface. Lubricating and cooling medium is the key factor that affects the processing result parameters (Danish et al., 2021). Some researchers have verified the unique advantages of MQL in cutting fiber composites (Senthilkumar et al., 2018; Giasin et al., 2016). MQL is a new cooling and lubrication technology (Li et al., 2017; Wang et al., 2018c; Zhang et al., 2012). High-pressure gas and minimal lubricating oil are mixed and atomized to form oil mist containing droplets, which are sprayed to the cutting area at high speed through the nozzle (Yang et al., 2017; Zhang et al., 2016c; Jia et al., 2022). Given the volatile property of MQL, it has no residue on the wall holes, which avoid the secondary cleaning of aviation parts after processing. MQL liquid forms lubricating oil film on the interface between the tool and workpiece, and high-pressure air mainly plays the role of cooling and chip removal (Jia et al., 2016). To realize green production from the source of machining, vegetable oil with high biodegradability and nontoxic is generally selected as the base oil (Wang et al., 2020f; Guo et al., 2017). This approach greatly reduces the harm of cutting fluid to the environment and human body. As shown in Figure 14, the literature review and analysis of MQL are presented, with data sourced from the Web of Science (retrieved Dec. 2025).

Figure 14. Literature review and analysis on MQL

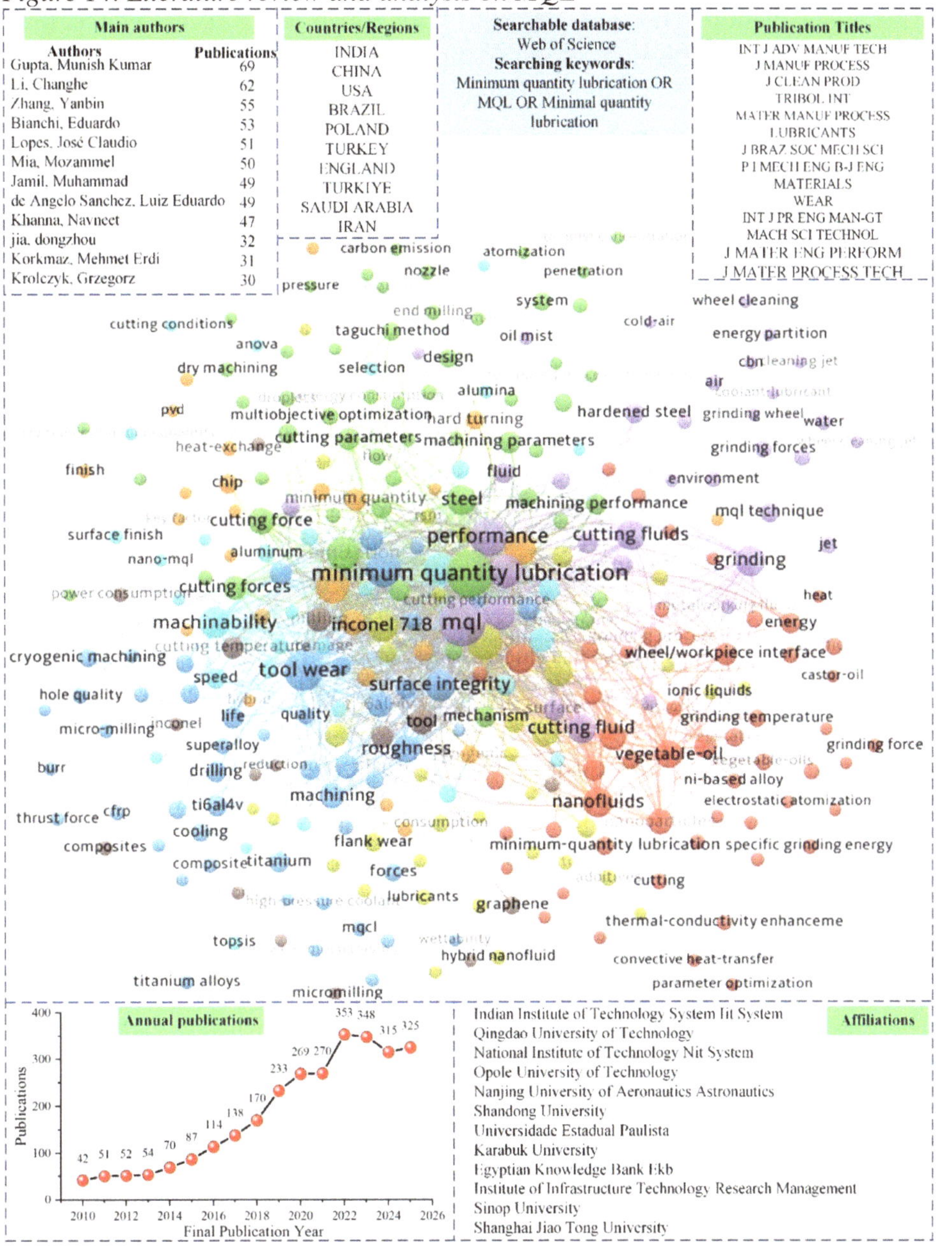

Yang et al. (2019a, 2019b) proved through experiments that MQL could solve the bottleneck of debris formation mechanism in the process of plastic removal of hard and brittle materials. The jet-flow condition of MQL has significant influence on the atomization effect and processing performance (Duan et al., 2021).

Cococcetta et al. (2020) showed that milling CFRP at a lower feed rate would lead to material adhesion to the tool, which leads to tool upsetting material rather than shearing material. MQL reduces the friction and tool wear between the tool tip and the workpiece. Dry machining results in a large number of burrs, which can be reduced significantly by MQL. Qu et al. (2020c) studied the effect of MQL on the grinding properties of unidirectional C/SiC composites. The grinding surface quality is optimal, and the grinding force is small when the nozzle angle is 15°, the air pressure is 5 bar (1 bar = 10^5 Pa), the flow rate is 100 ml/h, and the nozzle distance is 80 mm. Smooth fiber separation, fiber fracture, fiber outcrop, fiber pullout, and matrix crack are the main failure forms of the surface. In MQL grinding, water vapor takes away a lot of heat and reduces the grinding temperature significantly. An effective oil film is also formed in the contact area between the grains and the workpiece surface. The experimental results in Esmaeili et al. (2019) showed that brittle fracture was the main material removal mechanism of MQL grinding of C/SiC composites as shown in Figure 15(a). MQL can reduce the cutting force by 37.95%, reduce the surface roughness by 75.93%, and improve the grinding efficiency by 150% compared with dry grinding. The lubrication mechanism of MQL is analyzed, that is, the pressurized directional transportation of oil mist makes oil droplets that penetrate into the pore network on the workpiece and grinding wheel effectively. The G-ratio of MQL grinding is 115.38% higher than that of dry grinding (Adibi et al., 2018). Pervaiz et al. (2020) found different rules and believed that the MQL strategy in external configuration could not provide favorable performance. One possible reason is that part of the lubricant flows into the cavity during MQL-assisted CFRP machining, which leads to the formation of insufficient lubricating film. Through the CFRP milling experiment, Iskandar et al. (2014) found that the spray mode that combined high air flow and low oil flow promoted the breakup mechanism of droplets, and the best spray effect was obtained. It also produces a more coherent jet with less vortex formation and better penetration into the cutting area. Finally, the flank wear of MQL is reduced by 30% compared with pressurized air and 22% compared with dry and flood. Although MQL has been proved to have certain potential and advantages in machining of fiber composites, some studies also have different opinions. Helmy et al. (2018) found that MQL spray cooling increased cutting force compared with flood, because it did not adequately remove heat and dust in the cutting area. However, MQL has the advantage of reducing the pollution of cooling lubricant and the ability to provide the machining effect when it is close to the requirements of dry machining.

To further improve the heat transfer capacity and tribological characteristics of MQL, some scholars have proposed a new clean cooling and lubrication technology with low carbon, low consumption, and high efficiency nanofluid MQL (NMQL) (Li et al., 2020; Yang et al., 2021b; Said et al., 2021). The technical approach aims to add one or several nanoparticles with different physical and chemical properties to MQL base oil and fully disperse and mix the nanoparticles to prepare nanofluid (Cui et al., 2022; Zhang et al., 2022d; Said et al., 2022). Then, under the action of high-pressure air flow, the nanofluid is atomized and sprayed into the cutting area through the nozzle (Zhang et al., 2015). NMQL effectively improves the heat transfer capacity and lubrication performance of MQL in the grinding area (Dong et al., 2019; Bai et al., 2019; Yang et al., 2021c). Some studies show that NMQL achieves a cooling effect close to flood lubrication. In addition, nanoparticles help improve the lubrication characteristics of grinding wheel/workpiece and grain/chip interface because of their excellent anti-wear and antifriction properties (Cui et al., 2019; Gao et al., 2019; Zhang et al., 2022e). Gao et al. (2020b, 2021b) studied the grinding of CFRP with CNTs nanofluid. Compared with dry grinding, the results show that CNT nano-lubricant reduced maximum F_n and F_t by 20.07% and 26.81%, respectively. The friction coefficient under CNT nano-lubricant MQL is the smallest at $\mu = 0.141$. As shown in Figure 15(b), the grinding force reduction mechanism is explained in two points: (1) The strong constraint of the matrix material on the fiber under temperature reduction is conducive to the removal of fiber by grains; and (2) the excellent tribological properties of CNT nanoparticles at the interface between grains and fiber. The 2D and 3D surface roughness and fractal dimension under NMQL are the lowest. In terms of local morphology characteristics, the spectral width and spectral difference of multi-fractal spectrum of CNTs NMQL are reduced by 21.76% and 31%, respectively, compared with dry grinding. CNT NMQL significantly reduces resin coating, multi-fiber block pull-out, voids, and edge collapse. The nanoparticles filled in the debonding gap between the fiber and the matrix play a supporting role on the fiber, thereby easily removing the fiber (Figure 15(b)). The oil film between the grinding wheel and workpiece interface can effectively reduce or prevent the adhesion and blockage of resin matrix to grains and effectively improve the blockage of the grinding wheel. Qu et al. (2020b) carried out the grinding experiment of unidirectional carbon fiber-reinforced CMCs under different conditions, and the results are shown in Figure 15(c). In conclusion, the optimal parameters of carbon nanofluid MQL are: nanoparticle concentration of 5 g/L, air pressure of 7 bar, flow rate of 80 mL/h, and nozzle distance of 60 mm. The debonding depth between matrix and carbon fiber depends on the sharpness and lubrication state of grains. James and Nejadian (2020) compared the cutting performance of vegetable oil-based NMQL with different nanoparticles (e.g., CNT, Al_2O_3, Ni, and Al) to process different CFRP stacks. The experimental results show that 2.5%vol Al_2O_3 + MQL reduces the surface roughness of CFRP and titanium composites most than dry machining. However, 1%vol SWCNT + MQL has the optimal effect on improving the surface roughness of CFRP and Al composites. Table 3 lists the processing results of different MQL conditions and dry processing.

Figure 15. Lubrication mechanism, force reduction mechanism, and beneficial effects of MQL and NMQL

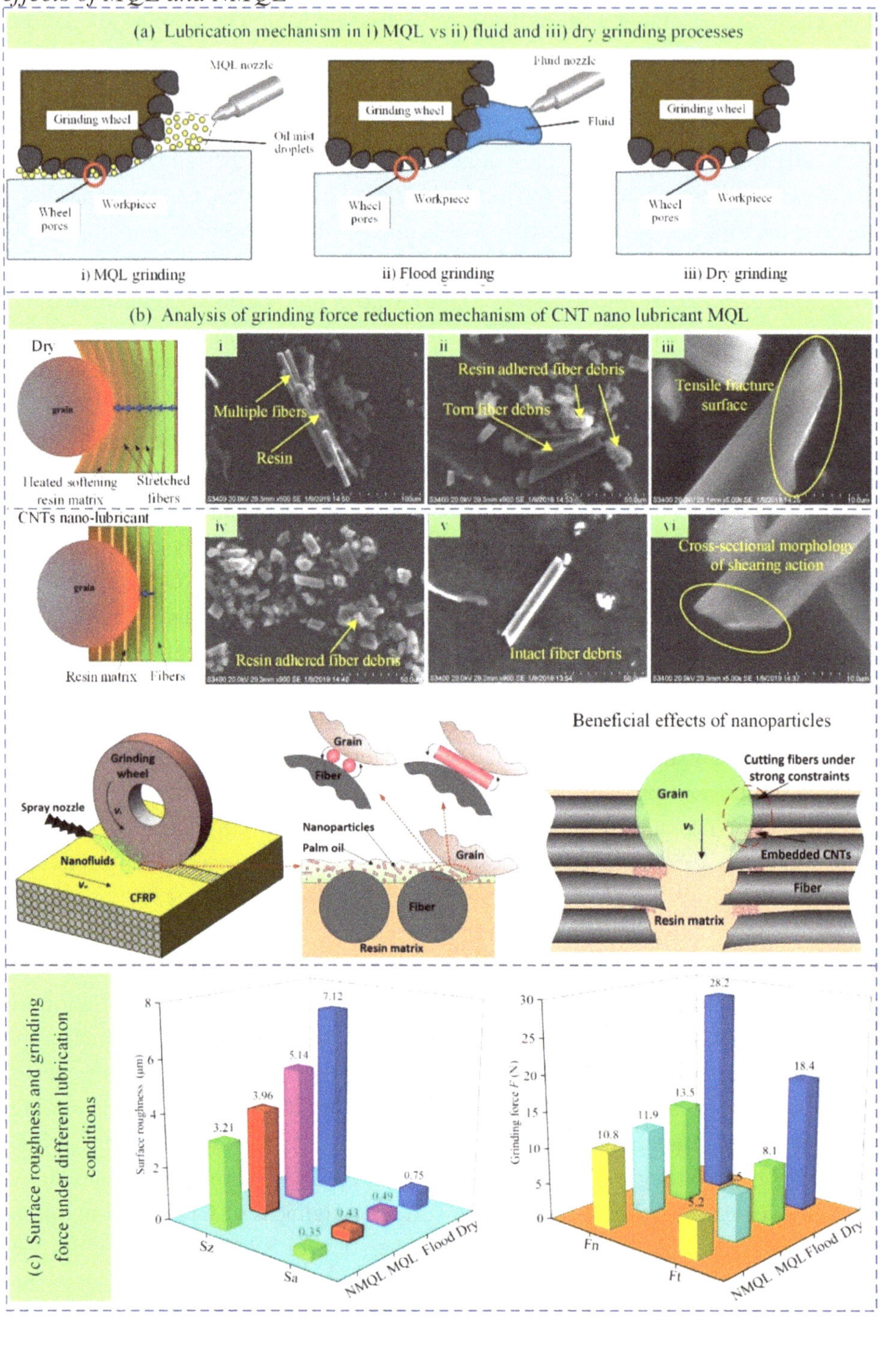

Table 3. MQL conditions and output parameters

Reference	Lubricant	Supply parameters			Composites	Output parameters (compared with dry)
		Flow rate / (mL h^{-1})	Air pressure / bar	Device		
Qu et al. (2020b)	Deionized water-based carbon nanofluid	40, 60, 80, 100	3, 5, 7, 9	KINS KS-2107	C/SiC	F_n: ↓ 62.5%; F_t: ↓ 71.7%; S_a: ≈ ↓ 53.3%; S_z: ≈ ↓ 54.9%
Qu et al. (2020c)	MQL oil	80, 100, 120	3, 5, 7	KINS KS-2107	C/SiC	F_n: ↓ 57.8%; F_t: ↓ 64.7%; S_a: ≈ ↓ 41.3%; S_z: ≈ ↓ 42.9%;
Danish et al. (2021)	Eco-friendly MQL oil	150	6	–	CFRP	Friction coefficient: ≈ ↓ 60%
Zou et al. (2021)	Cryogenic MQL: soluble vegetable oil	20	85–90	–	CFRP	Temperature: ↓ 70.3%
Cococcetta et al. (2021)	Coolube® 2210 oil-based cutting fluid	–	–	UNIST MQL system	CFRP	Slot milling: linear tool wear: ↓ 44%; tool wear area: ↓ 16%; edge milling: linear tool wear: ↓ 43%; tool wear area: ↓ 30%
Esmaeili et al. (2019)	Corn oil	6000	4	MQL system produced by Royal Dutch	C / SiC	Cutting force: ↓ 37.95%; power: ↓ 38.39%; R_a: ↓ 75.93%; G-ratio: ≈ ↑ 70%
Adibi et al. (2018)	Corn oil	6000	4	--	C / SiC	F_n: ↓ 31.16%; F_t: ↓ 38.88%; specific grinding energy: ↓ 41.77%; G-ratio: ≈ ↑ 115.38%; R_a: ↓ 75.26%; radial wheel wear: ≈ ↓ 50%–66.7%; grain flattening wear: ≈ ↓ 93.75%
Helmy et al. (2018)	Water-based cutting fluid	Two-nozzle system: 2220	4	NEX FLOW mist cooling system	CFRP	–

continued on following page

Table 3. Continued

Reference	Lubricant	Supply parameters			Composites	Output parameters (compared with dry)
		Flow rate / (mL h^{-1})	Air pressure / bar	Device		
Gao et al. (2020b, 2021a)	Palm oil-based CNTs nanofluids	60	6	KINS KS-2106	CFRP	R_a: ↓ 12.68%–17.7%; R_z: ↓ 20.78%–25.06%; average width of profile micro unevenness RS_m: ↓ 11.43%–25.4%; single grain F_n: ↓ 20.07%; single grain F_t: ↓ 26.81%; friction coefficient: ↓ 54%
James and Nejadian (2020)	CNT, Al_2O_3, Ni, and Al enhanced Mixture of castor oil and jojoba oil	13097	1.013	–	CFRP	–
Rodriguez et al. (2019)	Accu-Lube LB-1000 with chlorinated extreme pressure additives	100	8000	–	CFRP	–

1.6.3 Development Milestone

Figure 16 illustrates the developmental timeline of cooling/lubrication technologies and ultrasonic vibration-assisted grinding (UVAG) technologies. Regarding cooling and lubrication technologies, cutting fluids began to gain widespread adoption in the 1980s. As efforts to reduce cutting fluid usage advanced, researchers proposed MQL, which developed rapidly. This was followed by the emergence of NMQL, where nanoparticles are added to the lubricant, enhancing its anti-wear, friction-reduction, and heat dissipation capabilities. The use of green, biodegradable vegetable oils as base fluids further aligned the technology with environmental sustainability. Subsequently, scholars began investigating MQL assisted by various energy fields, such as electrostatic, ultrasonic, and magnetic fields, all of which contributed to improving MQL performance. Currently, with the growing adoption of nano-bio-lubricant MQL technology systems, the field is progressively advancing toward cleaner, more efficient, precise, and intelligent solutions. In recent years, the number of SCI publications, as well as ESI hot papers and highly cited papers in this area, has continued to increase. In the domain of UVAG, the first model for elliptical ultrasonic vibration machining was proposed in the 1960s. This was followed by early developments in rotary ultrasonic grinding, and subsequent exploration of one-dimensional ultrasonic vibration-assisted grinding, including axial, longitudinal, and tangential directions. Later, fixed two-dimensional ultrasonic vibration was introduced, and the adjustability of 2D multi-directional ultrasonic vibration further enhanced grinding performance. Currently, multi-directional, multi-mode, and adjustable composite ultrasonic vibration has gradually become a research focus.

When using ultrasonic vibration assisted grinding to process anisotropic materials such as carbon fiber reinforced composites, the interaction between the tool and the workpiece strongly depends on the fiber orientation angle (FOA). Previous studies have fully confirmed that using a fixed elliptical vibration trajectory can effectively reduce grinding force, suppress fiber tearing, and improve surface quality. However, the vibration direction angle of such fixed trajectories remains constant throughout the entire machining process and cannot match the local FOA that continuously changes with the surface contour or layer direction. When the vibration direction deviates from the optimal angle of fiber orientation, the material removal mechanism undergoes an unfavorable transformation, leading to an intensified bending effect of the cutting edge on the fibers, which in turn causes typical damages such as burrs and delamination. To overcome this limitation, this work proposes dynamically adjusting the elliptical vibration angle to form a synergistic match with the instantaneous FOA at the machining point, thereby maintaining a low damage material removal mode at all times. However, as described in this article, although the advantages of fixed elliptical trajectories have been repeatedly validated, the ability to adjust vibration angles in real-time to achieve synergy with specific FOAs has been almost unexplored in existing literature. This gap highlights the urgency of developing adaptive vibration strategies for anisotropic materials and opens up new directions for further improving the grinding quality of difficult to machine composite materials.

Figure 16. Evolution of cooling lubrication technology and UVAG technology

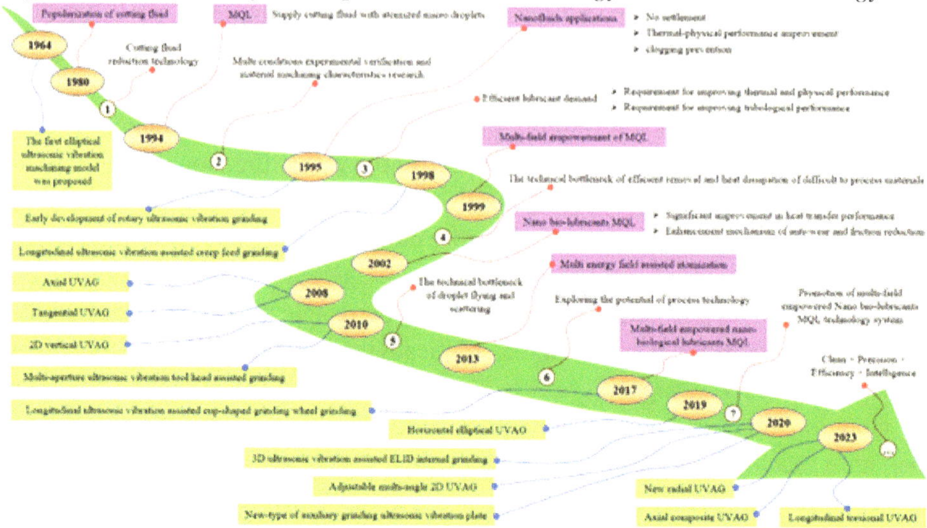

1.7 LOGICAL FRAMEWORK AND RESEARCH SIGNIFICANCE

Precision grinding is an essential machining process for achieving high surface integrity and dimensional accuracy in near-net-shaped aerospace CFRP components, particularly for obtaining precision mating surfaces that ensure assembly tolerances and alignment accuracy. Flood lubrication is unsuitable due to moisture absorption-induced swelling of CFRPs, which degrades their mechanical properties. Meanwhile, dry grinding under severe working conditions tends to induce severe thermo-mechanical damage. As a result, nanofluid minimum quantity lubrication (NMQL) has become an indispensable alternative to dry grinding. However, under extreme boundary conditions characterized by high temperature, pressure, and speed, it is challenging to simultaneously satisfy the dual requirements of adequate oil film thickness and effective droplet penetration. The insufficient effective flow rate of nano-lubricants often limits their cooling and lubrication performance.

To address these technical challenges, this study proposes a novel multi-directional two-dimensional ultrasonic-assisted grinding method with nanofluid MQL for heterogeneous and anisotropic CFRPs. The approach aims to enhance NMQL effectiveness by improving dynamic pressure lubrication at the wheel-workpiece interface, enhancing pumping effects in the grinding zone, and promoting the spreading and infiltration of lubricants on the CFRP grinding surface.

The logical framework and research roadmap of this book are shown in Figure 17. The book will analyze the kinematic behavior and cutting characteristics of abrasive grains under multi-directional two-dimensional ultrasonic vibration, clarify the mechanisms of lubricant penetration and transport through micro-channels in the ultrasonically activated wheel-workpiece interface, and reveal the cross-scale material removal mechanics in grinding CFRPs with varying fiber orientations. A grinding force model will be established for the multi-directional 2D ultrasonic-assisted NMQL grinding process. The formation and evolution of thermo-mechanical damage during grinding of CFRPs with different fiber orientations will be investigated, and experimental studies will be conducted on multi-parameter cooperative damage suppression strategies driven by multi-energy fields.

This research aims to overcome the critical challenge of severe thermo-mechanical damage in conventional CFRP grinding, achieve coordinated multi-parameter control of grinding damage in CFRPs for aerospace and marine engineering applications, and provide theoretical guidance and technical support for precision manufacturing of high-end aerospace equipment.

Figure 17. Logical framework and research roadmap

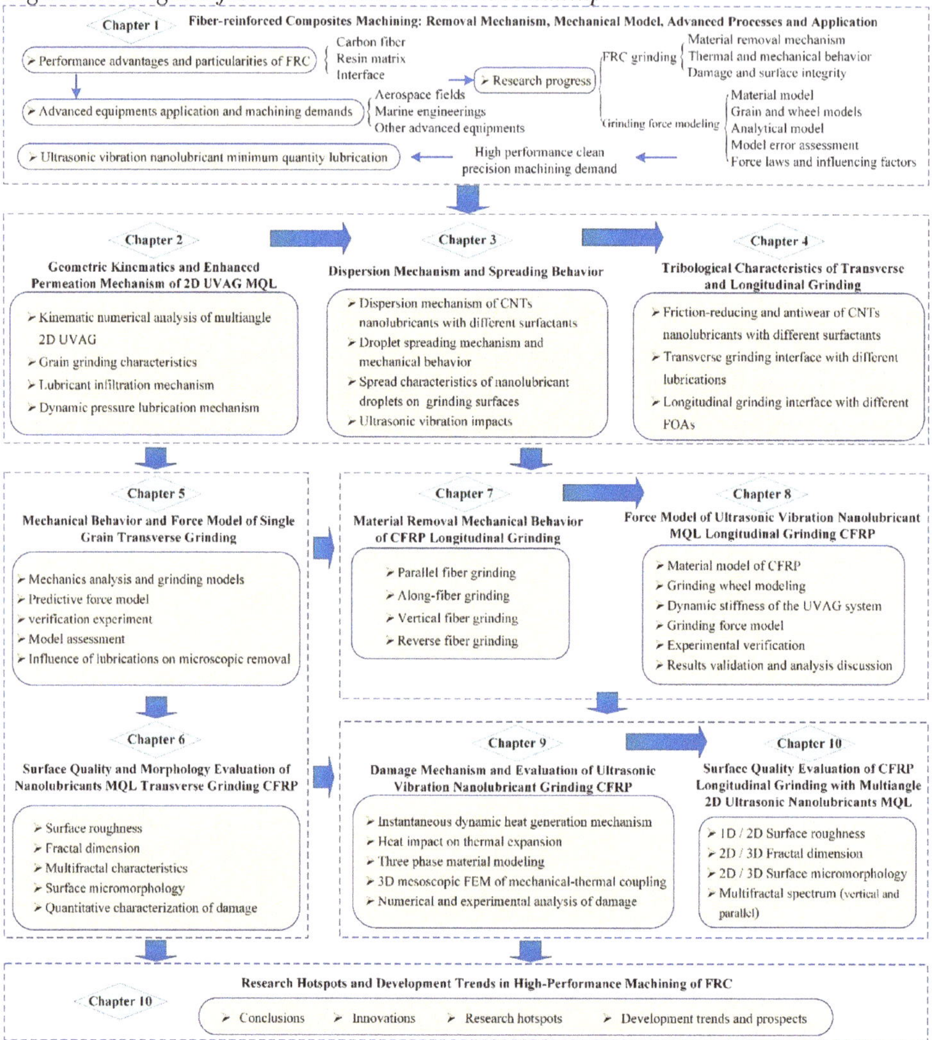

REFERENCES

Adibi, H., Esmaeili, H., & Rezaei, S. M. (2018). Study on minimum quantity lubrication (MQL) in grinding of carbon fiber-reinforced SiC matrix composites (CMCs). *International Journal of Advanced Manufacturing Technology*, *95*(9), 3753–3767. DOI: 10.1007/s00170-017-1464-x

Amin, M., Yuan, S., Israr, A., Zhen, L., & Qi, W. (2018b). Development of cutting force prediction model for vibration-assisted slot milling of carbon fiber reinforced polymers. *International Journal of Advanced Manufacturing Technology*, *94*(9), 3863–3874. DOI: 10.1007/s00170-017-1087-2

Amin, M., Yuan, S., Khan, M. Z., Wu, Q., & Zhu, G. (2017). Development of a generalized cutting force prediction model for carbon fiber reinforced polymers based on rotary ultrasonic face milling. *International Journal of Advanced Manufacturing Technology*, *93*(5), 2655–2666. DOI: 10.1007/s00170-017-0469-9

Amin, M., Yuan, S., Khan, M. Z., Zhang, C., & Wu, Q. (2018a). Development of cutting force prediction model for carbon fiber reinforced polymers based on rotary ultrasonic slot milling. *Machining Science and Technology*, *22*(3), 402–426. DOI: 10.1080/10910344.2017.1365895

Anand, P. S. P., Arunachalam, N., & Vijayaraghavan, L. (2018). Study on grinding of pre-sintered zirconia using diamond wheel. *Materials and Manufacturing Processes*, *33*(6), 634–643. DOI: 10.1080/10426914.2017.1364761

Asmael, M., Safaei, B., Zeeshan, Q., Zargar, O., & Nuhu, A. A. (2021). Ultrasonic machining of carbon fiber–reinforced plastic composites: A review. *International Journal of Advanced Manufacturing Technology*, *113*(11), 3079–3120. DOI: 10.1007/s00170-021-06722-2

Azarhoushang, B., & Tawakoli, T. (2011). Development of a novel ultrasonic unit for grinding of ceramic matrix composites. *International Journal of Advanced Manufacturing Technology*, *57*(9), 945–955. DOI: 10.1007/s00170-011-3347-x

Babbar, A., Sharma, A., Jain, V., & Jain, A. K. (2019). Rotary ultrasonic milling of C/SiC composites fabricated using chemical vapor infiltration and needling technique. *Materials Research Express*, *6*(8), 085607. DOI: 10.1088/2053-1591/ab1bf7

Bai, X., Li, C., Dong, L., & Yin, Q. (2019). Experimental evaluation of the lubrication performances of different nanofluids for minimum quantity lubrication (MQL) in milling Ti-6Al-4V. *International Journal of Advanced Manufacturing Technology*, *101*(9), 2621–2632. DOI: 10.1007/s00170-018-3100-9

Bie, W., Chen, F., Wang, X., Chen, S., Fu, Z., & Zhao, B. (2023). Longitudinal-torsional coupled rotary ultrasonic machining end surface grinding of SiCf/SiC composites: A mechanical model of cutting force. *International Journal of Advanced Manufacturing Technology*, *129*(3), 1227–1248. DOI: 10.1007/s00170-023-12360-7

Bin, L. I. N., Haoji, W. A. N. G., Jinhua, W. E. I., & Tianyi, S. U. I. (2021). Diamond wheel grinding characteristics of 3D orthogonal quartz fiber reinforced silica ceramic matrix composite. *Chinese Journal of Aeronautics*, *34*(5), 404–414. DOI: 10.1016/j.cja.2020.12.026

Bu, P., Ruan, W., & Liu, J. (2023). Effect of interface and matrix strength on transverse tensile strength of unidirectional CFRP composite. *Modelling and Simulation in Materials Science and Engineering*, *31*(6), 065005. DOI: 10.1088/1361-651X/acde93

Cao, J., Wu, Y., Lu, D., Fujimoto, M., & Nomura, M. (2014a). Material removal behavior in ultrasonic-assisted scratching of SiC ceramics with a single diamond tool. *International Journal of Machine Tools & Manufacture*, *79*, 49–61. DOI: 10.1016/j.ijmachtools.2014.02.002

Cao, X., Lin, B., Wang, Y., & Wang, S. (2014b). Influence of diamond wheel grinding process on surface micro-topography and properties of SiO2/SiO2 composite. *Applied Surface Science*, *292*, 181–189. DOI: 10.1016/j.apsusc.2013.11.109

Cao, X., Lin, B., & Zhang, X. (2013). A study on grinding surface waviness of woven ceramic matrix composites. *Applied Surface Science*, *270*, 503–512. DOI: 10.1016/j.apsusc.2013.01.069

Cao, X., Lin, B., & Zhang, X. (2015). Investigations on grinding process of woven ceramic matrix composite based on reinforced fiber orientations. *Composites. Part B, Engineering*, *71*, 184–192. DOI: 10.1016/j.compositesb.2014.11.029

Cao, Y., Zhao, B., Ding, W., Wu, J., Jia, X., Zhang, J., & Das, R. (2024). Intermittent cutting behavior and grinding force model in ultrasonic vibration-assisted grinding K4002 nickel-based superalloy. *International Journal of Advanced Manufacturing Technology*, *131*(5), 3085–3102. DOI: 10.1007/s00170-024-13053-5

Chen, J., An, Q., Ming, W., & Chen, M. (2020a). Investigation on machined surface quality in ultrasonic-assisted grinding of Cf/SiC composites based on fracture mechanism of carbon fibers. *International Journal of Advanced Manufacturing Technology*, *109*(5), 1583–1599. DOI: 10.1007/s00170-020-05739-3

Chen, J., Ming, W., An, Q., & Chen, M. (2020b). Mechanism and feasibility of ultrasonic-assisted milling to improve the machined surface quality of 2D Cf/SiC composites. *Ceramics International*, *46*(10), 15122–15136. DOI: 10.1016/j.ceramint.2020.03.047

Chen, J., Shen, J., Huang, H., & Xu, X. (2010). Grinding characteristics in high speed grinding of engineering ceramics with brazed diamond wheels. *Journal of Materials Processing Technology*, *210*(6-7), 899–906. DOI: 10.1016/j.jmatprotec.2010.02.002

Chen, T., Li, H., Ye, M., Xiang, C., & Tian, S. (2019). Experimental study on effects of structural characteristics of C/E composite laminates on grinding temperature. *Composites. Part B, Engineering*, *157*, 100–108. DOI: 10.1016/j.compositesb.2018.08.120

Chockalingam, P., & Kuang, K. C. (2015). Effects of abrasive types on grinding of chopped strand mat glass fiber-reinforced polymer composite laminates. *Machining Science and Technology*, *19*(2), 313–324. DOI: 10.1080/10910344.2015.1018534

Chockalingam, P., Kuang, K. C., & Vijayaram, T. R. (2013). Effects of grinding process parameters and coolants on the grindability of GFRP laminates. *Materials and Manufacturing Processes*, *28*(10), 1071–1076. DOI: 10.1080/10426914.2013.792411

Choudhary, A., Chakladar, N. D., & Paul, S. (2021). Identification and estimation of defects in high-speed ground C/SiC ceramic matrix composites. *Composite Structures*, *261*, 113274. DOI: 10.1016/j.compstruct.2020.113274

Cococcetta, N., Jahan, M. P., Schoop, J., Ma, J., Pearl, D., & Hassan, M. (2021). Post-processing of 3D printed thermoplastic CFRP composites using cryogenic machining. *Journal of Manufacturing Processes*, *68*, 332–346. DOI: 10.1016/j.jmapro.2021.05.054

Cococcetta, N. M., Pearl, D., Jahan, M. P., & Ma, J. (2020). Investigating surface finish, burr formation, and tool wear during machining of 3D printed carbon fiber reinforced polymer composite. *Journal of Manufacturing Processes*, *56*, 1304–1316. DOI: 10.1016/j.jmapro.2020.04.025

Cong, W. L., Pei, Z. J., Mohanty, N., Van Vleet, E., & Treadwell, C. (2011). Vibration amplitude in rotary ultrasonic machining: a novel measurement method and effects of process variables.

Cui, X., Li, C., Zhang, Y., Jia, D., Zhao, Y., Li, R., & Cao, H. (2019). Tribological properties under the grinding wheel and workpiece interface by using graphene nanofluid lubricant. *International Journal of Advanced Manufacturing Technology*, *104*(9), 3943–3958. DOI: 10.1007/s00170-019-04129-8

Cui, X., Li, C. H., Ding, W. F., Chen, Y., Mao, C., Xu, X. F., Liu, B., Wang, D. Z., Li, H. N., Zhang, Y. B., Said, Z., Debnath, S., Jamil, M., Ali, H. M., & Sharma, S. (2022). Minimum quantity lubrication machining of aeronautical materials using carbon group nanolubricant: From mechanisms to application. *Chinese Journal of Aeronautics*, *35*(11), 85–112. DOI: 10.1016/j.cja.2021.08.011

Danish, M., Gupta, M. K., Rubaiee, S., Ahmed, A., Mahfouz, A., & Jamil, M. (2021). Machinability investigations on CFRP composites: A comparison between sustainable cooling conditions. *International Journal of Advanced Manufacturing Technology*, *114*(11), 3201–3216. DOI: 10.1007/s00170-021-07073-8

Diaz, O. G., & Axinte, D. A. (2017). Towards understanding the cutting and fracture mechanism in ceramic matrix composites. *International Journal of Machine Tools & Manufacture*, *118*, 12–25. DOI: 10.1016/j.ijmachtools.2017.03.008

Ding, K., Fu, Y., Su, H., Cui, F., Li, Q., Lei, W., & Xu, H. (2017b). Study on surface/subsurface breakage in ultrasonic assisted grinding of C/SiC composites. *International Journal of Advanced Manufacturing Technology*, *91*(9), 3095–3105. DOI: 10.1007/s00170-017-0012-z

Ding, K., Fu, Y., Su, H., Xu, H., Cui, F., & Li, Q. (2017a). Experimental studies on matching performance of grinding and vibration parameters in ultrasonic assisted grinding of SiC ceramics. *International Journal of Advanced Manufacturing Technology*, *88*(9), 2527–2535. DOI: 10.1007/s00170-016-8977-6

Dong, L., Li, C., Bai, X., Zhai, M., Qi, Q., Yin, Q., Lv, X., & Li, L. (2019). Analysis of the cooling performance of Ti–6Al–4V in minimum quantity lubricant milling with different nanoparticles. *International Journal of Advanced Manufacturing Technology*, *103*(5), 2197–2206. DOI: 10.1007/s00170-019-03466-y

Dong, Z., Zhang, H., Kang, R., Ran, Y., & Bao, Y. (2022). Mechanical modeling of ultrasonic vibration helical grinding of SiCf/SiC composites. *International Journal of Mechanical Sciences*, *234*, 107701. DOI: 10.1016/j.ijmecsci.2022.107701

Du, J., Ming, W., Ma, J., He, W., Cao, Y., Li, X., & Liu, K. (2018). New observations of the fiber orientations effect on machinability in grinding of C/SiC ceramic matrix composite. *Ceramics International*, *44*(12), 13916–13928. DOI: 10.1016/j.ceramint.2018.04.240

Duan, Z. J., Li, C. H., Zhang, Y. B., Dong, L., Bai, X. F., Yang, M., Jia, D. Z., Li, R. Z., Cao, H. J., & Xu, X. F. (2021). Milling surface roughness for 7050 aluminum alloy cavity influenced by nozzle position of nanofluid minimum quantity lubrication. *Chinese Journal of Aeronautics*, *34*(6), 33–53. DOI: 10.1016/j.cja.2020.04.029

El-Hofy, M. H., & El-Hofy, H. (2019). Laser beam machining of carbon fiber reinforced composites: A review. *International Journal of Advanced Manufacturing Technology*, *101*(9), 2965–2975. DOI: 10.1007/s00170-018-2978-6

Esmaeili, H., Adibi, H., & Rezaei, S. M. (2019). An efficient strategy for grinding carbon fiber-reinforced silicon carbide composite using minimum quantity lubricant. *Ceramics International*, *45*(8), 10852–10864. DOI: 10.1016/j.ceramint.2019.02.163

Fan, B. P., Chen, Y., Chen, B. B., Liang, Y. H., & Sun, L. (2019). Finite element analysis on heat distribution ratio during grinding CFRP. *Diamond and Abrasives Engineering*, *39*(1), 66–71.

Fernando, P. K. S. C., Zhang, M., & Pei, Z. (2019). Rotary ultrasonic machining: Effects of tool natural frequency on ultrasonic vibration amplitude. *Machining Science and Technology*, *23*(4), 595–611. DOI: 10.1080/10910344.2019.1575404

Gao, C. Y., Xiao, J. Z., Zhang, L. C., & Ke, Y. L. (2017). On the static and dynamic properties of fiber-reinforced polymer composites: A three-phase constitutive model. *Journal of Thermoplastic Composite Materials*, *30*(11), 1560–1577. DOI: 10.1177/0892705716646418

Gao, T., Li, C., Jia, D., Zhang, Y., Yang, M., Wang, X., Cao, H., Li, R., Ali, H. M., & Xu, X. (2020b). Surface morphology assessment of CFRP transverse grinding using CNT nanofluid minimum quantity lubrication. *Journal of Cleaner Production*, *277*, 123328. DOI: 10.1016/j.jclepro.2020.123328

Gao, T., Li, C., Yang, M., Zhang, Y., Jia, D., Ding, W., Debnath, S., Yu, T., Said, Z., & Wang, J. (2021a). Mechanics analysis and predictive force models for the single-diamond grain grinding of carbon fiber reinforced polymers using CNT nano-lubricant. *Journal of Materials Processing Technology*, *290*, 116976. DOI: 10.1016/j.jmatprotec.2020.116976

Gao, T., Li, C., Zhang, Y., Yang, M., Jia, D., Jin, T., Hou, Y., & Li, R. (2019). Dispersing mechanism and tribological performance of vegetable oil-based CNT nanofluids with different surfactants. *Tribology International*, *131*, 51–63. DOI: 10.1016/j.triboint.2018.10.025

Gao, T., Xu, P., Wang, W., Zhang, Y., Xu, W., Wang, Y., An, Q., & Li, C. (2024). Force model of ultrasonic empowered minimum quantity lubrication grinding CFRP. *International Journal of Mechanical Sciences*, *280*, 109522. DOI: 10.1016/j.ijmecsci.2024.109522

Gao, T., Zhang, X., Li, C., Zhang, Y., Yang, M., Jia, D., Ji, H., Zhao, Y., Li, R., Yao, P., & Zhu, L. (2020a). Surface morphology evaluation of multi-angle 2D ultrasonic vibration integrated with nanofluid minimum quantity lubrication grinding. *Journal of Manufacturing Processes*, *51*, 44–61. DOI: 10.1016/j.jmapro.2020.01.024

Gao, T., Zhang, Y., Li, C., Wang, Y., An, Q., Liu, B., Said, Z., & Sharma, S. (2021b). Grindability of carbon fiber reinforced polymer using CNT biological lubricant. *Scientific Reports*, *11*(1), 22535. DOI: 10.1038/s41598-021-02071-y PMID: 34795390

Ge, J., Fu, G., Almeida, J. H. S.Jr, Jin, Y., & Sun, D. (2025). Thermal effect in CFRP machining: Temperature field characteristics, heat generation mechanism and thermal damage management. *Composite Structures*, *356*, 118845. DOI: 10.1016/j.compstruct.2025.118845

Geng, D., Zhang, D., Xu, Y., He, F., Liu, D., & Duan, Z. (2015). Rotary ultrasonic elliptical machining for side milling of CFRP: Tool performance and surface integrity. *Ultrasonics*, *59*, 128–137. DOI: 10.1016/j.ultras.2015.02.006 PMID: 25708349

Giasin, K., Ayvar-Soberanis, S., & Hodzic, A. (2016). Evaluation of cryogenic cooling and minimum quantity lubrication effects on machining GLARE laminates using design of experiments. *Journal of Cleaner Production*, *135*, 533–548. DOI: 10.1016/j.jclepro.2016.06.098

Guo, M., Tao, J., Wu, C., Luo, C., & Lin, Z. (2023). High-speed grinding fracture mechanism of Cf/SiC composite considering interfacial strength and anisotropy. *Ceramics International*, *49*(2), 2600–2612. DOI: 10.1016/j.ceramint.2022.09.240

Guo, S., Li, C., Zhang, Y., Wang, Y., Li, B., Yang, M., Zhang, X., & Liu, G. (2017). Experimental evaluation of the lubrication performance of mixtures of castor oil with other vegetable oils in MQL grinding of nickel-based alloy. *Journal of Cleaner Production*, *140*, 1060–1076. DOI: 10.1016/j.jclepro.2016.10.073

Haddad, M., Zitoune, R., Eyma, F., & Castanie, B. (2014). Study of the surface defects and dust generated during trimming of CFRP: Influence of tool geometry, machining parameters and cutting speed range. *Composites. Part A, Applied Science and Manufacturing*, *66*, 142–154. DOI: 10.1016/j.compositesa.2014.07.005

Hanasaki, S., Fujiwara, J., & Tashiro, T. (2003). 218 Study on Surface Grinding of Unidirectional Carbon Fiber Reinforced Plastics in Dry Method. In *Proceedings of International Conference on Leading Edge Manufacturing in 21st century: LEM21 2003* (pp. 297-302). The Japan Society of Mechanical Engineers. DOI: 10.1299/jsmelem.2003.297

Helmy, M. O., El-Hofy, M. H., & El-Hofy, H. (2018). Effect of cutting fluid delivery method on ultrasonic assisted edge trimming of multidirectional CFRP composites at different machining conditions. *Procedia CIRP*, *68*, 450–455. DOI: 10.1016/j.procir.2017.12.077

Hoang, Q. N., Lee, S., Lee, S., & Park, H. (2025). Characterization of CFRP interface properties with varying fiber surface roughness using AFM measurements and finite element modeling. *Composites. Part B, Engineering*, •••, 112990.

Hu, N. S., & Zhang, L. C. (2003). A study on the grindability of multidirectional carbon fibre-reinforced plastics. *Journal of Materials Processing Technology*, *140*(1-3), 152–156. DOI: 10.1016/S0924-0136(03)00704-0

Huang, B., Wang, W. H., Xiong, Y. F., Wu, X. F., Liu, J. T., Liu, C., & Wang, D. H. (2023). Investigation of force modeling in ultrasonic vibration-assisted drilling SiCf/SiC ceramic matrix composites. *Journal of Manufacturing Processes*, *96*, 21–30. DOI: 10.1016/j.jmapro.2023.04.040

Hussien, A. A., Qasem, I., Kataraki, P. S., Al-Kouz, W. G., & Janvekar, A. A. (2021). Studying the performance of cutting carbon fibre-reinforced plastic using an abrasive water jet technique. *Strojniški vestnik*, *67*(4), 125-141.

Inoue, H., & Kawaguchi, I. (1990). Study on the grinding mechanism of glass fiber reinforced plastics. *Journal of Engineering Materials and Technology*, *112*(3), 341–345. DOI: 10.1115/1.2903335

Iskandar, Y., Tendolkar, A., Attia, M. H., Hendrick, P., Damir, A., & Diakodimitris, C. (2014). Flow visualization and characterization for optimized MQL machining of composites. *CIRP Annals*, *63*(1), 77–80. DOI: 10.1016/j.cirp.2014.03.078

Islam, S., Yuan, S., & Li, Z. (2020). A cutting force prediction model, experimental studies, and optimization of cutting parameters for rotary ultrasonic face milling of C/SiC composites. *Applied Composite Materials*, *27*(4), 407–431. DOI: 10.1007/s10443-020-09815-5

James, S., & Nejadian, S. M. (2020). Experimental study on high-speed saw cutting of hybrid composite stacks using nanoparticle-enhanced minimum quantity lubrication. *International Journal of Advanced Manufacturing Technology*, *110*(11), 3077–3090. DOI: 10.1007/s00170-020-06008-z

Jebaratnam, J. M., & Hassan, M. H. (2025). Comprehensive review of drilling strategies for CFRP/Ti stacks in aircraft manufacturing. *Machining Science and Technology*, *29*(3), 334–438. DOI: 10.1080/10910344.2025.2475484

Jia, D., Li, C., Zhang, Y., Zhang, D., & Zhang, X. (2016). Experimental research on the influence of the jet parameters of minimum quantity lubrication on the lubricating property of Ni-based alloy grinding. *International Journal of Advanced Manufacturing Technology*, *82*(1), 617–630. DOI: 10.1007/s00170-015-7381-y

Jia, D., Zhang, Y., Li, C., Yang, M., Gao, T., Said, Z., & Sharma, S. (2022). Lubrication-enhanced mechanisms of titanium alloy grinding using lecithin biolubricant. *Tribology International*, *169*, 107461. DOI: 10.1016/j.triboint.2022.107461

Kodama, H., Okazaki, S., Jiang, Y., Yoden, H., & Ohashi, K. (2020). Thermal influence on surface layer of carbon fiber reinforced plastic (CFRP) in grinding. *Precision Engineering*, *65*, 53–63. DOI: 10.1016/j.precisioneng.2020.04.005

Larbi, S., Bensaada, R., Bilek, A., & Djebali, S. (2015). Hygrothermal ageing effect on mechanical properties of FRP laminates. In. AIP Conference Proceedings: Vol. 1653. *No. 1* (p. 020066). AIP Publishing LLC. DOI: 10.1063/1.4914257

Li, B., Li, C., Zhang, Y., Wang, Y., Jia, D., Yang, M., Zhang, N., Wu, Q., Han, Z., & Sun, K. (2017). Heat transfer performance of MQL grinding with different nanofluids for Ni-based alloys using vegetable oil. *Journal of Cleaner Production*, *154*, 1–11. DOI: 10.1016/j.jclepro.2017.03.213

Li, H., Lin, B., Wan, S., Wang, Y., & Zhang, X. (2016). An experimental investigation on ultrasonic vibration-assisted grinding of SiO2f/SiO2 composites. *Materials and Manufacturing Processes*, *31*(7), 887–895. DOI: 10.1080/10426914.2015.1090586

Li, L., Zhang, Y., Cui, X., Said, Z., Sharma, S., Liu, M., Gao, T., Zhou, Z., Wang, X., & Li, C. (2023). Mechanical behavior and modeling of grinding force: A comparative analysis. *Journal of Manufacturing Processes*, *102*, 921–954. DOI: 10.1016/j.jmapro.2023.07.074

Li, M., Yu, T., Zhang, R., Yang, L., Ma, Z., Li, B., Wang, X. Z., Wang, W., & Zhao, J. (2020). Experimental evaluation of an eco-friendly grinding process combining minimum quantity lubrication and graphene-enhanced plant-oil-based cutting fluid. *Journal of Cleaner Production*, *244*, 118747. DOI: 10.1016/j.jclepro.2019.118747

Li, Y., Ge, X., Wang, H., Hu, Y., Ning, F., Cong, W., & Ren, C. (2019a). Study of material removal mechanisms in grinding of C/SiC composites via single-abrasive scratch tests. *Ceramics International*, *45*(4), 4729–4738. DOI: 10.1016/j.ceramint.2018.11.165

Li, Y., Ren, C., Wang, H., Hu, Y., Ning, F., Wang, X., & Cong, W. (2019b). Edge surface grinding of CFRP composites using rotary ultrasonic machining: Comparison of two machining methods. *International Journal of Advanced Manufacturing Technology*, *100*(9), 3237–3248. DOI: 10.1007/s00170-018-2901-1

Liang, Y., Chen, Y., Chen, B., Fan, B., Yan, C., & Fu, Y. (2019). Feasibility of ultrasonic vibration assisted grinding for carbon fiber reinforced polymer with monolayer brazed grinding tools. *International Journal of Precision Engineering and Manufacturing*, *20*(7), 1083–1094. DOI: 10.1007/s12541-019-00135-8

Liang, Y., Chen, Y., Zhu, Y., Ji, J., & Ding, W. (2021). Investigations on tool clogging and surface integrity in ultrasonic vibration-assisted slot grinding of unidirectional CFRP. *International Journal of Advanced Manufacturing Technology*, *112*(5), 1557–1570. DOI: 10.1007/s00170-020-06364-w

Liu, H., Lin, J., Sun, Y., & Zhang, J. (2019b). Micro model of carbon fiber/cyanate ester composites and analysis of machining damage mechanism. *Chinese Journal of Mechanical Engineering*, *32*(1), 52. DOI: 10.1186/s10033-019-0364-4

Liu, M., Zhang, Y., Yang, M., Gao, T., Cui, X., Wang, X., & Sharma, S. (2023). Analysis of grinding mechanics and improved grinding force model based on randomized grain geometric characteristics. *Chinese Journal of Aeronautics*, *36*(7), 160–193. DOI: 10.1016/j.cja.2022.11.005

Liu, Q., Huang, G., Cui, C., Tong, Z., & Xu, X. (2019a). Investigation of grinding mechanism of a 2D Cf/C–SiC composite by single-grain scratching. *Ceramics International*, *45*(10), 13422–13430. DOI: 10.1016/j.ceramint.2019.04.041

Liu, Q., Huang, G., Fang, C., Cui, C., & Xu, X. (2017c). Experimental investigations on grinding characteristics and removal mechanisms of 2D–Cf/C-SiC composites based on reinforced fiber orientations. *Ceramics International*, *43*(17), 15266–15274. DOI: 10.1016/j.ceramint.2017.08.064

Liu, Q., Huang, G., Xu, X., Fang, C., & Cui, C. (2017a). A study on the surface grinding of 2D C/SiC composites. *International Journal of Advanced Manufacturing Technology*, *93*(5), 1595–1603. DOI: 10.1007/s00170-017-0626-1

Liu, Q., Huang, G., Xu, X., Fang, C., & Cui, C. (2018). Influence of grinding fiber angles on grinding of the 2D–Cf/C–SiC composites. *Ceramics International*, *44*(11), 12774–12782. DOI: 10.1016/j.ceramint.2018.04.083

Liu, S., Chen, T., & Wu, C. (2017b). Rotary ultrasonic face grinding of carbon fiber reinforced plastic (CFRP): A study on cutting force model. *International Journal of Advanced Manufacturing Technology*, *89*(1), 847–856. DOI: 10.1007/s00170-016-9151-x

Liu, Y., Liu, Z. B., Wang, X. B., & Huang, T. (2021). Experimental study on cutting force and surface quality in ultrasonic vibration-assisted milling of C/SiC composites. *International Journal of Advanced Manufacturing Technology*, *112*(7), 2003–2014.

Long, Y., Huang, D., Tao, N., Chen, G., He, Z., & Zhang, H. (2025). A review on laser drilling of CFRP laminates: Fundamental mechanisms, defect formation, strategies for high-quality processing. *Optics & Laser Technology*, *192*, 113842. DOI: 10.1016/j.optlastec.2025.113842

Luna, G. G., Axinte, D., & Novovic, D. (2020). Influence of grit geometry and fibre orientation on the abrasive material removal mechanisms of SiC/SiC Ceramic Matrix Composites (CMCs). *International Journal of Machine Tools & Manufacture*, *157*, 103580. DOI: 10.1016/j.ijmachtools.2020.103580

Ma, G., Kang, R., Dong, Z., Yin, S., Bao, Y., & Guo, D. (2020). Hole quality in longitudinal–torsional coupled ultrasonic vibration assisted drilling of carbon fiber reinforced plastics. *Frontiers of Mechanical Engineering*, *15*(4), 538–546. DOI: 10.1007/s11465-020-0598-y

Mei, J., Diaz, O. G., & Axinte, D. A. (2017). An approach on capturing the influence of the stochasticity of fibre distributions for modelling the variability of cutting forces in composite materials. *Composites. Part B, Engineering*, *125*, 27–38. DOI: 10.1016/j.compositesb.2017.05.056

Meng, Q. Y., Guo, B., Zhao, Q. L., Li, H. N., Jackson, M. J., Linke, B. S., & Luo, X. (2023). Mark, j., Barara, S., Luo, X.C. Modelling of grinding mechanics: A review [J]. *Chinese Journal of Aeronautics*, *36*(7), 25–39. DOI: 10.1016/j.cja.2022.10.006

Ning, F., Cong, W., Wang, H., Hu, Y., Hu, Z., & Pei, Z. (2017). Surface grinding of CFRP composites with rotary ultrasonic machining: A mechanistic model on cutting force in the feed direction. *International Journal of Advanced Manufacturing Technology*, *92*(1), 1217–1229. DOI: 10.1007/s00170-017-0149-9

Ning, F., Wang, H., & Cong, W. (2019). Rotary ultrasonic machining of carbon fiber reinforced plastic composites: A study on fiber material removal mechanism through single-grain scratching. *International Journal of Advanced Manufacturing Technology*, *103*(1), 1095–1104. DOI: 10.1007/s00170-019-03433-7

Pereira, A. C., Monteiro, S. N., da Silva, L. R. R., Binali, R., Kuntoğlu, M., Machado, A. R., & Pimenov, D. Y. (2025). Machinability and surface integrity of glass fiber reinforced plastic composite: A review. *Journal of Materials Research and Technology*, *35*, 6446–6467. DOI: 10.1016/j.jmrt.2025.02.215

Pervaiz, S., Kannan, S., Huo, D., & Mamidala, R. (2020). Ecofriendly inclined drilling of carbon fiber-reinforced polymers (CFRP). *International Journal of Advanced Manufacturing Technology*, *111*(7), 2127–2153. DOI: 10.1007/s00170-020-06203-y

Qian, M., Xiao, J., Wang, G., Huang, P., Chen, Z., & Han, G. (2020). Evaluation of heat generation using a microscopic cutting model with thermo-mechanical coupling for carbon fiber reinforced polymer composites. *Journal of Reinforced Plastics and Composites*, *39*(21-22), 793–804. DOI: 10.1177/0731684420931589

Qiao, G., Cheng, Z., Zheng, W., Yi, S., & Zhang, F. (2022). Grinding force model for longitudinal-torsional ultrasonic-assisted face grinding of ceramic matrix composites. *International Journal of Advanced Manufacturing Technology*, *120*(11), 7721–7733. DOI: 10.1007/s00170-022-09174-4

Qu, S., Gong, Y., Yang, Y., Sun, Y., Wen, X., & Qi, Y. (2020c). Investigating minimum quantity lubrication in unidirectional Cf/SiC composite grinding. *Ceramics International*, *46*(3), 3582–3591. DOI: 10.1016/j.ceramint.2019.10.076

Qu, S., Gong, Y., Yang, Y., Wang, W., Liang, C., & Han, B. (2020b). An investigation of carbon nanofluid minimum quantity lubrication for grinding unidirectional carbon fibre-reinforced ceramic matrix composites. *Journal of Cleaner Production*, *249*, 119353. DOI: 10.1016/j.jclepro.2019.119353

Qu, S., Gong, Y., Yang, Y., Xu, Y., Wang, W., Xin, B., & Pang, S. (2020a). Mechanical model and removal mechanism of unidirectional carbon fibre-reinforced ceramic composites. *International Journal of Mechanical Sciences*, *173*, 105465. DOI: 10.1016/j.ijmecsci.2020.105465

Qu, S., Yang, Y., Yao, P., Li, L., Sun, Y., & Chu, D. (2025). Fiber reinforced ceramic matrix composites: From the controlled fabrication to precision machining. *International Journal of Extreme Manufacturing*, *7*(6), 062004. DOI: 10.1088/2631-7990/adeee1

Qu, S., Yao, P., Gong, Y., Yang, Y., Chu, D., & Zhu, Q. (2022). Modelling and grinding characteristics of unidirectional C–SiCs. *Ceramics International*, *48*(6), 8314–8324. DOI: 10.1016/j.ceramint.2021.12.036

Ran, Y., Kang, R., Dong, Z., Jin, Z., & Bao, Y. (2023). Ultrasonic assisted grinding force model considering anisotropy of SiCf/SiC composites. *International Journal of Mechanical Sciences, 250*, 108311. DOI: 10.1016/j.ijmecsci.2023.108311

Ran, Y., Kang, R., Sun, J., Dong, Z., Wang, H., & Bao, Y. (2024). Insight into crack propagation induced by fiber orientations during single grain scratching of SiCf/SiC composites using FEM. *Composites. Part A, Applied Science and Manufacturing, 177*, 107928. DOI: 10.1016/j.compositesa.2023.107928

Rodriguez, R. L., Lopes, J. C., Mancini, S. D., de Ângelo Sanchez, L. E., de Almeida Varasquim, F. M. F., Volpato, R. S., de Mello, H. J., de Aguiar, P. R., & Bianchi, E. C. (2019). Contribution for minimization the usage of cutting fluids in CFRP grinding. *International Journal of Advanced Manufacturing Technology, 103*(1), 487–497. DOI: 10.1007/s00170-019-03529-0

Said, Z., Arora, S., Farooq, S., Sundar, L. S., Li, C., & Allouhi, A. (2022). Recent advances on improved optical, thermal, and radiative characteristics of plasmonic nanofluids: Academic insights and perspectives. *Solar Energy Materials and Solar Cells, 236*, 111504. DOI: 10.1016/j.solmat.2021.111504

Said, Z., Sharma, P., Sundar, L. S., Afzal, A., & Li, C. (2021). Synthesis, stability, thermophysical properties and AI approach for predictive modelling of Fe3O4 coated MWCNT hybrid nanofluids. *Journal of Molecular Liquids, 340*, 117291. DOI: 10.1016/j.molliq.2021.117291

Sambhav, K., Kumar, A., & Choudhury, S. K. (2011). Mechanistic force modeling of single point cutting tool in terms of grinding angles. *International Journal of Machine Tools & Manufacture, 51*(10-11), 775–786. DOI: 10.1016/j.ijmachtools.2011.06.007

Scida, D., Assarar, M., Poilâne, C., & Ayad, R. (2013). Influence of hygrothermal ageing on the damage mechanisms of flax-fibre reinforced epoxy composite. *Composites. Part B, Engineering, 48*, 51–58. DOI: 10.1016/j.compositesb.2012.12.010

Senthilkumar, M., Prabukarthi, A., & Krishnaraj, V. (2018). Machining of CFRP/ Ti6Al4V stacks under minimal quantity lubricating condition. *Journal of Mechanical Science and Technology, 32*(8), 3787–3796. DOI: 10.1007/s12206-018-0731-6

Sheikh-Ahmad, J., & Mohammed, J. (2014). Optimization of process parameters in diamond abrasive machining of carbon fiber-reinforced epoxy. *Materials and Manufacturing Processes, 29*(11-12), 1361–1366. DOI: 10.1080/10426914.2014.941869

Sheikh-Ahmad, J. Y., Almaskari, F., Hafeez, F., & Meng, F. (2019). Evaluation of heat partition in machining CFRP using inverse method. *Machining Science and Technology, 23*(4), 530–546. DOI: 10.1080/10910344.2019.1575401

Shi, H., Yuan, S., Zhang, C., Chen, B., Li, Q., Li, Z., Zhu, G., & Qian, J. (2020). A cutting force prediction model for rotary ultrasonic side grinding of CFRP composites considering coexistence of brittleness and ductility. *International Journal of Advanced Manufacturing Technology*, *106*(5), 2403–2414. DOI: 10.1007/s00170-019-04730-x

Shi, W., Ma, T., Li, J., Li, J., Zhang, X., Guo, B., Wen, Z., Li, J., Wang, L., & Cao, S. (2025). A review of research on hole machining technology for FRP/LA laminated materials: From mechanical drilling to non-traditional machining. *International Journal of Advanced Manufacturing Technology*, *137*(5), 2109–2156. DOI: 10.1007/s00170-025-15181-y

Shu, J., Liao, W., Zheng, K., & Hussein Youssef, A. M. F. (2021). Surface morphology on carbon fiber composites by rotary ultrasonic milling. *Machining Science and Technology*, *25*(5), 721–737. DOI: 10.1080/10910344.2021.1971705

Soo, S. L., Shyha, I. S., Barnett, T., Aspinwall, D. K., & Sim, W. M. (2012). Grinding performance and workpiece integrity when superabrasive edge routing carbon fibre reinforced plastic (CFRP) composites. *CIRP Annals*, *61*(1), 295–298. DOI: 10.1016/j.cirp.2012.03.042

Sun, J., Qin, F., Chen, P., & An, T. (2016). A predictive model of grinding force in silicon wafer self-rotating grinding. *International Journal of Machine Tools & Manufacture*, *109*, 74–86. DOI: 10.1016/j.ijmachtools.2016.07.009

Verma, G. C., & Pandey, P. M. (2019). Machining forces in ultrasonic-vibration assisted end milling. *Ultrasonics*, *94*, 350–363. DOI: 10.1016/j.ultras.2018.07.004 PMID: 30029793

Wang, D., Lu, S., Xu, D., & Zhang, Y. (2020d). Research on material removal mechanism of C/SiC composites in ultrasound vibration-assisted grinding. *Materials (Basel)*, *13*(8), 1918. DOI: 10.3390/ma13081918 PMID: 32325807

Wang, F., Jin, X., Wang, X., Jia, Z., & Zhang, B. (2022). Numerical investigation on the influences of processing conditions on damage in the CFRTP cutting using a novel Elastic–plastic damage model. *Applied Composite Materials*, *29*(5), 2063–2094. DOI: 10.1007/s10443-022-10041-4

Wang, F. J., Zhang, B. Y., Ma, J. W., Bi, G. J., & Hu, H. B. (2019a). Computation of the distribution of the fiber-matrix interface cracks in the edge trimming of CFRP. *Applied Composite Materials*, *26*(1), 159–186. DOI: 10.1007/s10443-018-9687-z

Wang, H., Cong, W., Ning, F., & Hu, Y. (2018a). A study on the effects of machining variables in surface grinding of CFRP composites using rotary ultrasonic machining. *International Journal of Advanced Manufacturing Technology*, *95*(9), 3651–3663. DOI: 10.1007/s00170-017-1468-6

Wang, H., Hu, Y., Cong, W., & Hu, Z. (2019c). A mechanistic model on feeding-directional cutting force in surface grinding of CFRP composites using rotary ultrasonic machining with horizontal ultrasonic vibration. *International Journal of Mechanical Sciences*, *155*, 450–460. DOI: 10.1016/j.ijmecsci.2019.03.009

Wang, H., Hu, Y., Cong, W., Hu, Z., & Wang, Y. (2020c). A novel investigation on horizontal and 3D elliptical ultrasonic vibrations in rotary ultrasonic surface machining of carbon fiber reinforced plastic composites. *Journal of Manufacturing Processes*, *52*, 12–25. DOI: 10.1016/j.jmapro.2020.01.027

Wang, H., Ning, F., Hu, Y., & Cong, W. (2018b). Surface grinding of CFRP composites using rotary ultrasonic machining: A comparison of workpiece machining orientations. *International Journal of Advanced Manufacturing Technology*, *95*(5), 2917–2930. DOI: 10.1007/s00170-017-1401-z

Wang, H., Ning, F., Hu, Y., Fernando, P. K. S. C., Pei, Z. J., & Cong, W. (2016a). Surface grinding of carbon fiber–reinforced plastic composites using rotary ultrasonic machining: Effects of tool variables. *Advances in Mechanical Engineering*, *8*(9), 1687814016670284. DOI: 10.1177/1687814016670284

Wang, H., Pei, Z. J., & Cong, W. (2020a). A mechanistic cutting force model based on ductile and brittle fracture material removal modes for edge surface grinding of CFRP composites using rotary ultrasonic machining. *International Journal of Mechanical Sciences*, *176*, 105551. DOI: 10.1016/j.ijmecsci.2020.105551

Wang, H., Pei, Z. J., & Cong, W. (2020b). A feeding-directional cutting force model for end surface grinding of CFRP composites using rotary ultrasonic machining with elliptical ultrasonic vibration. *International Journal of Machine Tools & Manufacture*, *152*, 103540. DOI: 10.1016/j.ijmachtools.2020.103540

Wang, H., Zhang, D., Li, Y., & Cong, W. (2020e). The effects of elliptical ultrasonic vibration in surface machining of CFRP composites using rotary ultrasonic machining. *International Journal of Advanced Manufacturing Technology*, *106*(11), 5527–5538. DOI: 10.1007/s00170-020-04976-w

Wang, J., Zhang, J., & Feng, P. (2017). Effects of tool vibration on fiber fracture in rotary ultrasonic machining of C/SiC ceramic matrix composites. *Composites. Part B, Engineering*, *129*, 233–242. DOI: 10.1016/j.compositesb.2017.07.081

Wang, X., Li, C., Zhang, Y., Ding, W., Yang, M., Gao, T., Cao, H., Xu, X., Wang, D., Said, Z., Debnath, S., Jamil, M., & Ali, H. M. (2020f). Vegetable oil-based nanofluid minimum quantity lubrication turning: Academic review and perspectives. *Journal of Manufacturing Processes*, *59*, 76–97. DOI: 10.1016/j.jmapro.2020.09.044

Wang, X., Wang, F., Gu, T., Jia, Z., & Shi, Y. (2021). Computational simulation of the damage response for machining long fibre reinforced plastic (LFRP) composite parts: A review. *Composites. Part A, Applied Science and Manufacturing*, *143*, 106296. DOI: 10.1016/j.compositesa.2021.106296

Wang, X., Xu, Q., Zheng, Q., Shao, Y., & Shen, J. (2025). Reviews of fiber-reinforced phenolic resin-based thermal protection materials for aircraft. *Energies*, *18*(4), 819. DOI: 10.3390/en18040819

Wang, Y., Li, C., Zhang, Y., Yang, M., Li, B., Dong, L., & Wang, J. (2018c). Processing characteristics of vegetable oil-based nanofluid MQL for grinding different workpiece materials. *International Journal of Precision Engineering and Manufacturing-Green Technology*, *5*(2), 327–339. DOI: 10.1007/s40684-018-0035-4

Wang, Y., Sarin, V. K., Lin, B., Li, H., & Gillard, S. (2016b). Feasibility study of the ultrasonic vibration filing of carbon fiber reinforced silicon carbide composites. *International Journal of Machine Tools & Manufacture*, *101*, 10–17. DOI: 10.1016/j.ijmachtools.2015.11.003

Wang, Y., Wang, H., Wei, J., Lin, B., Xu, J., & Fang, S. (2019b). Finite element analysis of grinding process of long fiber reinforced ceramic matrix woven composites: Modeling, experimental verification and material removal mechanism. *Ceramics International*, *45*(13), 15920–15927. DOI: 10.1016/j.ceramint.2019.05.100

Wei, J., Wang, H., Lin, B., Sui, T., Zhao, F., & Fang, S. (2019a). Acoustic emission signal of fiber-reinforced composite grinding: Frequency components and damage pattern recognition. *International Journal of Advanced Manufacturing Technology*, *103*(1), 1391–1401. DOI: 10.1007/s00170-019-03645-x

Wei, J., Wang, H., Lin, B., Sui, T., Zhao, F., & Fang, S. (2019b). A force model in single grain grinding of long fiber reinforced woven composite. *International Journal of Advanced Manufacturing Technology*, *100*(1), 541–552. DOI: 10.1007/s00170-018-2719-x

Wu, X., Li, C., Zhou, Z., Nie, X., Chen, Y., Zhang, Y., Cao, H., Liu, B., Zhang, N., Said, Z., Debnath, S., Jamil, M., Ali, H. M., & Sharma, S. (2021). Circulating purification of cutting fluid: An overview. *International Journal of Advanced Manufacturing Technology*, *117*(9), 2565–2600. DOI: 10.1007/s00170-021-07854-1 PMID: 34465936

Xie, Z., Liu, Z., Wang, B., Xin, M., Song, Q., & Jiang, L. (2021). Longitudinal amplitude effect on material removal mechanism of ultrasonic vibration-assisted milling 2.5 DC/SiC composites. *Ceramics International*, *47*(22), 32144–32152. DOI: 10.1016/j.ceramint.2021.08.106

Xu, J., Feng, P., Feng, F., Zha, H., & Liang, G. (2021). Subsurface damage and burr improvements of aramid fiber reinforced plastics by using longitudinal–torsional ultrasonic vibration milling. *Journal of Materials Processing Technology*, *297*, 117265. DOI: 10.1016/j.jmatprotec.2021.117265

Xu, J., Huang, X., Chen, M., & Paulo Davim, J. (2020). Drilling characteristics of carbon/epoxy and carbon/polyimide composites. *Materials and Manufacturing Processes*, *35*(15), 1732–1740. DOI: 10.1080/10426914.2020.1784935

Xu, J., Lin, T., Li, L., Ji, M., Davim, J. P., Geier, N., & Chen, M. (2022). Numerical study of interface damage formation mechanisms in machining CFRP/Ti6Al4V stacks under different cutting sequence strategies. *Composite Structures*, *285*, 115236. DOI: 10.1016/j.compstruct.2022.115236

Xu, W., Zhang, L. C., & Wu, Y. (2014). Elliptic vibration-assisted cutting of fibre-reinforced polymer composites: Understanding the material removal mechanisms. *Composites Science and Technology*, *92*, 103–111. DOI: 10.1016/j.compscitech.2013.12.011

Xue, F., Zheng, K., Liao, W., Shu, J., & Dong, S. (2021b). Investigation on fiber fracture mechanism of c/sic composites by rotary ultrasonic milling. *International Journal of Mechanical Sciences*, *191*, 106054. DOI: 10.1016/j.ijmecsci.2020.106054

Xue, F., Zheng, K., Liao, W., Shu, J., & Miao, D. (2021a). Experimental investigation on fatigue property at room temperature of C/SiC composites machined by rotary ultrasonic milling. *Journal of the European Ceramic Society*, *41*(6), 3341–3356. DOI: 10.1016/j.jeurceramsoc.2021.01.046

Yang, H., Zhao, G., Xia, H., Xin, L., Nian, Z., & Li, L. (2024). Effect of fiber orientation on material removal mechanisms in the machining of 3D Cf/SiC composite with a brazed diamond grinding rod: Experiment and simulation. *Journal of Manufacturing Processes*, *114*, 18–38. DOI: 10.1016/j.jmapro.2024.01.066

Yang, M., Li, C., Luo, L., Li, R., & Long, Y. (2021c). Predictive model of convective heat transfer coefficient in bone micro-grinding using nanofluid aerosol cooling. *International Communications in Heat and Mass Transfer*, *125*, 105317. DOI: 10.1016/j.icheatmasstransfer.2021.105317

Yang, M., Li, C., Said, Z., Zhang, Y., Li, R., Debnath, S., Ali, H. M., Gao, T., & Long, Y. (2021b). Semiempirical heat flux model of hard-brittle bone material in ductilc microgrinding. *Journal of Manufacturing Processes, 71*, 501–514. DOI: 10.1016/j.jmapro.2021.09.053

Yang, M., Li, C., Zhang, Y., Jia, D., Li, R., Hou, Y., & Cao, H. (2019a). Effect of friction coefficient on chip thickness models in ductile-regime grinding of zirconia ceramics. *International Journal of Advanced Manufacturing Technology, 102*(5), 2617–2632. DOI: 10.1007/s00170-019-03367-0

Yang, M., Li, C., Zhang, Y., Jia, D., Li, R., Hou, Y., Cao, H., & Wang, J. (2019b). Predictive model for minimum chip thickness and size effect in single diamond grain grinding of zirconia ceramics under different lubricating conditions. *Ceramics International, 45*(12), 14908–14920. DOI: 10.1016/j.ceramint.2019.04.226

Yang, M., Li, C., Zhang, Y., Wang, Y., Li, B., Jia, D., Hou, Y., & Li, R. (2017). Research on microscale skull grinding temperature field under different cooling conditions. *Applied Thermal Engineering, 126*, 525–537. DOI: 10.1016/j.applthermaleng.2017.07.183

Yang, Y., Qu, S., & Gong, Y. (2021a). Investigating the grinding performance of unidirectional and 2.5 DC/SiCs. *Ceramics International, 47*(4), 5123–5132. DOI: 10.1016/j.ceramint.2020.10.090

Yang, Z., Zhu, L., Zhang, G., Ni, C., & Lin, B. (2020). Review of ultrasonic vibration-assisted machining in advanced materials. *International Journal of Machine Tools & Manufacture, 156*, 103594. DOI: 10.1016/j.ijmachtools.2020.103594

Yao, L., Liu, Z., Song, Q., Wang, B., & Cai, Y. (2023b). Prediction modelling of cutting force in rotary ultrasonic end grinding 2.5 D woven SiO2f/SiO2 ceramic matrix composite. *Composite Structures, 304*, 116448. DOI: 10.1016/j.compstruct.2022.116448

Yao, L., Liu, Z., Song, Q., Wang, B., Cai, Y., & Zhao, J. (2023a). Numerical prediction of surface morphology and roughness in rotary ultrasonic face grinding SiO2f/SiO2 composite. *Journal of Materials Research and Technology, 25*, 5917–5937. DOI: 10.1016/j.jmrt.2023.07.044

Yin, J., Li, M., Xu, J., Ding, W., & Su, H. (2022). Edge chipping characteristics in grinding SiCf/SiC composite. *Ceramics International, 48*(5), 7126–7135. DOI: 10.1016/j.ceramint.2021.11.272

Yin, J., Xu, J., Ding, W., & Su, H. (2021). Effects of grinding speed on the material removal mechanism in single grain grinding of SiCf/SiC ceramic matrix composite. *Ceramics International*, *47*(9), 12795–12802. DOI: 10.1016/j.ceramint.2021.01.140

Yuan, J., Lyu, B., Hang, W., & Deng, Q. (2017a). Review on the progress of ultra-precision machining technologies. *Frontiers of Mechanical Engineering*, *12*(2), 158–180. DOI: 10.1007/s11465-017-0455-9

Yuan, S., Fan, H., Amin, M., Zhang, C., & Guo, M. (2016). A cutting force prediction dynamic model for side milling of ceramic matrix composites C/SiC based on rotary ultrasonic machining. *International Journal of Advanced Manufacturing Technology*, *86*(1), 37–48. DOI: 10.1007/s00170-015-8099-6

Yuan, S., Zhu, G., & Zhang, C. (2017b). Modeling of tool blockage condition in cutting tool design for rotary ultrasonic machining of composites. *International Journal of Advanced Manufacturing Technology*, *91*(5), 2645–2654. DOI: 10.1007/s00170-016-9911-7

Zan, Z., Guo, K., Sun, J., Lu, Y., Yang, B., Jia, X., Sivalingam, V., & Xi, J. (2022). Investigation on scratching force and material removal mechanism of 3D SiCf/C–SiC composites during single grain scratching. *Journal of the European Ceramic Society*, *42*(13), 5366–5379. DOI: 10.1016/j.jeurceramsoc.2022.05.079

Zhang, C., Liu, K., Cepero-Mejias, F., Curiel-Sosa, J. L., & Mao, C. (2024a). Numerical investigation on orthogonal cutting and damage response of CFRP/Ti6Al4V stacks. *Journal of Thermoplastic Composite Materials*, *37*(11), 3492–3515. DOI: 10.1177/08927057241234877

Zhang, C., Yuan, S., Amin, M., Fan, H., & Liu, Q. (2016b). Development of a cutting force prediction model based on brittle fracture for C/SiC in rotary ultrasonic facing milling. *International Journal of Advanced Manufacturing Technology*, *85*(1), 573–583. DOI: 10.1007/s00170-015-7894-4

Zhang, L., Ren, C., Ji, C., Wang, Z., & Chen, G. (2016a). Effect of fiber orientations on surface grinding process of unidirectional C/SiC composites. *Applied Surface Science*, *366*, 424–431. DOI: 10.1016/j.apsusc.2016.01.142

Zhang, L., Wang, S., Li, Z., Qiao, W., Wang, Y., & Wang, T. (2019). Influence factors on grinding force in surface grinding of unidirectional C/SiC composites. *Applied Composite Materials*, *26*(3), 1073–1085. DOI: 10.1007/s10443-019-09767-5

Zhang, L., Wu, Z., Wu, C., & Wu, Q. (2022a). On the numerical modelling of composite machining. *Composites. Part B, Engineering*, *241*, 110023. DOI: 10.1016/j.compositesb.2022.110023

Zhang, M., Shan, C., Xia, Z., Jia, F., & Luo, M. (2024b). Dynamic mechanical model in grinding C/SiC composites. *International Journal of Mechanical Sciences*, *268*, 109042. DOI: 10.1016/j.ijmecsci.2024.109042

Zhang, M., Xia, Z., Shan, C., & Luo, M. (2023). Analytical model of grinding force for ultrasonic-assisted grinding of Cf/SiC composites. *International Journal of Advanced Manufacturing Technology*, *126*(5), 2037–2052. DOI: 10.1007/s00170-023-11257-9

Zhang, S., Li, J. F., & Wang, Y. W. (2012). Tool life and cutting forces in end milling Inconel 718 under dry and minimum quantity cooling lubrication cutting conditions. *Journal of Cleaner Production*, *32*, 81–87. DOI: 10.1016/j.jclepro.2012.03.014

Zhang, X., Sun, G., Wang, C., Li, H., & Zhou, S. (2025). A review of structural topology optimization for fiber-reinforced composites. *Composites. Part B, Engineering*, *299*, 112393. DOI: 10.1016/j.compositesb.2025.112393

Zhang, Y., Li, C., Ji, H., Yang, X., Yang, M., Jia, D., Zhang, X., Li, R., & Wang, J. (2017). Analysis of grinding mechanics and improved predictive force model based on material-removal and plastic-stacking mechanisms. *International Journal of Machine Tools & Manufacture*, *122*, 81–97. DOI: 10.1016/j.ijmachtools.2017.06.002

Zhang, Y., Li, C., Jia, D., Li, B., Wang, Y., Yang, M., Hou, Y., & Zhang, X. (2016c). Experimental study on the effect of nanoparticle concentration on the lubricating property of nanofluids for MQL grinding of Ni-based alloy. *Journal of Materials Processing Technology*, *232*, 100–115. DOI: 10.1016/j.jmatprotec.2016.01.031

Zhang, Y., Li, C., Jia, D., Zhang, D., & Zhang, X. (2015). Experimental evaluation of the lubrication performance of MoS2/CNT nanofluid for minimal quantity lubrication in Ni-based alloy grinding. *International Journal of Machine Tools & Manufacture*, *99*, 19–33. DOI: 10.1016/j.ijmachtools.2015.09.003

Zhang, Y., Li, H. N., Li, C., Huang, C., Ali, H. M., Xu, X., Mao, C., Ding, W., Cui, X., Yang, M., Yu, T., Jamil, M., Gupta, M. K., Jia, D., & Said, Z. (2022d). Nano-enhanced biolubricant in sustainable manufacturing: From processability to mechanisms. *Friction*, *10*(6), 803–841. DOI: 10.1007/s40544-021-0536-y

Zhang, Z., Sui, M., Li, C., Zhou, Z., Liu, B., Chen, Y., Said, Z., Debnath, S., & Sharma, S. (2022e). Residual stress of grinding cemented carbide using MoS2 nanolubricant. *International Journal of Advanced Manufacturing Technology*, *119*(9), 5671–5685. DOI: 10.1007/s00170-022-08660-z

Zhang, Z., Yuan, S., Gao, X., Xu, W., Zhang, J., & An, W. (2022b). Analytical modelling of side grinding of orthogonal laminated SiCf/SiC composites based on effective elastic properties. *International Journal of Advanced Manufacturing Technology*, *120*(9), 6419–6434. DOI: 10.1007/s00170-022-09170-8

Zhang, Z., Yuan, S., Xu, W., Ouyang, X., Zhang, J., & An, W. (2022c). Damage behavior and removal mechanism of different yarn orientations 2.5 D SiCf/SiC composites under single-abrasive scratch test. *Ceramics International*, *48*(20), 30868–30883. DOI: 10.1016/j.ceramint.2022.07.041

Zhou, Y., Tian, C., Jia, S., Ma, L., Yin, G., & Gong, Y. (2023). Study on grinding force of two-dimensional ultrasonic vibration grinding 2.5 DC/SiC composite material. *Crystals*, *13*(1), 151. DOI: 10.3390/cryst13010151

Zou, F., Dang, J., Wang, X., Zhang, H., Sun, X., An, Q., & Chen, M. (2021). Performance and mechanism evaluation during milling of CFRP laminates under cryogenic-based conditions. *Composite Structures*, *277*, 114578. DOI: 10.1016/j.compstruct.2021.114578

Chapter 2
Geometric Kinematics and Enhanced Permeation Mechanism of Multiangle 2D Ultrasonic Vibration Nanolubricants Grinding

ABSTRACT

This chapter analyzes the geometric kinematics of multiangle 2D UVAG grains and the workpiece, establishing a model for their relative motion trajectory. The cutting characteristics and the surface generation mechanism are revealed, including multiple buffing of tangential ultrasonic vibration,furrow widening of axial ultrasonic vibration,repeated cutting of single grains,differential cutting of follow-up grains. Excellent gain from UVAG can be obtained by adjusting grinding and ultrasonic vibration parameters. Furthermore, the permeation mechanism at the grinding interface and the hydrodynamics lubrication laws in ultrasonic-assisted nanofluid MQL grinding are investigated. A model for the lubricant film thickness was established. By formulating the hydrodynamic pressure equation, the variations in hydrodynamic pressure, lubricant flow rate, and flow velocity were clarified. The introduction of the ultrasonic vibration velocity increases the lubricant flow velocity and flow rate.

2.1 INTRODUCTION

The variation in tool–workpiece contact states induced by ultrasonic vibration constitutes another important research domain. Tawakoli et al. (2009) reported that dry UVAG eliminates the negative environmental impact of cutting fluids, enabling

DOI: 10.4018/979-8-3373-9013-0.ch002

purely dry grinding. However, without cooling and lubrication media, surface damage cannot be effectively controlled. Considering the interrupted cutting mechanism in UVAM, Ni et al. (2018) demonstrated that four key parameters—excitation frequency, amplitude, cutting speed, and rotation angle—influence the tool–workpiece contact ratio. Through kinematic analysis, Wang et al. (2018) proposed a novel UVAG surface topography regulation and prediction method, capable of controlling surface microstructures and adjusting machining parameters. Ultrasonic vibration alters the trajectory and motion state of abrasive grains, thereby changing the mechanism of ground surface generation. Zhang et al. (2015a) investigated the critical conditions for achieving intermittent cutting and the influence of ballistic work through modeling and simulation of grain motion, respectively. They identified the following key conditions: the frequency coefficient is a positive integer, and the input amplitude ratio is 2. Yan et al. (2009) fully considered the influence of machining parameters on UVAG motion trajectories, derived the motion trajectory equations for adjacent grains during UVAG, and provided criteria for the interaction relationships between adjacent grains in the motion trajectory. However, ultrasonic vibration is typically confined to fixed directions and resulting vibration angles. Abdullah et al. (2012) noted that multiple discontinuous cuts occur in the contact zone between the grinding wheel and workpiece. Under conventional grinding, abrasive grains follow continuous circumferential motion, whereas under ultrasonic vibration-assisted cutting, the cutting path transforms into a semi-sinusoidal curve. Chen & Tang (2015) improved the roughness prediction method under UVAG conditions and established a grain trajectory time function considering workpiece vibration and grain shape. Furthermore, by meshing the workpiece and calculating the minimum value of all grains at each grid point, a simulation model for surface morphology was proposed.

MQL and nanoparticles have been extensively applied in many fields (Li & Ali, 2019; Gu et al., 2024; Zhang et al., 2017a). Zhang et al. (2015b) reported that palm oil-based nanofluids exhibit favorable lubrication performance under MQL conditions due to their high saturated fatty acid content and strong film-forming ability of carboxyl groups. Zhang et al. (2016) explored the influence of MoS_2 nanoparticles at different volume fractions on lubrication. Results for specific grinding energy and surface roughness indicated that satisfactory lubrication effects can be achieved using 2 vol% MoS_2 nanoparticles. To attain better cooling and lubrication effects, some scholars have begun blending nanoparticles in their experimental studies (Sharma et al., 2017). Zhang et al. (2015c) proposed a hybrid NMQL grinding scheme, which involves adding two or more types of nanoparticles to the base oil after physical or chemical modification, followed by mixing, before delivering the mixture to the grinding zone for lubrication and cooling. Zhang et al. (2017b) reported that hybrid NMQL enables complementary properties of multiple nanoparticles to enhance lubrication performance and heat exchange efficiency. Although numerous studies have been conducted on applying NMQL methods to machining operations, the issue of nanofluid fully penetrating the grinding zone has not been thoroughly explored.

UVAG is a precision machining process significantly different from conventional grinding. Li et al. (2012) achieved the best surface finish by using high frequency (11.4 kHz) and low feed rates. Additionally, MQL significantly improved tool life in UVAG. However, a gap exists between the parameters chosen by researchers and the strict requirements for generating ultrasonic vibration. Consequently, the superior machining effects of ultrasonic vibration may not have been fully realized. Jia et al. (2019) found that by combining multi-angle two-dimensional ultrasonic vibration with NMQL technology, adhesion and material spalling on zirconia ceramics were noticeably reduced. However, the machining mechanisms related to the three-dimensional trajectory of grains and the infiltration state of nanofluids were not analyzed. Rabiei et al. (2017) found that the combination of NMQL and UVAG could reduce grinding temperature from 254°C to 132°C compared to dry grinding, without causing any thermal damage. Molaie et al. (2016) demonstrated through research that the integrated application of UVAG and NMQL could reduce forces by approximately 60%. Yan et al. (2018) proposed an ultrasonic vibration-assisted continuous MQL system for near-dry machining and conducted experimental studies. They found that under the conditions of the ultrasonic vibration-assisted continuous MQL system, tool contact length was shortened, and chip morphology was optimized.

Currently, few studies have conducted simulation analyses on the relative motion trajectories of grains/workpieces under multi-angle two-dimensional UVAG conditions from a kinematic perspective to reveal the grain cutting characteristics of multi-angle 2D UVAG. In particular, the technique combining multi-angle 2D ultrasonic vibration with NMQL has not been fully explored. Most existing studies are limited to one-dimensional vibration or two-dimensional vibration with fixed directions, and there is a scarcity of research that reveals the grain motion trajectories and surface generation mechanisms of 2D multi-angle UVAG through simulation analysis from geometrical and kinematic perspectives. Furthermore, whether the coupling of multi-angle 2D ultrasonic vibration technology with nanolubricant MQL technology produces synergistic enhancement effects, how the cutting characteristics of grinding wheel grains and surface generation mechanisms change, has not been thoroughly investigated by scholars to date.

The objective of this chapter is to analyze the geometric kinematics of multi-angle 2D ultrasonic vibration grinding and reveal the mechanisms of nanolubricant infiltration and transport empowered by multi-angle 2D ultrasonic vibration. Through geometric kinematic analysis and motion trajectory modeling/simulation of grains/workpieces in multi-angle 2D UVAG, the effects of different two-dimensional ultrasonic vibration angles on the relative motion trajectory of the grinding wheel to the workpiece and the surface generation mechanism induced by abrasive grains are investigated. Furthermore, experiments are conducted to explore the mechanisms of multi-angle 2D ultrasonic vibration-empowered nanolubricant MQL grinding and the grinding performance resulting from the coupling of ultrasonic vibration and MQL technologies. Finally, the hydrodynamic lubrication mechanism at the grinding interface under ultrasonic-empowered nanolubricant MQL is elucidated, a lubricant oil film thickness model is established, and the variations in hydrodynamic pressure, lubricant flow rate, and flow velocity are clarified by formulating the hydrodynamic equation.

2.2 KINEMATIC NUMERICAL ANALYSIS

2.2.1 Kinematic Modeling and Simulation

The trajectory of the grains and the differential cutting effect of the adjacent subsequent grains were highlighted. The emphasis was on the degree of trajectory interference, which was independent of grain shape. The following assumptions on the grinding process and study object were made to facilitate calculation (Zhang & Li, 2025). (1) The grains on the surface of the grinding wheel are uniformly spread and have the same shape and size. (2) The frequency and amplitude of ultrasonic vibration in the grinding process are stable. (3) The grinding time is 0 when the grain first touches the workpiece, and the initial phases of tangential and axial ultrasonic vibrations are 0° and 90°, respectively. The motion diagram of the 2D UVAG system is shown in Figure 1. During 2D UVAG, when tangential and axial ultrasonic vibrations were exerted on the workpiece, the motion equation of workpiece was as follows:

$$\begin{cases} x = A_U cos(2\pi f_U t) \\ y = B_U cos(2\pi f_U t + \phi_U) \end{cases} \tag{2.1}$$

Figure 1. Schematic diagram of 2D UVAG motion and combined vibration angle θ_U

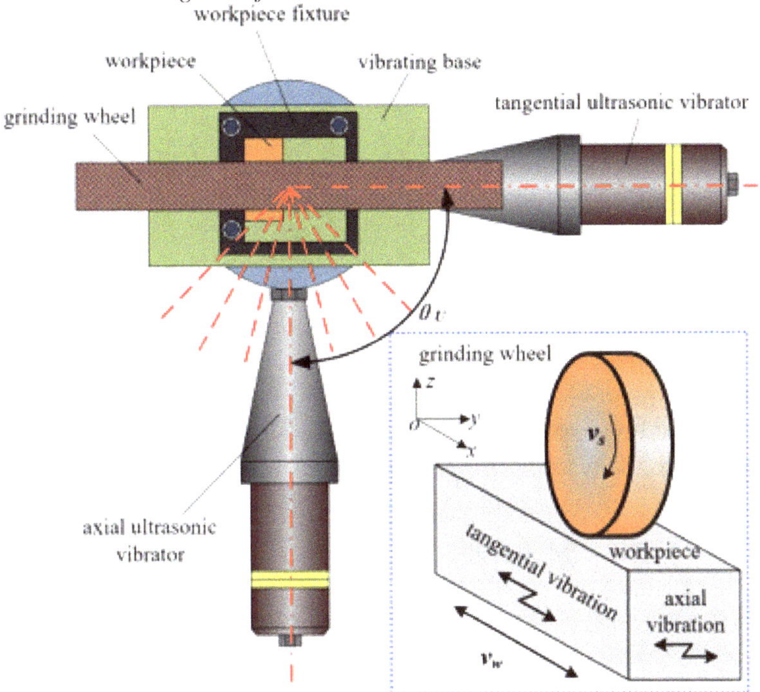

Through controlling the ultrasonic generator, different elliptical trajectories are obtained by fitting different elliptical trajectories with MATLAB software by changing the closing angle θ_U between tangential and axial ultrasonic vibrators. When the resultant vibration angle is θ_U, its component in the y direction can be obtained according to Formula (2.1).

$$y_\theta = y \cdot \sin\theta_U = B_U \cos(2\pi f_U t + \phi_U) \cdot \sin\theta_U \tag{2.2}$$

Its component in the x direction is,

$$x_\theta = y \cdot \cos\theta_U = B_U \cos(2\pi f_U t + \phi_U) \cdot \cos\theta_U \tag{2.3}$$

Given that the position of the tangential ultrasonic vibrator was unchanged, the geometric parameter equation of the relative motion of grains to the workpiece along the x direction after the angle of the axial ultrasonic vibrator was changed is as follows:

$$x_1 = x - x_\theta \tag{2.4}$$

The following can be obtained by substituting Formulas (2.1) and (2.3) into Formula (2.4).

$$x_1 = A_U \cos(2\pi f_U t) - B_U \cos(2\pi f_U t + \phi_U) \cdot \cos\theta_U \tag{2.5}$$

The component of axial ultrasonic vibration in the y direction is the geometric parameter equation of the relative motion of grains of the grinding wheel to the workpiece.

$$y_1 = y_\theta = B_U \cos(2\pi f_U t + \phi) \cdot \sin\theta_U \tag{2.6}$$

On the basis of workbench movement at fixed velocity v_w and the rotation of the grinding wheel at fixed velocity v_s, the relative motion equation of the grains of the grinding wheel to the workpiece was obtained as follows:

$$\begin{cases} x_2 = A_U cos(2\pi f_U t) - B_U cos(2\pi f_U t + \phi_U) \cdot cos\theta_U + v_w t + \frac{d_s}{2} sin\left(\frac{v_s}{d_s/2} t\right) \\ y_2 = B_U cos(2\pi f_U t + \phi_U) \cdot sin\theta_U \\ z = \frac{d_s}{2}\left[1 - \cos\left(\frac{v_s}{d_s/2} t\right)\right] \end{cases} \tag{2.7}$$

where v_w is the feed speed of the workbench, d_s is the diameter of the grinding wheel, and v_s is the circumferential velocity of the grinding wheel.

2.2.2 Motion Trajectory and Path Length

The relative motion trails of the grains to the workpiece in the time for the grain to complete the entire cutting arc were obtained via simulation with the kinematics model and are shown in Figure 2. The turning curvature of the relative motion trails of the grains to the workpiece initially increased and subsequently decreased as θ increased. A small turning curvature of the relative motion trail of grains to the workpiece equated to a small impact load exerted by the grains on the workpiece material through ultrasonic vibration. Relieving the impact of grains on the workpiece material helped relieve the surface/subsurface damage degree caused by grinding.

Figure 2. Relative motion trajectory of grain and workpiece under different θ_U of UVAG

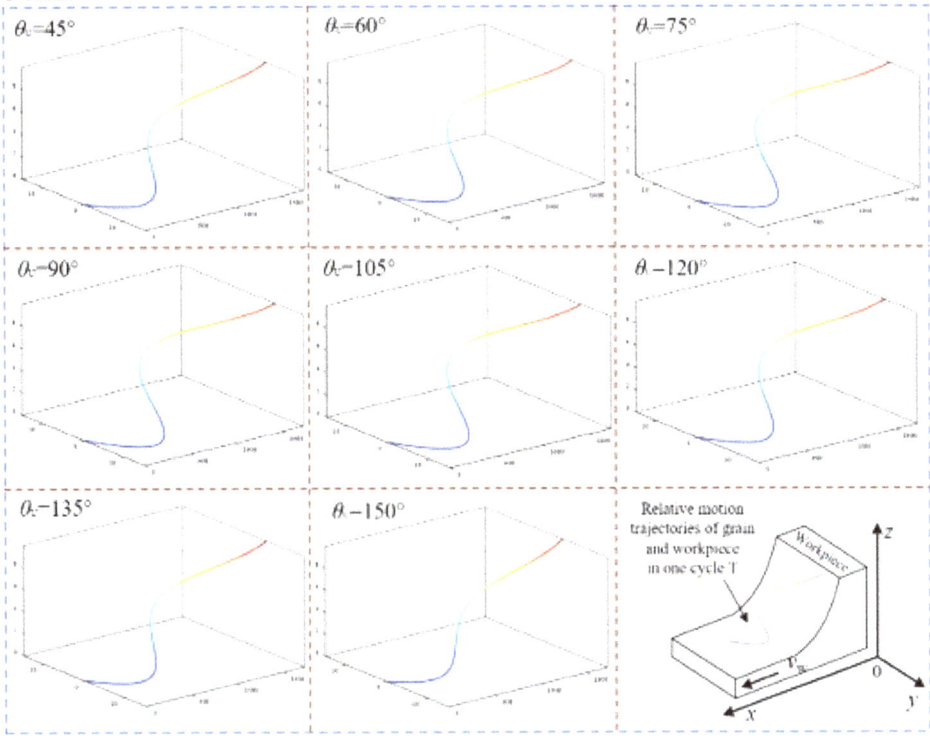

As shown in Figure 3, by analyzing the projection of the abrasive grain cutting motion path of ultrasonic assisted grinding on *xoz* surface, the abrasive grain intermittent cutting process (separation – cutting) in UVAG can be obtained. Several points are taken on the whole abrasive grain grinding contact arc length to analyze the interference state between abrasive grain and workpiece under different two-dimensional ultrasonic closing angles. Taking Figure 3 (a-d) as examples, the abrasive grain cutting states under different two-dimensional ultrasonic vibration angles are respectively marked and explained. The four contact interference states are different. It can be seen that the state of the abrasive grain is different under different vibration modes when the abrasive grain scratches the grinding area.

Figure 3. Grain cutting trajectory and intermittent cutting in UVAG

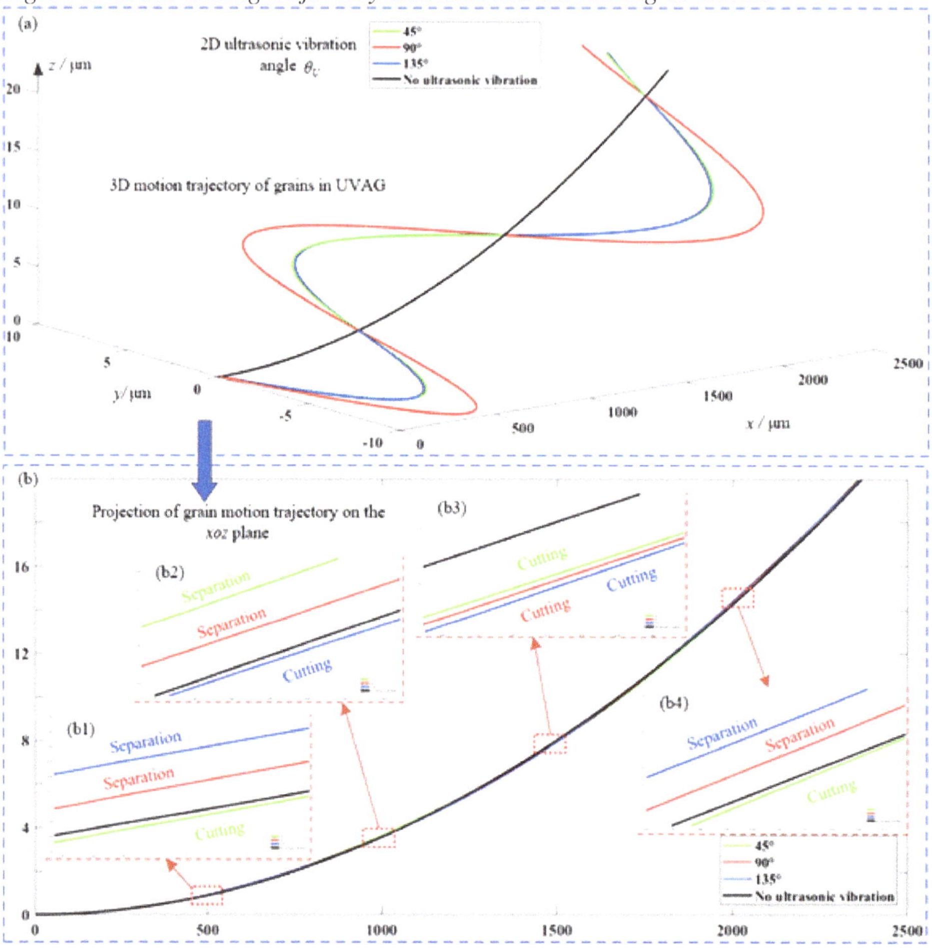

The path length of the motion trajectory can be calculated based on the simulation of grain trajectory and according to the curve integral of arc length.

$$\int_L f(x,y,z)ds = \int_\alpha^\beta f[x(t),y(t),z(t)]\sqrt{x'^2(t)+y'^2(t)+z'^2(t)}\,dt\ (\alpha\beta) \tag{2.8}$$

Figure 4 shows the variation of relative path length of abrasive grain under different grinding parameters and ultrasonic parameters in two-dimensional ultrasonic vibration. For the grinding wheel with low ultrasonic vibration frequency, small amplitude and high circumferential linear velocity, the path length variable can be ignored. It can be seen from Figure 4 that the cutting path length of abrasive grain decreases with the increase of ultrasonic vibration angle. The influence of tangential and axial ultrasonic vibration amplitude on the cutting path length of abrasive grain is opposite. The bigger the tangential ultrasonic vibration amplitude is, the smaller the cutting path length of abrasive grain is, while the axial ultrasonic vibration amplitude is opposite. The influence of the peripheral speed of grinding wheel on the cutting path length of abrasive grain is complicated. In the range of 15 m/s < v_s < 55 m/s, the cutting path length of abrasiv" gra'n Increases first and then decreases, and then increases and then decreases. Similarly, the influence of ultrasonic vibration frequency on the cutting path length of abrasive grain is sinusoidal.

Figure 4. Path length of relative motion trajectory between grinding wheel and workpiece in a single vibration cycle under different UVAG parameters

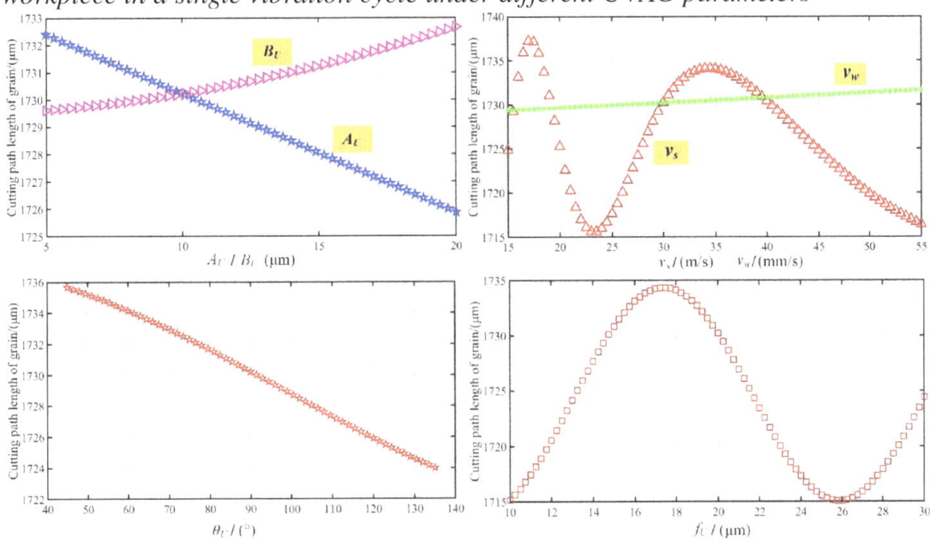

2.3 GRAIN GRINDING CHARACTERISTICS

2.3.1 Multiple Buffing Effect of Tangential Ultrasonic Vibration

The diagram of the relative motions of the grinding wheel and workpiece when tangential ultrasonic vibration alone was applied is shown in Figure 5. When the grinding wheel was at the cutting state at initial time 0, the workpiece moved along the horizontal direction driven by the workbench at velocity v_w. Bore tangential ultrasonic vibration on the same horizontal plane (vibration amplitude was A_U) and two ultrasonic vibration grinding effects occurred in the positive and negative directions of the displacement axis. When the grinding wheel was in the 0–$T/2$ period, namely, the tangential "ltra'onic vibration crest was in the positive direction of the displacement axis, the workpiece motion direction was contrary to that of tangential ultrasonic vibration. The workpiece displacement formula is

$$A_n = V_w \bullet \frac{n}{f_U} \tag{2.9}$$

where f_U is the frequency of ultrasonic vibration, n is the number of periods, and A_n is the displacement of the workpiece in the nth period. When tangential ultrasonic vibration amplitude A_U was greater than workpiece displacement A_1, the grains were separated from the cutting zone with a large separation amplitude. The buffing effect was again exerted on the machined surface. According to Figure 5, when ultrasonic vibration reached the $T/2$ period and satisfied $A > A_1/2$, the buffing effect was exerted on the machined surface twice, which effectively reduced the surface roughness value and improved the workpiece surface quality. When $A \leq A_1/2$, the separation amplitude of the grains of the grinding wheel from the cutting zone was small, and this kind of buffing effect by multiple times was diminished to a great degree. The ultrasonic vibration effect was not obvious.

When the grinding wheel was in the $T/2$- T period, namely, the tangential ultrasonic vibration crest was in the negative direction of the displacement axis, the moving direction of the workpiece was identical to the tangential ultrasonic vibration direction. In this phase, grains entered the cutting zone in advance under the effect of ultrasonic vibration, thus completing the cutting phase. Then, the workbench drove the workpiece toward the completed cutting zone to complete the buffing effect multiple times. Meanwhile, due to the high-frequency ultrasonic vibration in this phase, the action time of the grains on the workpiece was shortened, and the surface damage generated during hard alloy grinding was effectively reduced, thus improving the workpiece surface quality. One period of tangential ultrasonic vibration covered two phases, and the buffing effect occurred multiple times. As indicated in Figure 5, if the grinding wheel moves to position A, it must experience A/A_1 ultrasonic vibration periods. Thus, if the f_U and A_U are fixed and grinding efficiency is ensured, the v_w can be appropriately reduced, and the number of times of the buffing effect can be increased in the UVAG process.

Figure 5. Schematic diagram of relative motion between grinding wheel and workpiece under tangential ultrasonic vibration conditions

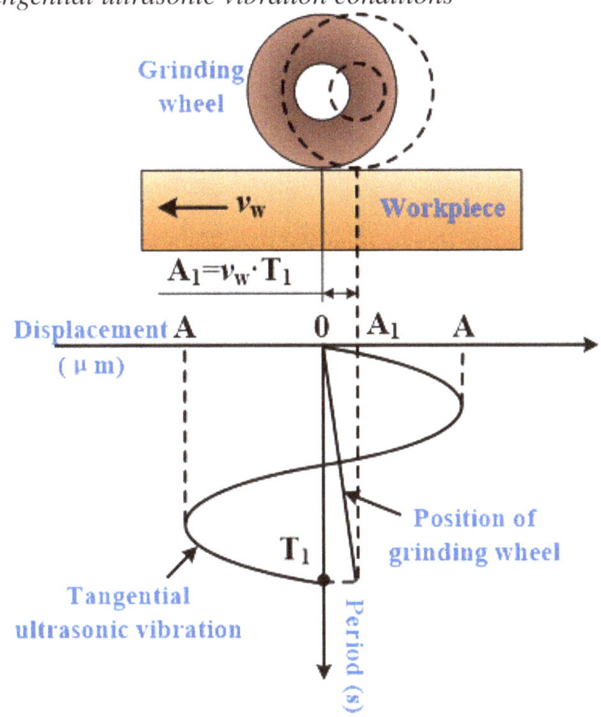

2.3.2 Furrow Widening Effect of Axial Ultrasonic Vibration

A diagram of the relative motions of the grinding wheel and workpiece when axial ultrasonic vibration alone was applied is shown in Figure 6. According to Figure 6, the axial ultrasonic vibration direction was vertical to the direction of workpiece motion. The effect of ultrasonic vibration on the relative motion of the grains and workpiece was mainly manifested by the furrow width on the grinding workpiece surface and the intermittent grinding traces. Bulges formed between grains because the area of ordinary grinding was high and wide, as shown by the triangular cross section with bottom D and height H in Figure 6. After axial ultrasonic vibration was applied, the positions of the grains on the dotted portion were formed. The width (D) and height (H) of the triangular cross section were reduced, thereby reducing the workpiece surface roughness and improving the workpiece surface integrity further. To eliminate the bulges between grains, the axial ultrasonic vibration amplitude should be greater than the average particle size of the grains of the grinding wheel.

Figure 6. Schematic diagram of relative motion between grains and workpiece under axial ultrasonic vibration conditions

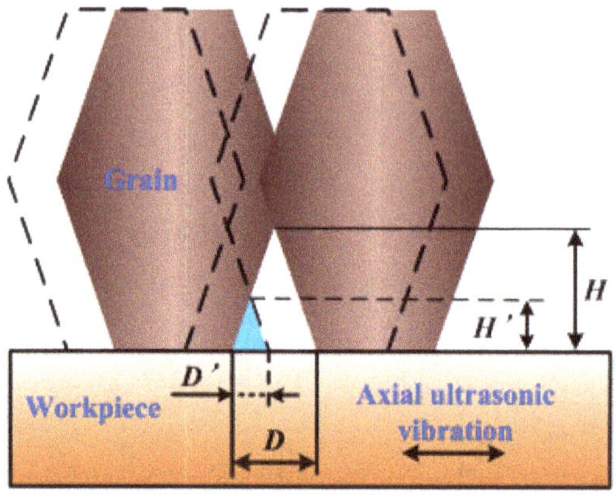

2.3.3 Repeated Cutting of Single Grains

With the motion diagram of 2D UVAG, the contact arc model of the cutting of single grains was obtained, as shown in Figure 7. The contact arc length formula is as follows:

$$l = \sqrt{a_p d_s} \tag{2.10}$$

where a_p is the cutting depth and d_s is the diameter of the grinding wheel. The time (Δt) needed by the grains to complete the cutting process can be obtained with Equation (2.10).

$$\Delta t = \frac{l}{v_s} \tag{2.11}$$

Given that the ultrasonic vibration period was $T=1/f$, at Δt, the number of ultrasonic vibrations is

$$m_U = \frac{\Delta t}{T} = f_U \Delta t = \frac{f_U \sqrt{a_p d_s}}{v_s} \tag{2.12}$$

According to Formula (2.12), to obtain the optimized effect of ultrasonic vibration, Δt must be greater than T. Only in this way can the grains complete ultrasonic vibration once or multiple times within Δt. When $a_p = 10$ μm, $f_U = 20$ kHz, $v_s = 30$

m/s, and $d_s = 0.3$ m, the number of ultrasonic vibrations completed by the grains in the cutting process was calculated to be $m_U = 1.2$, that is, time Δt was greater than T, and 1.2 times of vibration were completed. Grains can vibrate multiple times in the grinding zone by controlling the a_p and v_s of the grinding wheel for an improved grinding surface.

Figure 7. Schematic diagram of contact arc length for grain grinding

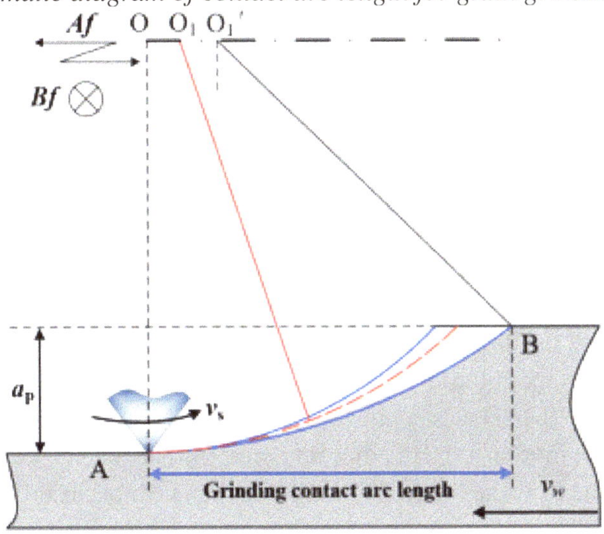

2.3.4 Differential Cutting of Follow-Up Grains

Figure 8 shows the follow-up differential cutting diagram of adjacent grains. The blue line represents the cutting path of the previous grain, and the red line represents the cutting path of the following grain. The grains exerted single-side extrusion cutting on the workpiece material during the multi-angle 2D UVAG process, which was different from bi-directional extrusion applied by grains in the ordinary grinding process. Bi-directional extrusion can distribute the workpiece material to the two sides of the grinding groove, and this mode of acting force would aggravate the cracking of the machined surface. However, single-side extrusion cutting can relieve the tearing effect of the grains on the workpiece surface, reduce the quantity of meso-position cracks on the subsurface, and relieve the subsurface damage degree and workpiece strength loss after machining. Furthermore, the differential cutting effect also increased the quantity of effective grains on the grinding wheel participating in the cutting process and thus improved the machining efficiency.

Figure 8. Differential cutting of adjacent subsequent grains

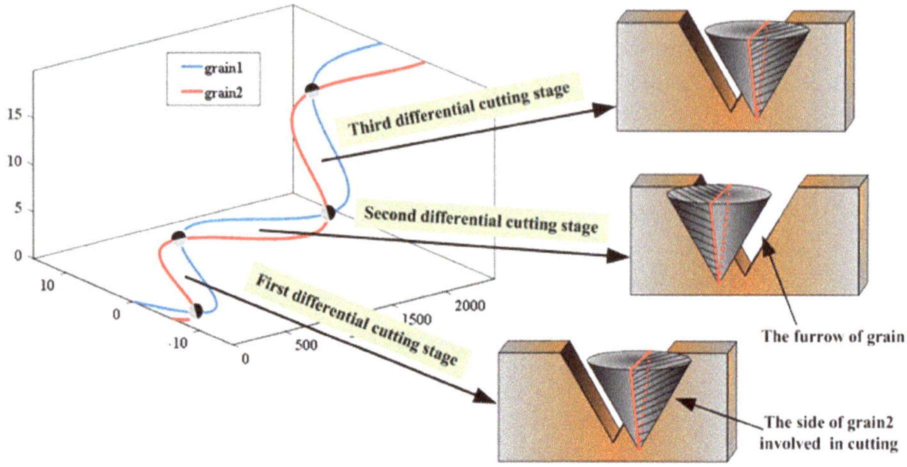

Figure 9 shows the differential cutting under different resultant vibration angles as reflected by the motion trails of the follow-up adjacent grains in one vibration period. When θ_U in the 2D ultrasonic vibration varied from 45° to 135°, the interference degree of the motion trails of adjacent grains initially decreased then increased with θ_U. When the θ_U values were 45° and 135°, the effect of the differential cutting motion of the follow-up grains was the strongest, and the interference degree of the motion trails of the multiple follow-up grains was the highest. The stronger the differential cutting motional effect of follow-up grains is, the shorter the formed cutting debris is. This condition can expand the plastic machining region of crisp and hard materials and reduce the damage degree of the machined surface. However, when θ_U value is 90°, the interference range is the widest, and the action surface of differential cutting effect of subsequent abrasive grain is larger.

Figure 9. Differential cutting under different 2D combined ultrasonic vibration angles

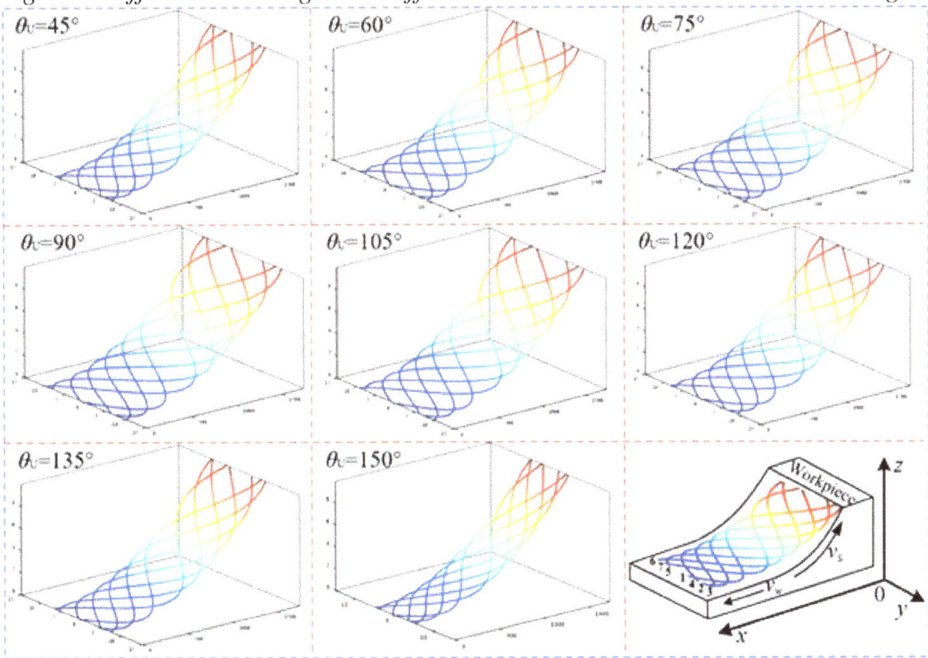

Different processing parameters were tested in the simulation program to explore the impacts of different processing parameters on the differential cutting effect. The contrast diagrams of differential cutting effect under different parameters are shown in Figure 10. Six sets of processing parameters were selected, one of which was the control group. Figure 10 shows that f, v_s, and a_p had a considerable influence on the differential cutting effect. However, the differential cutting effect was minimally affected by the changes in ultrasonic amplitude, grain size, and v_w. Therefore, the differential cutting effect can be improved by increasing f, reducing v_s, and increasing a_p properly to obtain an enhanced grinding performance. In one ultrasonic vibration period (T), the number (n) of continuously adjacent cut grains is the ratio of grain motion stroke to grain size, as follows:

$$n_g = \frac{v_s \cdot 10^6 \cdot (1/f_U)}{d_{grain}} \qquad (2.13)$$

where d_{grain} is the nominal diameter of the grain.

Figure 10. Effect of grinding and ultrasonic vibration parameters on differential cutting results

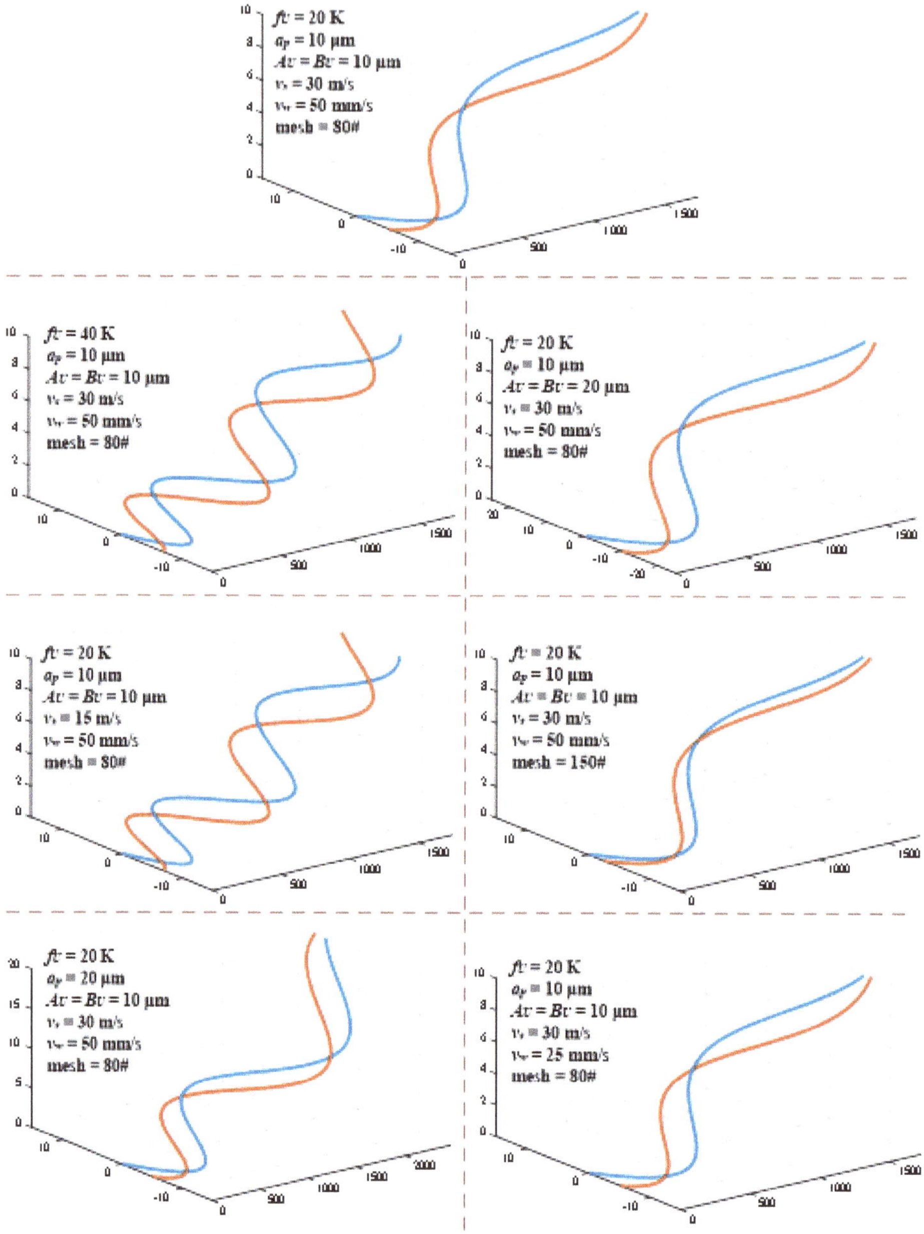

2.4 LUBRICANT INFILTRATION MECHANISM

As shown in Figure 11, when one side of the subsequent adjacent abrasive grain is involved in the cutting process, the other side is completely wetted and cooled by the nano lubricant, thus improving the effective action rate of the nano lubricant. When entering the next differential cutting stage, the cutting state on both sides of the abrasive grain changes. Therefore, the integrated coupling of the two technologies enhances the processing advantages of each other compared to the single NMQL technology and ultrasonic vibration grinding technology. Differences in the state of infiltration lead to significant differences in dynamic characteristics. As shown in Figure 11, droplets on the rough surface have two wetting states, Casie and Wenzel. When nanolubricant droplets infiltrate rough surface microstructures in Wenzel state, they show strong adhesion, resulting in good cooling and lubrication effects on the workpiece surface. While the droplets in Casie state do not completely wet the surface microstructure and suspend on the rough surface, so the cooling and lubrication effect is poor. The lubricant droplets on rough surfaces are usually between the full Cassie and Wenzel states, not completely immersed in the microstructure but rather form an air zone at the bottom. The vertical and horizontal oscillations promote the Cassie-Wenzel wettability transition of droplets on a microscale porous substrate, and the driving force for the transition is the pressure jump caused by the droplet oscillations. As shown in Figure 11, ultrasonic vibration can lead to a transition from Casie state to Wenzel state of the nanolubricant on the machined surface and in the grinding zone of the workpiece. It also provides better penetration Into the workpiece and grinding areas and provides more cooling and lubrication. Tangential ultrasonic vibration is perpendicular to the wetted surface of the droplets, i.e. in the direction of lubricant penetration into the grinding area. When the θ_U is 45 °, the effect of the combined vibration is the strongest In the direction of the penetration Into the grinding zone, and the ultrasonic vibration shows the best contribution to the infiltration of micro droplets.

Figure 11. Coupling effect of ultrasonic energized MQL and wetting state transition of rough surfaces

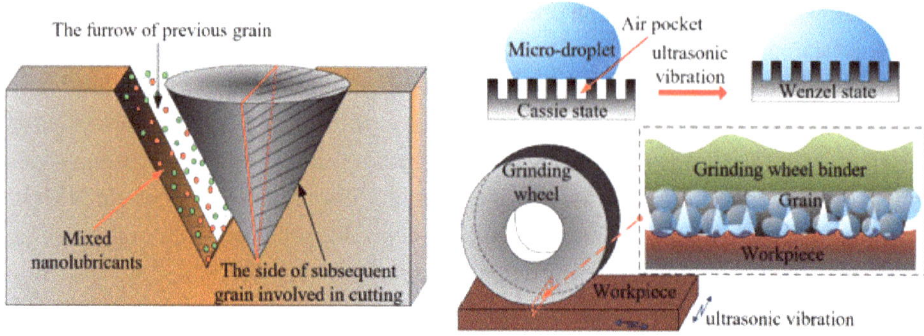

2.5 DYNAMIC PRESSURE LUBRICATION MECHANISM

2.5.1 Formation and Thickness Analysis of Dynamic Pressure Oil Film

Under the influence of ultrasonic vibration, the workpiece in UVAG separates from and contacts the grinding wheel, producing a regular impact effect that facilitates the removal of surface material from the workpiece. Generally, this is often overlooked due to the workpiece feed rate being significantly lower than the circumferential speed of the grinding wheel. As illustrated in Figure 12 (a), during the down-grinding process, a separation zone forms under the action of ultrasonic vibration, allowing for the regular penetration of the nanolubricant into the grinding zone. In UVAG, ultrasonic vibrations generate rapidly varying gaps between the grinding wheel and the workpiece. Consequently, nanolubricants inevitably undergo flow and dynamic pressure when subjected to impact compression within this fluctuating gap. Based on the ultrasonic vibration parameters and motion characteristics, the separation distance between the grinding wheel and the workpiece can be determined as follows:

$$h_{nd} = d_s \cdot \left(\sin\theta_{h2} - \sin\theta_{h1}\right) \tag{2.14}$$

where,

$$\theta_{h1} = \arccos\frac{x_{nd}}{d_s} \tag{2.15}$$

$$\theta_{h2} = \arccos \frac{x_{nd} - 2A_U \sin(2\pi ft + \phi_U)}{d_s} \tag{2.16}$$

Among them, A_U is the amplitude of ultrasonic vibration, f_U is the frequency of ultrasonic vibration, and ϕ_U is the phase angle of ultrasonic vibration.

As shown in Figure 12(b), during the up-grinding process, the nanolubricant is continuously pumped into the grinding zone, specifically in the wedge-shaped area between the grinding wheel and the workpiece. Generally, this dynamic is overlooked because the workpiece feed rate is significantly lower than the circumferential speed of the grinding wheel. The thickness of the oil film in this context is represented as follows:

$$h_{nu} = d_s \cdot (1 - \sin\theta_{h3}) \tag{2.17}$$

where,

$$\theta_{h3} = \arccos \frac{x_{nu}}{d_s} \tag{2.18}$$

At the origin of the coordinate system, the film thickness is denoted as h_0 in Figure 12, where the finished surface makes contact with the grinding wheel. It is assumed that the grinding wheel ceases to remove material from the CFRP workpiece at this point, resulting only in microscopic contact characterized by elastic deformation. The micro protrusions experience elastic deformation when the contact pressure is low. The stress distribution at the micro convex tips within the elliptical contact area follows the Hertz contact principle. The relationship between the average pressure during the el"stic'deformation phase and the depth of indentation h_0 is obtained as follows (Gao et al., 2025),

$$h_0 = \left(\frac{3\pi f_{ge}}{4E}\right)^2 \cdot r_e \tag{2.19}$$

where, h_0 is the depth of indentation, f_{ge} is the average pressure during the elastic contact stage of the grains on CFRP, E is the equivalent elastic modulus, and r_e is the curvature radius of the contact point.

According to the theory of contact mechanics, when $f_{ge} = \sigma$ (where σ is the yield strength of the material), the workpiece material begins to undergo plastic deformation. So, the critical indentation depth h_0 corresponding to the critical point of plastic deformation of the workpiece material is,

$$h_0 = \left(\frac{3\pi\sigma}{4E}\right)^2 \cdot r_e \tag{2.20}$$

The geometric dimensions of a single micro convex surface contour can be determined by the following equation,

$$y\left(x\right) = h_p \cdot \cos\left(\frac{\pi x}{l_p}\right) -\frac{l_p}{2} < x < \frac{l_p}{2} \tag{2.21}$$

The curvature radius of the vertex of the contour line of the semi-periodic cosine curve, representing the rotational body shape, can be derived from the expression for the micro convex body contour.

$$re = \frac{l_p^2}{h_p \pi^2} \tag{2.22}$$

Figure 12. Formation process and film thickness geometric model of dynamic pressure oil film

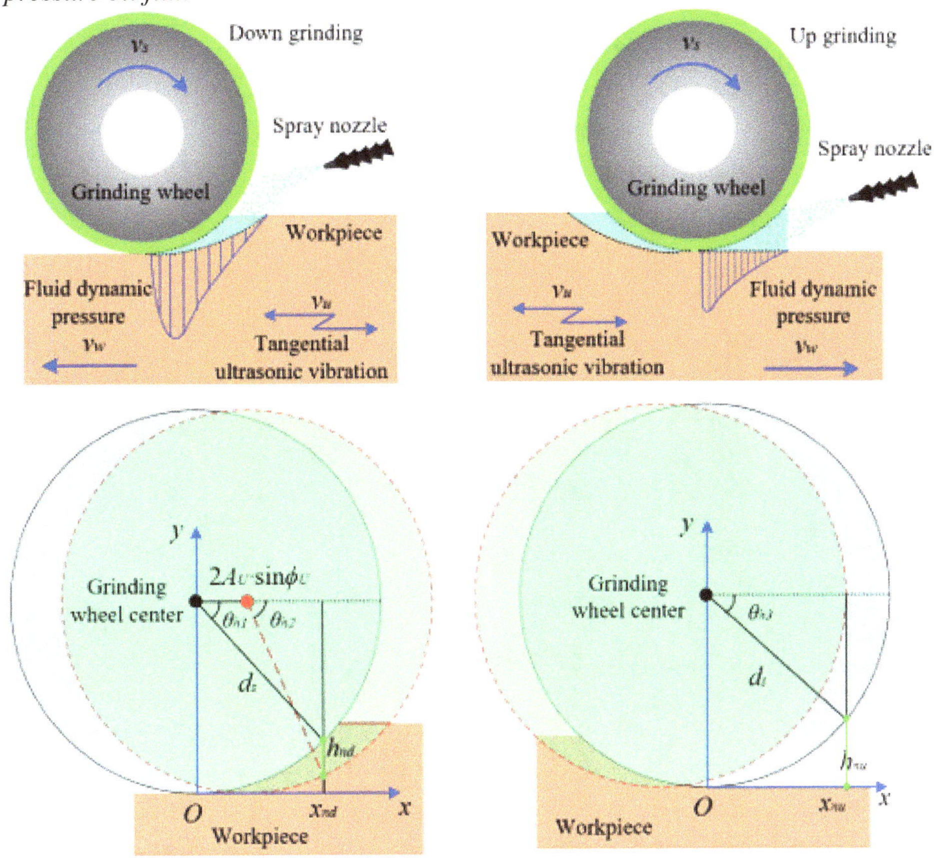

2.5.2 Fluid Dynamic Pressure and Flow Velocity Analysis

The schematic diagram illustrating the dynamic pressure analysis of the lubricating oil film is presented in Figure 13. According to fluid dynamics theory, the general form of the Reynolds equation for the dynamic pressure lubrication process of nanolubricants in the grinding zone is [21],

$$\frac{\partial}{\partial x}\left(\frac{\rho h^3}{12\eta_{hl}}\frac{\partial p_{hl}}{\partial x}\right) + \frac{\partial}{\partial z}\left(\frac{\rho h^3}{12\eta_{hl}}\frac{\partial p_{hl}}{\partial z}\right) = \frac{\partial(\rho h)}{\partial x}\frac{v_s - v_u \cdot \sin\phi_U}{2} \tag{2.23}$$

At a certain point $h = h_m$ where $\frac{\partial p_{hl}}{\partial x} = 0$, the above equation is simplified as follows,

$$\frac{\partial p_{hl}}{\partial x} = \frac{6\eta_{hl}(v_s + v_u \sin\phi_U)}{h^3} \cdot (h - h_m) \tag{2.24}$$

According to Newton's law of friction for viscous fluids,

$$\frac{\partial p_{hl}}{\partial x} = \eta_{hl} \cdot \frac{\partial^2 u}{\partial y^3} \tag{2.25}$$

Figure 13. Schematic diagram of dynamic pressure analysis of lubricating oil film fluid

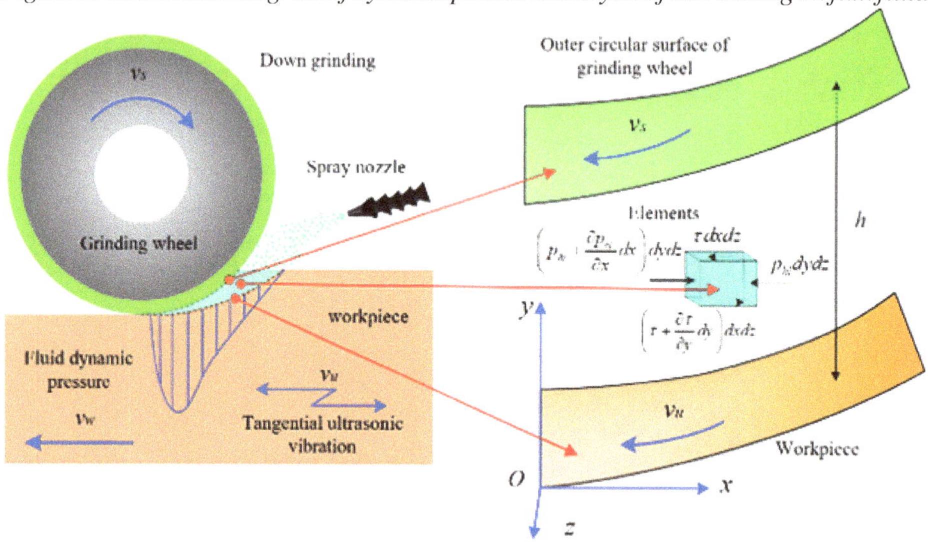

Integrate y to obtain the flow velocity of the lubricant in the x direction,

$$u = \frac{1}{2\eta_{hl}} \frac{\partial p_{hl}}{\partial x} y(y - h) + \frac{v_s}{h} y + \frac{v_u \cdot \sin\phi_U}{h}(y - h) \tag{2.26}$$

Integrating the system between the grinding wheel and the workpiece allows for the determination of the nanolubricant flow rate across any cross-section.

$$q_{hl} = \int_0^h \left[\frac{1}{2\eta_{hl}} \frac{\partial p_{hl}}{\partial x} y(y - h) + \frac{v_s}{h} y + \frac{v_u \cdot \sin\phi_U}{h}(y - h)\right] dy \tag{2.27}$$

To obtain the integral,

$$q_{hl} = \frac{h^3}{12\eta_{hl}} \frac{\partial p_{hl}}{\partial x} + \frac{h}{2}(v_s + v_u \cdot \sin\phi_U) \tag{2.28}$$

According to equations (2.26) and (2.28), it can be seen that the introduction of ultrasonic vibration v_u increases the lubricant velocity of flow u and volume of flow q_{hl}.

2.6 SUMMARY

In this chapter, the multi-angle two-dimensional UVAG geometry and transport mechanism of nano lubricant in the grinding interface and workpiece surface are studied. Through the kinetic modeling and simulation of the abrasive cutting path, the abrasive cutting characteristics of different two-dimensional UVAG parameters are studied. On this basis, the infiltration mechanism of nano lubricant is disclosed, the film thickness model of dynamic pressure lubricating oil is established, and the influence of ultrasonic vibration on the flow and flow rate in the grinding area of nano lubricant is analyzed.

(1) Tangential ultrasonic vibration will produce multiple polishing effects on the machined surface. Ultrasonic vibration reduces the contact and cutting time of abrasive particles to materials, and can effectively improve the surface integrity of workpiece. Proper reduction of v_w can improve the multiple polishing effect of UVAG process. The furrow widening effect of axial ultrasonic vibration can eliminate the residual material bulge caused by the abrasive grain gap, thereby reducing the grinding surface roughness, improving the processing quality, and the furrow widening effect is more significant under the condition of high vibration amplitude. In order to obtain a better ultrasonic vibration gain effect, the grinding parameters and ultrasonic vibration parameters are adjusted so that the ultrasonic vibration period is shorter than the time when the abrasive grain completes the cutting process in the grinding area, so that the abrasive grain can realize the repeated cutting effect of the abrasive grain. By adjusting a_p and v_s, it is possible to improve the grinding surface by improving the repetitive cutting effect of abrasive particles.
(2) The two-dimensional UVAG subsequent abrasive grain differential cutting effect changes the two-way extrusion cutting to unilateral extrusion, which can relieve the tearing effect of abrasive grain on the workpiece surface and inhibit the subsurface crack and subsurface damage. In addition, the abrasive grain differential cutting effect also increases the number of effective abrasive grains participating in the cutting process on the grinding wheel, and improves the processing efficiency. Ultrasonic vibration frequency, peripheral speed of grinding wheel and grinding depth have great influence on differential cutting effect, while ultrasonic amplitude, abrasive grain size and workbench moving

speed have little influence on differential cutting effect. Therefore, on the premise of ensuring the grinding efficiency and precision requirements, the differential cutting effect can be improved by appropriately increasing the ultrasonic vibration frequency, reducing the peripheral speed of the grinding wheel and increasing the grinding depth, so as to improve the grinding performance.

(3) Ultrasonic vibration promotes a transition from Casie state to Wenzel state of nanolubricant on the grinding surface and grinding wheel workpiece interface, fully wetting the penetrating grinding area and providing more cooling lubrication. When θ_U of 2D ultrasonic vibration is 45°, the effect is the strongest in the direction of penetration into the grinding area, and ultrasonic vibration has the best promotion effect on the infiltration of micro droplets.

(4) Based on the theory of fluid dynamics, the hydrodynamic pressure of lubricating oil film is analyzed, and the dynamic lubrication mechanism of the grinding interface of ultrasonic enabled nano lubricant MQL is disclosed. A dynamic pressure lubricating oil film model is established. The influence of ultrasonic vibration on the lubricant flow and flow rate is studied. The introduction of ultrasonic vibration v_u improves the lubricant flow rate u and flow rate q_{hl}.

REFERENCES

Abdullah, A., Farhadi, A., & Pak, A. (2012). Ultrasonic-assisted dry creep-feed up-grinding of superalloy Inconel738LC. *Experimental Mechanics*, *52*(7), 843–853. DOI: 10.1007/s11340-011-9557-1

Chen, H., & Tang, J. (2015). A model for prediction of surface roughness in ultrasonic-assisted grinding. *International Journal of Advanced Manufacturing Technology*, *77*(1-4), 643–651. DOI: 10.1007/s00170-014-6482-3

Gao, T., Liu, J., Sun, X., Zhang, Y., Yang, M., Liu, M., Xu, W., An, Q., Wang, D., Xu, P., & Li, C. (2025). Enhanced permeation mechanism and tribological assessment of ultrasonic vibration nanolubricants grinding CFRP. *Tribology International*, *204*, 110494. DOI: 10.1016/j.triboint.2024.110494

Gu, G., Wang, D., Wu, S., Zhou, S., & Zhang, B. (2024). Research status and prospect of ultrasonic vibration and minimum quantity lubrication processing of nickel-based alloys. *Intelligent and Sustainable Manufacturing*, *1*(1), 10006. DOI: 10.35534/ism.2024.10006

Jia, D., Li, C., Zhang, Y., Yang, M., Zhang, X., Li, R., & Ji, H. (2019). Experimental evaluation of surface topographies of NMQL grinding ZrO2 ceramics combining multiangle ultrasonic vibration. *International Journal of Advanced Manufacturing Technology*, *100*(1-4), 457–473. DOI: 10.1007/s00170-018-2718-y

Li, C., & Ali, H. M. (Eds.). (2019). *Enhanced heat transfer mechanism of nanofluid MQL cooling grinding*. IGI Global, 43-60.

Li, K. M., Hu, Y. M., Yang, Z. Y., & Chen, M. Y. (2012). Experimental study on vibration-assisted grinding. *Journal of Manufacturing Science and Engineering*, *134*(4), 041009. DOI: 10.1115/1.4007102

Molaie, M. M., Akbari, J., & Movahhedy, M. R. (2016). Ultrasonic assisted grinding process with minimum quantity lubrication using oil-based nanofluids. *Journal of Cleaner Production*, *129*, 212–222. DOI: 10.1016/j.jclepro.2016.04.080

Ni, C., Zhu, L., Liu, C., & Yang, Z. (2018). Analytical modeling of tool-workpiece contact rate and experimental study in ultrasonic vibration-assisted milling of Ti–6Al–4V. *International Journal of Mechanical Sciences*, *142*, 97–111. DOI: 10.1016/j.ijmecsci.2018.04.037

Rabiei, F., Rahimi, A. R., Hadad, M. J., & Saberi, A. (2017). Experimental evaluation of coolant-lubricant properties of nanofluids in ultrasonic assistant MQL grinding. *International Journal of Advanced Manufacturing Technology*, *93*(9-12), 3935–3953. DOI: 10.1007/s00170-017-0774-3

Sharma, A. K., Singh, R. K., Dixit, A. R., & Tiwari, A. K. (2017). Novel uses of alumina-MoS2 hybrid nanoparticle enriched cutting fluid in hard turning of AISI 304 steel. *Journal of Manufacturing Processes*, *30*, 467–482. DOI: 10.1016/j.jmapro.2017.10.016

Tawakoli, T., Azarhoushang, B., & Rabiey, M. (2009). Ultrasonic assisted dry grinding of 42CrMo4. *International Journal of Advanced Manufacturing Technology*, *42*(9), 883–891. DOI: 10.1007/s00170-008-1646-7

Wang, Y., Fu, Z., Dong, Y., Liu, S., & Lu, S. (2018). Research on surface generating model in ultrasonic vibration-assisted grinding. *International Journal of Advanced Manufacturing Technology*, *96*(9-12), 3429–3436. DOI: 10.1007/s00170-018-1744-0

Yan, L., Zhang, Q., & Yu, J. (2018). Effects of continuous minimum quantity lubrication with ultrasonic vibration in turning of titanium alloy. *International Journal of Advanced Manufacturing Technology*, *98*(1-4), 827–837. DOI: 10.1007/s00170-018-2323-0

Yan, Y., Zhao, B., & Liu, J. (2009). Ultraprecision surface finishing of nano-ZrO2 ceramics using two-dimensional ultrasonic assisted grinding. *International Journal of Advanced Manufacturing Technology*, *43*(5-6), 462–467. DOI: 10.1007/s00170-008-1732-x

Zhang, J., Zhao, Y., Tian, F., Zhang, S., & Guo, L. (2015a). Kinematics and experimental study on ultrasonic vibration-assisted micro end grinding of silica glass. *International Journal of Advanced Manufacturing Technology*, *78*(9-12), 1893–1904. DOI: 10.1007/s00170-014-6761-z

Zhang, X., Li, C., Zhang, Y., Wang, Y., Li, B., Yang, M., Guo, S., Liu, G., & Zhang, N. (2017b). Lubricating property of MQL grinding of Al2O3/SiC mixed nanofluid with different particle sizes and microtopography analysis by cross-correlation. *Precision Engineering*, *47*, 532–545. DOI: 10.1016/j.precisioneng.2016.09.016

Zhang, Y., & Li, C. (2025). *Hybrid-Energy Sustainable Machining: Mechanism and Processability*. Springer Nature, 458-459.

Zhang, Y., Li, C., Ji, H., Yang, X., Yang, M., Jia, D., Zhang, X., Li, R., & Wang, J. (2017a). Analysis of grinding mechanics and improved predictive force model based on material-removal and plastic-stacking mechanisms. *International Journal of Machine Tools & Manufacture*, *122*, 81–97. DOI: 10.1016/j.ijmachtools.2017.06.002

Zhang, Y., Li, C., Jia, D., Li, B., Wang, Y., Yang, M., Hou, Y., & Zhang, X. (2016). Experimental study on the effect of nanoparticle concentration on the lubricating property of nanofluids for MQL grinding of Ni-based alloy. *Journal of Materials Processing Technology*, *232*, 100–115. DOI: 10.1016/j.jmatprotec.2016.01.031

Zhang, Y., Li, C., Jia, D., Zhang, D., & Zhang, X. (2015b). Experimental evaluation of MoS_2 nanoparticles in jet MQL grinding with different types of vegetable oil as base oil. *Journal of Cleaner Production*, *87*, 930–940. DOI: 10.1016/j.jclepro.2014.10.027

Zhang, Y., Li, C., Jia, D., Zhang, D., & Zhang, X. (2015c). Experimental evaluation of the lubrication performance of MoS2/CNT nanofluid for minimal quantity lubrication in Ni-based alloy grinding. *International Journal of Machine Tools & Manufacture*, *99*, 19–33. DOI: 10.1016/j.ijmachtools.2015.09.003

Chapter 3 Dispersion Mechanism and Spreading Behavior of Ultrasonic CNTs Nanolubricants on Grinding Surface

ABSTRACT

This chapter first investigates the dispersion performance of different surfactants, comparing the influence of their molecular structures and physicochemical properties on dispersion mechanism. The dispersion mechanisms of various surfactants for CNTs nanofluids were studied from the perspectives of space barriers, reversed micellization, hydrocarbon chain, london force, and synergism. Subsequently, the spreading and diffusion mechanisms of nanolubricant droplets on different FOA CFRP grinding surfaces and the influencing patterns on droplet contact angle were investigated. The mechanical behavior of the droplet spreading contact line and the anisotropic spreading mechanism are revealed. Finally, the influence of ultrasonic vibration on the spreading and diffusion was analyzed, and the mapping relationship between surface roughness and spreading rate was explored. This chapter provides a theoretical foundation and data support for the subsequent in-depth study of the tribological behavior and material removal mechanics in MQL UVAG of CFRP.

3.1 INTRODUCTION

Owing to the hygroscopic expansion characteristics of resin-based CFRP, flood cooling cannot be applied in cutting and grinding processes. Furthermore, the substan-

DOI: 10.4018/979-8-3373-9013-0.ch003

tial consumption of grinding fluids in flood cooling poses significant environmental threats, failing to meet the requirements of green manufacturing. Currently, dry processing serves as the primary machining method, however, its inadequate friction reduction and heat transfer capabilities often lead to interfacial cracking between high-strength fibers and the relatively weak matrix under high grinding forces and thermal loads, resulting in interlaminar damage and other defects. Grinding heat can further soften the matrix and exacerbate crack propagation. Particularly when the processing temperature reaches the glass transition temperature, the matrix's support for fibers is severely compromised, leading to degraded mechanical properties that adversely affect load-bearing capacity and service safety. Additionally, dry processing generates considerable dust pollution, endangering workers' health. Consequently, both dry grinding and flood grinding of CFRP exhibit significant limitations that restrict their applications.

To address the technical shortcomings of dry and flood grinding, MQL has emerged as an indispensable and ideal alternative to dry processing (Gu et al., 2024). The addition of solid nanoparticles to base oils for preparing nanofluids can overcome the technical bottleneck of poor heat exchange performance in traditional MQL grinding by leveraging the exceptional heat transfer enhancement properties of nanoparticles (Hadad et al., 2012). At ambient temperature, carbon nanotubes (CNTs) exhibit a thermal conductivity of 3000 W/m·K, approximately one thousand times that of water. For solid materials of identical type and mass, smaller volume corresponds to finer grain size, larger specific surface area, and superior thermal conductivity (Choi et al., 2011; Hwang et al., 2006). CNTs possess small dimensions, large specific surface area, and high aspect ratio. Their surface energy and tension are also significantly higher than those of conventional materials. CNTs tend to entangle and aggregate, forming large clusters due to strong electrostatic and van der Waals forces between surface molecules or atoms. Such aggregation can lead to incomplete oil film formation on friction pairs and substantial entanglement accumulation, preventing effective rolling behavior and hindering their function as "micro-bearings" akin to spherical nanoparticles. These characteristics ultimately limit the friction-reduction capability of CNTs (Shen & Shih, 2009). Xie et al. (2011) incorporated various nanoparticles into base fluids to prepare nanofluids, investigating the effects of particle size, volume concentration, pretreatment processes, additives, and physical properties of base fluids on thermal conduction. They proposed promising methods for optimizing the thermal conductivity of nanofluids. Hu et al. (1998) added sorbitan monostearate and copper stearate as dispersants into 500SN oil containing amorphous iron oxide nanoparticles, finding that dispersants play a crucial role in improving tribological performance. The stability and homogeneity of nanofluids significantly enhance thermal conductivity. Stable and uniform dispersion of CNTs is essential for suppressing entanglement and aggregation while optimizing their performance (Yamamoto et al., 2011).

Substantial research has been conducted to explore the dispersive effects of various surfactants on CNTs. Duan et al. (2011) employed molecular mechanics simulations from an energy perspective to investigate the dispersion of CNTs with sodium dodecyl sulfate (SDS), revealing the aggregation morphology of SDS on CNT surfaces and the influence of CNT diameter on adsorption density. Moore et al. (2003) suspended single-walled carbon nanotubes (SWCNTs) in aqueous media using various anionic, cationic, and nonionic surfactants and polymers, comparing the performance of surfactant-SWCNT suspensions and the quality of absorption and fluorescence spectra. The results indicated that the ionic surfactant SDBS exhibited optimal spectral resolution characteristics, while nonionic surfactants with high molecular weights could suspend substantial amounts of SWCNTs and achieve remarkable spectral properties. Rastogi et al. (2008) reported a comparative analysis of multi-walled carbon nanotube (MWCNT) dispersion with four surfactants, Triton X-100, Tween 20, Tween 80, and SDS. Among these, Triton X-100 and SDS provided the maximum and minimum dispersion stability, respectively. The observed trend in surfactant dispersion capability aligned with their chemical structures. Estellé et al. (2014) measured the viscosity and thermal conductivity of water-based CNT nanofluids stabilized by lignin as a surfactant. Compared with SDBS, lignin reduced the viscosity and shear-thinning behavior of nanofluids at high volume fractions without compromising thermal conductivity enhancement.

Researchers have also investigated the performance of other nanofluids under surfactant conditions. Mao et al. (2014) examined the effect of SDBS on the suspension stability of Al_2O_3 nanofluids and analyzed the dispersion morphology of Al_2O_3 nanoparticles in the dispersion system. They found that large aggregates formed when no surfactant was applied, whereas Al_2O_3 nanoparticles were uniformly dispersed upon surfactant addition. Guo et al. (2017) observed that phosphinate ionic liquids, when used as surfactants mixed with ceramic nanoparticles, maintained long-term stability. Behera et al. (2017) investigated the correlation between spreading behavior, surface tension, and surface energy of Al_2O_3 nanofluids with different surfactants. The optimal wetting behavior was achieved with nonionic surfactants.

Droplet spreading on solid substrates is ubiquitous in nature and industry. The wetting of CFRP surfaces by nanolubricants and the spreading dynamics of droplets critically influence lubrication and cooling performance. Microscopically, the final state of droplet spreading is determined by the combined effects of substrate attraction and friction at solid–liquid interfaces. Solid–liquid interactions play a crucial role in spreading kinetics. The infiltration and spreading of nanolubricant droplets on workpiece surfaces significantly affect their cooling and lubrication efficacy (Liu & Li, 2024). Studies have demonstrated that ultrasonic and micro-vibrations positively promote droplet infiltration, penetration, and spreading on surfaces and interfaces. Rahimzadeh and Eslamian (2017) elucidated the force balance and pinning/depinning behavior in pristine and excited sessile droplet systems, demonstrating that applied vibrations influence droplet stick-slip behavior. Zhang et al. (2020) investigated the dynamic behavior of droplets impacting ultrasonically vibrated surfaces, finding that ultrasonic vibration can control droplet spreading, with higher amplitudes leading to larger average droplet sizes. Wu et al. (2024) discovered that ultrasonic excitation significantly reduces droplet contact angles. Moreover, the wetting state of droplets undergoes permanent changes under ultrasonic action; even after cessation of ultrasonic excitation, the final contact angle remains substantially smaller than the initial value. Furthermore, ultrasonic cavitation effects also positively contribute to lubricant infiltration and penetration. Guo et al. (2018) reported that the oil film pressure in ultrasonic vibration honing alternates between positive and negative pulses, inducing secondary expansion of bubbles and resulting in dual micro-jets during bubble collapse. With increasing ultrasonic amplitude, the amplitude of bubble expansion increases and the oscillation interval prolongs. Naskar et al. (2021) studied the influence of fluid lubricity on abrasive–workpiece interactions in a hybrid UVAG + MQL grinding method. More lubricating fluids offered less resistance to vibration transmission into the grinding zone, thereby increasing the effective vibration amplitude with enhanced fluid lubricity. Additionally, more lubricating vegetable oils yielded the lowest grinding forces and higher compressive residual stresses, with UVAG enhancing compressive residual stress by at least 20%.

Researchers have observed macroscopic dispersion effects of CNTs and analyzed the particle size of CNT aggregates using laser particle size analyzers and other instruments. The influence of different surfactants on CNT dispersion stability has been investigated via SEM and TEM. However, the mechanisms by which different surfactants affect CNTs vary due to differences in their chemical molecular structures and physical properties. Currently, only a few studies have compared the tribological performance of CNT nanofluids with different types of surfactants and analyzed the corresponding mechanisms. Consequently, it remains challenging to identify surfactants that promote optimal CNT dispersion, determine the surfactant type that yields CNT nanofluids with the best tribological performance, and establish fundamental relationships with dispersion stability. Therefore, this chapter analyzes the influence of different surfactants on CNT dispersion mechanisms and evaluates dispersion stability via static observation methods. Furthermore, it investigates the spreading characteristics of CNT nanolubricant droplets on CFRP grinding surfaces with different FOAs under multi-angle ultrasonic vibration conditions, and explores the mapping relationship between surface roughness and spreading rate.

3.2 DISPERSION MECHANISM OF DIFFERENT SURFACTANTS

CNTs are coaxial circular tubes composed of a monolayer to multilayer carbon atoms in a hexagonal arrangement. Carbon atoms in CNTs hybridized by SP^2 have a larger S orbit than those hybridized by SP^3, thus exhibiting high modulus and strength. CNTs possess favorable lubrication performance and wearing resistance considering their unique tube structure and extremely high tenacity. However, CNTs are characterized by their tubular, easy-winding, and conglobation characteristics, which limit their applications. The addition of surfactants to CNT nanolubricants can relieve the winding and conglomeration of CNTs effectively, thereby improving the CNT dispersion effect. Different surfactants have various dispersing mechanisms to CNTs.

3.2.1 Space Barriers of Interfaces

The main function of surfactants as dispersants is to prevent the reaggregation of dispersed solid particles. Due to the large surface area and high surface energy of dispersed systems, they are thermodynamically unstable systems. The adsorption of surfactants on the CNTs palm oil interface will reduce the interfacial tension of CNTs palm oil, thus greatly reducing the surface free energy and decreasing their tendency to aggregate with each other. In addition, due to the adsorption of surfactants, the surface adsorption layer of CNTs thickens, forming a spatial barrier that hinders particles from approaching each other. If the surfactant is ionic, its adsorption will also make its surface charged, creating an electrostatic repulsion barrier, which will hinder the particles from approaching each other and thus prevent re aggregation. In non-aqueous dispersion systems, due to the low dielectric constant, the electric repulsion barrier is not effective, and the stability of dispersed particles mainly relies on the spatial repulsion barrier.

The surface of CNTs belongs to high-energy surfaces, and palm oil contains a large amount of triglycerides with good polarity. Therefore, palm oil cannot spread well on the surface of CNTs, and CNTs themselves are prone to entanglement and aggregation, which will cause CNTs to be unevenly dispersed in palm oil, as shown in Figure 1a. After adding surfactants, due to the special functional groups on the surfactant molecules, they have strong adsorption properties on the surface of CNTs or palm oil. As a result, the surfactant molecules are tightly adsorbed on the surface of CNTs, forming a spatial barrier and evenly dispersing CNTs in palm oil, as shown in Figure 1.

Figure 1. Space barrier of surfactant adsorption

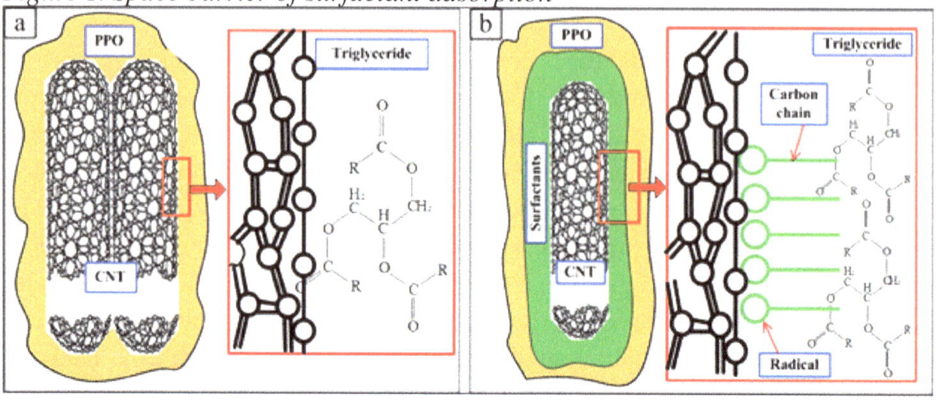

3.2.2 Reversed Micellization

The micelle formation of surfactants in the solution closely resembles the surfactant adsorption at an interface. The state of surfactant molecules will change when a certain concentration is reached. Under this circumstance, a certain number of surfactant molecules in the solution will create an aggregation in the form of a polar group proceeding to opposite directions, and the aggregation is a micelle, which can effectively increase the space repulsion barriers (Gao et al., 2019). The number of micelles increases with the further increase in surfactant concentration while keeping the size of each micelle and the number of molecules in each micelle constant, and the entire process is exhibited in Figure 2. A certain surfactant concentration is required to form micelles, and the certain concentration is described as a critical micelle concentration (CMC). The micelle formation will cause sudden changes In the surfactant properties. The CMC is an important property of surfactant and can be used as an evaluation index of the surface activity of the surfactant. A low CMC indicates a low concentration required to form micelles and a low concentration for the interface to reach the saturated adsorption of the monomolecular layer. Therefore, a low concentration can significantly change the internal and surface properties of the system; that is, surfactant performs effectively under low micelle concentration (Naskar et al., 2021).

Figure 2. Formation process of micelles

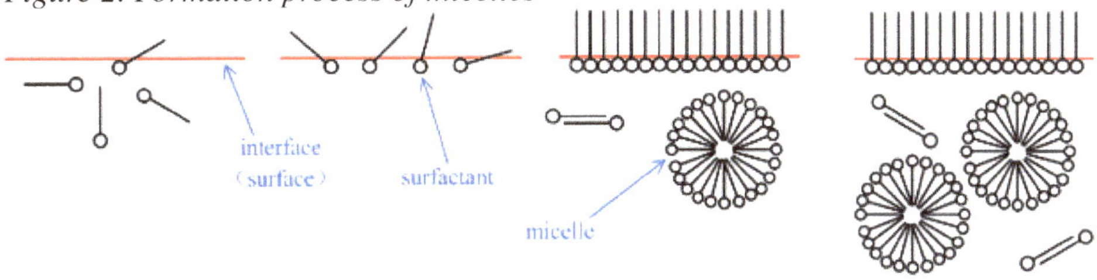

Surfactants can also form micelles in a non-aqueous solution. Such aggregate, with hydrophilic groups as the internal kernel, is called reversed micelle (W/O micelle) and has opposite structures with micelles in the aqueous solution. The association of surfactant in a nonpolar solvent is the consequence of forming a reversed micelle through dipole–dipole between two molecules and ion interaction. Reversed micellization is different from micelles formed by surfactant in water through hydrophobic interaction; This process is complicated, as displayed in Figure 3. The reversed micellization further increases the space barrier and improves the dispersion stability of CNTs. Several special groups and chemical bonds in surfactant can influence the CMC differently. The -O- and -OH polar groups in the molecular hydrocarbon chain of surfactant will decrease the critical reversed micelle concentration of surfactant significantly (Gao et al., 2019).

Figure 3. Space barrier of reversed micelles

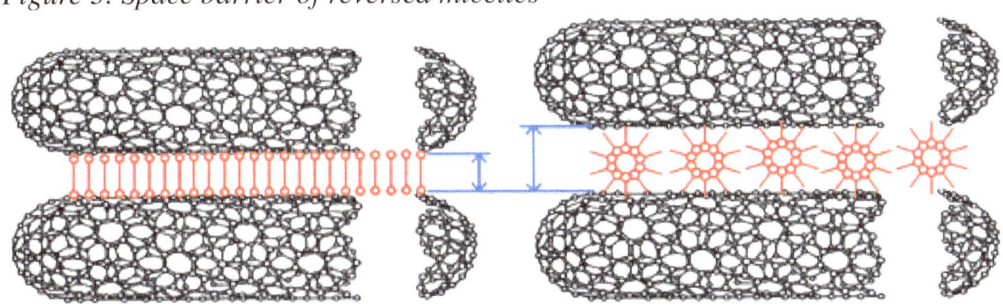

3.2.3 Nature of the Hydrocarbon Chain

The hydrophilicity of the hydrophilic group is negatively related to the length of the hydrocarbon chain. A strong relative hydrophilicity of hydrophilic groups implies a strong tendency to escape from oil, easiness to be adsorbed onto the solid surface, and a high adsorption capacity. Surfactants with significantly different properties also conform to the abovementioned law. For surfactants that belong to homolog, the surface adsorption capacity will increase slightly with the growth of methyne -CH_2 in the hydrocarbon chain length (Attwood & Florence, 1983).

The hydrocarbon chain length of io"Ic s'rfactant can influence its hydrophilicity and lipophilicity; that is, a long hydrocarbon chain leads to a strong lipophilicity. However, such a relationship is sensitive to the hydrophilicity of hydrophilic groups. Given strong lipophilicity, the adsorption capacity on CNT–PPO interface enhances, and the absorption stability improves. However, the long chain will be curled up when the carbon number in the straight alkane chain of the lipophilic group increases up to 16, thus also increasing the single-molecule cross-sectional area of the surfactant. Consequently, the saturated adsorption capacity of the surfactant decreases, and the adsorption layer becomes loose, thereby indicating the declining adsorption capacity of the surfactant (Gao et al., 2019).

3.2.4 London Force

An instantaneous relative displacement between electron cloud and atomic nucleus frequently occurs given the ceaseless movement of electrons and the continuous vibration of an atomic nucleus in molecules; thus, an instantaneous dipole is produced. The instantaneous dipole exists briefly but continuously. The molecular interaction force generated by the instantaneous dipole is called the London force, and London force constantly occurs on the surfactant and solid surface. The significance of London force lies not only in its independent adsorption mechanism but also in its supplementation mechanism of other adsorption types (Gao et al., 2019). London force increases with the molecular weight of surfactant. London force is also related to molecular deformability, ionization potential, and molecular distance.

3.2.5 Synergism

Mixed surfactant possesses excellent performance, is prepared by two or several types of surfactants, and has an excellent performance (Holland & Rubingh, 1992). The physicochemical properties of the mixed surfactant solution change significantly compared with those of the independent surfactant, which is called synergism or called synergistic effect. The surface activity of mixed surfactant is superior to that of each component, and several of its properties may even be beyond the capability of the original components. The mixed surfactant prepared by homolog can be approximately viewed as the ideal solution, and micelle is viewed as the ideal micelle (Clint, 1975).

3.3 COMPARISON OF DIFFERENT SURFACTANTS EFFECTS

3.3.1 Dispersants and Experiment Details

Six lubricating fluids (six types of surfactants + CNT nanolubricants) were used in the frictional test, and the PPO was selected as the base oil, as shown in Table 1. The control variate method of the test was adopted. In each test, only the type of lubricant was changed, whereas other conditions remained the same. Each frictional test was performed thrice to assure a high test accuracy. After each test, the white alundum ball and disc were cleaned by acetone and absolute ethyl alcohol to eliminate impurity residues (e.g., lubricating fluid and cutting debris) on the sample for the convenience of further measurements. Surfactants with different basic groups and hydrocarbon chain lengths were applied. The main properties of the surfactants are summarized in Table 2.

Table 1. Conditions of different nanolubricants

Sample	Nanolubricants composition
1	PPO+2 wt% (CNTs)+3 wt% (APE-10)
2	PPO+2 wt% (CNTs)+3 wt% (CTAB)
3	PPO+2 wt% (CNTs)+3 wt% (OP-10)
4	PPO+2 wt% (CNTs)+3 wt% (SDBS)
5	PPO+2 wt% (CNTs)+3 wt% (SDS)
6	PPO+2 wt% (CNTs)+3 wt% (TTAB)

Table 2. Properties of surfactant

Surfactant	APE-10	OP-10	SDBS	SDS	CTAB	TTAB
Mr		646	348.5	288.4	364.5	336.4
HLB		8~8.6	10.6	40	15.8	16.75
Type	nonionic	nonionic	anionic	anionic	cationic	cationic

continued on following page

Table 2. Continued

Surfactant	APE-10	OP-10	SDBS	SDS	CTAB	TTAB
Physical state	liquid	liquid	liquid	granular	powdery	powdery
Chemical molecular formula	80% NPE 15% OPE 1% DPE 1% DNPE	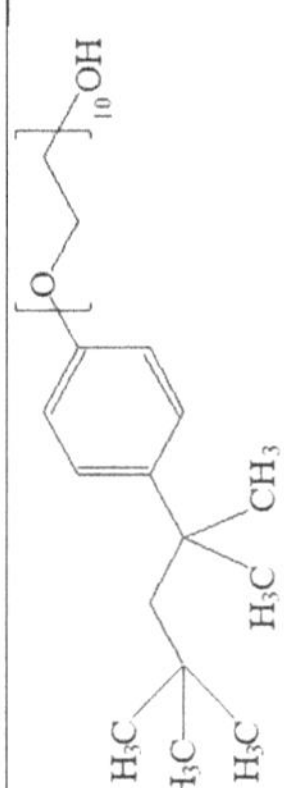	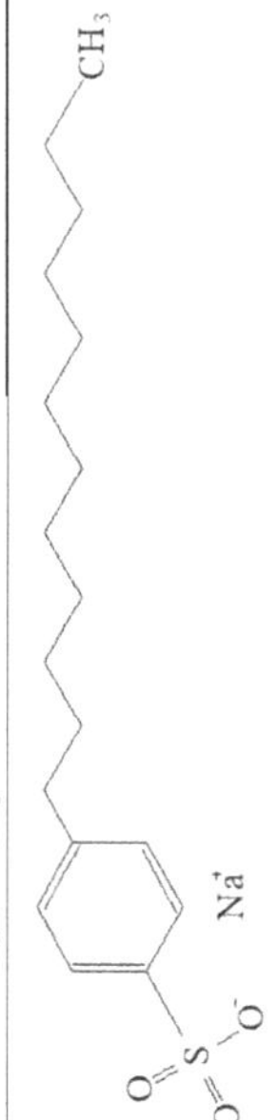	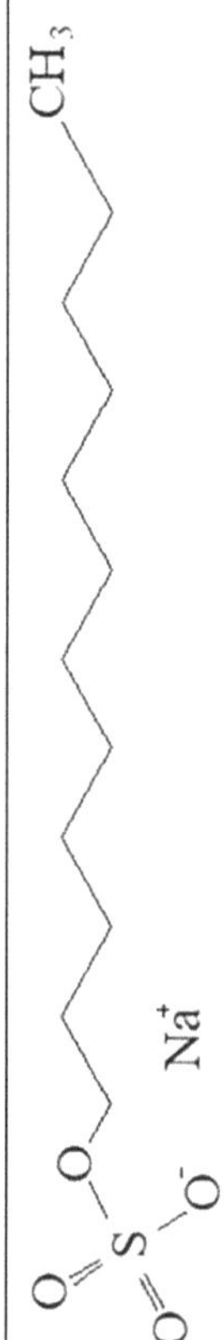	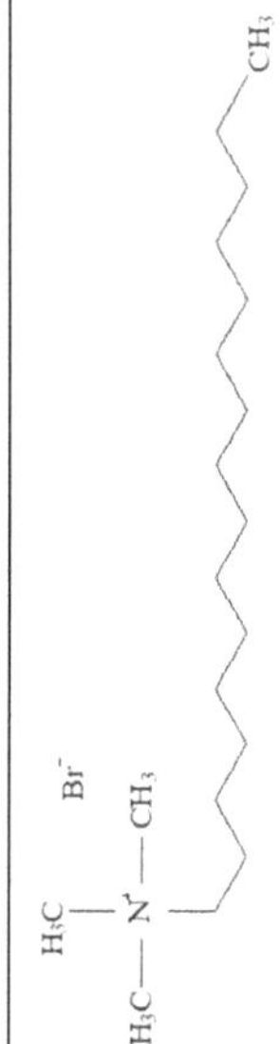	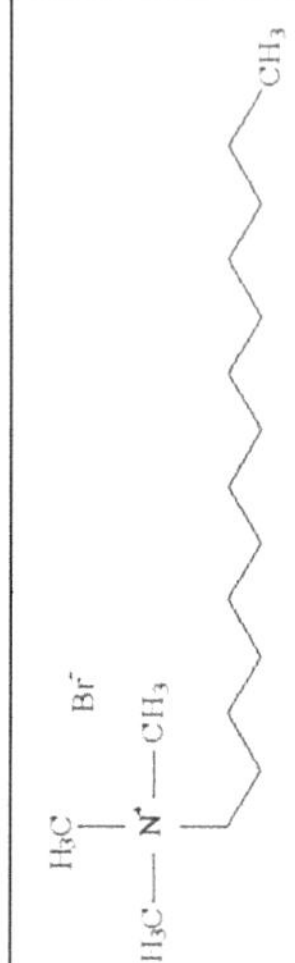

3.3.2 Stability Persistence

Variations in CNT nanolubricants with different surfactants under static conditions were observed and recorded, as depicted in Figure 4. Prepared CNT nanolubricants with different surfactants were processed through a 10 min ultrasonic vibration in an ultrasonic cleaning machine to ensure the same initial dispersion state. After putting static for 1 h, the CNT nanolubricants with SDS showed obvious layering, their bottom layer was the CNT aggregation layer, the middle was the PPO layer with few CNTs, and the top was the SDS particulate layer. SDS is a particulate surfactant with high HLB and poor oil solubility, and abundant SDS particles are accumulated on the top layer rather than dispersing (Gao et al., 2019). After 3 h of being static, the CNT nanolubricants with SDBS began to demonstrate CNT aggregation settlement, and the color lightened up from the bottom to the top of the tube. The CNT nanolubricants with CTAB began to develop a flocculent CNT aggregation system in the upper part and abundant CNT aggregation settlement in the lower part. After 5 h of being static, the CNT nanolubricants with SDS, SDBS, and CTAB continued to aggregate and deposit, and the layering phenomenon intensified. However, no obvious changes were observed in the other groups of nanolubricants. After 21 h of being static, the CNT nanolubricants with APE-10 and OP-10 developed extremely thin transparent layers on the liquid surface; these transparent layers might be a small amount of mixture of surfactants and palm oil. No obvious changes occurred in the other groups of nanolubricants. After 45 h of being static, none of the nanolubricants had developed marked changes. Therefore, APE-10(>45 h), OP-10(>45 h), and TTAB(>45 h) were the surfactants with a high dispersion stability under extended time statics.

Figure 4. Static observation results of nanolubricants with different surfactants

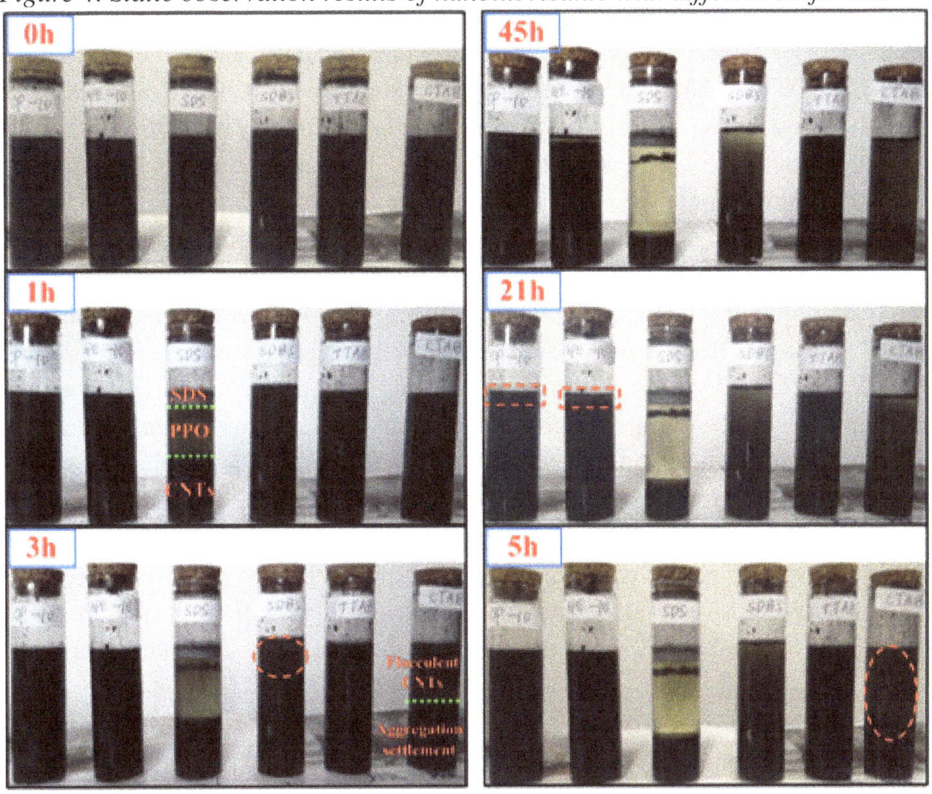

3.3.3 Effects of Surfactants on the Viscosity

The dispersing mechanisms of the different surfactants in the CNT nanolubricants were discussed by static observation and viscosity comparison. The prepared CNT nanolubricants with different surfactants were dispersed through a 10 min ultrasonic vibration in the ultrasonic cleaning machine, and viscosity was measured using the DV2TLV digital viscometer. The test parameters were set as follows: 75% torque, 1 min measurement time, and viscosity data collected every 1 s. Finally, the viscosity average throughout the entire measurement was obtained.

The viscosities of the CNT nanolubricants added with different surfactants under room temperature are demonstrated in Figure 5, which illustrates that adding surfactant can increase the viscosity of the CNT nanolubricants. Different surfactants increased the viscosity of the CNT nanolubricants to different extents. The order of viscosity of lubricating fluids was (PPO+CNTs+APE-10) > (PPO+CNTs+CTAB) > (PPO+CNTs+SDS) > (PPO+CNTs+TTAB) > (PPO+CNTs+OP-10) > (PPO+CNTs+SDBS) > (PPO+CNTs) > (PPO). The comparison of the measurement results with frictional test and static observation showed that the nanolubricants with APE-10 and CTAB exhibit improved dispersion stability and tribological properties. APE-10 increased the viscosity of nanolubricants significantly because it is a mixed surfactant; that is, different surfactant molecules had high-efficiency "synergism" and were adsorbed onto the CNT–PPO interface.

CTAB could increase the viscosity of nanolubricants more significantly than the TTAB; this phenomenon could be explained by the following aspects. CTAB contains two hydrophilic groups (amidogens and halogen -Br). With strong hydrophilicity, CTAB shows a high tendency to escape from oil and is easy to be adsorbed onto the CNT surface. Thus, the adsorption capacity is high, which is conducive to a uniform CNT dispersion and a high nanolubricant viscosity. In addition, the carbon chain is longer in the lipophilic groups of CTAB than in TTAB, and the longer chain will not be curled up in a short time. CTAB is adsorbed onto CNTs, thereby increasing the distance between CNTs compared with that under TTAB adsorption and improving the CNT dispersion effect. Nanolubricants with SDS achieved the third highest viscosity but also poorer dispersion stability. The reasons for this result might be as follows: SDS is a powdery surfactant with a high HLB value; therefore, the oil solubility of SDS is poor, and the dispersion effect of CNT nanolubricants after adding SDS deteriorates. Moreover, serious layering phenomena, namely, CNT deposited at the bottom of the lubricating fluid, PPO occupied the middle layer, and SDS particles accumulated at the top layer, was developed, thus resulting in high viscosity. The low viscosity of the nanolubricants with OP-10 might be caused by the weak London force of OP-10 molecules on the CNT surface. The rotor rotation of viscometer in the measurement significantly damaged the adsorption of OP-10 considering the strong external disturbance and the weak London force of OP-10 molecules on the CNT surface.

Figure 5. Viscosity of nanolubricants with different surfactants

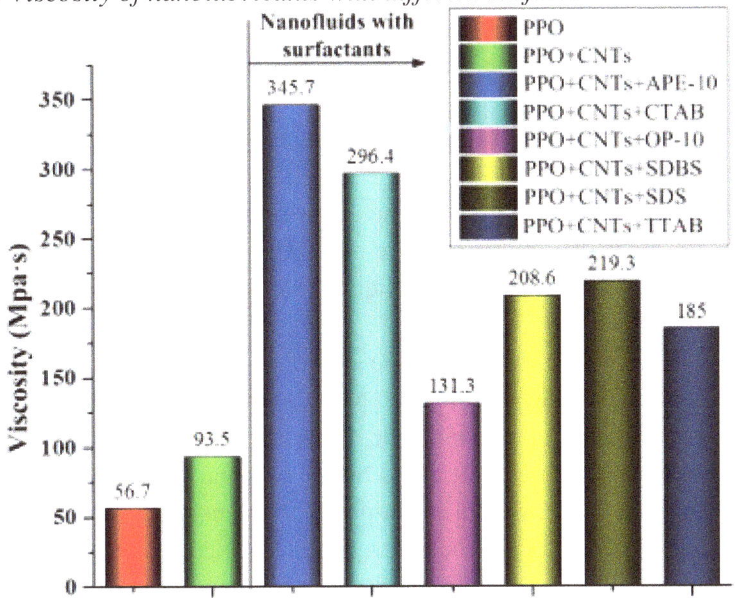

3.4 DROPLET SPREADING MECHANISM AND MECHANICAL BEHAVIOR

3.4.1 Mechanical Properties of Droplet Spreading Contact Line

The entire boundary D of the nanolubricant droplets in contact with the CFRP grinding surface can be divided into three regions (D_s, D_l, D_g), as illustrated in Figure 6. Here, D_s represents the boundary portion corresponding to the liquid in contact with the substrate, excluding the contact line. D_l denotes the contact line, while D_g signifies the remaining portion of the liquid phase boundary, which corresponds to the free surface of the droplet. For two-dimensional problems, the contact line is simplified to two triple points. Assuming that the droplet is surrounded by its saturated liquid gas phase and neglecting evaporation—thereby asserting that it exerts no mechanical influence on the droplet—the droplet as an isolated entity can be analyzed. When the droplet makes stable contact with the CFRP grinding surface, a static contact angle θ_{sc} is established, as shown in Figure 6. In this situation, the lubricant surface tension γ_{lv}, CFRP surface energy γ_{sv}, and interfacial tension γ_{iv} achieve an equilibrium state. Upon normalization, these forces transform into the interfacial tension among the three phases (solid/liquid/gas). The Young's equation is derived by projecting the equilibrium force onto the CFRP solid plane.

$$\cos\theta_{sc} = \frac{\gamma_{sv} - \gamma_{iv}}{\gamma_{lv}} \tag{3.1}$$

At this equilibrium state, the contact line velocity is $v_l = 0$. Thus, the boundary conditions applied to the contact line are as follows: if the contact angle remains within a specified range, the contact line is considered fixed, a condition referred to as a fixed contact line. The minimum and maximum contact angles that facilitate fixed contact lines depend on the liquid and solid surfaces, and are typically determined experimentally.

When the droplet begins to spread, the aforementioned equation ceases to apply, necessitating the use of slip boundary conditions for the velocity at the contact line. However, this leads to an unrealistic evolution of the contact angle due to the contact line speed not being constrained by the physical viscous dissipation at the solid-liquid interface, resulting in a non-physical extension of the moving contact line. Conversely, the velocity of the contact line can be correlated with the Young's stress acting upon it (Ren & e, 2007; Buscaglia & Ausas, 2011),

$$u \propto \gamma\left(\cos\theta_{sc} - \cos\theta_{dc}\right) \tag{3.2}$$

Among them, u_{cl} is the tangential component of the velocity vector on the contact line, that is $u_{cl} = v_{cl} \bullet t$, θ_{dc} is the dynamic contact angle. The proportionality coefficient between contact line velocity and Young's stress is the effective slip boundary condition (Ren & e, 2007),

$$\beta_{es} u \propto \gamma\left(\cos\theta_{sc} - \cos\theta_{dc}\right) \tag{3.3}$$

In the formula, β_{es} is the effective slip coefficient, defined as (Mahrous et al., 2020),

$$\beta_{es} = \zeta_{dcec} + \beta_s + \beta_n \tag{3.4}$$

Among them, ζ_{dcec} is the dynamic capillary effect coefficient, β_s is the Navier slip coefficient, and β_n is the normal stress coefficient. The capillary effect coefficient is expressed as (Buscaglia & Ausas, 2011; Manservisi & Scardovelli, 2009),

$$\zeta_{dcec} = \frac{\gamma}{u} R(Ca) \tag{3.5}$$

Furthermore, $R(Ca)$ is a function of the capillary number calculated based on the contact line velocity, with its formulation typically derived from empirical data fitting. For droplets driven by capillary forces, the contact line velocity is comparable to the droplet impact velocity, for which the following expression is applied (Jiang et al., 1979),

$$R(Ca) = (\cos\theta_{sc} + 1)\tanh(4.96Ca^{0.702}) \tag{3.6}$$

Among them, $Ca = u\mu/\gamma$ is the number of capillaries calculated based on the contact line velocity. Therefore, ζ_{dcec} is,

$$\zeta_{dcec} = \frac{\gamma}{u}(\cos\theta_{sc} + 1)\tanh(4.96Ca^{0.702}) \tag{3.7}$$

The Navier slip coefficient β_s is a function of shear stress,

$$\beta_s = \frac{\mu\ \nabla\ (v \cdot t) \cdot n}{u} \tag{3.8}$$

In addition, with the increase of speed, a significant normal stress jump is introduced on the contact line. By adding the coefficient β_n to balance the normal stress jump across the contact line,

$$\beta_n = \frac{\mu\ \nabla\ (v \cdot t) \cdot t}{u} \tag{3.9}$$

After considering the contributions of capillary effect, navigator slip, and normal stress in equation (3.4), the total dissipated force can be obtained using the coefficient β_{es}.

In scenarios where a viscous fluid interfaces with a solid substrate in a stationary state, no-slip boundary conditions should be applied. However, the no-slip condition observed in single wetting phenomena contradicts the physical dissipation observed at the solid-liquid interface, necessitating the introduction of Navier slip boundary conditions. The slip coefficient at the solid-liquid interface, denoted as D_s, is defined as follows,

$$\beta_{slc} = \frac{\mu\ \nabla\ (v \cdot t) \cdot n}{u} \tag{3.10}$$

Figure 6. Schematic diagram of droplet contact angle and contact line mechanics analysis

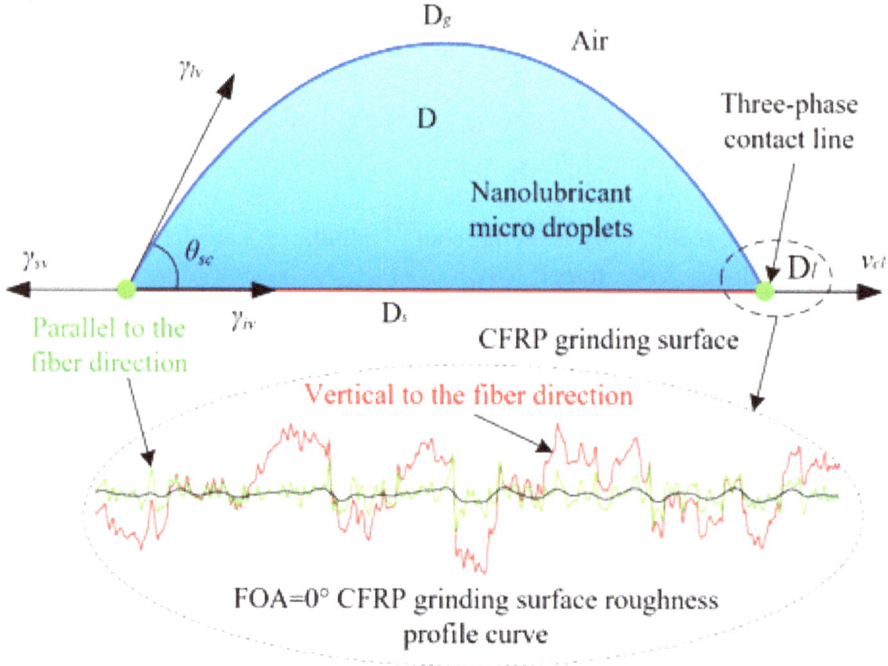

3.4.2 Anisotropic Spreading Mechanism and Capillary Action of Droplets

Droplets exhibit anisotropic infiltration on asymmetric barrier surfaces, with their deformation occurring asymmetrically due to unbalanced forces. This results in varying contact angles in different directions and directional movement of the liquid, driven by surface structural asymmetry and material heterogeneity. When lubricant droplets spread and move across a solid surface, the liquid closest to the surface first penetrates the microstructure due to capillary forces, increasing the solid-liquid contact area. Subsequently, the liquid located at the upper portion of the droplet extends outward under the Laplace pressure difference within the droplet. As the droplet spreads, the liquid at the forefront gradually infiltrates the microstructure, which increases the curvature of the liquid surface and enhances the Laplace pressure. This process empowers the liquid between adjacent voids, accelerating the spreading speed of the contact line and reducing the contact angle (Yuan & Zhao, 2013; Yuan et al., 2014). The droplet spreading on the CFRP grinding surface with FOA=0° exhibits the most pronounced anisotropic infiltration. The spreading speed of droplets along the parallel fiber direction is considerably higher than that in the vertical fiber direction. This disparity arises from the gap grooves created by missing fibers, the inherent fiber gap grooves in the parallel fiber direction, and the natural micro grooves on the surface of the carbon fibers. A schematic diagram illustrating droplet permeation on the CFRP grinding surface with FOA=0° and the microstructure of carbon fiber surfaces is presented in Figure 7. The capillary action of the furrow capillaries is relatively weak, as the number of furrow capillaries is significantly lower than that of fiber capillaries, and their scale and operational mode differ from those of the capillaries on the carbon fiber surface.

Figure 7. Capillary interaction between 0° carbon fibers and surface microstructure of carbon fibers

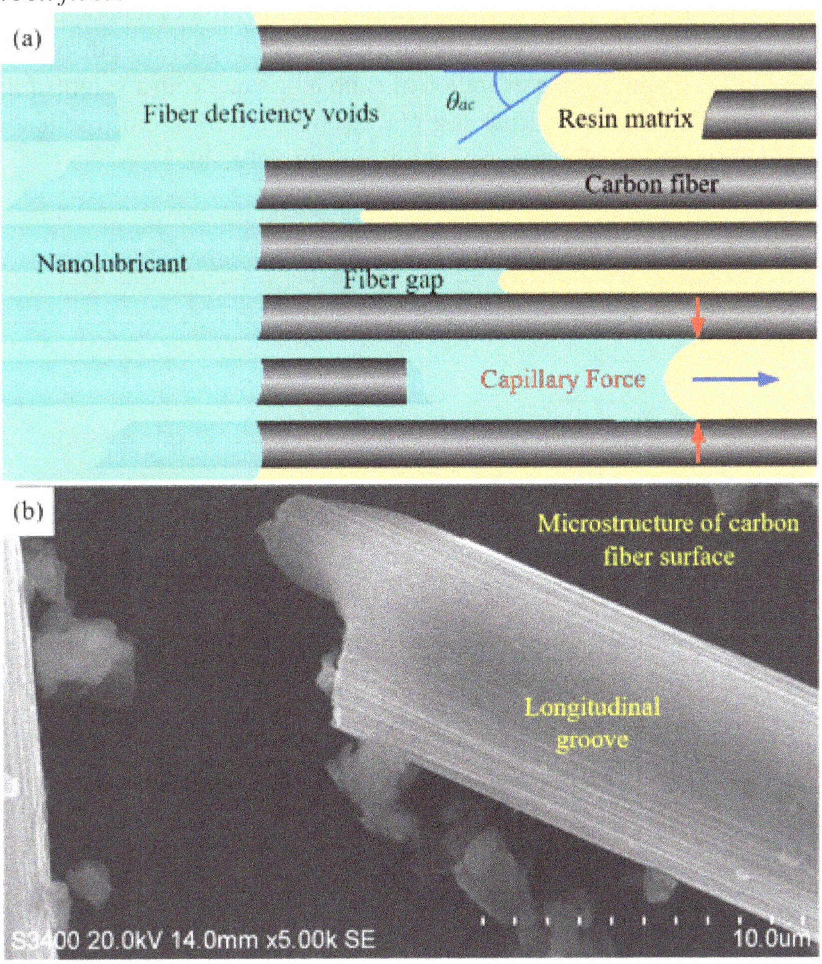

3.5 EXPERIMENT AND TESTING

3.5.1 Preparation of Dispersion Stable Nanolubricants

Carbon nanotubes (CNTs) have been selected as nanoparticles due to their high thermal conductivity and ability to effectively transfer heat. In addition, adding an appropriate amount of dispersant during the preparation of nanolubricants can solve the problem of CNTs aggregation and precipitation affecting the anti friction heat transfer performance. Pure palm oil (PPO) is selected as the base oil for nano lubricants. Due to the presence of saturated fatty acids with excellent lubrication properties and strong intermolecular binding forces, PPO forms a more stable lubricating film, further improving the anti-wear and friction reducing effect. The specific physicochemical properties of nanolubricants are shown in Table 3. The preparation of nano lubricants adopts a two-step method, in which CNTs nanoparticles are dispersed into PPO base oil to complete the preparation of nano lubricants with a volume fraction of 3%. The prepared nano lubricants are placed on a CNC ultrasonic oscillator and subjected to 20 minutes of ultrasonic vibration (Cui & Li, 2025).

Table 3. Physical and chemical properties of CNTs nanolubricants

CNTs nanoparticles	
Particle size	50 nm
Purity	99.0%
Colour	Black
Density	2.1 g/cm^3
PPO	
Colour	Faint yellow
Density (20 ° C)	0.93 g/ml^3
Viscosity (20 ° C)	56.7 mpa·s
Flash point	165 °C
Acid value	0.16 mg/g
Saponification value	195.07mg/g
Main components of fatty acids	42.02% oleic acid, 40.24% palmitic acid, 12.32% linoleic acid, 4.27% stearic acid

3.5.2 Lubricant Infiltration Spreading Experiment and Contact Angle Measurement

The CFRP grinding surface with the parameters v_s=30m/s, v_w=10mm/s and a_p =30μm was selected as the experimental object. As illustrated in Figure 8(a), the dynamic macroscopic spreading process of nanolubricant droplets on the CFRP surface, under different ultrasonic vibration directions and FOAs was observed and recorded using an optical microscope. The micromorphology of the spreading droplet contact line was captured using a Qlympus laser confocal microscope, as shown in Figure 8(b). Additionally, contact angles in different directions (perpendicular and parallel to the grinding direction) on the CFRP surface were measured with the JC2000C1B contact angle measuring instrument, as shown in Figure 8(c).

Figure 8. Droplet spreading state characterization and contact angle measuring instrument

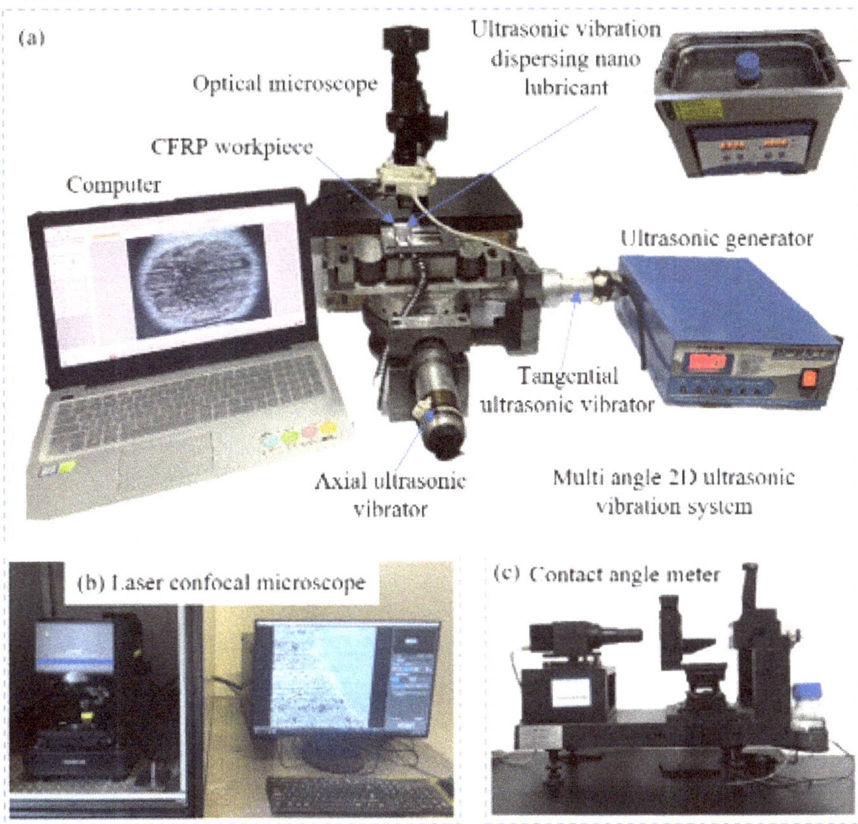

3.6 SPREAD CHARACTERISTICS OF NANOLUBRICANT DROPLETS ON CFRP GRINDING SURFACES

3.6.1 The Influence of Fiber Orientation Angle on Droplet Contact Angle

As illustrated in Figure 9, the droplet contact angles of CFRP grinding surfaces with varying FOA were measured in both vertical and parallel grinding directions, with the average of three measurements recorded for each condition. The contact angle observed in the parallel grinding direction is smaller than that in the vertical direction. Specifically, the contact angles measured perpendicular to the grinding direction for FOAs of 0°, 45°, 90°, and 135° are 20.09%, 20.98%, 16.02%, and 5.95% higher, respectively, than those measured parallel to the grinding direction. This finding indicates that droplet spreading demonstrates significant anisotropy, with the exception of the 135° FOA. The pinning effect of grinding traces impedes the spreading of droplets in the vertical grinding direction, resulting in an increased contact angle. The difference in contact angle between the two directions at the 135° FOA is not significant, which may be attributed to the deterioration of grinding surface quality, disruption of grinding texture, or considerable surface damage. For varying FOAs, the contact angle of CNT nanolubricant on the CFRP ground surface is smallest at a 0° FOA and largest at a 135° FOA, with differences of 59.62% and 40.83% in the vertical and parallel grinding directions, respectively. Therefore, the nanolubricant exhibits optimal spreading characteristics on the CFRP grinding surface oriented at 0° FOA, which facilitates enhanced cooling and lubrication effects.

Figure 9. Comparison of droplet contact angles on CFRP grinding surfaces with different FOA in vertical and parallel grinding directions

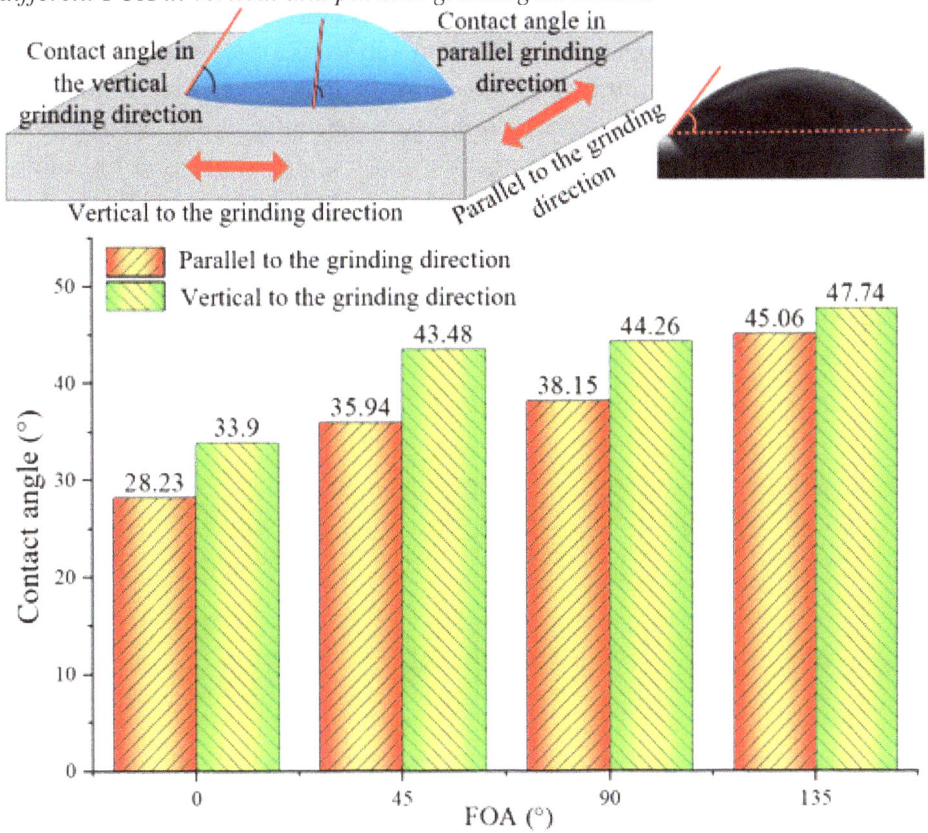

3.6.2 Evolution and Stability of Nanolubricant Infiltration Spreading

The droplet spreading behavior on the ground surface of CFRP with different FOAs is illustrated in Figure 10. In this figure, the horizontal axis represents the grinding direction, which aligns with the fiber direction of the 0° FOA CFRP. The order of droplet spreading speeds observed across different surfaces is as follows: 0° > 90° > 135° > 45°. Notably, the spreading speed of CNTs nanolubricant droplets on the 0° FOA grinding surface is the highest, achieving a stable state at 163 seconds. In contrast, the 45° FOA exhibits the slowest spreading rate, stabilizing at 252 seconds. The strong pinning effect on the surface results in the smallest spreading diameter, indicating that the droplet has not yet fully infiltrated or spread. The aspect ratio of droplet spreading serves as a measure of droplet deformation on anisotropic surfaces. A larger aspect ratio signifies a more pronounced anisotropic spreading behavior. In Figure 10, the vertical measurement of the droplet aligns parallel to the grinding feed direction, while the horizontal measurement is perpendicular to it. The aspect ratio for droplets on the 0° FOA surface is 1.87, reflecting the most significant anisotropic infiltration observed. Furthermore, compared to the 135° FOA CFRP grinding surface, the anisotropic spreading effect is enhanced by 29%. Consequently, this analysis indicates that the anti-friction heat transfer rate of nanolubricant droplets on the 0° FOA CFRP surface is the highest, and the cooling and lubrication area of a single droplet is maximized. This result also confirms the theoretical analysis in the third section. The capillary action on the surface of carbon fiber and the grinding surface leads to this result.

Figure 10. Droplet spreading stability time and aspect ratio of CFRP grinding surfaces with different FOAs

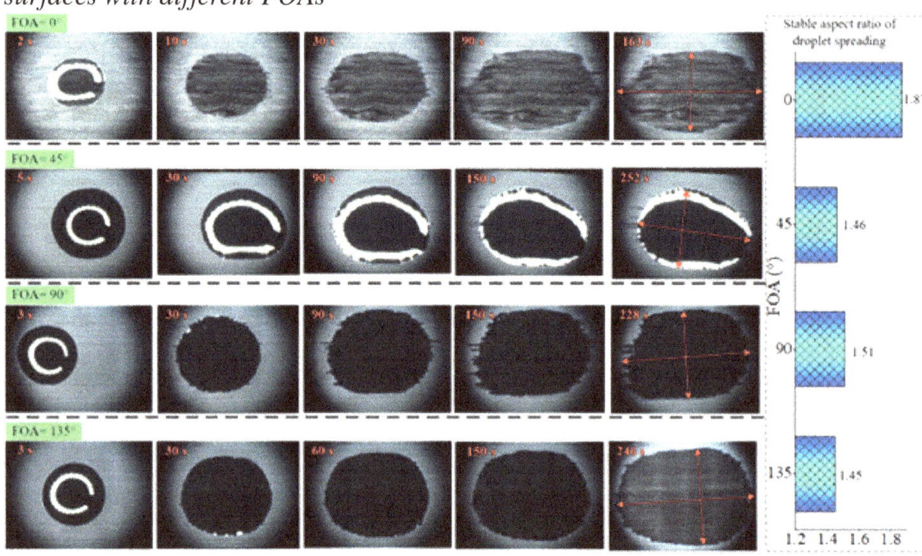

As depicted in Figure 11, the dynamic spreading diameter of nanolubricant droplets on CFRP grinding surfaces, under varying FOAs and two different grinding directions, can be represented by the slope of the curve. For all FOAs examined, the final droplet spreading radius in the direction parallel to the grinding is larger and the spreading speed is greater than that observed in the perpendicular direction. This enhanced spreading behavior is attributable to the microtraces aligned with the grinding direction, which facilitate droplet spreading, while the plowing grooves impede diffusion and spreading in the vertical grinding direction. Due to the strongest capillary action along the fiber direction, the spreading speed for FOA = 0° is the highest. In the case of FOA = 45°, the spreading rates in both the parallel and perpendicular directions to the grinding direction are at their lowest, particularly in the vertical grinding direction where stability in spreading is reached more quickly. Conversely, the fastest spreading in the direction perpendicular to the grinding occurs at FOA = 135°, while the slowest spreading is observed at FOA = 0° and 45°.

Figure 11. Variation of dynamic spreading diameter of nanolubricant droplets on grinding surface of CFRP with different FOAs with time

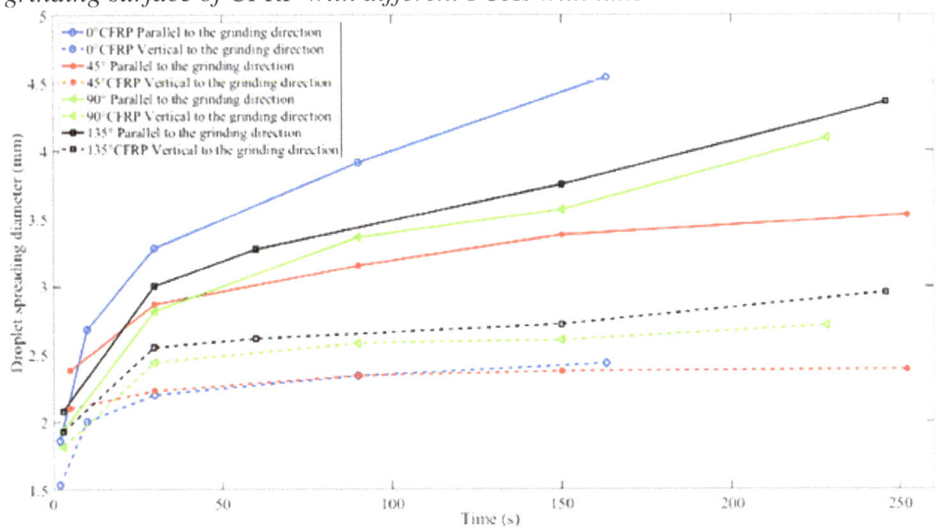

In Figure 12, the microstructure of the droplet spreading contact lines on CFRP surfaces subjected to grinding at various FOAs reveals notable differences when comparing parallel and perpendicular grinding directions, particularly for the 0° CFRP orientation. Notably, there is little to no significant pinning effect observed on the FOA=0° CFRP grinding surface. The contour of the droplet's spreading contact line is less distinct, attributable to the microscopic capillary action of the longitudinal grooves present on the carbon fiber surface. In the direction of parallel grinding (fiber orientation), this capillary action is more pronounced, owing to the increased number of capillaries formed by the alignment of the fibers. Conversely, in the vertical grinding direction (fiber), the movement of the droplet's spreading contact line is impeded by the influence of the fiber's orientation. For FOA=90°, the numerous and prominent grinding grooves facilitate a capillary effect; however, they also introduce a degree of pinning effect perpendicular to the grinding direction, which restricts the spreading of the droplet. In the cases of FOA=45° and 135°, the grinding marks are comparatively larger in size and fewer in number than the carbon fibers, leading to a less pronounced capillary action.

Figure 12. Micromorphology of droplet spreading contact line on CFRP grinding surfaces with different FOAs

3.6.3 The Mapping Relationship of Surface Roughness and Spreading Speed

The correlation between the stable time of droplet spreading on the grinding surface of CFRP and the surface roughness value is presented in Figure 13. A relationship is observed between droplet spreading speed and surface roughness, particularly with respect to the one-dimensional surface roughness measured perpendicular to the grinding direction. As the surface roughness increases, the stability time of the droplet decreases, indicating that droplets reach a stable state more quickly. This phenomenon may be attributed to the predominant hindering effect of the surface structure. Nonetheless, this mapping relationship is influenced by fiber orientation and exhibits a degree of specificity. Particularly on the FOA=0° CFRP grinding surface, the capillary action facilitated by the microchannels between the fibers and the longitudinal grooves on the carbon fiber surface is more significant than the capillary action resulting from the grinding texture.

Figure 13. Mapping relationship between droplet spreading stability time and surface roughness on CFRP grinding surface

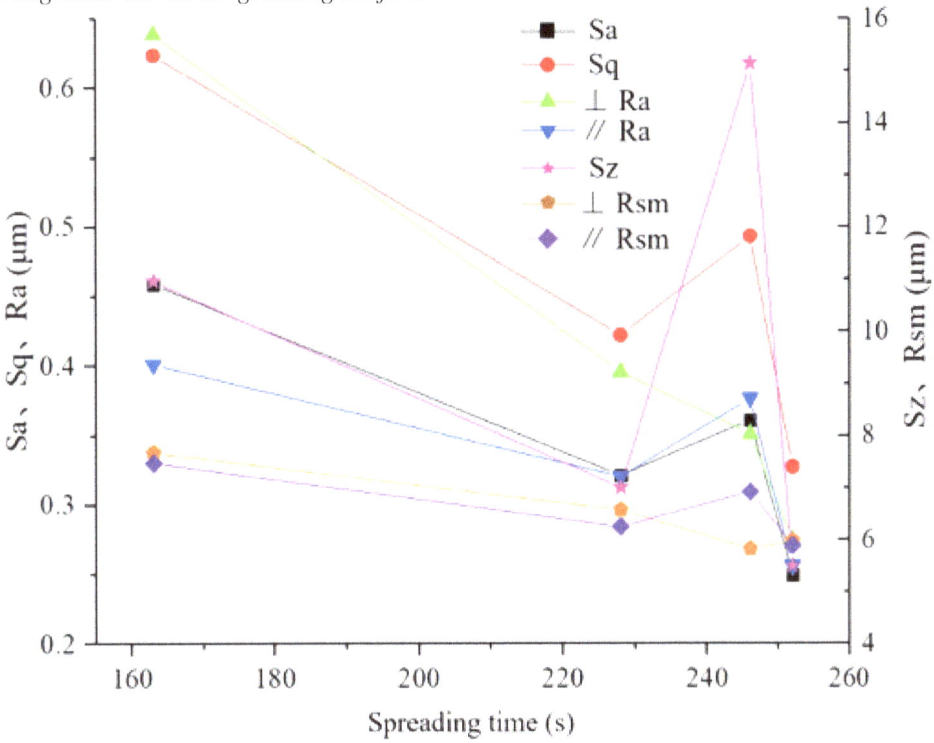

3.7 ULTRASONIC VIBRATION IMPACTS ON SPREADING BEHAVIOR

3.7.1 Stable Aspect Ratio

Currently, scholars have demonstrated that ultrasonic vibration enhances the wetting and spreading of droplets at interfaces (Morozov & Manor, 2018; Friend & Yeo, 2011; Ding et al., 2013), This section examines CFRP grinding surfaces with FOA=0 and 45° as experimental subjects to investigate the effects of varying ultrasonic vibration directions on droplet spreading behavior. Notably, the aspect ratio of droplet spreading on the CFRP grinding surface with an FOA of 0° is the highest, indicating the most significant anisotropic spreading. In contrast, the droplet spreading stability time on the CFRP grinding surface at an FOA of 45° is the longest, while the spreading rate is the lowest. The study also explores the impact of different coupled ultrasonic vibration angles, denoted as θ_U, on droplet spreading behavior. As illustrated in Figure 14, the stable aspect ratio of droplet spreading on the CFRP grinding surface under various θ_U conditions is presented. The application of 2D ultrasonic vibration reduces the stable droplet aspect ratio, leading to more uniform droplet spreading. Furthermore, the ultrasonic vibration effect at $\theta_U = 90°$ is more pronounced than at $\theta_U = 45°$. Compared to conditions without ultrasonic vibration, the aspect ratio of droplets on the surface of FOA = 0° CFRP decreased by 28.34%, while the aspect ratio on the surface of FOA = 45° CFRP decreased by 18.49%. This suggests that ultrasonic vibration at $\theta_U = 90°$ is more effective in reducing the anisotropic spreading of droplets, promoting uniform spreading, and expanding the wetting area. Consequently, this results in a larger cooling and lubrication area for the lubricant. The increased significance of anisotropic spreading of droplets under ultrasonic vibration at $\theta_U = 90°$ may be attributed to the maximum curvature of grain cutting motion, the widest interference range of subsequent grains, and a larger surface area affected by the differential cutting effect. Specifically, this relates to the spreading of liquid droplets on workpieces experiencing prolonged half-axes in elliptical motion.

Figure 14. Stable aspect ratio of droplet spreading on CFRP grinding surface under ultrasonic vibration in different directions

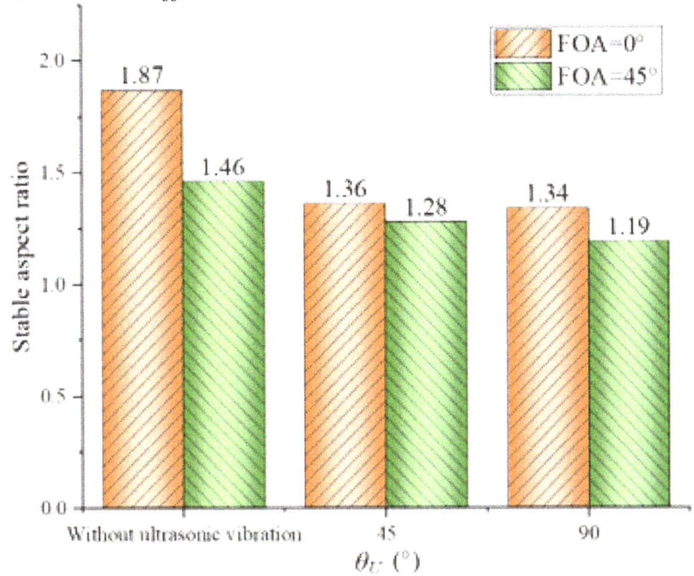

3.7.2 Dynamic Spreading Diameter

The rate at which lubricant droplets spread on the CFRP surface significantly influences the cooling and lubrication efficiency at the interface during high-temperature, high-pressure, and high-speed grinding processes. Figure 15 illustrates the stabilization time of droplet spreading, aligned with the grinding direction on the CFRP surface subjected to ultrasonic vibration at varying angles (θ_U). In this figure, the droplet spreading diameter is represented on the *y*-axis as the maximum diameter achieved. It is evident that the application of ultrasonic vibration accelerates droplet spreading, allowing them to reach a stable state more quickly. Notably, droplets on CFRP surfaces with FOA=45° exhibit a smaller spreading diameter in the grinding direction compared to those with an FOA of 0°. Additionally, while maintaining the same FOA, ultrasonic vibration at (θ_U = 45°) yields a slightly higher spreading rate and a marginally larger spreading diameter in the grinding direction than at (θ_U = 90°). This phenomenon can be attributed to the lower resonance angle at (θ_U = 45°), which facilitates improved coupling between the two directions.

Figure 15. Variation of dynamic spreading diameter of droplets over time under different directions of ultrasonic vibration

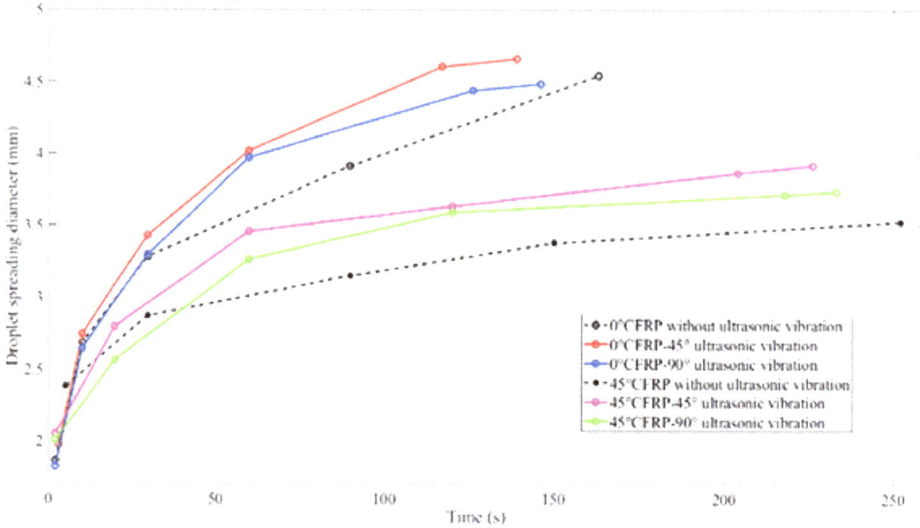

3.8 SUMMARY

In this chapter, the dispersing properties of different types of surfactants were revealed. The molecular structures and physicochemical properties of different surfactants were analyzed, and their dispersing mechanisms were studied. The spreading characteristics of CNTs nanolubricant droplets on the CFRP grinding surface with different FOA under different θ_U are studied, the mechanical behavior of droplet spreading contact line is revealed, and the anisotropic spreading mechanism and capillary action mechanism of droplets are analyzed. Through the observation of droplet spreading process and the measurement of contact angle, the influence mechanism of ultrasonic vibration on the spreading and diffusion of nanolubricant droplets on the grinding surface of CFRP with different fiber orientation was analyzed. The influence rule of fiber orientation on the contact angle of droplets was explored, and the mapping relationship between surface roughness and spreading speed was clarified.

(1) Among the six surfactants, the CNT nanolubricants with APE-10 obtained the longest time without deposition (>45 h) and the highest viscosity (345.7 MPa·s), thereby demonstrating the most outstanding dispersion stability.

(2) The order of droplet spreading speed on CFRP surfaces with different FOAs is as follows: 0° > 90° > 135° > 45°. The stable aspect ratio of droplets on the 0° FOA CFRP surface is 1.87, indicating that anisotropic spreading is most pronounced in this configuration.
(3) In the vertical grinding direction, the contact angles are 20.09%, 20.98%, 16.02%, and 5.95% higher than those observed in the parallel grinding direction for FOAs of 0°, 45°, 90°, and 135°, respectively. Except for the 135° FOA, which shows minimal anisotropy in droplet spreading, the contact angle of CNT nanolubricant on the ground surface of CFRP is smallest at 0° FOA and largest at 135° FOA. Therefore, the nanolubricant demonstrates superior spreading characteristics on the grinding surface of CFRP at 0° FOA, which enhances its cooling and lubrication effects.
(4) The application of 2D ultrasonic vibration significantly reduced the stable aspect ratio of nanolubricant droplets. When compared to conditions without ultrasonic vibration, the aspect ratio of droplets on the surface of CFRP at FOA = 0° decreased by 28.34%, while at FOA = 45°, it decreased by 18.49%. The use of ultrasonic vibration accelerates droplet spreading, allowing droplets to reach a stable state more quickly. Additionally, 2D ultrasonic vibration at θ_U = 45° resulted in a slightly higher spreading speed compared to θ_U = 90°.

REFERENCES

Attwood, D., & Florence, A. T. (1983). *Surfactant Systems: Their Chemistry, Pharmacy and Biology*. Chapman and Hall. DOI: 10.1007/978-94-009-5775-6

Behera, B. C., Setti, D., Ghosh, S., & Rao, P. V. (2017). Spreadability studies of metal working fluids on tool surface and its impact on minimum amount cooling and lubrication turning. *Journal of Materials Processing Technology*, *244*, 1–16. DOI: 10.1016/j.jmatprotec.2017.01.016

Buscaglia, G. C., & Ausas, R. F. (2011). Variational formulations for surface tension, capillarity and wetting. *Computer Methods in Applied Mechanics and Engineering*, *200*(45-46), 3011–3025. DOI: 10.1016/j.cma.2011.06.002

Choi, C., Jung, M., Choi, Y., Lee, J., & Oh, J. (2011). Tribological properties of lubricating oil-based nanofluids with metal/carbon nanoparticles. *Journal of Nanoscience and Nanotechnology*, *11*(1), 368–371. DOI: 10.1166/jnn.2011.3227 PMID: 21446457

Clint, J. H. (1975). Micellization of mixed nonionic surface active agents. Journal of the Chemical Society, Faraday Transactions 1. *Physical Chemistry in Condensed Phases*, *71*, 1327–1334.

Cui, X., & Li, C. H. (2025). *Magnetic Traction Nanolubricant Grinding: Force Model and Mechanisms*. Nova Science Publishers, 33-44. DOI: 10.52305/PVGY4472

Ding, X., Li, P., Lin, S. C. S., Stratton, Z. S., Nama, N., Guo, F., Slotcavage, D., Mao, X., Shi, J., Costanzo, F., & Huang, T. J. (2013). Surface acoustic wave microfluidics. *Lab on a Chip*, *13*(18), 3626–3649. DOI: 10.1039/c3lc50361e PMID: 23900527

Duan, W. H., Wang, Q., & Collins, F. (2011). Dispersion of carbon nanotubes with SDS surfactants: A study from a binding energy perspective. *Chemical Science (Cambridge)*, *2*(7), 1407–1413. DOI: 10.1039/c0sc00616e

Estellé, P., Halelfadl, S., & Maré, T. (2014). Lignin as dispersant for water-based carbon nanotubes nanofluids: Impact on viscosity and thermal conductivity. *International Communications in Heat and Mass Transfer*, *57*, 8–12. DOI: 10.1016/j.icheatmasstransfer.2014.07.012

Friend, J., & Yeo, L. Y. (2011). Microscale acoustofluidics: Microfluidics driven via acoustics and ultrasonics. *Reviews of Modern Physics*, *83*(2), 647–704. DOI: 10.1103/RevModPhys.83.647

Gao, T., Li, C., Zhang, Y., Yang, M., Jia, D., Jin, T., Hou, Y., & Li, R. (2019). Dispersing mechanism and tribological performance of vegetable oil-based CNT nanofluids with different surfactants. *Tribology International, 131*, 51–63. DOI: 10.1016/j.triboint.2018.10.025

Gu, G., Wang, D., Wu, S., Zhou, S., & Zhang, B. (2024). Research status and prospect of ultrasonic vibration and minimum quantity lubrication processing of nickel-based alloys. *Intelligent and Sustainable Manufacturing*, *1*(1), 10006. DOI: 10.35534/ism.2024.10006

Guo, C., Zhu, X., Liu, J., & Zhang, D. (2018). Cavitation bubble dynamics induced by hydrodynamic pressure oil film in ultrasonic vibration honing. *Journal of Tribology*, *140*(4), 041707. DOI: 10.1115/1.4039409

Guo, J., Barber, G. C., Schall, D. J., Zou, Q., & Jacob, S. B. (2017). Tribological properties of ZnO and WS2 nanofluids using different surfactants. *Wear*, *382*, 8–14. DOI: 10.1016/j.wear.2017.04.003

Hadad, M. J., Tawakoli, T., Sadeghi, M. H., & Sadeghi, B. (2012). Temperature and energy partition in minimum quantity lubrication-MQL grinding process. *International Journal of Machine Tools & Manufacture*, *54*, 10–17. DOI: 10.1016/j.ijmachtools.2011.11.010

Holland, P. M., & Rubingh, D. N. (1992). Mixed surfactant systems: An overview. *ACS Symposium Series*, 501, 2-30. DOI: 10.1021/bk-1992-0501.ch001

Hu, Z. S., Dong, J. X., & Chen, G. X. (1998). Study on antiwear and reducing friction additive of nanometer ferric oxide. *Tribology International*, *31*(7), 355–360. DOI: 10.1016/S0301-679X(98)00042-5

Hwang, Y., Park, H. S., Lee, J. K., & Jung, W. H. (2006). Thermal conductivity and lubrication characteristics of nanofluids. *Current Applied Physics*, *6*, 67–71. DOI: 10.1016/j.cap.2006.01.014

Jiang, T. S., Soo-Gun, O. H., & Slattery, J. C. (1979). Correlation for dynamic contact angle. *Journal of Colloid and Interface Science*, *69*(1), 74–77. DOI: 10.1016/0021-9797(79)90081-X

Liu, M., & Li, C. (2024). Thermodynamic mechanism of cryogenic air minimum quantity lubrication grinding. IGI Global, 136-176. DOI: 10.4018/979-8-3693-3378-5

Mahrous, E., Jarauta, A., Chan, T., Ryzhakov, P., Weber, A. Z., Roy, R. V., & Secanell, M. (2020). A particle finite element-based model for droplet spreading analysis. *Physics of Fluids*, *32*(4), 042106. DOI: 10.1063/5.0006033

Manservisi, S., & Scardovelli, R. (2009). A variational approach to the contact angle dynamics of spreading droplets. *Computers & Fluids*, *38*(2), 406–424. DOI: 10.1016/j.compfluid.2008.05.001

Mao, C., Zou, H., Zhou, X., Huang, Y., Gan, H., & Zhou, Z. (2014). Analysis of suspension stability for nanofluid applied in minimum quantity lubricant grinding. *International Journal of Advanced Manufacturing Technology*, *71*(9), 2073–2081. DOI: 10.1007/s00170-014-5642-9

Moore, V. C., Strano, M. S., Haroz, E. H., Hauge, R. H., Smalley, R. E., Schmidt, J., & Talmon, Y. (2003). Individually suspended single-walled carbon nanotubes in various surfactants. *Nano Letters*, *3*(10), 1379–1382. DOI: 10.1021/nl034524j

Morozov, M., & Manor, O. (2018). Vibration-driven mass transfer and dynamic wetting. *Current Opinion in Colloid & Interface Science*, *36*, 37–45. DOI: 10.1016/j.cocis.2017.12.002

Naskar, A., Choudhary, A., & Paul, S. (2021). Surface generation in ultrasonic-assisted high-speed superabrasive grinding under minimum quantity cooling lubrication with various fluids. *Tribology International*, *156*, 106815. DOI: 10.1016/j.triboint.2020.106815

Rahimzadeh, A., & Eslamian, M. (2017). Experimental study on the evaporation of sessile droplets excited by vertical and horizontal ultrasonic vibration. *International Journal of Heat and Mass Transfer*, *114*, 786–795. DOI: 10.1016/j.ijheatmasstransfer.2017.06.099

Rastogi, R., Kaushal, R., Tripathi, S. K., Sharma, A. L., Kaur, I., & Bharadwaj, L. M. (2008). Comparative study of carbon nanotube dispersion using surfactants. *Journal of Colloid and Interface Science*, *328*(2), 421–428. DOI: 10.1016/j.jcis.2008.09.015 PMID: 18848704

Ren, W., & e, W. (2007). Boundary conditions for the moving contact line problem. *Physics of Fluids*, *19*(2), 022101. DOI: 10.1063/1.2646754

Shen, B., & Shih, A. J. (2009). Minimum quantity lubrication (MQL) grinding using vitrified CBN wheels. *Transactions of NAMRI/SME, 37*, 129-136.

Wu, X., Gong, J., Gao, T., Wang, Y., Huang, R., & Hou, J. (2024). Experimental study on dynamic wetting behavior of liquid droplets on the aluminum surfaces excited by ultrasonic waves. *Applied Thermal Engineering*, *248*, 123209. DOI: 10.1016/j.applthermaleng.2024.123209

Xie, H., Yu, W., Li, Y., & Chen, L. (2011). Discussion on the thermal conductivity enhancement of nanofluids. *Nanoscale Research Letters*, *6*(1), 124. DOI: 10.1186/1556-276X-6-124 PMID: 21711638

Yamamoto, T., Noda, S., & Kato, M. (2011). A simple and fast method to disperse long single-walled carbon nanotubes introducing few defects. *Carbon*, *49*(10), 3179–3183. DOI: 10.1016/j.carbon.2011.03.040

Yuan, Q., Huang, X., & Zhao, Y. P. (2014). Dynamic spreading on pillar-arrayed surfaces: Viscous resistance versus molecular friction. *Physics of Fluids*, *26*(9), 092104. DOI: 10.1063/1.4895497

Yuan, Q., & Zhao, Y. P. (2013). Wetting on flexible hydrophilic pillar-arrays. *Scientific Reports*, *3*(1), 1944. DOI: 10.1038/srep01944 PMID: 23736041

Zhang, H., Zhang, X., Yi, X., He, F., Niu, F., & Hao, P. (2020). Dynamic behaviors of droplets impacting on ultrasonically vibrating surfaces. *Experimental Thermal and Fluid Science*, *112*, 110019. DOI: 10.1016/j.expthermflusci.2019.110019

Chapter 4

Tribological Characteristics of Transverse and Longitudinal Grinding CFRP With Nanolubricants

ABSTRACT

This chapter analyzes the tribological characteristics of transverse and longitudinal grinding of CFRP surfaces. Firstly, a contrast analysis on the coefficient of friction (COF) of different CNT nanofluids and the wearing loss of discs is performed to compare the dispersion mechanism. Secondly, the tribological characteristics in transverse grinding under different lubrications are analyzed, focusing on the COF and wheel blockage. Finally, the tribological behavior of the nanolubricant on longitudinal grinding surfaces of different FOA is studied, by evaluating the coefficient of friction, wear scar profiles, and wear volume. Compared to dry grinding, the COF of nanolubricant under four FOAs is reduced by 17.09%, 34.87%, 19.59%, and 27.13%, respectively. Notably, the lowest COF is observed at an FOA of 90°. Similarly, the anti-wear performance at 90° FOA is enhanced by 48.29% compared to the dry condition. This chapter provides a theoretical foundation and essential data to support further investigation into the material removal mechanics in MQL UVAG of CFRP.

DOI: 10.4018/979-8-3373-9013-0.ch004

4.1 INTRODUCTION

Nanofluid minimum quantity lubrication (NMQL) represents an emerging precision machining technology characterized by high efficiency, low consumption, clean operation, and low carbon footprint (da Silva et al., 2007). The addition of solid nanoparticles to base oils for preparing nanofluids contributes to enhancing frictional and lubrication performance at the wheel/workpiece and wheel/chip interfaces in MQL processes (Li & Ali, 2019). Nanoparticles can further improve the tribological properties within the grinding zone (Gu et al., 2024; Alves et al., 2013). Ge et al. (2015) explored the tribological behavior of novel greases incorporating lipophilic TiO_2 and SiO_2 nanoparticles. These researchers found that deposited Ti and Si or metal oxides form a protective oil film that reduces friction and improves tribological performance. Qu et al. (2016) demonstrated that copper microparticles can enhance the friction-reducing and anti-wear properties of sunflower seed oil, exhibiting a synergistic lubricating effect that explains the lubrication mechanism. Zhang et al. (2015) investigated the lubrication performance of NMQL by adding hybrid nanoparticles (MoS_2/CNTs). The results indicated that MoS_2/CNT hybrid nanoparticles (with an optimal mixing ratio of 2:1) provide better lubrication than single nanoparticles. Vegetable oil, as a non-toxic, biodegradable, and environmentally friendly renewable resource (Malkin and Guo, 2007), causes minimal or no harm to the environment and operators while offering excellent lubrication performance (Wickramasinghe et al., 2020). Consequently, vegetable oils can replace conventional mineral oils as base oils for lubricating fluids (Rahim and Sasahara, 2011; Lawal et al., 2012). Guo et al. (2017) used castor oil as a base oil and mixed it individually with six other vegetable oils at a 1:1 ratio; each mixture was then used as the base oil for MQL grinding. The results showed that the blended oils exhibited superior comprehensive lubrication performance compared to castor oil alone, with soybean/castor oil blend demonstrating the best performance and surface quality. Wang et al. (2016) explored the lubrication performance of seven vegetable oils in MQL grinding by comparing them with conventional flood lubrication. These researchers found that vegetable oils possess excellent lubrication properties, as evidenced by low friction coefficients and reduced wheel wear. The results indicated that palm oil offers the most outstanding lubrication performance. Mughal et al. (2023) found that a 1% MoS_2 nanolubricant formed a tribofilm on the machined surface, reducing wear and adhesion. Virdi et al. (2021) experimentally demonstrated that a palm oil-based Al_2O_3 nanolubricant with a concentration of 0.5 wt% achieved lower surface roughness, attributed to the ball-bearing and polishing effects of nanoparticles at the wheel–workpiece interface.

Regarding frictional interfaces, ultrasonic vibration also exhibits encouraging effects in improving tribological behavior. Leus et al. (2023) found that variations in the tangential contact stiffness coefficient significantly affect friction values in sliding motion, particularly under the influence of longitudinal tangential vibrations introduced into the contact area. A reduction in surface roughness leads to increased contact stiffness, enhancing the effectiveness of friction reduction under longitudinal tangential vibration. Liu et al. (2020) observed that ultrasonic vibration significantly reduces the friction coefficient, which slightly increases with vibration frequency but markedly decreases with increasing vibration velocity amplitude. Ali et al. (2025) discovered that longitudinal ultrasonic vibration-assisted milling surfaces exhibit superior tribological behavior when milling under high feed rates, applied loads, and sliding speeds. Sun et al. (2021) proposed an analytical method based on the Coulomb static friction model to calculate the friction coefficient in ultrasonic vibration-assisted side milling. The results indicated that increasing ultrasonic power reduces the friction coefficient, thereby lowering cutting forces. Shen and Zhang (2013) found that the scale-like surfaces generated in ultrasonic vibration-assisted machining (UVAM) show reduced fluctuations in both friction coefficient curves and average friction coefficient values, along with significantly improved wear resistance and load-bearing capacity.

There is a general consensus that the machining performance and surface quality of CFRP are affected by its tribological behavior. Liang and Wu (2019) identified two wear types in carbon fibers: fiber fracture and fiber polishing, which primarily depend on the direction angle of friction. Moreover, the anisotropy leads to chaotic wear tracks on the ball. Seward et al. (2022) proposed a novel experimental open-loop tribology testing method to generate a dynamic finite element friction model for CFRP machining simulation. The newly developed dynamic friction model is based on functions of fiber angle, contact pressure, and slip rate, and has been validated through experimental results and constant COF finite element simulations. Li et al. (2018) found that fretting amplitude, frequency, temperature, and fretting mode collectively influence the evolution of the coefficient of friction (COF). Fiber polishing, matrix shear fracture, and fatigue cracking are the dominant fretting wear mechanisms under low fretting amplitudes. Birleanu et al. (2022) investigated the tribological behavior of CFRP under boundary lubrication, observing a slight reduction in stick-slip effects and static/kinetic friction coefficients with increasing normal load on carbon fiber-reinforced polymer composites. Wang et al. (2019) demonstrated that the COF along the direction parallel to 0° fibers is lower than that perpendicular to 0° fibers. Fergani et al. (2017) proposed a model based on the Reynolds equation to predict the contact friction coefficient of surfaces produced under different grinding parameters. Increasing wheel speed was found to affect surface roughness and the tribological behavior of surfaces. Surfaces generated at higher wheel speeds ($v_s > 90$ m/s) exhibited lower friction coefficients.

Having established the enhanced infiltration and spreading behavior of nanolubricants under ultrasonic excitation in Chapters 2 and 3, this chapter now turns to an experimental investigation of the resulting tribological characteristics at both the grinding interface and the CFRP workpiece surface. The analysis encompasses both transverse and longitudinal grinding orientations, laying the foundation for understanding how nanolubricant properties translate into frictional and wear responses in CFRP machining. Currently, only a limited number of studies have compared the tribological performance of carbon nanotube (CNT) nanofluids incorporating different types of surfactants and analyzed the corresponding mechanisms. Consequently, it remains unclear which surfactant promotes optimal CNT dispersion, which type of surfactant yields CNT nanofluids with the best tribological performance, and whether fundamental relationships exist between surfactant type and dispersion stability. Therefore, this chapter evaluates the tribological properties of nanofluids with six different surfactants and analyzes the friction-reduction and anti-wear mechanisms using a tribological testing machine. Furthermore, the tribological characteristics of CNT nanolubricants on CFRP grinding surfaces with different fiber orientations under varying angles of ultrasonic vibration are investigated. Friction and wear tests on CFRP with different fiber orientations under nanolubricant conditions were conducted. By comparing the coefficient of friction, wear profiles, wear volume, and other aspects, the tribological effects of nanolubricants on CFRP grinding surfaces with different fiber orientations were systematically explored. These findings provide a theoretical foundation and data support for further research into the material removal mechanics in ultrasonic-assisted nanolubricant MQL grinding of CFRP.

4.2 FRICTION AND WEAR EXPERIMENTAL

To analyze the frictional properties of CFRP grinding surfaces with varying FOAs, friction and wear tests were conducted using the UMT-3 multifunctional friction and wear tester under different FOAs and lubrication conditions, as shown in Figure 1. This experiment employed a reciprocating motion, point-surface contact, and a pin-on-disk model. To ensure the reliability of the test, conduct three tests for each operating condition and take the average. The friction and wear experimental parameters, such as load and frequency, were determined based on specific grinding conditions, as detailed in Table 1.

Figure 1. Multifunctional friction and wear testing machine UMT

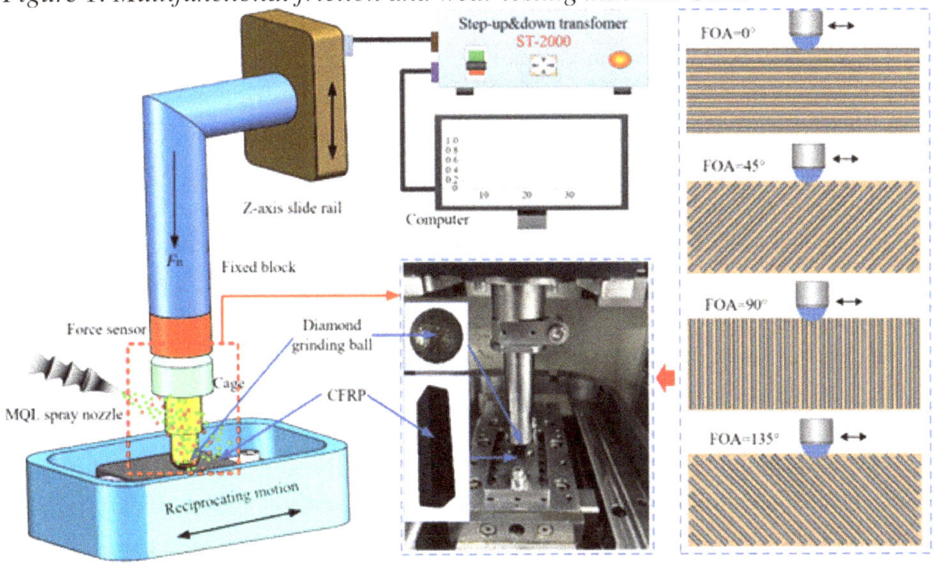

Table 1. Friction test parameters

Parameters	Value
Temperature (°C)	22±5
Test time (s)	900
Load (N)	10
Frequency (Hz)	50

4.3 FRICTION-REDUCING AND ANTIWEAR OF CNTS NANOLUBRICANTS WITH DIFFERENT SURFACTANTS

4.3.1 Coefficient of Friction

The utilization of surfactant could increase the dispersion stability of CNTs in PPO. However, different surfactants caused various friction coefficients. Therefore, the possible reasons and mechanisms of actions related to the friction-reduction effects of CNT nanolubricants with different surfactants must be discussed. The experimental results of the friction coefficient and viscosity were highly correlated.

CNT nanolubricants with surfactant possessed high viscosity, thus, a low friction coefficient was obtained. With the increase in viscosity, the colloidal force, Brownian movement, and viscosity force among nanolubricant molecules increased accordingly, thereby benefitting the formation of oil membrane on the friction surface and the increase in the thickness and strength of the adsorption membrane. Consequently, the lubrication performance increased slightly. An oil membrane with sufficient thickness was difficult to form on the high-temperature friction surface when the lubricating fluid had low viscosity. The oil membrane had the low bearing capacity and was easily damaged under loads. The lubrication in the friction pair was inadequate, and the friction coefficient increased accordingly.

The friction coefficients of the different lubricating fluids are exhibited in Figure 2(a). The addition of different surfactants to CNT nanolubricants caused different friction coefficients, and the fluctuation in the friction coefficients significantly changed with time. Given the entire experiment, the friction coefficients of nanolubricants with a non-ionic surfactant (APE-10 and OP-10) increased with time. The friction coefficient of nanolubricants with OP-10 increased from 0.135 to 0.157 after 200 s because APE-10 and OP-10 exhibited high molecular mass and London force, thereby playing the dominant role in several dispersing mechanisms. Therefore, nanolubricants showed a high dispersion stability in a static observation. The lubricating fluid was disturbed by the rotation sliding of the white alundum ball, and the stable system formed by London force, which was generated by an instantaneously induced dipole, was damaged throughout the frictional test, thus weakening the dispersion of CNTs and intensifying the aggregation of CNTs gradually. Consequently, a stable adsorption film was impossible to form, and the continuity of the adsorption film was destroyed; this result contradicted the lubrication performance.

The friction coefficient of nanolubricants with a cationic surfactant (CTAB and TTAB) slightly decreased with time. The friction coefficients of nanolubricants with CTAB decreased from 0.15 to 0.135, whereas those with TTAB decreased from 0.17 to 0.15. The reason for these results might be as follows: with the continued friction and wear process, the friction surface gradually smoothened, the wear changed from the violent wearing state to the stable wearing state gradually, and the destroyed friction film was updated and supplemented timely, thereby contributing to the continuously satisfying lubrication performances. The friction coefficient curves of nanolubricants with anionic SDS and SDBS remained constant throughout the experiment, except for a slight fluctuation.

Fig. 4.2(b) illustrates that the average friction coefficient of the CNT nanolubricants with APE-10 was the lowest (0.121), thereby indicating the optimal friction-reduction performance. The overall order of average friction coefficient was APE-10 (0.121) < CTAB (0.144) < OP-10 (0.148) < SDBS (0.156) < TTAB (0.160) < SDS

(0.169). On the basis of the average friction coefficient, the comparative analysis was as follows:

APE-10 versus OP-10: The friction coefficient of the CNT nanolubricants added with APE-10 was the lowest, and the friction coefficients fluctuated slightly. APE-10 is a mixed surfactant, and the surface active molecules of APE-10 have high-efficiency synergism; in addition, APE-10 can adsorb onto the CNT–PPO interface, and a high-efficiency synergistic effect exists between APE-10 and CNTs (Lu and Somasundaran, 2008; Alargova et al., 2001). The specific performance might be as follows: different types of surfactants formed reversed micelles with different sizes and shapes, which filled different spatial spaces and formed stable dispersions. The nanolubricants with OP-10 showed a favorable dispersion effect in the static observation. A possible reason was that the surfactant concentration reached the critical reversed micelle concentration. The other possible reason is that OP-10 molecules contain -O- and -OH polar groups, which will decrease the critical reversed micelle concentration of OP-10 sharply. Therefore, a low concentration could form the reversed micelles and reach the saturated adsorption of molecular monolayer. The low concentration could change the internal and interface performances of the system significantly; that is, the OP-10 surfactant could develop high performances under a low concentration. The frictional test reported that the friction coefficient under nanolubricants with OP-10 was not the lowest, and the disc surface quality analysis demonstrated the poor friction-reduction and antiwear performances, which were caused by the instability of the dominant London force in the frictional tests.

CTAB versus TTAB: Figure 2(b) displays that the average friction coefficient of nanolubricants with CTAB was lower than that of nanolubricants with TTAB. The possible reason was that the carbon chain of the lipophilic group of CTAB is longer than that of TTAB, and the distance among CNTs when CTAB was adsorbed onto CNTs is longer than that under CTAB. Thus, CNTs could uniformly disperse and easily form the uniform friction membrane on the friction pair surface, thus contributing favorable friction-reduction and antiwear performances. Moreover, CTAB exhibited strong lipophilicity and low critical reversed micelle concentration in palm oil and performed efficiently under a low concentration, thereby causing it to have a high adsorption capacity on the CNT–PPO interface and improved adsorption stability. The experimental results of the static observation were opposite to the abovementioned analysis. Nanolubricants with CTAB required a shorter time for layering deposition than those with TTAB, thus indicating a favorable dispersion effect of CNT nanolubricants with TTAB. One possible reason is that TTAB possesses shorter hydrocarbon chain and stronger hydrophilicity of hydrophilic groups; thus, TTAB is apt to escape from oil and can easily be adsorbed onto the CNT surface, thereby contributing to the increase in adsorption capacity. The other possible explanation was as follows: The carbon number on the straight alkane chain of the lipophilic group exceeds 16, and the static observation time is longer than that of the frictional test, thus causing the curling up of the long chain; the cross-sectional area of single surfactant molecules was further increased. Ultimately, the saturated adsorption capacity of surfactant and the loosening of the adsorption layer were decreased.

SDBS versus SDS: Nanolubricants with SDBS are superior to those with SDS in terms of the average friction coefficient and static dispersion performance; the possible reasons are that SDS has a large HLB value that leads to the poor oil solubility, and the solubility of SDS powder in palm oil is relatively small, thereby resulting in a small number of micelles formed by SDS molecules; consequently, the dispersion effect is the poorest (Gao et al., 2019). Macroscopically, the experimental results of a static observation showed that SDS is layered up in the lubricating fluid, and abundant SDS is no longer dissolved in the lubricating fluid and cannot effectively disperse.

Certain information is displayed in Figure 2(b). First, the friction coefficients of nanolubricants with nonpolar ionic surfactant OP-10 and cationic surfactants CTAB and TTAB varied extensively. The extensive variation in the friction coefficient of nanolubricants with OP-10, as presented in Figure 2(a), was due to the friction coefficient curve steadily increased after 200 s, accompanied by a small fluctuation. Second, the friction coefficients of nanolubricants with cationic surfactants CTAB and TTAB fluctuated significantly throughout the experiment, and the maximum friction coefficient influenced the mean considerably, thereby indicating a mutual corroboration as illustrated in Figure 2(a). The violent fluctuation might be the consequence of the accumulation and exfoliative discharge of cutting debri

Figure 2. (a) Friction coefficient, (b) Box plot of friction coefficient

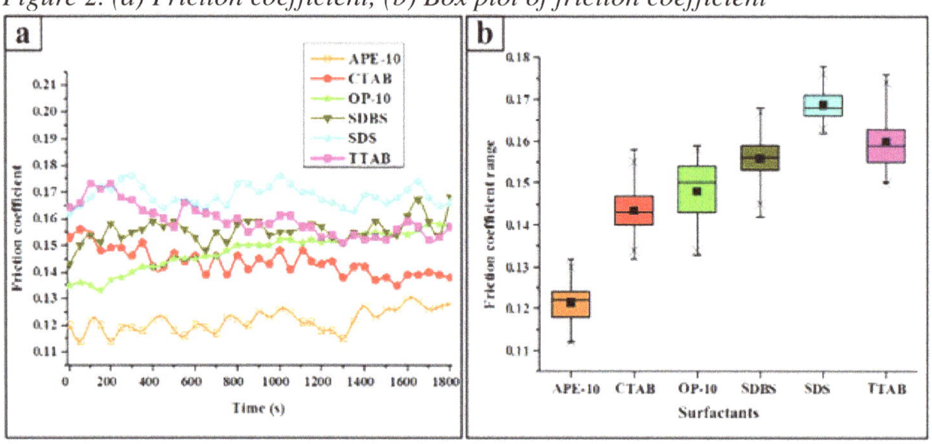

4.3.2 Wear Depth

MATLAB program was used to generate the scratch profile curves based on the original roughness data for fully reflecting the grinding performance of white alundum balls and the antiwear performances of different nanolubricants with surfactants. Figure 3 displays that the maximum scratch depth order under different nanolubricants was APE-10 (7.024 μm) < TTAB (7.790 μm) < SDBS (8.469 μm) < OP-10 (9.238 μm) < CTAB (9.608 μm) < SDS (10.057 μm). The nanolubricants with APE-10 contributed the optimal antiwear performance. The dispersion effect of CNTs under CNT nanolubricants with APE-10 was stable, and CNTs easily formed a stable friction film, thereby leading to the reduction in surface roughness. A low surface roughness enhanced the capacity of film forming and the thickness of the oil film. A virtuous circle was formed in this way. The optimal antiwear performance was achieved under nanolubricants with APE-10.

Figure 3. Roughness profile curve and wear depth (a): PPO+2 wt% (CNTs)+3 wt% (APE-10); (b): PPO+2 wt% (CNTs)+3 wt% (CTAB); (c): PPO+2 wt% (CNTs)+3 wt% (OP-10); (d): PPO+2 wt% (CNTs)+3 wt% (SDBS); (e):PPO+2 wt% (CNTs)+3 wt% (SDS); (f): PPO+2 wt% (CNTs)+3 wt% (TTAB)

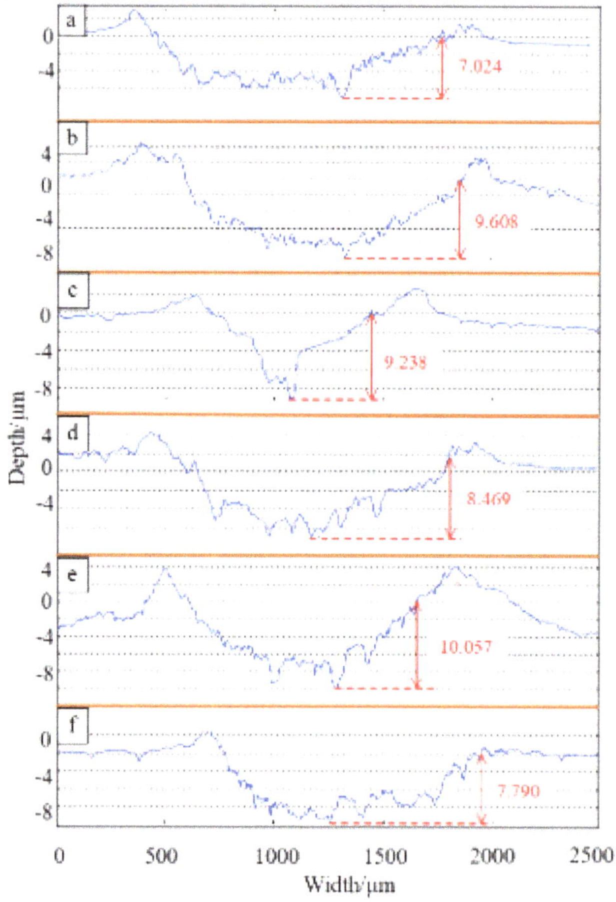

4.3.3 Material Wear Rate

Material removal rate is the material volume on the disc removed by the grinding ball per unit time. The antiwear performance of different CNT nanolubricants can further be analyzed by material removal rate. Area ratio = $(S_1 + S_2) / S_3$ can reflect the material removal rate, and it is negatively correlated with the material removal rate. where S_1 (μm^2) and S_2 (μm^2) are the cross-sectional areas of plastic accumulation, and S_3 (μm^2) is the cross-sectional area of the furrows produced by the white alundum ball cutting of the disc. The cross section of a scratch is illustrated in Figure

4. The values of S_1, S_2, and S_3 were calculated by MATLAB software. In Figure 4, the order of area ratio of the six disc groups in terms of material removal rate is TTAB<SDBS<SDS<APE-10<OP-10<CTAB. Therefore, the order of nanolubricants in the antiwear performance is CTAB>OP-10>APE-10>SDS>SDBS>TTAB. Figure 3 demonstrates that the disc scratches under nanolubricants with OP-10 presented obvious narrow gully and high maximum scratch depth, this result obviously disagreed with the white alundum ball profile. This result might be due to the large or several cutting debris piled up in the grinding zone to cut the disc as the "third body" under the extrusion of white alundum balls (Yu et al., 2013).

Figure 4. The wear marks profile and Material wear rate

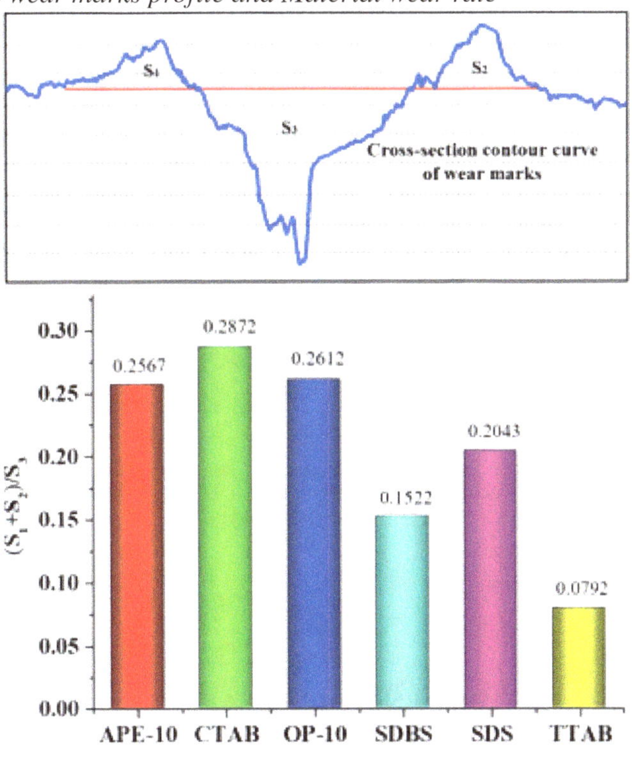

4.3.4 Roughness of Frictional Wear Surface

The surface roughness value of the disc samples was measured after the frictional test to discuss the frictional wear of samples. Roughness measurement and mean were gained from 3 points to 5 points of each disc sample, and the point that is closest to the roughness mean was selected for evaluation. The roughness values of disc samples under different nanolubricants are depicted in Figure 5. The roughness value order under different lubrication conditions was APE-10 (0.48 μm) < OP-10 (0.518 μm) < SDBS (0.533 μm) < TTAB (0.552 μm) < CTAB (0.691 μm) < SDS (0.839 μm). The minimum roughness value was achieved by CNT nanolubricants with APE-10; therefore, they achieved the optimal lubrication and tribological properties. Scratches were only one part of the disc-sampling section for roughness measurement, and the roughness value could not comprehensively reflect the surface quality and lubrication performance of nanolubricants.

Figure 5. Roughness value of CNT nanolubricants with different surfactants

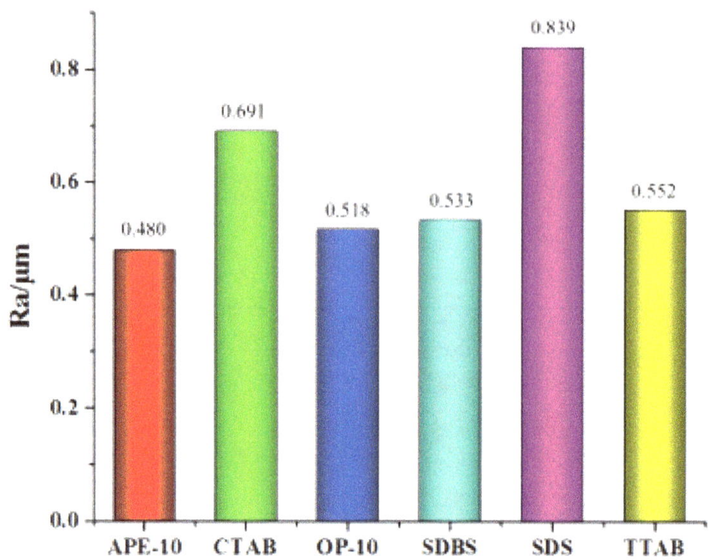

4.4 TRANSVERSE GRINDING INTERFACE WITH DIFFERENT LUBRICATIONS

4.4.1 Coefficient of Friction

The variation curve of friction coefficient with time under different lubrication conditions was shown in Figure 6. On the whole, the average friction coefficient of MQL (0.156) and CNTs nano lubricant (0.141) was significantly lower than that of dry condition (0.303), decreased by 48.51% and 53.47% respectively. For dry condition: in the initial stage, the friction coefficient decreased sharply and then increased sharply. This was because the surface of CFRP plate was smooth at first, but then the diamond grinding ball destroyed the surface of CFRP plate and worsened the contact condition. In the second stage (antifriction stage), the friction coefficient decreased gradually, which was due to the fact that the frictional removal process of the diamond ball on the CFRP plate tended to be stable gradually, and the removed fiber debris acted as a solid lubricant between the interfaces. In the third stage (stable stage), the contact state between the friction pairs tended to be stable, and the diamond grinding ball no longer had cutting effect on the CFRP plate. For MQL and CNTs nano-lubricants conditions: in the initial stage, the friction coefficient decreased sharply under the action of lubricant. After that, the lubricant had entered the stable stage, and the lubricant played a full role in lubrication between the friction pairs. The difference was that with the continuous grinding of diamond ball and CFRP plate, the friction coefficient of MQL increased slowly in the third stage, and the frictional contact condition became deteriorated. This was due to the fact that bio-lubricants were insufficient to form an effective lubricant film. On the contrary, the friction coefficient curve of CNTs nano lubricant condition was stably, indicating that the frictional contact state was stable all the time. The reason was that CNTs nanoparticles played an excellent role in anti-wear and friction reduction between the friction pairs, improved the tribological properties and the lasting ability of friction reduction of biological lubricants.

Figure 6. Friction coefficient curve under different lubrication conditions

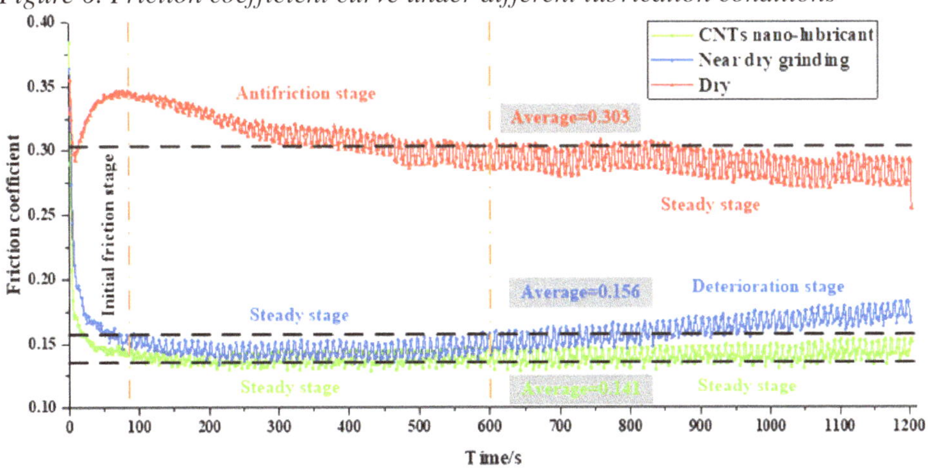

4.4.2 Grinding Wheel Blockage

The blockage of grinding wheel is the main factor affecting the grinding performance of grinding wheel, and different lubrication conditions have a direct impact on the blockage of grinding wheel. Excellent lubrication method and grinding fluid are very important to improve the blockage of grinding wheel. The types of grinding wheel blockage mainly include adhesive type, attached type, embedded type and mixed type. Figure 7 shows the schematic diagram of different types of grinding wheel blockage. Sticking grinding wheel blockage refers to the adhesion of the grinding chips around the protruding cutting edge or on the binding agent after the grinding chips melt. Because the resin matrix is sensitive to temperature, it is easy to adhere after being softened by heat, or a large number of carbon fiber chips are adhered to the grinding wheel surface. The attached grinding wheel blockage refers to the mechanical blockage of abrasive chips on the back cutting surface of abrasive cutting edge due to temporary power, which mainly comes from internal and external factors such as electricity, heat and physics. Plugging of embedded grinding wheel refers to the blocking condition of grinding chips embedded in the air hole on the grinding wheel surface. Mixed grinding wheel blockage refers to the aggregation and collection of the above three types at a certain place.

Figure 7. Diagram of grinding wheel blocking type

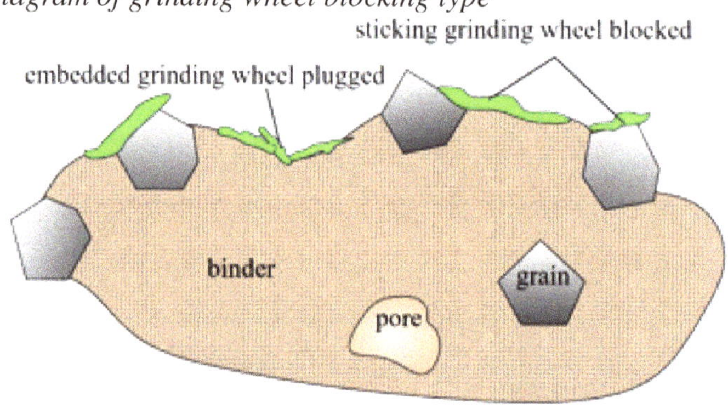

The comparison of wheel clogging and oil film on the grinding wheel surface during grinding CFRP under different conditions is shown in Figure 8. Serious grinding wheel clogging in dry grinding was observed. The grinding wheel surface of MQL and CNT biological lubricant was relatively clean, no adhesion blockage of a large number of fiber debris occurred, and obvious oil film spreading and covering were found on the grinding wheel surface, which was conducive to lubricating the grinding wheel workpiece interface and reducing the grinding wheel blockage.

The reason for this condition is that the high grinding temperature of dry grinding leads to the softening of the resin matrix. The softened resin matrix adheres to the grains and bonds with the fiber debris, which would accumulate with the grinding process. In addition, under the dry grinding condition, the grain wear on the grinding wheel was serious, and the contact area between the grinding wheel and the workpiece increased, thereby increasing the difficulty of removing the debris and, in turn, worsening the adhesion of the debris. In MQL and CNT biological lubricant grinding, the oil film of MQL oil and nano-lubricant between the grinding wheel and workpiece interface can effectively reduce and avoid the blockage of grinding chips on the grinding wheel. Moreover, the grinding temperature of MQL and CNT biological lubricant conditions was low, which prevented the thermal softening of resin matrix, and few fiber debris adhered to the grinding wheel surface. The high-pressure gas from the nozzle also had a certain cleaning effect on the grinding wheel surface, which could remove the debris blocked by the attached and embedded grinding wheels, further reducing the blockage on the surface of the grinding wheel and improving the service life of the grinding wheel.

Figure 8. Comparisons of grinding wheel blockage

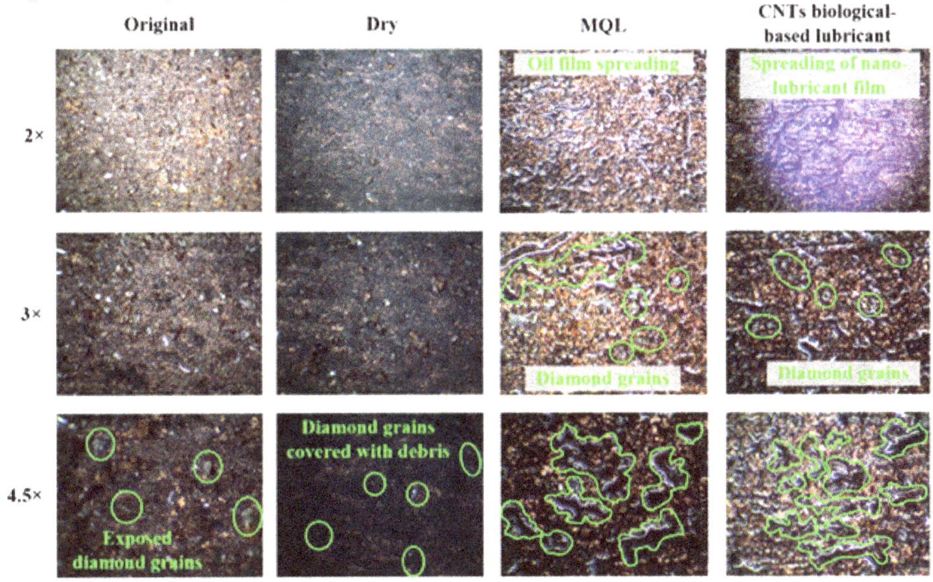

4.5 LONGITUDINAL GRINDING INTERFACE WITH DIFFERENT FIBER ORIENTATION ANGLES

4.5.1 Coefficient of Friction

The COF curves for CFRP grinding surfaces at various FOAs under both dry and nanolubricant conditions are presented in Figure 8. Unlike traditional homogeneous isotropic materials, the COF curve of CFRP exhibits complex variations and fluctuations over the course of friction and wear. At an FOA of 0°, the COF under dry conditions gradually increases throughout the friction and wear experiments, displaying significant fluctuations prior to the 400-second mark. This behavior may be attributed to unstable contact conditions between the friction pairs due to the removal of CFRP material during the initial stages of friction. In contrast, the curve under nanolubricant conditions remains relatively stable. When the FOA is 45°, the friction coefficient curve under dry conditions experiences the most fluctuation, stabilizing only after 700 seconds. Conversely, under nanolubricant conditions, the trend reverses, demonstrating initial stability followed by later fluctuations. This phenomenon is likely due to the lubricant's insufficient anti-wear and anti-friction properties, as well as the removal of CFRP material during the later stages of the friction process. At an FOA of 90°, the friction coefficient curves for both dry and nanolubricant conditions exhibit relative stability. However, for an FOA of 135°, the friction coefficient curve under dry conditions shows significant fluctuations after 300 seconds, persisting until the end of the experiment, while the nanolubricant conditions also reveal continuous fluctuations after 500 seconds. The palm oil-based nanolubricant forms a stable lubricating oil film between the diamond ball and CFRP interface, significantly improving the anti-wear and friction reduction properties and reducing grinding force.

Figure 9. Friction coefficient curve of CFRP grinding surface with different FOAs

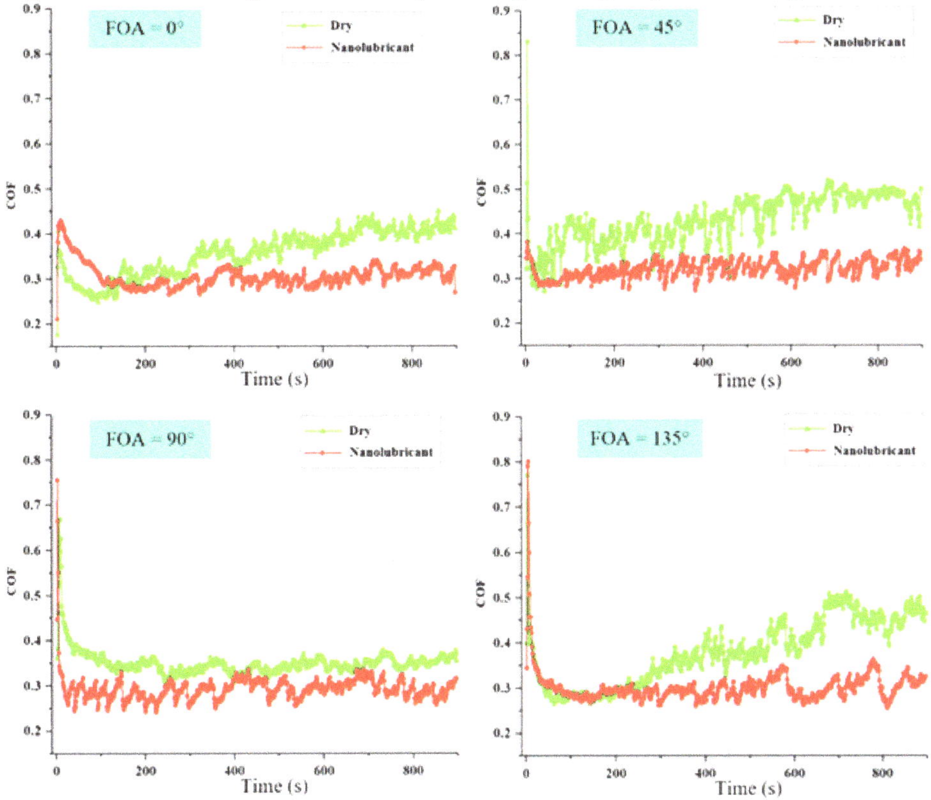

As illustrated in Figure 10, the average friction coefficients of CFRP grinding surfaces at different FOAs were compared under dry and nanolubricant conditions. Notably, the use of nanolubricants significantly reduces the friction coefficient by 17.09%, 34.87%, 19.59%, and 27.13% for the four FOAs, respectively. Furthermore, there are notable differences in the friction coefficients of various FOAs under identical conditions. The friction coefficients are lower for FOAs of 0° and 90°, while the highest values are observed at an FOA of 45° in both lubrication conditions. During the reciprocating friction tests involving FOAs of 45° and 135° on CFRP, the reciprocating friction effect of grains on these two surfaces is equivalent; however, there is a significant difference in the average friction coefficient observed between the two conditions. Given that the CFRP workpiece used in the friction and wear tests consists of surfaces that have been ground, this strongly suggests that the grinding process exerts differing effects on the microstructure, surface quality, integrity, and subsurface properties of these two FOA CFRP surfaces, particularly under dry conditions where the effects are more pronounced.

Figure 10. Mean friction coefficient of CFRP with different FOA under different lubrication conditions

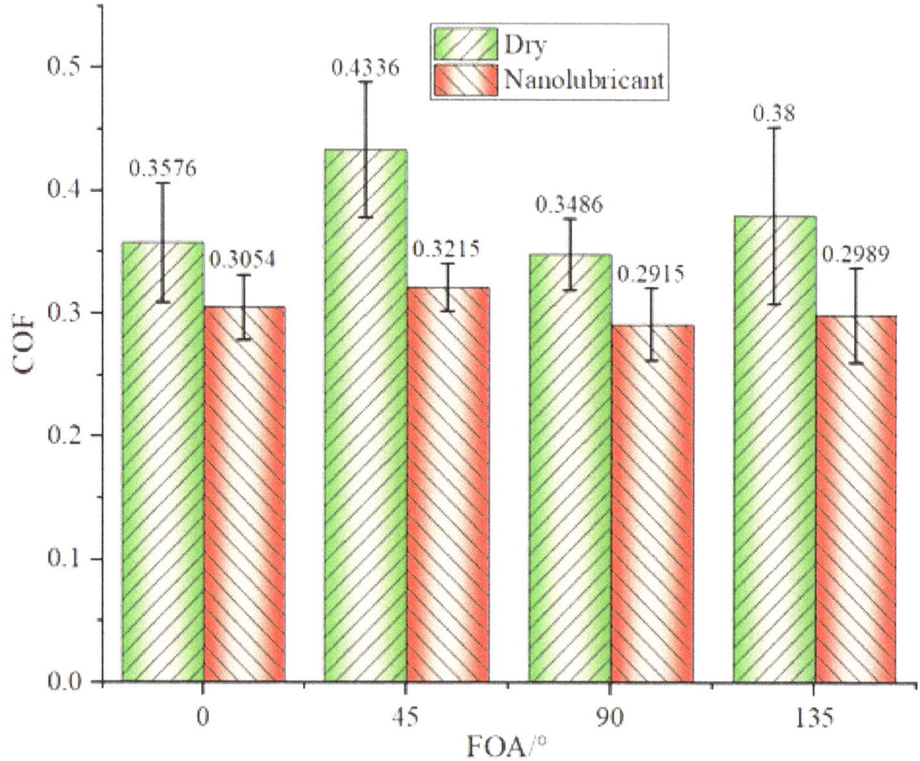

4.5.2 Wear Scar Contour and Area

While the preceding analysis has focused on workpiece surface integrity and wheel loading, the progressive wear of the abrasive grains themselves warrants further attention, as it directly governs process stability and overall machining cost. Under the present UVAG MQL conditions, two primary wear modes are of interest: attritious wear, which involves the gradual blunting of grain cutting edges due to mechanical rubbing and thermal softening, and grain fracture, where micro-cracks propagate within the abrasive grain until small fragments detach. The protective nanolubricant film formed by the MQL medium is expected to reduce the coefficient of friction at the grain–workpiece interface, thereby mitigating attritious wear and suppressing the temperature rise that often accelerates grain degradation. Simultaneously, the intermittent contact characteristic of ultrasonic vibration superimposes cyclic impact loads on the abrasive grains. This may promote controlled micro-fracture and self-sharpening of the grains, maintaining cutting efficiency, but excessive impact energy could also lead to premature grain pull-out or macro-fracture, negatively affecting wheel life and process consistency. A systematic investigation of these coupled wear mechanisms—preferably through in-situ observation of grain morphology evolution and grinding force signatures—would help establish a predictive model for wheel wear under hybrid UVAG MQL conditions. Such efforts would complement the current findings on hole quality and directly address the economic and industrial viability of the proposed process.

The smaller the wear area and the volume of material removed, the more effective the lubrication condition. As illustrated in Figure 11, the profiles and unit surface area (within the observation area of laser confocal microscopy) of CFRP wear scars at various FOAs under different lubrication conditions are compared. It is evident that, under the influence of nanolubricants, the CFRP wear scar area for all FOAs is smaller than that observed under dry conditions, with the maximum reduction reaching 25.13% for FOA = 0°. Additionally, FOA = 0° exhibits the lowest wear area across both cooling lubrication conditions. The overall differences in wear area among the various FOAs under dry conditions are not significant. Nanolubricants have effectively demonstrated their lubricating capabilities on CFRP grinding surfaces, particularly at FOA = 0° and 135°. This enhanced lubrication is attributed to the optimal infiltration and spreading of the lubricant on the CFRP surface at FOA = 0°, along with the sliding of the diamond grinding ball friction pair along the fiber direction, which promotes the nanolubricant's infiltration effect at the interface. Conversely, when FOA = 135°, the intense friction and wear processes may lead to the removal of fiber blocks and the extraction of fibers, resulting in gaps where nanolubricants can be stored, thereby enhancing lubrication.

Figure 11. Wear scar contour and area of grinding CFRP with different FOAs under different lubrication conditions

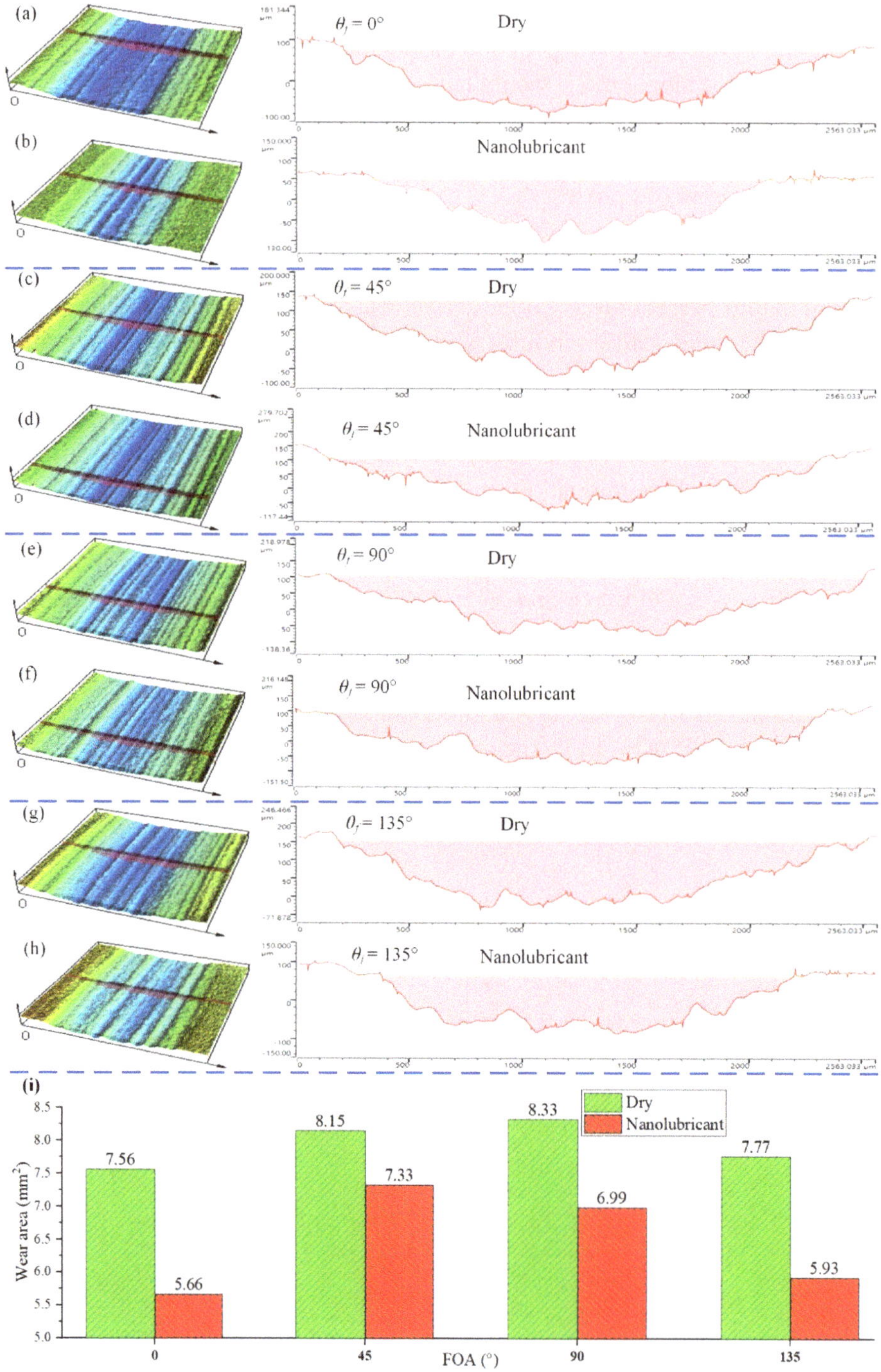

4.5.3 Wear Volume

The wear volume of the wear scar within the observation area (unit area) of the laser confocal microscope is defined as the unit wear volume of the wear scar. As illustrated in Figure 12, the comparison of the unit wear volume of CFRP wear scars with different FOAs under various lubrication conditions aligns with the wear area observed. Under dry conditions, the wear volume of CFRP wear scars across different FOAs shows no significant difference. The maximum wear volume for a 45° FOA is 0.6559 mm^3, while the minimum wear volume for a 0° FOA CFRP is 0.6277 mm^3, resulting in a minimal difference of only 4.5%. This similarity may be attributed to the severe friction experienced under dry conditions, leading to substantial material removal and the accumulation of debris between the friction pairs, which acts as a solid lubricant. Furthermore, the debris coating on the surface of CFRP wear scars diminishes the influence of the FOA. In contrast, the wear volume under nanolubricant conditions is considerably smaller than that under dry conditions, with significant variations in wear volume observed across different FOAs. This finding suggests that the anti-wear and friction-reducing effects of CNTs nanolubricants differ when interacting with diamond grinding balls and CFRP grinding surfaces. The order of wear volume across different FOAs under nanolubricant conditions is as follows: 45° > 0° > 135° > 90°. Notably, when the FOA is 90°, the nanolubricant demonstrates the most effective anti-friction and anti-wear performance, achieving a 48.29% improvement compared to dry conditions. The filling of gaps and surface pits caused by fiber loss with nanoparticles in nanolubricants can also have an anti-wear effect and reduce wear.

Figure 12. Wear volume of CFRP with different FOAs under different lubrication conditions

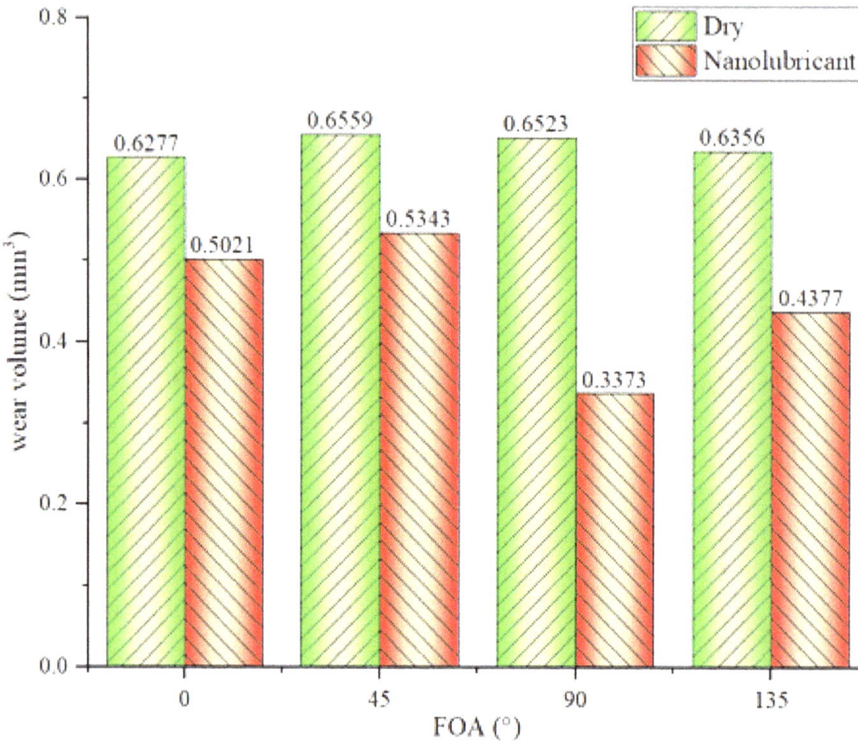

4.6 SUMMARY

A contrast analysis on the friction coefficient of the lubrication of different CNT nanolubricants and the wearing loss of discs was performed to compare the dispersion mechanism. The frictional behavior of the new lubrication method of CNT biological lubricant were evaluated by performing friction-wear tests under different lubrication conditions. Additionally, the frictional behavior of nanolubricants on the grinding surface of CFRP with different FOAs was also studied.

(1) The CNT nanolubricants with APE-10 gained the lowest friction coefficient (0.121), lowest maximum scratch depth (7.024 μm), lowest roughness value (0.48 μm), second lowest material removal rate, and favorable frictional surface morphology, thereby indicating the excellent tribological property, antiwear, and friction-reduction performance. The order of viscosity and the average

friction coefficient of nanolubricants with different surfactants were strongly correlated. The high viscosity of nanolubricants implied a stable oil film, thus exerting a favorable friction-reduction effect.

(2) Compared with the dry condition, MQL and CNT biological lubricant significantly reduced the friction coefficient by 48.51% and 53.47%, respectively. In addition, CNT nano-lubricants showed optimal and durable antifriction properties. Dry grinding caused severe wheel blockage. MQL and CNT biological lubricants can effectively improve the surface condition of the grinding wheel through temperature control, chip removal, and oil film formation. The addition of CNTs nanoparticles is conducive to the formation and spread of oil film.

(3) In comparison to dry conditions, the application of nanolubricants led to reductions in the friction coefficients for the four FOAs of 17.09%, 34.87%, 19.59%, and 27.13%, respectively. Among these, the lowest friction coefficient was observed at FOA = 90°, while the highest was noted at FOA = 45°. For FOAs of 0°, 45°, and 90°, the optimal reduction effect of the 90° UVAG MQL on grinding force ratio was a maximum decrease of 20.14%. Meanwhile, the 45° UVAG MQL demonstrated the most effective reduction for FOA = 135°, achieving a maximum reduction of 37.66%. Correspondingly, nanolubricants significantly decreased both the wear area and volume of wear scars, with a maximum reduction of 32.05% compared to dry conditions recorded at FOA = 90°. The order of wear volume under nanolubricant conditions is as follows: 45° > 0° > 135° > 90°. At FOA = 90°, nanolubricants exhibited the best anti-wear performance, with a 48.29% improvement compared to dry lubricants. While this study has demonstrated significant mitigation of wheel loading, a detailed investigation into the synergistic effect of the hydrodynamic lubricant film and ultrasonic impact on the progressive micro-wear of diamond abrasive grains would further advance the predictability of tool life in this process.

REFERENCES

Alargova, R. G., Kochijashky, I. I., Sierra, M. L., Kwetkat, K., & Zana, R. (2001). Mixed micellization of dimeric (gemini) surfactants and conventional surfactants: II. CMC and micelle aggregation numbers for various mixtures. *Journal of Colloid and Interface Science*, *235*(1), 119–129. DOI: 10.1006/jcis.2000.7311 PMID: 11237450

Ali, S., Kurniawan, R., Xu, M., Ko, T. J., & Gautam, R. K. S. (2025). Effect of High-Speed Longitudinal Ultrasonic Vibration-Assisted Milling on Tribological Characteristics of a Plane Surface. *Journal of Materials Engineering and Performance*, *34*(1), 870–884. DOI: 10.1007/s11665-024-09737-4

Alves, S. M., Barros, B. S., Trajano, M. F., Ribeiro, K. S. B., & Moura, E. J. T. I. (2013). Tribological behavior of vegetable oil-based lubricants with nanoparticles of oxides in boundary lubrication conditions. *Tribology International*, *65*, 28–36. DOI: 10.1016/j.triboint.2013.03.027

Birleanu, C., Pustan, M., Pop, G., Cioaza, M., Popa, F., Lazarescu, L., & Contiu, G. (2022). Experimental investigation of the tribological behaviors of carbon fiber reinforced polymer composites under boundary lubrication. *Polymers*, *14*(18), 3716. DOI: 10.3390/polym14183716 PMID: 36145857

Fergani, O., Yousfi, M., Ding, Z., Welo, T., & Liang, S. Y. (2017). A physics-based approach to relate grinding process parameters to tribological behavior of ground surfaces. *International Journal of Advanced Manufacturing Technology*, *91*(9), 4151–4161. DOI: 10.1007/s00170-017-0111-x

Gao, T., Li, C., Zhang, Y., Yang, M., Jia, D., Jin, T., Hou, Y., & Li, R. (2019). Dispersing mechanism and tribological performance of vegetable oil-based CNT nanofluids with different surfactants. *Tribology International*, *131*, 51–63. DOI: 10.1016/j.triboint.2018.10.025

Ge, X., Xia, Y., & Cao, Z. (2015). Tribological properties and insulation effect of nanometer TiO2 and nanometer SiO2 as additives in grease. *Tribology International*, *92*, 454–461. DOI: 10.1016/j.triboint.2015.07.031

Gu, G., Wang, D., Wu, S., Zhou, S., & Zhang, B. (2024). Research status and prospect of ultrasonic vibration and minimum quantity lubrication processing of nickel-based alloys. *Intelligent and Sustainable Manufacturing*, *1*(1), 10006. DOI: 10.35534/ism.2024.10006

Guo, S., Li, C., Zhang, Y., Wang, Y., Li, B., Yang, M., Zhang, X., & Liu, G. (2017). Experimental evaluation of the lubrication performance of mixtures of castor oil with other vegetable oils in MQL grinding of nickel-based alloy. *Journal of Cleaner Production*, *140*, 1060–1076. DOI: 10.1016/j.jclepro.2016.10.073

Lawal, S. A., Choudhury, I. A., & Nukman, Y. (2012). Application of vegetable oil-based metalworking fluids in machining ferrous metals—A review. *International Journal of Machine Tools & Manufacture*, *52*(1), 1–12. DOI: 10.1016/j.ijmachtools.2011.09.003

Leus, M., Gutowski, P., & Bachtiak-Radka, E. (2023). The effect of contact compliance of sliding pair on friction force reduction at longitudinal tangential vibrations. *Tribology International*, *187*, 108701. DOI: 10.1016/j.triboint.2023.108701

Li, C., & Ali, H. M. (Eds.). (2019). *Enhanced heat transfer mechanism of nanofluid MQL cooling grinding*. IGI Global, 298-307.

Li, J., Zhang, K., Liu, P., & Li, Y. (2018). Tribological Behavior and Microstructure of Carbon Fiber–Reinforced Polymer against Ti6Al4V Alloy in Fretting Contact. *Tribology Transactions*, *61*(2), 256–268. DOI: 10.1080/10402004.2017.1313471

Liang, X., & Wu, D. (2019). Tribological properties of carbon-fibre-reinforced plastic against tungsten carbide under dry condition. *Tribology International*, *134*, 118–128. DOI: 10.1016/j.triboint.2019.01.043

Liu, W., Ni, H., Wang, P., & Chen, H. (2020). Investigation on the tribological performance of micro-dimples textured surface combined with longitudinal or transverse vibration under hydrodynamic lubrication. *International Journal of Mechanical Sciences*, *174*, 105474. DOI: 10.1016/j.ijmecsci.2020.105474

Lu, S., & Somasundaran, P. (2008). Tunable synergism/antagonism in a mixed nonionic/anionic surfactant layer at the solid/liquid interface. *Langmuir*, *24*(8), 3874–3879. DOI: 10.1021/la703233d PMID: 18324849

Malkin, S., & Guo, C. (2007). Thermal analysis of grinding. *CIRP Annals*, *56*(2), 760–782. DOI: 10.1016/j.cirp.2007.10.005

Mughal, K., Mughal, M. P., Farooq, M. U., Anwar, S., & Ammarullah, M. I. (2023). Using nano-fluids minimum quantity lubrication (NF-MQL) to improve tool wear characteristics for efficient machining of CFRP/Ti6Al4V aeronautical structural composite. *Processes (Basel, Switzerland)*, *11*(5), 1540. DOI: 10.3390/pr11051540

Qu, M., Yao, Y., He, J., Ma, X., Liu, S., Feng, J., & Hou, L. (2016). Tribological performance of functionalized ionic liquid and Cu microparticles as lubricating additives in sunflower seed oil. *Tribology International*, *104*, 166–174. DOI: 10.1016/j.triboint.2016.08.035

Rahim, E. A., & Sasahara, H. (2011). A study of the effect of palm oil as MQL lubricant on high speed drilling of titanium alloys. *Tribology International*, *44*(3), 309–317. DOI: 10.1016/j.triboint.2010.10.032

Seward, O., Cepero-Mejías, F., Fairclough, J. P. A., & Kerrigan, K. (2022). Development of a novel friction model for machining simulations in unidirectional composite materials. *Polymers*, *14*(5), 847. DOI: 10.3390/polym14050847 PMID: 35267670

Shen, X., & Zhang, J. (2013). Studies on friction and wear properties of surface produced by ultrasonic vibration-assisted milling. *International Journal of Advanced Manufacturing Technology*, *67*(1), 349–356. DOI: 10.1007/s00170-012-4488-2

Silva, L. R., Bianchi, E. C., Fusse, R. Y., Catai, R. E., Franca, T. V., & Aguiar, P. R. (2007). Analysis of surface integrity for minimum quantity lubricant—MQL in grinding. *International Journal of Machine Tools & Manufacture*, *47*(2), 412–418. DOI: 10.1016/j.ijmachtools.2006.03.015

Sun, Y. J., Gong, H., Wang, Y., & Ni, H. (2021). Tribological viewpoint of the cutting force in rotary ultrasonic vibration-assisted side milling (RUVSM). *International Journal of Advanced Manufacturing Technology*, *114*(9), 2839–2848. DOI: 10.1007/s00170-021-06966-y

Virdi, R. L., Chatha, S. S., & Singh, H. (2021). Experimental investigations on the tribological and lubrication behaviour of minimum quantity lubrication technique in grinding of Inconel 718 alloy. *Tribology International*, *153*, 106581. DOI: 10.1016/j.triboint.2020.106581

Wang, X., Wang, C., Shen, X., & Sun, F. (2019). Tribological behaviors of the diamond films sliding against the T800/X850 CFRP laminates. *Wear*, *418*, 191–200. DOI: 10.1016/j.wear.2018.12.007

Wang, Y., Li, C., Zhang, Y., Yang, M., Li, B., Jia, D., Hou, Y., & Mao, C. (2016). Experimental evaluation of the lubrication properties of the wheel/workpiece interface in minimum quantity lubrication (MQL) grinding using different types of vegetable oils. *Journal of Cleaner Production*, *127*, 487–499. DOI: 10.1016/j.jclepro.2016.03.121

Wickramasinghe, K. C., Sasahara, H., Abd Rahim, E., & Perera, G. I. P. (2020). Green Metalworking Fluids for sustainable machining applications: A review. *Journal of Cleaner Production*, *257*, 120552. DOI: 10.1016/j.jclepro.2020.120552

Yu, H., Xu, Y., Shi, P., Wang, H., Wei, M., Zhao, K., & Xu, B. (2013). Microstructure, mechanical properties and tribological behavior of tribofilm generated from natural serpentine mineral powders as lubricant additive. *Wear*, *297*(1-2), 802–810. DOI: 10.1016/j.wear.2012.10.013

Zhang, Y., Li, C., Jia, D., Zhang, D., & Zhang, X. (2015). Experimental evaluation of the lubrication performance of MoS2/CNT nanofluid for minimal quantity lubrication in Ni-based alloy grinding. *International Journal of Machine Tools & Manufacture*, *99*, 19–33. DOI: 10.1016/j.ijmachtools.2015.09.003

Chapter 5
Mechanical Behavior and Force Model of Single Grain Transverse Grinding CFRP With Nanolubricants MQL

ABSTRACT

This chapter explores the CFRP grinding mechanism of single-grain of CNT nanolubricant MQL. A mechanical model is established based on the unique geometric interactions between the grain and fibers, which arise from random fiber arrangements and the grain topography under varying undeformed chip thicknesses. Firstly, four distinct contact and grinding models are developed, corresponding to different interaction states between the grain and fiber throughout the grinding arc length. Subsequently, the influence of different grinding parameters on the geometry of the cutting zone is elucidated. Then, a mechanical model for single-grain grinding is established, which incorporates the homogeneous randomness of fiber distribution across the entire grinding arc length. Finally, the model is numerically simulated and the trend of force along the entire grinding arc length is obtained. Experimental verifications demonstrate the approach for predicting the grinding force have acceptable accuracy and can successfully capture the mechanics of CFRPs.

5.1 INTRODUCTION

The majority of CFRP components are manufactured via near-net shaping processes, requiring secondary machining to meet assembly requirements and surface

DOI: 10.4018/979-8-3373-9013-0.ch005

precision (Hosokawa et al., 2014). High-quality, low-damage machining of CFRP is essential for its widespread application in aerospace components (Geier et al., 2019). Qiu et al. (2018) investigated the influence of twist-drill drilling behavior on delamination and found that step drilling yields better entry-hole quality than twist drilling. Ning et al. (2017) established an analytical model indicating that thermal properties can be correlated with fiber fracture and delamination. Voss et al. (2019) developed a force model comprising four sub-models that account for the effects of fiber orientation, tool wear, and geometry. Jia et al. (2018) proposed a principle for suppressing cutting damage at appropriate temperatures and revealed the influence of temperature on CFRP cutting quality. Additionally, a machining technology featuring negative-pressure reverse cooling and self-air-cooling chip removal was invented. Furthermore, based on the principles of micro-element removal and reverse shearing, a series of drilling tools were developed to achieve low-damage and high-quality machining of CFRP (Jia et al., 2016).

Among various machining methods, grinding is categorized as a finishing process, characterized by high machining accuracy and minimal material removal. Consequently, grinding is widely employed in the secondary machining of CFRP (Cao et al., 2015). However, the abrasive grains on engineered grinding wheels are small in size, with random shapes, dimensions, and distributions. The complex contact conditions and material removal behaviors between abrasive grains and the workpiece during grinding pose significant research challenges. From a microscopic perspective, grinding can be regarded as the cumulative cutting action of a large number of randomly distributed abrasive grains on the wheel surface (Luna et al., 2020). Therefore, utilizing single-grit cutting as a means to reveal the complex material removal mechanism in CFRP grinding represents an effective approach (Li et al., 2019).

Grinding force is an important indicator of material removal process, which directly affects machining efficiency, surface quality, and tool life (Cui and Li, 2025). The force model is the most direct and comprehensive mathematical expression for describing the mechanical behavior of material removal, and plays an important role in the application of grinding in various fields (Kong et al., 2025). The mechanical modeling of CFRP grinding is the key to effectively controlling the occurrence of machining damage. The grinding force, defined as the combined effect of cutting forces and frictional forces generated during grinding, significantly influences the formulation of grinding parameters and the extent of surface and subsurface damage. It also serves as a crucial indicator for evaluating grinding performance (Zhang et al., 2019). The calculation and measurement of grinding forces contribute to uncovering grinding mechanisms. Since the grinding force is the summation of forces acting on all effective cutting edges of abrasive grains within the wheel–workpiece contact zone, coupled with the multiphase and anisotropic nature of CFRP composites, modeling the cutting forces during single-grit machining of unidirectional CFRP composites is a fundamental and critical prerequisite for studying grinding forces and performance. Machining force modeling is one of the most effective methods for investigating and capturing material removal mechanisms and mechanical behaviors during machining processes (Zhang et al., 2020). Particularly for heterogeneous and anisotropic materials like CFRP, establishing high-precision mechanical models is challenging, as they are often treated as equivalent homogeneous materials in modeling, and CFRP specimens may contain uncertain pore defects. Zhang et al. (Zhang et al., 2021) developed a mechanical model capable of reflecting deformation mechanisms and predicting cutting forces, with maximum errors of 37% and 27% for vertical and horizontal forces, respectively. Nayak et al. (2005) analyzed the influence of fiber orientation, tool geometry, and depth of cut on cutting forces, reporting a force prediction error of 38%. Based on the assumption of material removal via brittle fracture, Ning et al. (2017) established a surface grinding force model for rotary ultrasonic machining with a single parameter, namely the fracture volume factor of the workpiece material. Qu et al. (2020) conducted grinding experiments on unidirectional carbon fiber-reinforced CMCs under various conditions, finding that the depth of debonding between the matrix and carbon fibers depends on grain sharpness and lubrication state.

Building on the macroscopic tribological behavior identified in Chapter 4, this chapter narrows the focus to the single-grain scale to deconvolve the complex tool–workpiece interaction. By isolating the mechanical response of an individual abrasive grain during transverse grinding under nanolubricant MQL, the study establishes a mechanistic force model that bridges the lubricant's interfacial effects with the fundamental material removal process. This chapter considers the random distribution of carbon fibers and establishes corresponding mechanical models based on the distinct contact states and material removal behaviors between abrasive grains and carbon fibers during different cutting stages. The cutting forces at various points along the cutting arc length are calculated. Accounting for the influence of lubrication conditions on frictional forces, friction and wear tests using diamond materials and CFRP composites as friction pairs were conducted. The friction coefficients under different lubrication conditions required for the mechanical model were obtained, enabling the determination of frictional forces between a single abrasive grain and CFRP at different stages. Finally, the cutting and frictional forces of a single abrasive grain are decomposed to obtain the tangential and normal forces, and the single-grit cutting force model is validated under identical machining conditions. The influence of MQL and nanofluid MQL (NMQL) on the material removal mechanisms during single-grit cutting of CFRP, as well as the role of nanoparticles between abrasive grains and carbon fibers, are analyzed.

5.2 MECHANICS ANALYSIS AND GRINDING MODELS

The predictive force model allows different modes of interaction between the tool grain and the work material, particularly the carbon fiber at different grinding stages and grain edge conditions (Figure 1). Here, a mechanical analysis is initially conducted to develop the force models under different cases before the predictive force model can be fully investigated. Different grinding sub-models for the whole grinding arc length are established according to the contact status, as the grain and the fiber may come in contact with one another because of the randomness of the fiber placements and the grain edge shapes under various undeformed chip thickness conditions.

Figure 1. Interaction modes between the single grain and CFRPs under different a_g values

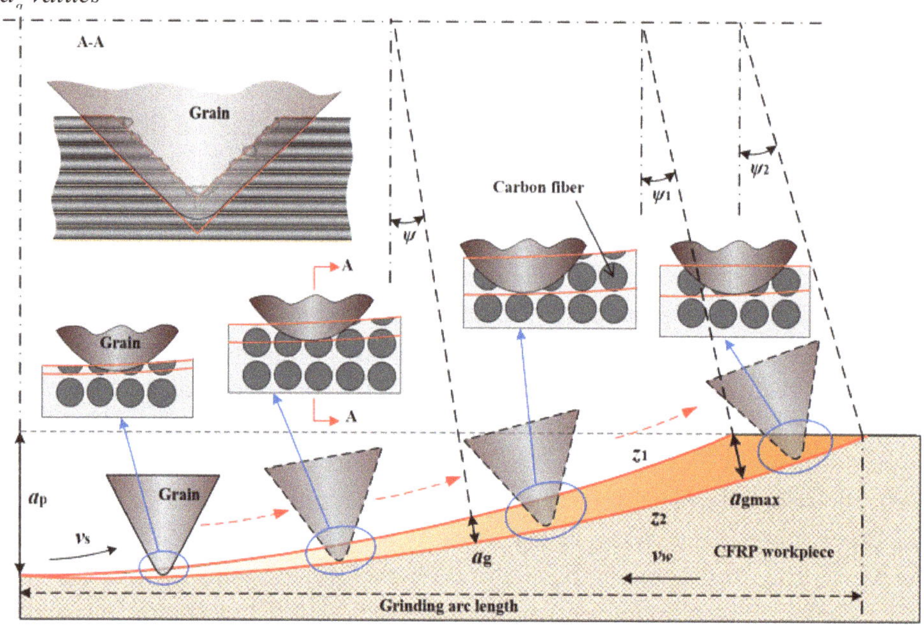

5.2.1 Contact Model Between Blunt Cone and Fibers

In the initial stage of the single-grain grinding process, the grain tip (blunt cone) produces extrusion pressure in the normal direction on the CFRP due to the low thickness of the undeformed chip. Figure 2 shows a deformation area, with a 2D blunt cone acting on a flat surface. The normal pressure distribution on the conical surface of the contact area between the 2D blunt cone and the plane is expressed as (Johnson, 1987),

$$P_{bc}(r) = \frac{1}{2}arch\left(\frac{r_{bc}}{r}\right)E^{*}\cot\theta, \tag{5.1}$$

$$r_{bc} = \sqrt{r_g^2 - \left(r_g h\right)^2}, \tag{5.2}$$

where

$$\cot\theta = \frac{r_{bc}}{r_g - h}. \tag{5.3}$$

The resultant force of the contact area is given by

$$F_{bc} = \frac{1}{2}\pi r_{bc}^2 E^{*}\cot\theta. \tag{5.4}$$

In the above equations, E^* is the equivalent modulus of elasticity, which is calculated according to (Zhang and Li, 2025),

$$\frac{1}{E^{*}} = \frac{1-\upsilon_f^2}{E_a} + \frac{1-\upsilon_g^2}{E_g}, \tag{5.5}$$

where E_a is the off-axis elastic modulus of the carbon fiber, E_g is the elastic modulus of the grain, v_f is the Poisson's ratio of the carbon fiber, and v_g is the Poisson's ratio of the grain. When grinding is performed at different angles, the off-axis modulus changes along the grinding direction. The calculation method according to Gao et al. (2021) is,

$$\frac{1}{E_a} = \frac{1}{E_l}\cos^4\alpha + \left(\frac{1}{G_f} - \frac{2\upsilon_f}{E_l}\right)\sin^2\alpha\cos^2\alpha + \frac{1}{E_f}\sin^4\alpha, \tag{5.6}$$

where E_l is the longitudinal modulus of the fiber, E_f is the transverse modulus of the fiber, G_f is the shear modulus of the fiber, and α is the angle between the fiber direction and the grinding direction. In the case of 90° grinding, the off-axis modulus along the grinding direction is close to the transverse modulus of the CFRPs. The interaction between normal and tangential forces, which is caused by the difference between the grain and solid elastic constant of the workpiece, is not considered when the grain slides along the grinding direction on the workpiece surface. On the basis of Amontons' friction law, the tangential force can be obtained from

$$f_{bc} = \frac{3\mu F_{bc}}{2\pi r_{bc}^3}\left(r_{bc}^2 r^2\right)^{1/2}. \tag{5.7}$$

Figure 2. Contact status analysis between the blunt cone on top of the grain and the carbon fiber

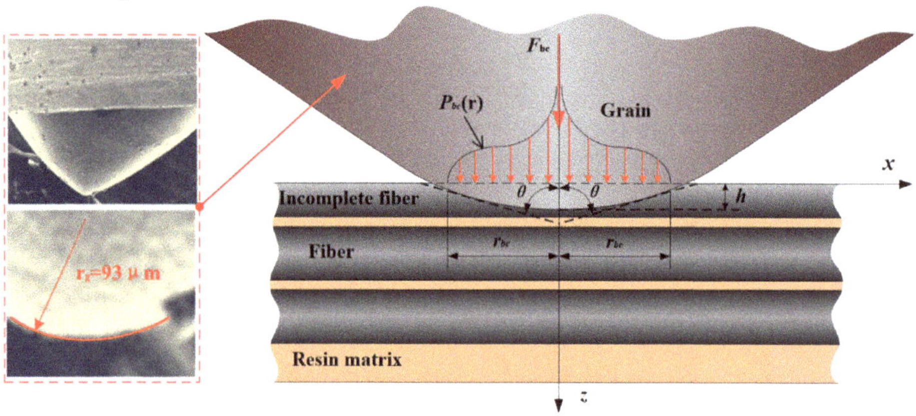

5.2.2 Local Contact of the Elliptical Region

The shape of the contact area formed by the grain tip is an ellipse when grinding with a single fiber (Figure 3). The long axis along the x-axis has a length of $2a_{II}$, whereas the short axis along the y-axis has a length of $2b_{II}$. The eccentricity of the ellipse is calculated as

$$k = \sqrt{1 - \frac{b_{II}^2}{a_{II}^2}}, \tag{5.8}$$

where,

$$a_{II} = \sqrt{r_g^2 - \left(r_g h\right)^2} \tag{5.9}$$

$$\cos\xi = \frac{r_g^2 + \left(r_f + r_g - h\right)^2 - r_f^2}{2 \cdot r_g \cdot \left(r_f + r_g - h\right)} \tag{5.10}$$

$$l^2 = \left(r_g - h\right)^2 + r_g^2 - 2 \cdot r_g\left(r_g - h\right) \cdot \cos\xi \tag{5.11}$$

$$\cos\eta = \frac{l^2 + r_g^2 - \left(r_g - h\right)^2}{2 \cdot l \cdot r_g} \tag{5.12}$$

$$b = 2l\sin(\xi + \eta) \tag{5.13}$$

Figure 3. Local contact between the grain and the carbon fiber

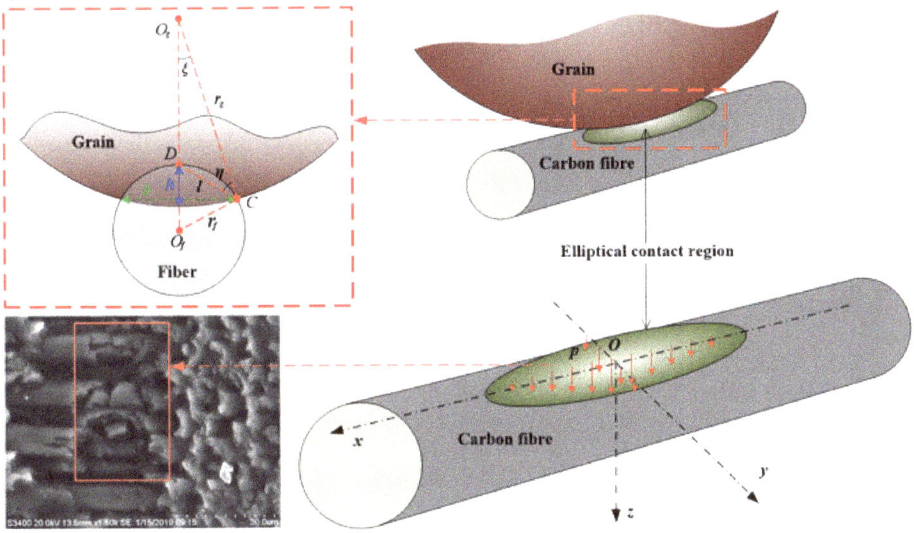

According to the point load theory of elastic half space, the pressure distribution in the elliptical contact region between the grain and the workpiece is

$$p = p_{\max}\left[1\left(x/a_{II}\right)^2\left(y/b_{II}\right)^2\right]^{1/2}. \tag{5.14}$$

According to the theory of maximum extension line strain in material mechanics, the fracture criterion of the fiber is

$$\sigma_z - \upsilon\left(\sigma_x + \sigma_y\right) = \sigma. \tag{5.15}$$

The stress along the z axis is calculated as

$$\frac{\sigma_x}{p_{II\max}} = \frac{2b_{II}}{k^2 a_{II}}\left(\Omega_x + \upsilon\Omega'_x\right) \tag{5.16}$$

$$\frac{\sigma_y}{p_{II\max}} = \frac{2b_{II}}{k^2 a_{II}}\left(\Omega_y + \upsilon\Omega'_y\right) \tag{5.17}$$

$$\frac{\sigma_z}{p_{II\max}} = -\frac{b_{II}}{k^2 a_{II}}\left(\frac{1-T^2}{T}\right) \tag{5.18}$$

where

$$\Omega_x = -\frac{1}{2}\left(1-T\right) + \xi\left[F\left(\phi,k\right) - E\left(\phi,k\right)\right] \tag{5.19}$$

$$\Omega'_x = 1 - \frac{a_{II}^2 T}{b_{II}^2} + \xi\left[\frac{a_{II}^2}{b_{II}^2}E\left(\phi,k\right) - F\left(\phi,k\right)\right] \tag{5.20}$$

$$\Omega_y = \frac{1}{2} + \frac{1}{2T} - \frac{a_{II}^2 T}{b_{II}^2} + \xi\left[\frac{a_{II}^2}{b_{II}^2}E\left(\phi,k\right) - F\left(\phi,k\right)\right] \tag{5.21}$$

$$\Omega'_y = -1 + T + \xi\left[F\left(\phi,k\right) - E\left(\phi,k\right)\right] \tag{5.22}$$

$$T = \sqrt{\frac{b_{II}^2 + h^2}{a_{II}^2 + h^2}} \tag{5.23}$$

$$\xi = \frac{z_{II}}{a_{II}} = \cot\phi \tag{5.24}$$

$$\phi = \operatorname{arccot}\left(\frac{z_{II}}{a_{II}}\right) \tag{5.25}$$

The elliptic integral of $F(\phi, k)$ and $E(\phi, k)$ are obtained from the lookup table of Carlson (1977, 1987).

The total load acting on the elliptical contact area is

$$F_e = \frac{2\pi a_{II} b_{II} p_{II\max}}{3}. \tag{5.26}$$

The friction force between the grain bottom and CFRP is

$$f_e = \mu F_e. \tag{5.27}$$

5.2.3 Grinding Model of the Single Fiber

In determining the fracture morphology, the grinding behavior of the whole carbon fiber is studied by observing its chip morphology by means of SEM. At present, the understanding of fiber tensile failure process has reached the mature stage. The fracture mode of a fiber mainly occurs as a tensile fracture (Figure 4). The fracture failure section of the fiber can be divided into five areas, namely, (A) origination area, (B) extension area, (C), slippage, (D), area of rapid increase of cracks, and (E) ultimate destruction area.

The carbon fiber, as the reinforcing phase in the meso-composition of CFRP, is the main body removed by grinding. The micro-performance of CFRP grinding is equivalent to the deformation and fracturing of a single carbon fiber under the action of grinding force or the cracking process of resin or the interface surrounding the fiber. In the process of mesofailure, the resin is wrapped around the carbon fiber due to the manufacturing process of CFRP, which has a restraining effect on the carbon fiber. Furthermore, the continuous distribution of resin in the material suggests that the final performance of the binding effect will affect the material surrounding the carbon fiber. As the resin and interface may have weak links, their respective cracking behaviors in the grinding process need to be determined accurately.

The behaviors are described in meso-scale in this study. A representative volume element of the material to be cut is selected as the research object based on the concept of representative elements in composite mechanics, including the single carbon fiber and its interface and equivalent homogeneous material. The physical description of grinding in the meso-scale by using a single carbon fiber is shown in Figure 4. The single carbon fiber in the representative element is affected by grinding force and constrained and supported by the surrounding materials, and it has a bonding effect on the interface. The surrounding materials are regarded to be equivalent homogeneous materials composed of resin and carbon fiber. Moreover, the stress state of the single carbon fiber is similar to that of the elastic foundation beam. The deflection of the single carbon fiber under grinding force is solved by the two-parameter elastic foundation beam analysis as a means of describing in practical terms the deformation of the carbon fiber under the grinding action.

For convenience of research, this study defines the coordinate system of x–y in the representative element of the single carbon fiber, as illustrated in Figure 4. The x axis represents the coordinate along the fiber direction, and w is the deflection of the fiber in this direction. k_m is the first parameter of the surrounding material support and used to express the normal constraint degree of the surrounding material when the fiber is bent; it is solved by Biot's equation (Biot, 1937), which is given by

$$k_m = \frac{1.23E}{C(1 - v_m^2)}\left[\frac{Ed_f^4}{C(1 - v_m^2)E_f I_f}\right]^{0.11}, \tag{5.28}$$

where C is the coefficient of action mode (uniform pressure $C = 1$; uniform deflection C = 1.13). A value of C = 1.13 is selected in this study.

Figure 4. Single-carbon fiber grinding model

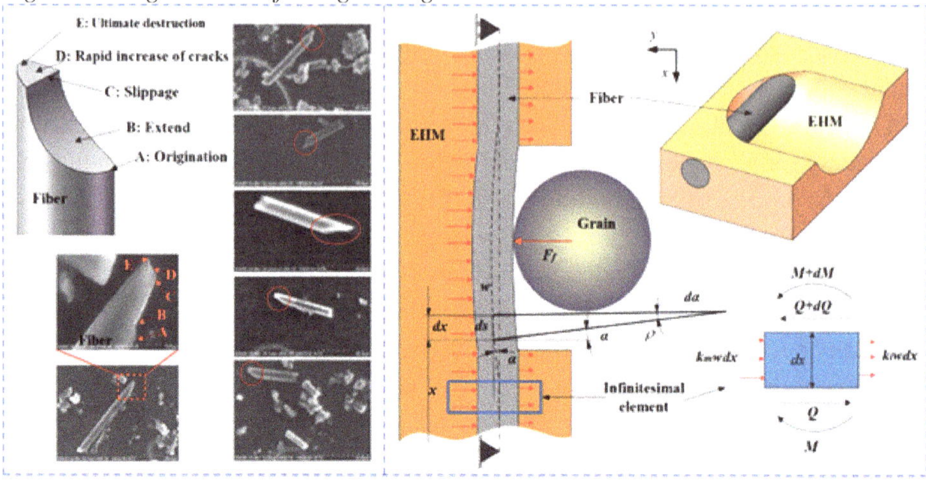

The equilibrium equation of the force and bending moment of carbon fiber micro-element with a length of dx is expressed as

$$\begin{cases} \sum Q = 0 \\ \sum M = 0 \end{cases}$$

$$\begin{aligned} Q - (Q + dQ) + \left(k_m + k_i\right)wdx &= 0 \\ M - (M + dM) + (Q + dQ)dx + \left(k_m + k_i\right)dx^2 &= 0 \end{aligned} \tag{5.29}$$

where Q is shear force, and M is bending moment. This study ignores the second-order differential variable in the above formula, i.e.,

$$\begin{cases} dQ = \left(k_m + k_i\right)wdx \\ \quad dM = Qdx \end{cases}. \tag{5.30}$$

In bending deformation, the angle in which the cross section of the beam turns to its original position is denoted by the section rotating angle α (Figure 4). According to the plane assumption, the section perpendicular to the fiber axis (x-axis) before bending continues to be perpendicular to the deflection curve after deformation. Therefore, the section rotating angle α, which is the angle between the y-axis and the normal of the deflection curve, should be equal to the angle of inclination of the deflection curve, i.e., the angle between the x-axis and the tangent of the deflection curve.

$$\tan\alpha = \frac{dw}{dx} \tag{5.31}$$

In the case of pure bending, the relationship between bending moment and the curvature is expressed as

$$\frac{1}{\rho} = \frac{M}{EI}. \tag{5.32}$$

By taking the differential arc *ds* amplification of the fiber (Figure 4), the intersection of normal at both ends of *ds* is selected as the center of the curvature. The approach can help to determine the radius of the curvature.

$$\frac{1}{\rho} = \frac{d\alpha}{ds} = \frac{M}{EI} \tag{5.33}$$

From Eq. (5.31),

$$\frac{d\alpha}{ds} = \frac{\frac{d^2w}{dx^2}}{1+\left(\frac{dw}{dx}\right)^2}\frac{dx}{ds} \tag{5.34}$$

Where $ds = \left[1\left(\frac{dw}{dx}\right)^2\right]^{1/2} dx$. Thus,

$$\frac{d\alpha}{ds} = \frac{\frac{d^2w}{dx^2}}{1+\left(\frac{dw}{dx}\right)^{3/2}}. \tag{5.35}$$

Substituting Eq. (5.33) obtains

$$\frac{\frac{d^2w}{dx^2}}{\left[1\left(\frac{dw}{dx}\right)^2\right]^{3/2}} = \frac{M}{EI}. \tag{5.36}$$

In the case of small deformation, if the deflection of the carbon fiber is far less than the span, and the angle α is extremely small, then

$$\alpha \approx \tan\alpha = \frac{dw}{dx}, \tag{5.37}$$

Where $\frac{d^2w}{dx^2}$ can be omitted compared with 1.

$$\frac{d^2w}{dx^2} = \frac{M}{E_f I_f} \tag{5.38}$$

The deformation control equation of the elastic foundation beam of the carbon fiber is

$$E_f I_f \frac{d^4 w}{dx^4} + \left(k_m + k_i\right) w = 0. \tag{5.39}$$

The general solution of Eq. (5.44) is

$$w = e^{-\lambda x}(A\cos\lambda x + B\sin\lambda x) + e^{\lambda x}(C\cos\lambda x + D\sin\lambda x), \tag{5.40}$$

where $\lambda = \sqrt[4]{\frac{k_m + k_i}{4EI}}$. Constants A, B, C and D are determined on the basis of the boundary conditions. In this model, the single carbon fiber appears as an infinite beam under the action of concentrated load F, and the stress point is taken as the coordinate origin. Only the right half of the fiber is studied using its symmetry. In the qualitative analysis, when $x \to \infty$ and $w(x) \to 0$ and when $x \to \infty$, then $e^{\lambda x} \to \infty$ and $e^{-\lambda x} \to 0$. From Eq. (5.40),

$$\infty \cdot (C\cos\lambda x + D\sin\lambda x) \to 0. \tag{5.41}$$

Eq. (5.41) only holds when $C = D = 0$. The expression of the carbon fiber flexure function is

$$w = e^{-\lambda x}(A\cos\lambda x + B\sin\lambda x). \tag{5.42}$$

The first derivative function is given by

$$\frac{dw}{dx} = -\lambda e^{-\lambda x}[A(\cos\lambda x + \sin\lambda x) + B(\sin\lambda x - \cos\lambda x)]. \tag{5.43}$$

On the basis of the boundary conditions,

If $w''(0) = 0$,then $A = B$.

Subsequently,

$$w = Ae^{-\lambda x}(\cos\lambda x + \sin\lambda x). \tag{5.44}$$

Eq. (5.44) is differentiated three times, i.e.,

$$\frac{d^3 w}{dx^3} = 4\lambda^3 A e^{-\lambda x}\cos\lambda x. \tag{5.45}$$

By means of symmetry, if $x = 0$, then $\alpha = 0$, and $Q = -\frac{F_f}{2}$.

$$Q = -E_f I_f \frac{d^3 w}{dx^3} = -\frac{F_f}{2} \tag{5.46}$$

The following solution is obtained:

$$A = \frac{F_f}{8\lambda^3 E_f I_f}. \tag{5.47}$$

When $x > 0$,

$$w = \frac{F_f}{8\lambda^3 E_f I_f} e^{-\lambda x}\left(\cos\lambda x + \sin\lambda x\right) \tag{5.48}$$

$$\frac{dw}{dx} = -\frac{F_f}{4\lambda^2 E_f I_f} e^{-\lambda x} \sin\lambda x \tag{5.49}$$

$$\frac{d^2 w}{dx^2} = \frac{F_f}{4\lambda E_f I_f} e^{-\lambda x}\left(\sin\lambda x - \cos\lambda x\right) \tag{5.50}$$

The relationship between stress and deflection caused by fiber deformation is given by

$$\sigma_f = E_f \eta w'', \tag{5.51}$$

where η is the distance from any point on the fiber to the neutral axis. If the point is on the left side of the neutral axis, then η is positive. If the point is on the right side of the neutral axis, then η is negative. If η is r_f and $-r_f$, then the absolute value of deformation stress has reached the maximum. As the compressive strength of the fiber is greater than the tensile strength, when $\eta = r_f$ and $x = 0$, the stress on the fiber is the largest. When the stress reaches the tensile strength σ, the fiber begins to fracture. Then,

$$E_f r_f w''(0) = \sigma, \tag{5.52}$$

$$F_f = \frac{4\lambda I_f \sigma}{r_f}. \tag{5.53}$$

The friction force produced by the grinding of the grain can be obtained as

$$f = \mu F_f. \tag{5.54}$$

The corresponding components of friction force in the normal direction and tangent direction are

$$f_n = f\sin\gamma_0, \tag{5.55}$$

$$f_t = f\cos\gamma_0. \tag{5.56}$$

The grinding force decomposition diagram of the single-grain grinding carbon fiber is shown in Figure 5. For convenience of calculation, the center point of the contact length l_g of grain grinding is taken in the formula of the decomposition of single-fiber grinding force in the tangential and normal directions.

$$\cos\gamma_1 = \frac{(r_g - z_2)}{r_g}, \tag{5.57}$$

$$\gamma_0 = \gamma_1 / 2. \tag{5.58}$$

The contact length of grain grinding is

$$l_g = r_g \cdot \gamma_1. \tag{5.59}$$

Figure 5. Schematic of grinding force decomposition

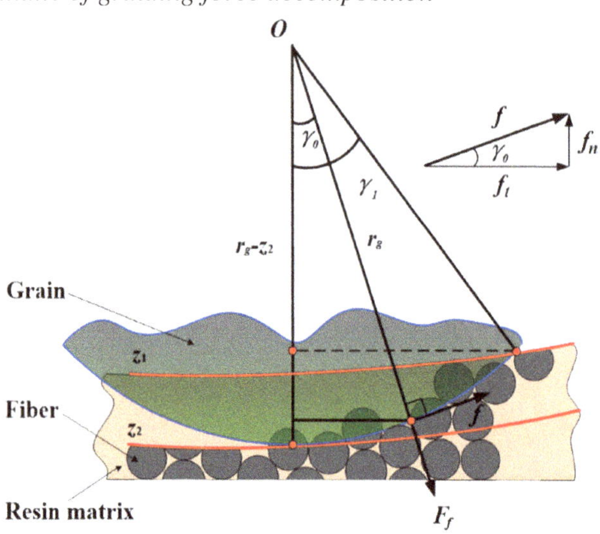

In the grinding process, the grain is in a grinding state on the incomplete carbon fiber. According to the calculation of internal force via the section method of material mechanics, internal force is related to external force, it is unrelated to material property, and it is related to the section where it is located. Moreover, the internal forces of different sections differ from one another. Therefore, the grinding force of the incomplete fiber can be calculated on the basis of the cross-sectional area of the incomplete fiber and the grinding force of the whole fiber.

The grinding of the incomplete carbon fiber by a single grain can be divided into two cases, as shown in Figure 6 (a) and (b). The cross-sectional area s is calculated according to the following cases: "+" for (a) and "-" for (b).

$$S = \left(r_f^2\beta \pm lh_i\right)/2, \tag{5.60}$$

where β is the center angle of the incomplete carbon fiber cross section, and h_i is the distance from the center of the carbon fiber cross section to the incomplete section. These two parameters are determined by the state of random arrangement of the carbon fiber. ψ is defined as the ratio of the cross-sectional area of the incomplete fiber to the cross-sectional area of the whole fiber.

$$\psi = \frac{S}{\pi r_f^2}, \tag{5.61}$$

$$F_{lli} = \psi F_f = \psi \frac{4\lambda I_f \sigma}{r_f} = \frac{4S\lambda I_f \sigma}{\pi r_f^3}. \tag{5.62}$$

Figure 6. Cross section of incomplete carbon fibers

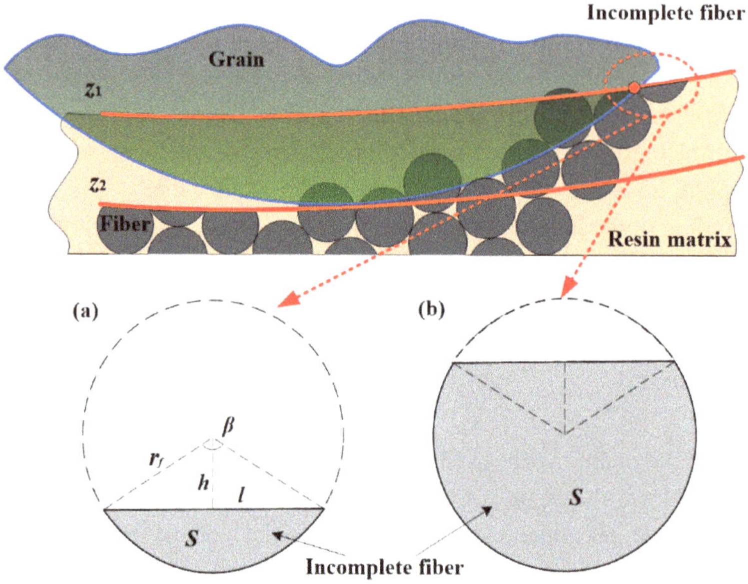

5.2.4 Model for Extrusion and Shearing

In the grinding process of the single grain, the carbon fiber cut by the last circular movement of the grain will be further fractured due to the extrusion of the two sides of the grain. The bending failure theory of carbon fibers is applied to solve the extrusion model and determine the mechanical behavior of the single carbon fiber under extrusion (Reifsnider and Case, 2002). With the influence of the initial bending degree of the cut carbon fiber on its bending failure, the bending deformation of the cut carbon fiber is thus introduced. Figure 7 shows the single-fiber extrusion model with the single-grain grinding CFRP.

Figure 7. Extrusion model of a single grain on the carbon fiber fracture surface

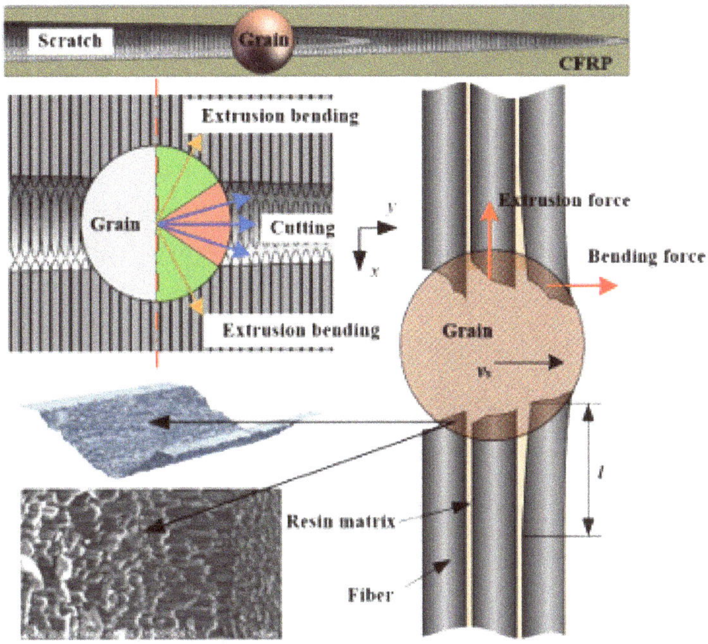

In the whole process of grain grinding, both sides of the grain extrude on the fracture fiber section. The fracture of the fiber under grain extrusion can be divided into two types, namely, the fracture under shear or bending stress. These two stress types are expressed as

$$\sigma_{shear} = \frac{G_{unit}}{1 + \pi\left(\frac{f_0}{l}\right)\frac{G_{unit}}{\sigma_\tau}} \tag{5.63}$$

$$\sigma_{bending} = \frac{G_{unit}}{1 + \pi^2\left(\frac{f_0}{2l^2}\right)\frac{d_f E_f}{\sigma_c}} \tag{5.64}$$

where f_0 is the initial maximum deformation of the deformed carbon fiber, i.e., $f_0 = w(0)$; σ_τ is the ultimate shear strength of the carbon fiber; σ_c is the ultimate compressive strength of the carbon fiber; d_f is the diameter of the carbon fiber; l is the length of the deformed carbon fiber below the contact point between the carbon fiber and the grain; and E_f is the elastic modulus of the carbon fiber. G_{unit} is the in-plane shear modulus of the composite material in the compression shear zone expressed in carbon fiber units (l in length and d_f in width), which can be solved using the strain energy balance equation that considers the different deformation lengths of the carbon fiber.

$$G_{unit} = \frac{E_f V_f}{2}\left(\frac{d_f}{l}\right)^2\left[1 + \left(\frac{d_f}{l}\right)^2 \frac{E_f}{4G_f}\right] \tag{5.65}$$

By considering the abovementioned two fracture forms, the ultimate value of ultimate stress can be used to calculate the extrusion pressure of the single carbon fiber as follows:

$$F_{shear} = \sigma_{shear}\frac{\pi d_f^2}{4} \tag{5.66}$$

$$F_{bending} = \sigma_{bending}\frac{\pi d_f^2}{4} \tag{5.67}$$

The friction force between the grain side and the carbon fiber section is

$$F_{side} = \mu\left(F_{shear} + F_{bending}\right). \tag{5.68}$$

5.2.5 Mechanical Properties

The size of the CFRP workpiece is 70 mm × 30 mm × 6 mm; the material properties are shown in Table 1. The material properties of the diamond grain are shown in Table 2. The MQL lubricant is a CNT nano-lubricant with 2 vol% as the volume fraction of the nanoparticles. The base oil is palm oil. CNTs have constraints (winding and conglobations) that hinder tribological performance. The constraints can be solved by applying surfactants. Therefore, the surfactant APE-10 is added to CNT nano-lubricant.

Table 1. Material properties of CFRP

CFRP	
Density (g/cm^3)	1.575
Hardness (HRB)	68-72
Transverse elasticity modulus (GPa)	9.65
Interface	
Interfacial bonding strength (MPa)	30
Interfacial equivalent modulus k_b (GPa/m)	115

continued on following page

Table 1. Continued

Carbon fiber (70%.vol)	
Density (g/cm^3)	1.8
Young modulus (GPa)	230
Transverse elasticity modulus (GPa)	19.6
Axial tensile strength (GPa)	3.45
Shear strength (GPa)	0.38
Poisson ratio	0.3
Fracture toughness (J/m^2)	2
Elongation at break (%)	1.5
Monofilament diameter (μm)	7.0
Vinyl resin (30%.vol)	
Density (g/cm^3)	1.05
Young modulus (GPa)	3.4
Axial tensile strength (MPa)	85
Shear strength GPa)	1.02
Poisson ratio	0.4
Fracture toughness (J/m^2)	500
Elongation at break (%)	5.5
Heat deflection temperature (HDT) (°C)	101

Table 2. Material properties of polycrystalline diamond grain

Properties	**Value**
Density (×10^3 kg/m^3)	4.12
Hardness (GPa)	50.00
Young modulus (GPa)	776.00
Poisson ratio	0.07
Transverse fracture strength (GPa)	1.20

5.3 PREDICTIVE FORCE MODEL

5.3.1 Geometric Analysis

The impact of the shape and size of the grinding zone on the contact state and grinding behavior of the grain and carbon fiber is critical. Thus, the grinding zone is geometrically analyzed using different grinding parameters. The time *t* for a grain to pass through the whole grinding area is

$$t = \frac{\sqrt{a_p \cdot d_s}}{v_s}. \tag{5.69}$$

The rotation time of a single grain on the grinding wheel is

$$t_0 = \frac{\pi d_s}{v_s}. \tag{5.70}$$

The mathematical expression of the grinding path of grain is

$$\begin{cases} y = v_w t + \frac{d_s}{2} sin\left(\frac{v_s}{d_s/2} t\right) \\ z = \frac{d_s}{2}\left[1 - \cos\left(\frac{v_s}{d_s/2} t\right)\right] \end{cases}. \tag{5.71}$$

The undeformed chip thickness a_g, is

$$a_g(\psi)\begin{cases} \pi \cdot D \cdot \frac{v_\omega}{v_s} \cdot \sqrt{2 \cdot (1 - cos\psi)}, 0 \leq \psi \leq \psi_1 \\ \frac{a_p - \frac{D}{2}(1 - cos\psi)}{\cos\psi}, \psi_1 \leq \psi \leq \psi_2 \end{cases}. \tag{5.72}$$

Figure 8 shows the input variables of a_p, v_s, and v_w, with the grinding zone's shape and undeformed chip thickness taken as the output results. Two groups of grinding parameters were selected to obtain the other effects. The influence of depth of cut on the geometry of the grinding area is obvious. The grinding area becomes longer and thicker with the increase in depth of cut, and the change trend of undeformed chip thickness is similar before reaching the maximum undeformed chip thickness. With the increase in the peripheral speed of the grinding wheel, the grinding area becomes shorter and thinner, and the undeformed chip becomes thick. Moreover, with the uniform increase in the peripheral speed of the grinding wheel, the change rates of the length and thickness of the grinding area also decrease; that is, the influence of the grinding wheel's peripheral speed on the grinding area's shape and size is weakened. Before reaching the maximum thickness of the undeformed chip, the peripheral speed of grinding wheel increases, and the curve slope of the undeformed chip's thickness changes as the grinding process is reduced. However, after reaching the maximum undeformed chip thickness, the slope remains the same. The length and thickness of the grinding area increase with the increase in workpiece feed speed, but the shape of the grinding area does not change. Before the maximum chip thickness is reached, the curve slope of the undeformed chip's thickness decreases with the increase in workpiece feed speed. After the maximum chip thickness is reached, the slope does not change.

Figure 8. Effect of grinding parameters on the grinding zone

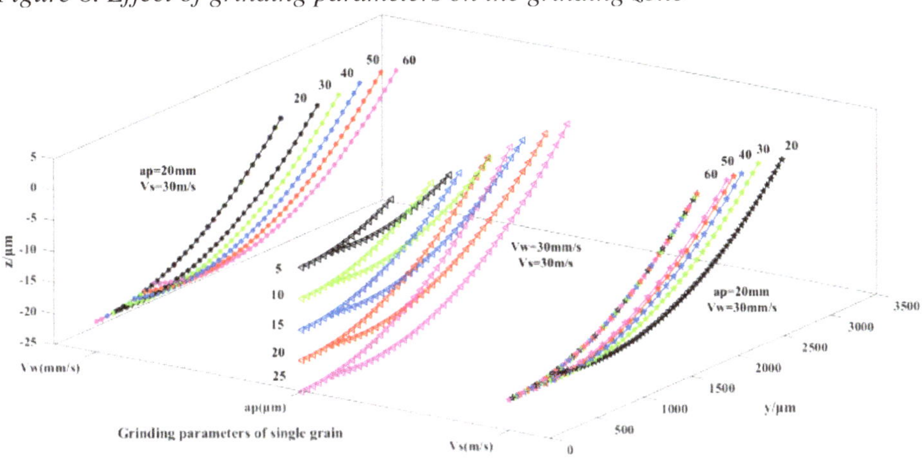

5.3.2 Generating Random Fibers Distributions

The minimum value denoted by $l_{min} = 0$ and the maximum value denoted by $l_{max} = d_f$ of inter-fiber distances are controlled to avoid fiber interference and limit the inter-fiber distances caused by the large fiber volume fraction of the CFRP. The generation principle and process of fibers' random distribution model are shown in Figure 9. First, a random coordinate point (x_1, y_1) is produced in a square area with a side length of l_r. Then, a second random point representing the second fiber is created. The second random coordinate point is limited by the distance l_1 and the orientation angle θ_{r1} from the first point (x_1, y_1), as shown in Figure 9 (b). Here, l_1 is assigned as a random value ranging between l_{min} and l_{max}, while θ_{r1} is a random angle ranging between 0 and 2π. The second step is repeated to generate a new fiber around the first fiber until no more fibers can be inserted, as shown in Figure 9 (d). Finally, the same algorithm shifts to the second fiber and repeats the second and third steps. In using this method, the above algorithm is carried out for each fiber until the square area is filled.

Figure 9. Schematic of the principle and process of generating the fibers' random distribution model

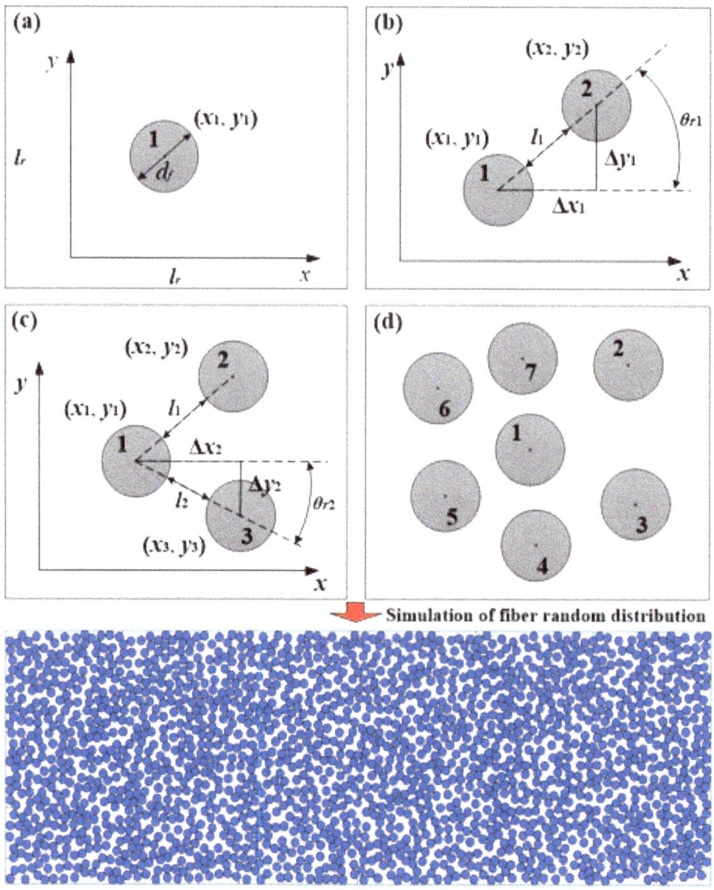

5.3.3 Frictional Force

The friction coefficient must be determined to be able to accurately determine the friction force in the single-grain grinding process. The UMT-3 friction (CETR Company) is used to solve the friction coefficient under different lubrication conditions, as shown in Figure 10. The friction test parameters are shown in Table 3.

The reciprocating motion and the point-to-face contact of the ball–disk model are both selected in the experiment. The collected data of vertical load and horizontal friction force are fed back and recorded in real time by a computer through the top dual channel force sensor. The ratio of friction force to load represents the friction coefficient of the CFRPs, then the relationship curve of the friction coefficient with time is obtained (Figure 10). According to the friction test results shown in Figure 11, the friction coefficients of the dry, MQL, and CNT nano-lubricants are 0.303, 0.156 and 0.141, respectively. Two reasons may explain why the CNT nano-lubricant has obtained the lowest friction coefficient. On the one hand, CNTs remain tubular under large load due to their high-strength and high-hardness characteristics, which may have played important roles, such as reducing the friction coefficient for the grinding interface and improving lubrication performance, similar to those of a bearing. On the other hand, the addition of CNTs increases the viscosity of the lubricant, consequently increasing the thickness and stability of the lubricant film at the grinding interface.

Table 3. Friction test parameters

Test conditions	Value
Test temperature (°C)	22±5
Test duration (s)	1200
Load (N)	10
Frequency (Hz)	50

Figure 10. UMT-3 multifunctional friction testing results

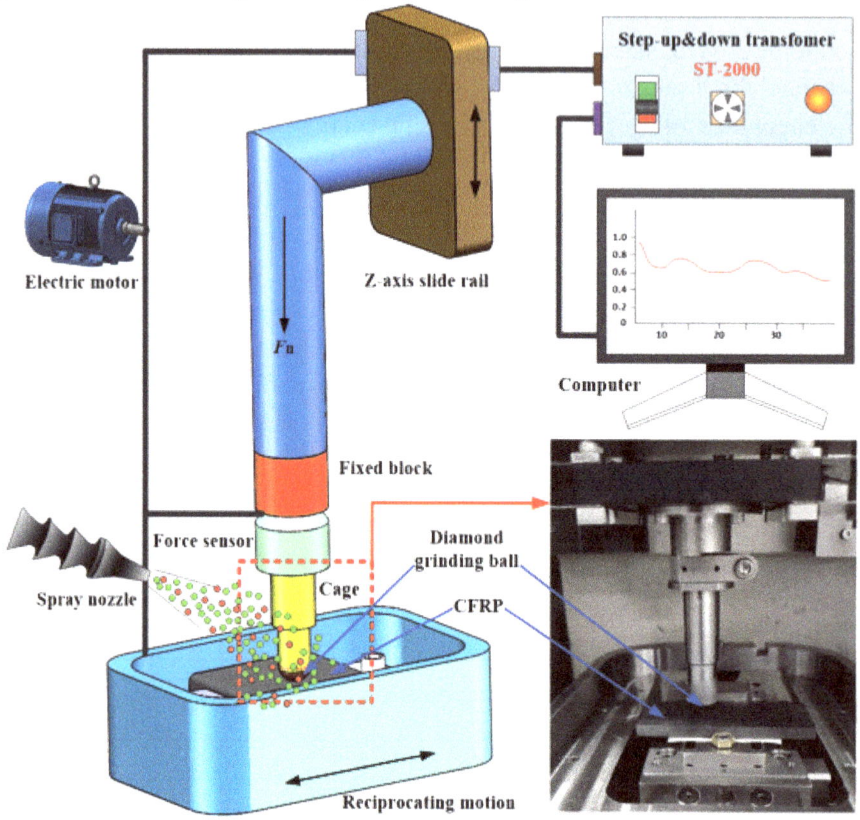

Figure 11. Friction coefficient curves and average friction coefficient under different lubrication conditions

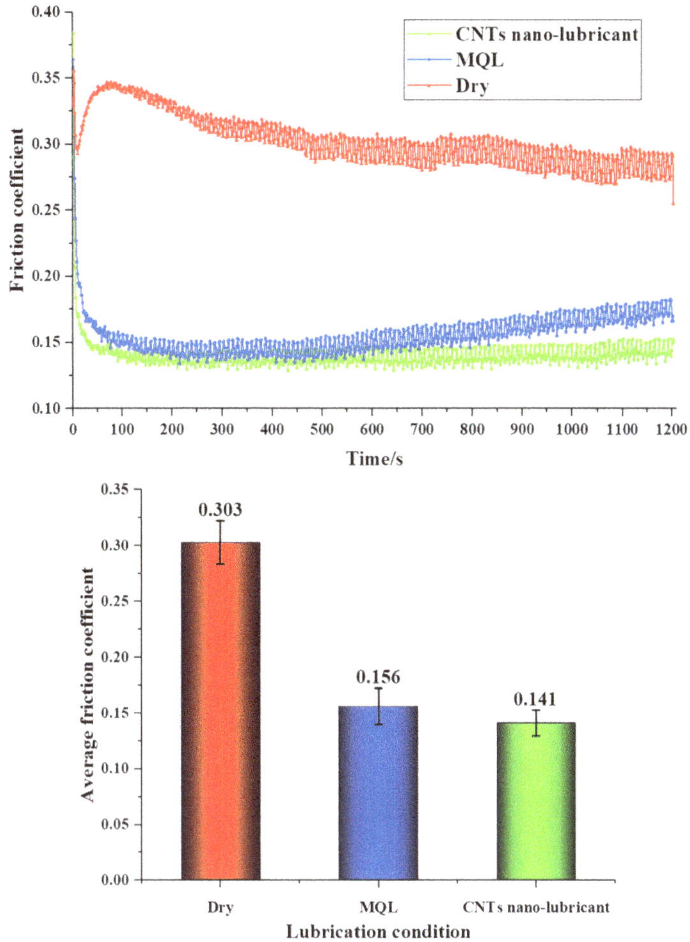

5.3.4 Model Implementation

As shown in Figure 12, when the grinding parameters (a_p, v_s, and v_w) are known, the position of the grain on the grinding arc length can be determined. Given the mechanical properties of CFRP and the geometric parameters of the grain, the carbon fiber arrangement can be implemented. Then, the number of carbon fibers cut by the grain and the contact state and removal behavior of the carbon fibers can also be determined. The total force of the grain on the carbon fiber is calculated on the basis of the different material removal mechanisms and contact stress distributions. Furthermore, the friction coefficient μ under different lubrication conditions is introduced in the calculation of the grinding force of the single grain. Finally, the single is decomposed in the tangential and normal directions, and the grinding force and normal force are obtained.

Figure 12. Flowchart of the grinding force modeling of the single grain

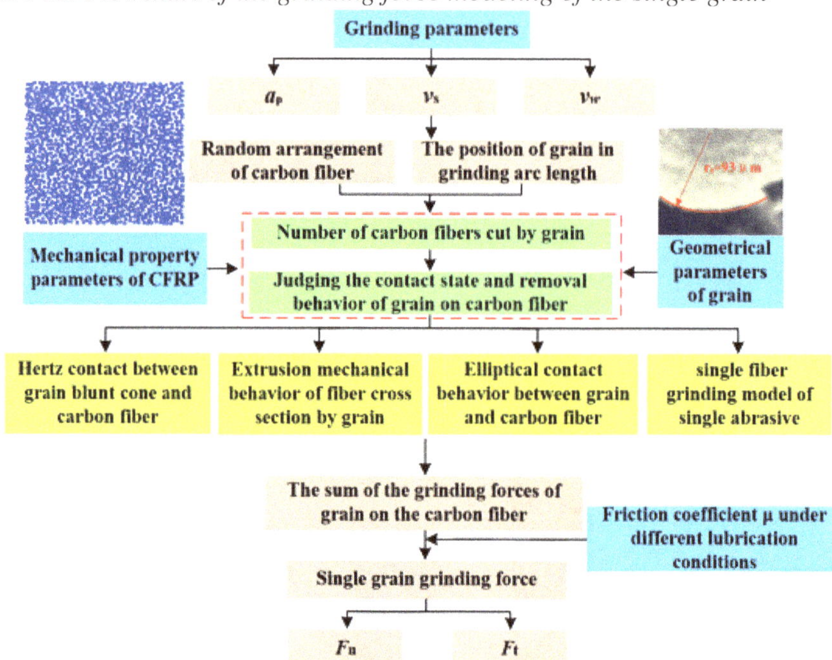

5.4 EXPERIMENTS

5.4.1 Setups

The grinding experimental platform of the single grain is shown in Figure 13, where the *x* and *y* axes are plotted by the Shinano stepping motor (resolution = 1 μm; model: Y110TA100), while the *z* axis is plotted by the Panasonic servo brake motor (resolution = 0.1 μm; model: DZY90TA50-ZS). A 5.5 KW high-speed grinding motorized spindle (rated speed = 18000 rad/min; model: CHM6068B-18-5.5SCF) is also used. The three-phase dynamometer, which is clamped on the platform, is used to collect force signals during the grinding process. A diamond abrasive grain is affixed to the insertion of the grinding wheel disc by means of brazing. The grain size is approximately 550 μm. The grinding force of each sampled group is measured in five sets, and the average value is taken. The MQL device adopted in the experiment is a KINS KS-2106 MQL system. The system uses a pulse generator to adjust the CNT nano-lubricant and the compressed gas in the MQL system, thus controlling their flow. A proper gas–liquid ratio can be selected by the system for processing. The compressed gas and the CNT nano-lubricant are mixed and atomized, then they are ejected from the nozzle into the grinding area with the assistance of high-pressure gas. The MQL system also includes adjustable flow pumps, compressed air flow control devices, and spray-assisted air atomizing nozzles. For the data acquisition, the YDM-III99 force measuring instrument is used to measure the grinding force.

Figure 13. Schematic of the experimental setup

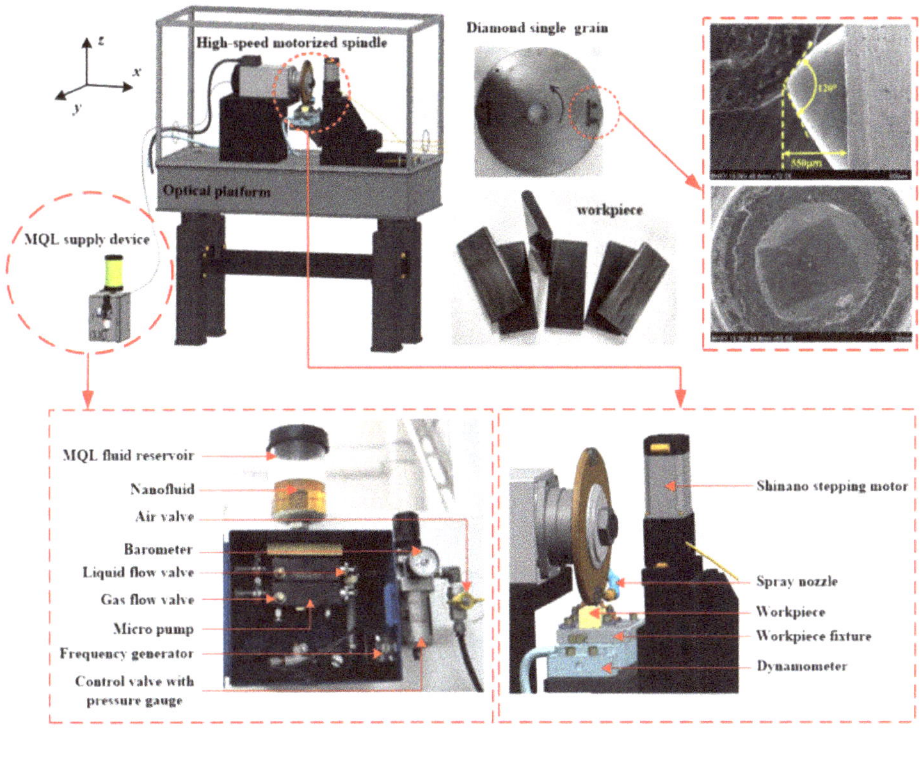

5.4.2 Scheme

The experiment is carried out in three lubri-cooling modes, namely, dry, MQL, and CNT nano-lubricant MQL modes. The experimental scheme is listed in Table 4.

Table 4. Experimental scheme

Lubricating conditions	**Lubricant**
Dry grinding	n/a
MQL	Pure palm oil
CNT nano-lubricant MQL	Pure palm oil based CNT nano-lubricant (particle size 50 nm)
Grinding process parameters	Value
Peripheral speed of grinding wheel v_s (m/s)	30
Depth of cut a_p (μm)	20
Feed speed v_w (mm/s)	30
Grinding wheel diameter (mm)	200
Gas pressure (bar)	6.0
Gas–liquid ratio	3.0
MQL flow rate (mL/h)	60
MQL nozzle distance (mm)	12
MQL nozzle angle (°)	15

5.5 MODEL ASSESSMENT

5.5.1 Contrastive Analysis of Prediction and Experiments

Figure 14 shows the predicted grinding force of the single grain under different lubrication conditions. The predicted grinding force of the single grain is compared with those in the experiments. The predicted grinding force is highly correlated with the experimental force (Table 5). Then, the estimation error between the predictions and the experimental of the single-grain grinding force are calculated and analyzed. The estimation errors are 14.76% and 16.14% for the normal force and the tangential force, respectively. The obtained errors are smaller than the estimation errors of previously reported force models (Zhang et al., 2001; Nayak et al., 2005; Voss et al., 2019). However, this difference is understandable because the experimental measurements are affected by a number of factors during CFRP specimen fabrication. For example, the fibers do not line up perfectly in the desired unidirectional arrangement or are not evenly distributed throughout the specimen. In addition, the equivalent homogeneous material is used to deal with the surrounding material of the cut fiber in the mechanical modeling, which can be the main reason for the prediction error. Another reason is that the ideal properties of the materials considered in the mechanical modeling have no inherent defects, such as voids in actual materials, during the processing. The calculation results compared with those of other force models indicate that the proposed prediction model can attain good prediction results.

Compared with dry grinding, the MQL and CNT nano-lubricant MQL techniques can effectively reduce the grinding forces, with the maximum F_n values reduced by 14.81% and 20.07% and the maximum F_t values reduced by 27.03% and 26.81%, respectively. With the grinding of a single grain for the CFRP, if the carbon fiber has a strong constraint parallel to the grinding direction, then a local contact between the grain and carbon fiber will be formed. As a result, the local contact stress in the elliptical contact area of the carbon fiber will reach the ultimate fracture strength and then be cut off. When the support constraint parallel to the grinding direction is weak, the carbon fiber to be cut will bend, resulting in the pile extrusion of multiple fibers, further leading to the fracturing of the carbon fiber under the joint action of local contact stress and bending stress. Finally, the distribution length of the carbon fiber's bending deformation area in the grinding feed direction will increase. The grinding deformation area and the grinding resistance will also both increase, thus easily leading to the cracking of the interface between the carbon fiber and the resin matrix.

Figure 14. Prediction and experimental results of grinding force under different lubrication conditions

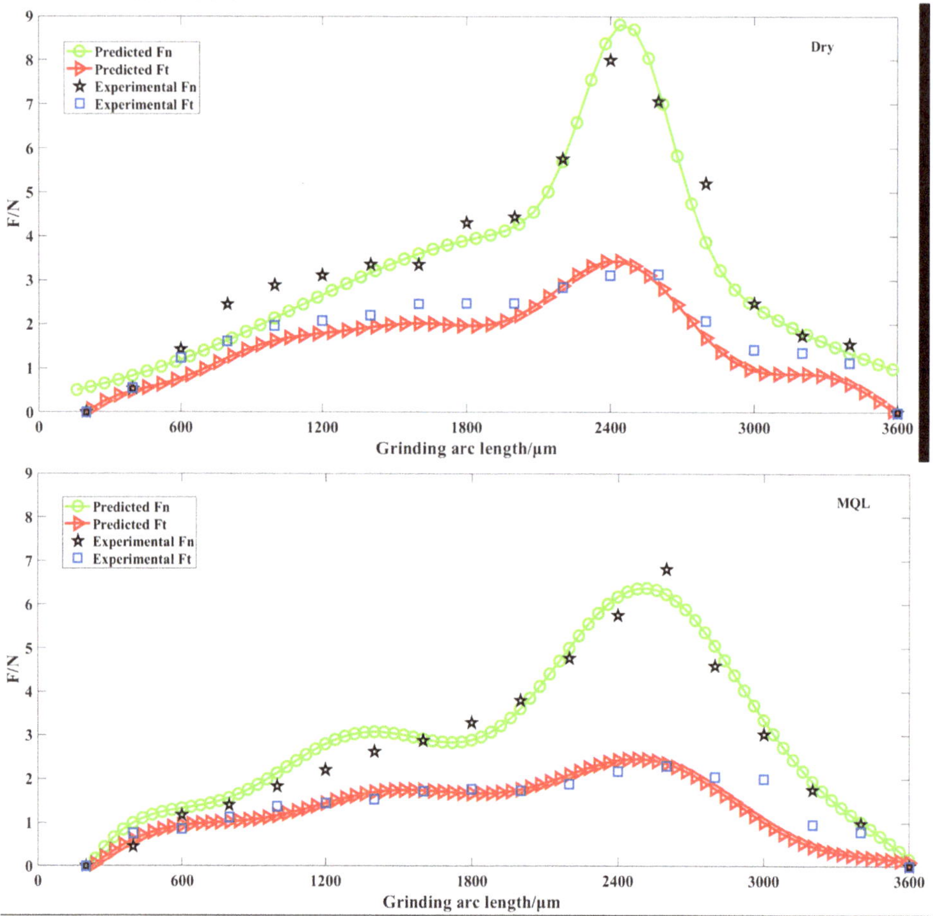

Table 5. Correlation coefficient and estimation error between predicted and experimental results

Lubrication	Force	Estimation error (%)
Dry	F_n	12.32
	F_t	17.05
MQL	F_n	14.60
	F_t	15.44
CNT nano-lubricant	F_n	17.36
	F_t	15.92

5.5.2 Reduction Mechanism Analysis of Single Grain Grinding Force

The schematic diagram of the reduction mechanism of the single grain's grinding force is shown in Figure 15. The figure also shows the weakened constraint strength of the softened resin matrix on the carbon fiber resulting from the increase in grinding temperature. Under the condition of weak constraint, many carbon fibers appear to be in a crowding effect, which is equivalent to the grinding of multiple carbon fibers by the grain. Furthermore, the fracture mode of the carbon fibers under this condition will appear to have a drawing fracture. In addition, when the material is compressed and deformed, it needs more energy to absorb. Therefore, compared with the MQL and CNT nano-lubricant MQL techniques, when the resin matrix softened by dry grinding and the carbon fiber in a weak constraint condition are cut by the grain, the material needs more energy and greater grinding force.

Figure 15. Schematic of the grinding force reduction mechanism

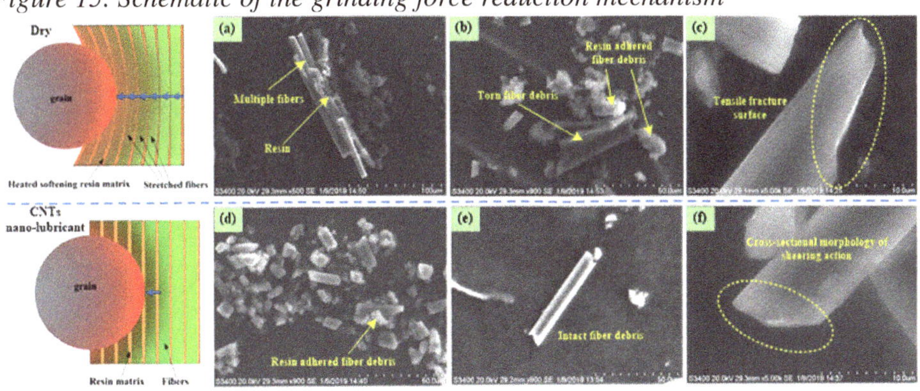

5.6 THE INFLUENCE OF LUBRICATIONS ON THE MICROSCOPIC REMOVAL

5.6.1 Micromorphology of Chips

As shown in Figure 16, the chip morphologies of CFRP under different lubrication conditions are presented. The chips generated under dry cutting consist of multi-fiber bulk debris (Figure 16a) and a significant amount of fiber debris adhered by resin (Figure 16b). In contrast, while the chips produced under MQL and NMQL conditions also exhibit some resin adhesion, the extent is considerably less compared to dry cutting, and no multi-fiber bulk debris is observed (Figure 16e and i). The fiber debris under MQL and NMQL conditions predominantly appears as intact rods (Figure 16f and j), whereas dry cutting results in fractured fibers (Figure 16b). Shear fracture surfaces are generated under all three lubrication conditions (Figure 16c, g, k, and l). Additionally, tensile fracture and tensile-shear mixed fracture modes are identified in both dry and MQL cutting (Figure 16d and h).

Figure 16. Morphology of grinding debris under different lubrication conditions.

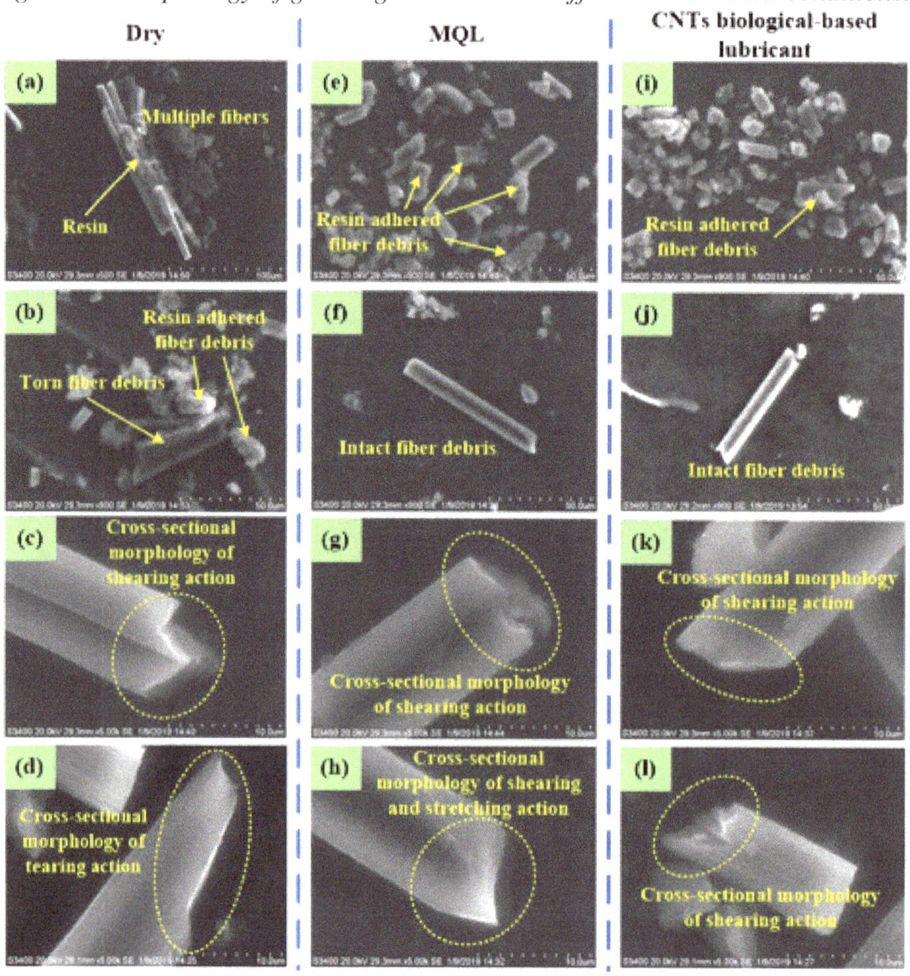

5.6.2 Micromorphology of Grooves

Figure 17 presents the three-dimensional topography of the scratch from a single-grain scratching experiment. A clear distinction in the groove morphology under the three lubrication conditions is evident. Under the dry condition, the scratch cross-section is heavily coated with resin matrix and grinding debris. This phenomenon is attributed to elevated grinding temperatures in dry cutting, which soften the resin. The softened resin, adhering to fiber debris, is subsequently smeared and adhered onto the scratched surface under the extrusion of the abrasive grain. Furthermore, the groove edges under the dry condition are irregular, a consequence of carbon fiber pull-out. In contrast, the groove profiles under both MQL and NMQL conditions are distinctly visible without any coating phenomenon. Notably, the groove edges under NMQL are more regular than those under MQL.

In addition, Figure 17 shows the SEM micrographs of the grooves generated by a single abrasive grain under different lubrication conditions. Under the dry condition, extensive resin coating is observed on both sides of the groove (Figure 17a, b, and c). This occurs because the resin, being more thermally sensitive, softens upon heating and adheres to the groove sides or is transferred and smeared onto the machined surface after adhering to the abrasive grain. Fiber pull-out is also evident at the groove edges in dry cutting (Figure 17a). This is caused by the temperature rise reducing the constraining capability of the resin matrix on the fibers. Weakened interfacial bonding diminishes or eliminates the support from the surrounding matrix and adjacent fibers to the carbon fiber being cut, ultimately leading to fiber pull-out. When observed at the same magnification, the grooves under both MQL and NMQL conditions show no signs of resin coating or fiber pull-out. However, the groove edges under MQL exhibit chipping phenomena (Figure 17e), whereas the groove edges and fiber fracture surfaces under NMQL appear more regular and neat (Figure 17d, g, and h). Additionally, although no resin coating occurs under MQL, fiber fragmentation is observed (Figure 17f). In contrast, the fiber shear fracture surfaces under NMQL are relatively more complete and orderly (Figure 17i). The reasons for the aforementioned phenomena can be attributed to the following two aspects. First, the CNTs nanoparticles in the NMQL enhance the convective heat transfer capability of the cutting fluid, removing more heat compared to MQL and thereby reducing the softening effect on the resin. This helps maintain stronger constraining forces from the resin matrix and surrounding material on the carbon fibers being cut, mitigating or preventing chipping and fiber pull-out. Second, the nanoparticles may become embedded in the micro-cracks between the carbon fibers and the resin matrix, which similarly enhances the constraining effect on the fibers being removed. This facilitates more effective shearing of the carbon fibers by the abrasive grain, consequently preventing fiber pull-out and edge chipping.

Figure 17. Micromorphology comparison of grinding groove under different lubrication conditions

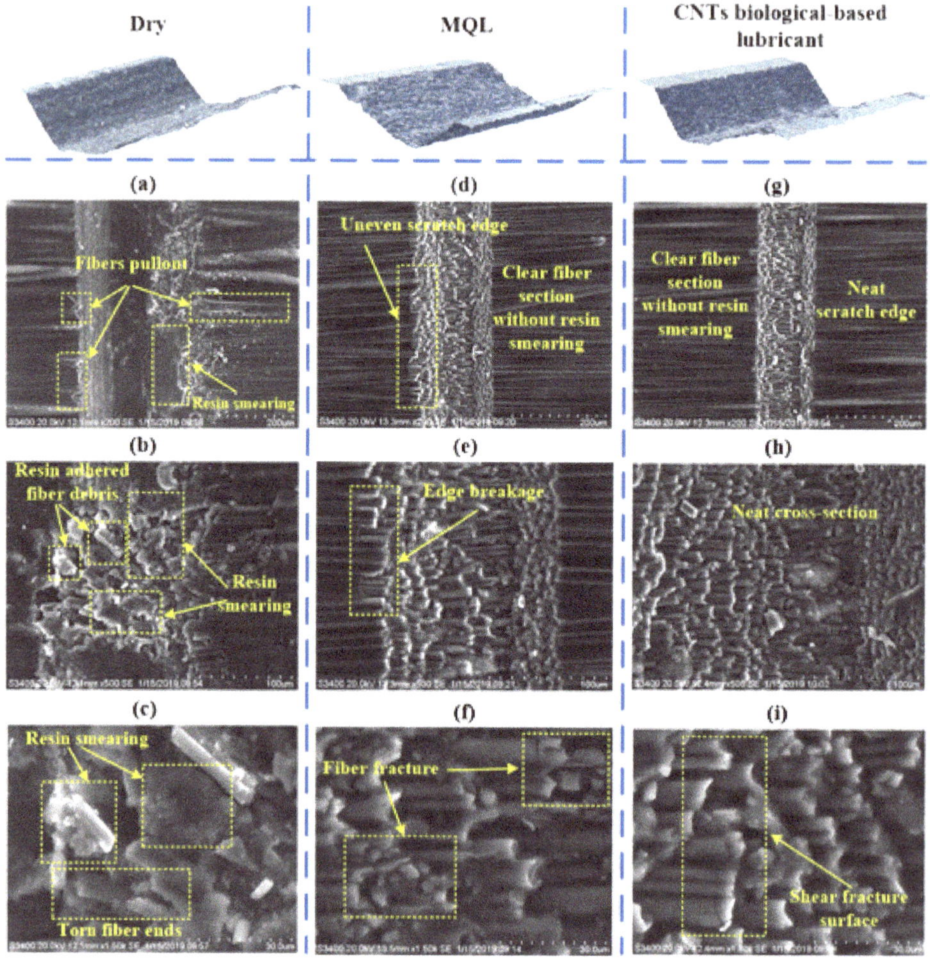

5.7 SUMMARY

In view of solving the problems of complex contact state and mechanical behavior between the grain and the CFRP workpiece in the grinding process, a single-grain grinding CFRP is used to reveal its material removal mechanism and mechanical behavior. The contact and grinding models of the grain and carbon fiber are established on the basis of the uniform and random arrangements of the carbon fiber in the CFRPs and the various contact states of the grain and carbon fiber in different stages at the whole grinding arc length. Considering the different friction coefficients between the grain and the CFRP under different lubrication conditions, the mechanical model of the single-grain grinding CFRP is established and verified experimentally, and the grinding force reduction mechanism of the grinding CFRP with CNT nanolubricants MQL is analyzed. The specific conclusions are as follows:

(1) MQL and CNT nano-lubricant MQL can effectively reduce the grinding forces. Compared with dry grinding, their corresponding maximum F_n values can be reduced by 14.81% and 20.07%, while their maximum F_t values can be reduced by 27.03% and 26.81%, respectively.
(2) The friction coefficient is introduced into the grinding force prediction model of the single grain to characterize the different lubrication conditions. The addition of CNT nanoparticles can effectively improve the tribological properties of the interface between the grain and the CFRP workpiece. The friction coefficient under CNT nano-lubricant MQL is the smallest at $\mu = 0.141$.
(3) The predicted and experimental grinding forces, which are dependent on the lubrication condition and the contact state and grinding behavior of the grain/fiber in this study, are in good agreement with each other based on the comparative results. The average value of the estimation error under the three lubrication conditions for F_n is 14.76% and that for F_t is 16.14%.
(4) Based on the comparative analysis of the SEM surface morphology of single-grain scratch grooves and the microscopic morphology of carbon fiber chips, the influence of different lubrication conditions on the mesoscopic material removal of carbon fiber composites was investigated. The results indicate that both MQL and NMQL promote a more favorable removal mechanism for carbon fibers compared to dry cutting, characterized by effective material removal under enhanced mechanical support or strong constraints.

REFERENCES

Biot, M. A. (1937). Bending of an infinite beam on an elastic foundation. *Journal of Applied Mechanics*, *4*(1), A1–A7. DOI: 10.1115/1.4008739

Cao, X., Lin, B., & Zhang, X. (2015). Investigations on grinding process of woven ceramic matrix composite based on reinforced fiber orientations. *Composites. Part B, Engineering*, *71*, 184–192. DOI: 10.1016/j.compositesb.2014.11.029

Carlson, B. C. (1977). Elliptic integrals of the first kind. *SIAM Journal on Mathematical Analysis*, *8*(2), 231–242. DOI: 10.1137/0508016

Carlson, B. C. (1987). A table of elliptic integrals of the second kind. *Mathematics of Computation*, *49*(180), 595–606. DOI: 10.1090/S0025-5718-1987-0906192-1

Cui, X., & Li, C. H. (2025). *Magnetic Traction Nanolubricant Grinding: Force Model and Mechanisms*. Nova Science Publishers. 255-291. DOI: 10.52305/PVGY4472

Gao, T., Li, C., Yang, M., Zhang, Y., Jia, D., Ding, W., Debnath, S., Yu, T., Said, Z., & Wang, J. (2021). Mechanics analysis and predictive force models for the single-diamond grain grinding of carbon fiber reinforced polymers using CNT nano-lubricant. *Journal of Materials Processing Technology*, *290*, 116976. DOI: 10.1016/j.jmatprotec.2020.116976

Geier, N., Davim, J. P., & Szalay, T. (2019). Advanced cutting tools and technologies for drilling carbon fibre reinforced polymer (CFRP) composites: A review. *Composites. Part A, Applied Science and Manufacturing*, *125*, 105552. DOI: 10.1016/j.compositesa.2019.105552

Hosokawa, A., Hirose, N., Ueda, T., & Furumoto, T. (2014). High-quality machining of CFRP with high helix end mill. *CIRP Annals*, *63*(1), 89–92. DOI: 10.1016/j.cirp.2014.03.084

Jia, Z., Fu, R., Niu, B., Qian, B., Bai, Y., & Wang, F. (2016). Novel drill structure for damage reduction in drilling CFRP composites. *International Journal of Machine Tools & Manufacture*, *110*, 55–65. DOI: 10.1016/j.ijmachtools.2016.08.006

Jia, Z., Fu, R., Wang, F., Qian, B., & He, C. (2018). Temperature effects in end milling carbon fiber reinforced polymer composites. *Polymer Composites*, *39*(2), 437–447. DOI: 10.1002/pc.23954

Johnson, K. L. (1987). *Contact mechanics*. Cambridge university press.

Kong, X., Li, C., Li, Z., Yang, M., Cui, X., Liu, M., Li, B., Zhang, Y., Ma, X., & Li, C. (2025). Biological bone and replacement materials in grinding: Force model and processing capability. *Intelligent and Sustainable Manufacturing*, *2*(1), 10003. DOI: 10.70322/ism.2025.10003

Li, Y., Ge, X., Wang, H., Hu, Y., Ning, F., Cong, W., & Ren, C. (2019). Study of material removal mechanisms in grinding of C/SiC composites via single-abrasive scratch tests. *Ceramics International*, *45*(4), 4729–4738. DOI: 10.1016/j.ceramint.2018.11.165

Luna, G. G., Axinte, D., & Novovic, D. (2020). Influence of grit geometry and fibre orientation on the abrasive material removal mechanisms of SiC/SiC Ceramic Matrix Composites (CMCs). *International Journal of Machine Tools & Manufacture*, *157*, 103580. DOI: 10.1016/j.ijmachtools.2020.103580

Nayak, D., Bhatnagar, N., & Mahajan, P. (2005). Machining studies of UD-FRP composites part 2: Finite element analysis. *Machining Science and Technology*, *9*(4), 503–528. DOI: 10.1080/10910340500398183

Ning, F., Cong, W., Wang, H., Hu, Y., Hu, Z., & Pei, Z. (2017). Surface grinding of CFRP composites with rotary ultrasonic machining: A mechanistic model on cutting force in the feed direction. *International Journal of Advanced Manufacturing Technology*, *92*(1), 1217–1229. DOI: 10.1007/s00170-017-0149-9

Ning, Z., Liu, R., Elhajjar, R. F., & Wang, F. (2017). Micro-modeling of thermal properties in carbon fibers reinforced polymer composites with fiber breaks or delamination. *Composites. Part B, Engineering*, *114*, 247–255. DOI: 10.1016/j.compositesb.2017.01.036

Qiu, X., Li, P., Li, C., Niu, Q., Chen, A., Ouyang, P., & Ko, T. J. (2018). Study on chisel edge drilling behavior and step drill structure on delamination in drilling CFRP. *Composite Structures*, *203*, 404–413. DOI: 10.1016/j.compstruct.2018.07.007

Qu, S., Gong, Y., Yang, Y., Wang, W., Liang, C., & Han, B. (2020). An investigation of carbon nanofluid minimum quantity lubrication for grinding unidirectional carbon fibre-reinforced ceramic matrix composites. *Journal of Cleaner Production*, *249*, 119353. DOI: 10.1016/j.jclepro.2019.119353

Reifsnider, K. L., & Case, S. W. (2002). *Damage tolerance and durability of material systems*.

Voss, R., Seeholzer, L., Kuster, F., & Wegener, K. (2019). Analytical force model for orthogonal machining of unidirectional carbon fibre reinforced polymers (CFRP) as a function of the fibre orientation. *Journal of Materials Processing Technology*, *263*, 440–469. DOI: 10.1016/j.jmatprotec.2018.08.001

Zhang, L., Wang, S., Li, Z., Qiao, W., Wang, Y., & Wang, T. (2019). Influence factors on grinding force in surface grinding of unidirectional C/SiC composites. *Applied Composite Materials*, *26*(3), 1073–1085. DOI: 10.1007/s10443-019-09767-5

Zhang, L. C., Zhang, H. J., & Wang, X. M. (2001). A force prediction model for cutting unidirectional fibre–reinforced plastics. *Machining Science and Technology*, *5*(3), 293–305. DOI: 10.1081/MST-100108616

Zhang, X., Yu, T., & Zhao, J. (2020). An analytical approach on stochastic model for cutting force prediction in milling ceramic matrix composites. *International Journal of Mechanical Sciences*, *168*, 105314. DOI: 10.1016/j.ijmecsci.2019.105314

Zhang, Y., & Li, C. (2025). *Hybrid-Energy Sustainable Machining: Mechanism and Processability*. Springer Nature, 596-609.

APPENDIX

Table 6. Nomenclature

CFRP	—	Carbon fiber reinforced polymer
MQL	—	Minimum quantity lubrication
NMQL	—	Nanofluid minimum quantity lubrication
$P_1(r)$	N	Normal pressure distribution on blunt cone surface in contact zone
Ψ	°	Swing angle of a single grain
a_1	μm	Radius of circular contact zone of blunt conical surface
r_g	μm	Arc radius of blunt cone at grain tip
h	μm	Width of grinding zone
θ	°	Half-angle of blunt cone
E^*	GPa	Equivalent elasticity modulus of CFRP
E_f	GPa	Elasticity modulus of carbon fibers
E_a	GPa	Biaxial modulus of elasticity of carbon fibers
E_g	GPa	Elastic modulus of grain
v_f	—	Poisson's ratio of carbon fibers
v_g	—	Poisson's ratio of grain
E_l	GPa	Longitudinal modulus of fibers
E_f	GPa	Transverse modulus of fibers
G_f	GPa	Shear modulus of fibers
α	°	Angle between fiber direction and grinding direction
f_0	μm	Initial maximum deformation of self-deformed fibers
σ_τ	GPa	Shear ultimate strength of fibers
d_f	μm	Diameter of carbon fiber
σ_c	GPa	Ultimate compressive strength of fibers
r_f	μm	Radius of carbon fiber
l	μm	Length of deformed fibers below the contact point between fiber and grain
G_{unit}	GPa	In-plane shear modulus of CFRP with fiber elements in compression-shear zone
V_f	—	Volume ratio of carbon fiber
k_{II}	—	Centrifugal rate in elliptical contact zone of grain and carbon fiber
a_{II}	μm	Long axis length in elliptical contact zone of grain and carbon fiber
b_{II}	μm	Short axis length in elliptical contact zone of grain and carbon fiber
P	N	Total load of grain in elliptic region
P_{max}	N	Maximum load of abrasive particles acting on elliptic domain
w	μm	Deflection of fibers in grinding direction

continued on following page

Table 6. Continued

k_m	—	First parameter of surrounding material support
C	—	Coefficient of action mode
a_p	μm	Depth of cut
a_g	μm	Undeformed chip thickness
a_{gmax}	μm	Maximum undeformed chip thickness
d_s	mm	Diameter of grinding wheel
v_s	m/s	Peripheral speed of grinding wheel
v_w	mm/s	Feed speed of workpiece
t	s	Time of abrasive grinding through the whole grinding arc
t_0	s	Time for grinding wheel to rotate round
F_{1n}	N	Normal force acting on blunt circumference
Q	N	Stress of carbon fiber's micro-element
M	kN·m	Bending moment of carbon fiber's micro-element
I_f	m^4	Interfacial moment of inertia of carbon fiber
k_i	GPa	Equivalent modulus of interface
μ	—	Friction coefficient
F_{shear}	N	Shear force of grain on cut carbon fibers
$F_{bending}$	N	Bending force of grain on cut carbon fibers
σ_{shear}	GPa	Shear ultimate strength of carbon fibers
$\sigma_{bending}$	GPa	Ultimate compressive strength of carbon fibers
η	μm	Distance from any point on carbon fibers to the neutral axis
σ	GPa	Tensile strength of carbon fiber
F_f	N	Tensile fracture force of carbon fiber
f	N	Friction between grain and carbon fibers

Chapter 6
Surface Quality and Morphology Evaluation of Nanolubricants Minimum Quantity Lubrication Transverse Grinding

ABSTRACT

This chapter investigates the grinding performance of CFRP composites using NMQL, with a focus on the characterization and analysis of grinding surface integrity and machining-induced damage. Initially, a comparative analysis of surface roughness parameters was conducted. Subsequently, the fractal dimension was employed to quantitatively characterize the overall fractal features across different dimensions. Furthermore, multifractal spectrum with its width and difference was utilized to analyze the local topographic characteristics. Then, SEM was used for qualitative analysis of the surface micro-morphology, detailed machining defects, and damage under different conditions. Moreover, the energy distribution ratio of wavelet decomposition were applied to examine the high-frequency texture variations in both surface roughness signals and SEM images across different directions, enabling a quantitative characterization of grinding defects. Finally, the surface of grinding wheel under different conditions is characterized, and the diffusion state of lubricating oil film is analyzed.

DOI: 10.4018/979-8-3373-9013-0.ch006

6.1 INTRODUCTION

Owing to the significant disparity in mechanical properties between the resin matrix and fibers, carbon fiber-reinforced polymer (CFRP) composites exhibit poor machinability. This often results in various machining-induced defects in processed CFRP workpieces, including fiber fracture and pull-out, matrix cracking, resin smearing, fiber-matrix interfacial debonding, and delamination (Davim et al., 2007; Wang and Zhang, 2003). With technological advancements, the aerospace industry has seen a growing demand for ultra-smooth surfaces with high machining-induced surface integrity (Li et al., 2020; Liao et al., 2019).

Current research efforts are predominantly focused on mitigating machining damage in difficult-to-machine CFRP composites. The resin matrix in these composites is temperature-sensitive, exhibiting a decline in strength and performance with increasing temperature. The low thermal conductivity of the resin combined with the high abrasion resistance of carbon fibers leads to elevated cutting forces and rapid tool wear. Fu et al. (2018) demonstrated that the temperature field in CFRP drilling assumes an elliptical distribution, with the eccentricity determined by the drilling depth and fiber layup sequence. Yashiro et al. (2013) found that high-speed cutting is suitable for CFRP milling. Wang et al. (2018) confirmed that rotary ultrasonic machining with an end-face tool can achieve reduced surface roughness and smaller chip widths. Sheikh-Ahmad et al. (2012) identified low feed rates and high rotational speeds as the optimal cutting conditions. Voss et al. (2016) compared the integrity of holes produced by orbital drilling using diamond-coated tools with that from conventional drilling. Su (2019) proposed a fiber deformation theory and observed that the affected zone decreases with increasing cutting speed. In contrast to conventional grinding, Geng et al. (2019) reported that rotary ultrasonic helical machining reduces the internal hole roughness Ra by 53.2%. Shi et al. (2019) discovered that the arithmetic mean height exhibits a banded distribution along the feed direction.

Traditional flood lubrication grinding consumes substantial amounts of cutting fluid, which not only escalates manufacturing costs but also poses potential risks to human health and the environment (Debnath et al., 2014). Under conditions of intense impact, high-temperature evaporation, and high-speed wheel rotation, cutting fluids generate significant oil mist and PM2.5 suspended particulates. Moreover, costs associated with cutting fluids account for 18%–21% of the total manufacturing expenses, representing 3–5 times the cost of tooling. It is noteworthy that conventional grinding of CFRPs still faces technical bottlenecks (Soo et al., 2012). CFRPs are prone to moisture absorption and swelling in hygrothermal environments, leading to degradation of mechanical properties (Cheng et al., 2011; Almeida Jr et al., 2016). This hygroscopic behavior primarily stems from the diffusion of water molecules within the resin and capillary action, rendering cutting fluids, which constitute wet cutting, unsuitable. Additionally, dry machining generates considerable amounts of hazardous dust (Santiuste et al., 2011). The absence of cooling and chip evacuation by cutting fluids further deteriorates machining conditions, making it challenging to control machining quality. Surface damage such as fiber fracture and pull-out, resin smearing, fiber-matrix debonding, and delamination can significantly compromise the service life and load-bearing capacity of CFRP components.

MQL is an emerging green machining technology, delivers atomized mixtures of minute lubricant oil and gas at specific pressures to the grinding zone to provide lubrication and cooling (Dogra et al., 2018; Gu et al., 2024; Yang and Li, 2025; Mia et al., 2018). The high-pressure gas primarily facilitates cooling and chip removal (Wu et al., 2019). Zhu et al. (2019) systematically investigated the combined effects of flow field characteristics, particle trajectories, airflow rate, and spindle speed on droplet penetration. Vegetable oils, as environmentally friendly and renewable resources, are biodegradable and non-toxic, positioning them as viable alternatives to conventional lubricants (Lawal et al., 2012; Zhang et al., 2015; Wickramasinghe et al., 2020). These resources pose minimal risk to the environment and operators while offering favorable lubrication performance. Adibi et al. (2018) found that MQL grinding of CFRPs yields superior surface quality compared to dry and flood conditions. Based on slotting experiments on CFRPs, Iskandar et al. (2014) reported significantly reduced flank wear under MQL compared to flood and dry machining. Molaie et al. (2016) demonstrated that the synergistic application of ultrasonic vibration and MQL reduces grinding forces and improves surface quality. Zhang et al. (2017) observed that under constant grinding parameters, MQL grinding leads to a concurrent decrease in frictional force and total grinding force. Rodriguez et al. (2019) integrated MQL with a wheel-cleaning jet technique, effectively removing adhered debris from the grinding wheel and reducing clogging. However, MQL technology has yet to overcome the technical bottleneck associated with its cooling effectiveness (Hadad et al., 2012), which severely limits its broader application. The high-pressure gas entering the grinding zone lacks sufficient cooling capacity, potentially causing heat accumulation on the workpiece surface, which can result in workpiece burn, reduced wheel life, or even wheel damage. The inadequate heat transfer performance of MQL has prompted research efforts aimed at its enhancement. The incorporation of nanoparticles into MQL lubricants has led to the development of NMQL technology, which not only effectively addresses heat transfer issues in the grinding zone but also enhances lubrication performance (Qiao et al., 2025).

Having characterized the single-grain forces and material removal mechanisms in Chapter 5, the investigation now returns to the workpiece side to evaluate the resulting surface quality. This chapter systematically assesses surface morphology, roughness, and defect characteristics produced under nanolubricant MQL transverse grinding, thereby linking the process mechanics established earlier to the final machined surface integrity. CFRP components under flood lubrication are susceptible to mechanical property degradation due to resin moisture absorption and swelling, whereas dry machining leads to poor surface integrity and problematic dust generation, limiting their application in the aerospace sector. Concurrently, the atomized mixture of lubricant and gas delivered by MQL still exhibits deficiencies in heat transfer and friction reduction. This study aims to investigate the grinding performance of CFRPs under CNT-based NMQL in comparison to dry and conventional MQL conditions, with a focus on surface integrity and machining-induced damage. The research methodology involves: firstly, comparing the roughness parameters *R*a, *R*z, and *R*Sm, which characterize overall surface texture features and performance; secondly, conducting qualitative analysis of detailed machining defects and damage morphology on ground CFRP surfaces; subsequently, employing fractal dimension and multifractal spectrum to quantitatively characterize global fractal attributes and local topographic characteristics; and finally, quantitatively studying high-frequency texture features via the energy distribution ratio derived from wavelet decomposition. Specific defects characterized include fiber pull-out, blocky pull-out, fiber fracture, resin matrix smearing, fiber-matrix debonding, and machining-induced pits. This study is anticipated to provide a feasibility basis for the partial or complete substitution of dry machining with NMQL in the manufacturing of precision CFRP components.

6.2 EVALUATION PARAMETERS

6.2.1 Fractal Dimension

The engineering surface presents non-stationary randomness, disorder, self-similarity, self-affine, and multiscale characteristics (Brown et al., 1993). The fractal characteristics and simulation of rough surfaces are also studied. Non-uniqueness is observed in the traditional surface morphology characterization parameters because of the unstable self-similarity of random rough surfaces. For example, the arithmetic mean deviations of contour, slope, and curvature variance, which are all related to the sampling length of samples and the measuring instrument resolution. Fractal dimension, a parameter used to characterize fractal characteristics quantitatively, is also a scale-free parameter. Theoretically, the fractal dimension can be used to study surface phenomena at any scale.

The machined surface of the workpiece has non-smooth and non-differentiable characteristics everywhere. With the improvement of measurement resolution, the machined surface will show fractal characteristics, and the fractal dimension is a parameter independent of scale, which can be used to study the change in surface morphology characteristics at different scales. The results of different fractal dimension calculation methods occasionally demonstrate considerable differences. The calculation results are accurate and the programming is convenient due to the clear concept of the box dimension method. Thus, the box-counting dimension method is used to calculate the two-dimensional fractal dimension of the CFRP machined surface.

The box-counting dimension is one of the most widely used dimensions. The basic definition of this method is as follows. Let F be an arbitrary bounded subset of non-empty R^n, and $N_\delta(F)$ be the minimum number of sets whose diameter is the largest and can cover F. The upper and lower box-counting dimensions of F are then defined as follows:

$$\underline{Dim_B}F = \underline{\lim}_{\delta \to 0} \frac{\log N_\delta(F)}{-\log \delta}, \tag{6.1}$$

$$\overline{Dim_B}F = \overline{\lim}_{\delta \to 0} \frac{\log N_\delta(F)}{-\log \delta}. \tag{6.2}$$

If the upper and the lower box dimensions are equal, then box-counting dimension with the same value F is called

$$Dim_B F = \lim_{\delta \to 0} \frac{\log N_\delta(F)}{-\log \delta}. \tag{6.3}$$

The differential box-counting (DBC) method is an improvement of the box-counting dimension method (Sarkar and Chaudhuri, 1992). The DBC method is effective in the dynamic range of fractal dimension and calculation efficiency. In addition, this method has good accuracy and generalization. First, the size of the SEM image is converted to 256 × 256, as shown in Figure 1. Then, the fractal algorithm is realized by programming, and the fitting line is shown in Figure 1. The fractal dimension of the SEM image D = 2.4842 and the correlation coefficient R = 0.9943 are obtained, thus demonstrating the good correlation of the fitting line. This finding proves that the image has good fractal characteristics in the calculation scale. Therefore, characterizing the morphological characteristics of the machined surface of CFRP by using the fractal dimension is feasible.

Figure 1. SEM image of machined CFRP surface and fitting line of fractal dimension

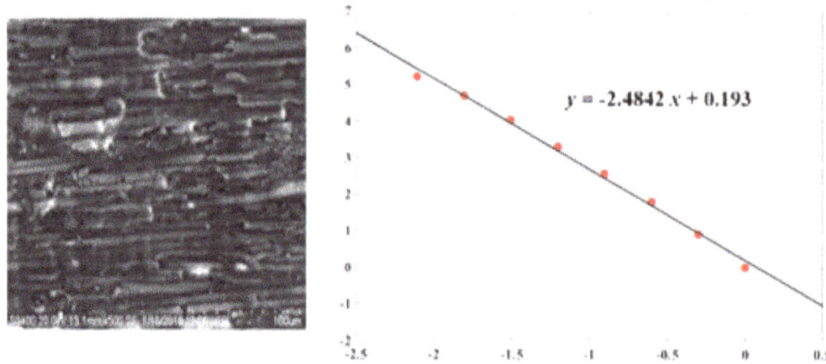

6.2.2 Multifractal Spectrum

Fractal dimension D can only characterize the overall morphology of the machined surface. Therefore, this paper introduces a multifractal method to explore the local morphology of the CFRP machined surface. The surface roughness signal is divided into N different regions s_i ($i = 1, 2, \ldots, N$), which are set as the scale of the i th region, and p_i is the generation probability (physical quantity) of the region s_i. Various s_i and p_i are also different, which can be expressed by different scaling indexes α_i. Thus, the following relationship is satisfied.

$$p_i(\varepsilon) \in \varepsilon_i^{\alpha_i}(i = 1,2,\cdots,N), \tag{6.4}$$

where α is a singularity index, which reflects the singularity degree of each small area in the fractal dimension. Therefore, the value of α is related to the probability of generating the region. Different generation probabilities form different subsets of the probability distribution. In practice, the size of each region ε_i is chosen as the same value. The fractal dimension of subsets with the same α value is defined as $f(\alpha)$, which is explained by the fractal dimension:

$$N_\alpha(\varepsilon) \sim \varepsilon^{-f(\alpha)}, \tag{6.5}$$

where N denotes the number of probability subsets corresponding to α under the scale of ε. $f(\alpha)$ is generally called multiple typing spectrum. $f(\alpha)$ and α are a set of parameters for describing multifractals.

A partition function $\chi_q(\varepsilon)$, which weights the probability $P_i(\varepsilon)$ with the q power, is defined. q is called the weight factor. The mathematical expression is as follows:

$$\chi_q(\varepsilon) = \sum P_i(\varepsilon)^q = \varepsilon^{\tau(q)}. \tag{6.6}$$

If the equation after the above formula holds, that is, the partition function has a power function relationship with ε, then the slope of $\ln\chi_q \in \ln\varepsilon$ curve can be obtained as follows:

$$\tau\left(q\right) = \frac{\ln\chi_q(\varepsilon)}{\ln\varepsilon}\left(\varepsilon \to 0\right). \tag{6.7}$$

$\tau(q)$ is generally called the mass index. The generalized fractal dimension D_q is obtained from $\tau(q)$:

$$D\left(q\right) = \frac{\tau_q}{q-1} = \frac{\ln\chi_q(\varepsilon)}{(q-1)\ln\varepsilon}\left(\varepsilon \to 0\right). \tag{6.8}$$

The singular exponent α and the multifractal spectrum $f(\alpha)$ can be obtained by Legendre transformation of $\tau(q)$ and q:

$$\alpha = \frac{d\tau(q)}{dq}, \tag{6.9}$$

$$f(\alpha) = \alpha q - \tau(q). \tag{6.10}$$

Formula (6.6) is substituted for formula (6.7) to obtain the following:

$$\tau\left(q\right) = \frac{\ln\sum P_i(\varepsilon)^q}{\ln\varepsilon}\left(\varepsilon \to 0\right). \tag{6.11}$$

Formula (6.11) is substituted into (6.10) to derive q:

$$\alpha = \frac{\sum\ln P_i(\varepsilon)\cdot P_i(\varepsilon)^q}{\sum P_i(\varepsilon)^q}. \tag{6.12}$$

Formulas (6.11) and (6.12) are substituted into (6.10) to obtain $f(\alpha)$:

$$f\left(\alpha\right) = \frac{\dfrac{\sum\ln P_i(\varepsilon)\cdot P_i(\varepsilon)^q\cdot q}{\sum P_i(\varepsilon)^q} - \ln\sum P_i(\varepsilon)^q}{\ln\varepsilon}. \tag{6.13}$$

Let the molecular terms of expressions (6.12) and (6.13) of α and $f(\alpha)$ be αq and fq, respectively, that is:

$$\alpha q = \frac{\sum\ln P_i(\varepsilon)\cdot P_i(\varepsilon)^q}{\sum P_i(\varepsilon)^q}, \tag{6.14}$$

$$fq = \frac{\sum \ln P_i(\varepsilon) \cdot P_i(\varepsilon)^q \cdot q}{\sum P_i(\varepsilon)^q} - \ln \sum P_i(\varepsilon)^q. \quad (6.15)$$

The expressions of α and $f(\alpha)$ can then be respectively expressed as functions of q:

$$\alpha = \frac{\alpha q}{\ln \varepsilon}, \quad (6.16)$$

$$f(\alpha) = \frac{fq}{\ln \varepsilon}. \quad (6.17)$$

Multifractal calculates the probability of each region according to the actual roughness profile value, counts all the probability differences, and divides the surface into many alpha subsets by weighting. Each subset represents a probability distribution, that is, $P(\varepsilon) \in \varepsilon^{\alpha}$, and the width of the fractal spectrum is defined as (Sun et al., 2003):

$$\Delta\alpha = \alpha_{max} - \alpha_{min}. \quad (6.18)$$

According to $P(\varepsilon) \in \varepsilon^{\alpha}$, further derivation can be obtained as follows:

$$\Delta\alpha = \frac{\ln(P_{\max}/P_{\min})}{\ln(1/\varepsilon)}. \quad (6.19)$$

The above formula concludes that the width of the multifractal spectrum $\Delta\alpha$ reflects the size of the probability distribution range. When the probability distribution is uneven, the corresponding $f(\alpha)$ curve is wide. That is, when $\Delta\alpha$ is large, the converse is small.

The shape of the multifractal spectrum reflects the characteristics of various probability distributions. The spectral difference Δf is defined as follows:

$$\Delta f = f(\alpha_{min}) - f(\alpha_{max}), \quad (6.20)$$

where $\alpha_{\min}$ and $\alpha_{\max}$ respectively correspond to the maximum and the minimum probability subsets. According to $N_{\alpha}(\varepsilon) \sim \varepsilon^{-f(\alpha)}$, further derivation can be obtained as follows:

$$\Delta f = \frac{\ln(N_{P\max}/N_{P\min})}{\ln \varepsilon}. \quad (6.21)$$

The Δf can be used to calculate the proportion between the number of maximum and minimum probabilistic units. $N_{P\max}$ and $N_{P\min}$ represent the number of maximum and minimum probabilistic subsets, respectively. Based on its application to the machined surface roughness profile curve, when $\Delta f < 0$, then $N_{P\max} > N_{P\min}$, and the multifractal spectrum is in the right-hook state. This finding shows that the large height of the roughness profile curve occupies the main position, the corresponding machined surface topography height is high, and the probability of the machined surface peak is large. When $\Delta f > 0$, then $N_{P\max} < N_{P\min}$, and the multifractal spectrum is left hook state. This finding indicates that the probability of a small height of the roughness profile curve is high, the corresponding height of the machined surface is low, and the probability of valleys appearing on the machined surface is high. When $f = 0$, the probability of peaks and valleys on the machined surface is equal.

6.2.3 Energy Distribution Proportion of Wavelet Decomposition

Low roughness does not guarantee the absence of defects in the surface. Defects, such as fiber pull-out, block pull-out, fiber breakage, resin matrix coating, and debonding between fiber and matrix, easily occur on the CFRP surface. Considering the surface morphology of CFRP, the surface of CFRP has abundant information on roughness and texture change in the fiber direction or at an angle with the fiber direction. The characteristics of these changes are mainly reflected in the high-frequency texture information part of the processing surface signal.

Let $\psi(t) \in L^2(R)$, $L^2(R)$ denote the square-integrable function space whose Fourier transformation is $\hat{\psi}(\omega)$. When $\hat{\psi}(\omega)$ satisfies the allowable conditions

$$C_\psi = \int_R \frac{|\hat{\psi}(\omega)|}{|\omega|} d\omega < \infty, \tag{6.22}$$

Then $\hat{\psi}(\omega)$ is called a basic or mother wavelet. A wavelet sequence is obtained after scaling and translating the mother wavelet. When stretching and translating continuously, the obtained wavelet sequence is as follows:

$$\psi_{a,b}(t) = \frac{1}{\sqrt{|a|}}\psi\left(\frac{t-b}{a}\right) \quad a, b \in R, a \neq 0, \tag{6.23}$$

where a is the scale expansion factor, and b is the translation factor.

For any function of $f(t) \in L^2(R)$, continuous wavelet transform is defined as:

$$\left(W_\psi f\right)(a,b) = \langle f, \psi_{a,b}\rangle = |a|^{-1/2} \int_R f(t)\psi^*\left(\frac{t-b}{a}\right)dt. \tag{6.24}$$

The superscript represents conjugation, and its inverse transformation is defined as

$$f(t) = \frac{1}{C_\psi}\int_{R^\cdot}\int_R \frac{1}{a^2}\left(W_\psi f\right)(a,b)\psi\left(\frac{t-b}{a}\right)dadb, \tag{6.25}$$

where a affects not only the position of the window on the frequency axis but also the shape of the window, while b only affects the position of the window on the phase-plane time axis. The time-domain sampling step of the wavelet transform is adjustable for different frequencies. That is, the time resolution of the wavelet transform is low at low frequency, while the frequency resolution is high at high frequency. In high frequency, the time resolution of the wavelet transform is high, but its frequency resolution is low. This condition is consistent with the characteristics of the slow and fast transformation of low-and high-frequency signals, respectively.

The construction of sym wavelet is similar to that of db wavelet family. The difference between the two wavelets lies in the improved symmetry, suitability for image processing, and reduction of the phase shift in the reconstruction of the sym wavelet. Daubechies adopted the idea of improving the symmetry of the sym wavelet while maintaining the simplicity of the db wavelet. Considering the m_0 function used to construct db wavelet, if this function is regarded as the function W of $z = e^{iw}$, then different wavelets can be obtained through various ways in the process of W construction. W is defined in the following form:

$$W(z) = U(z)\overline{U\left(\frac{1}{z}\right)} \quad z = e^{iw}, \tag{6.26}$$

where W can be regarded as a "minimum phase shift" filter. If the modulus square of the root of $U(z)$ is less than 1, then the db wavelet is obtained. If the root of $U(z)$ is larger than 1, then the sym wavelet is obtained.

The support range of the symN wavelet is 2N-1 and the vanishing moment is N, which also has good regularity. Compared with the dbN wavelet, this wavelet is consistent with the dbN wavelet in continuity and support and filter lengths. However, the symN wavelet has improved symmetry. That is, the symN wavelet can reduce the phase distortion in signal analysis and reconstruction to a certain extent. Therefore, the sym2 wavelet was chosen for the five-level wavelet decomposition. Taking MQL grinding of CFRP as an example, the signal maps of each layer after wavelet decomposition of the surface roughness profile curve are shown in Figure 2. Among the extracted multilayer signals, a5 is the approximate component of the decomposed signal, and d1–d5 are the detail components of the decomposed signal.

Figure 2. Wavelet decomposition signal maps of MQL grinding surface roughness profiles

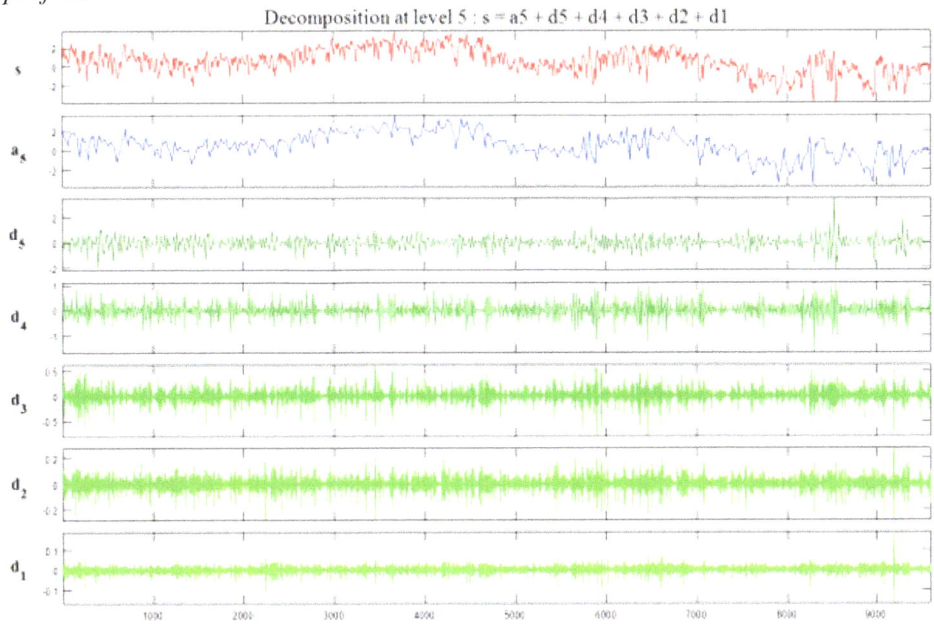

6.3 EXPERIMENTAL DETAILS

6.3.1 Experimental Equipment and Measuring Instruments

The experiments were conducted on a K-P36 numerical control precision surface grinder, and its configuration is shown in Figure 3. The grinding wheel material comprised diamond and ceramic bond due to the excellent thermal conductivity and wear resistance of diamond. The size, shape, and morphology of diamond grains slightly change in the grinding process. The sharpened diamond grinding wheel can maintain the micro-cutting edge of the grains for a long time, thus contributing to its suitability for high-precision machining. The MQL-supply device in the experiment is KINS KS-2106 minimum quantity oil supply system. The system uses a pulse generator to adjust the lubricating oil and compressed gas of the MQL system to control its flow. The compressed gas can be ejected from the nozzle into the processing area by high-pressure gas after the atomization of compressed gas and minimum quantity oil. The MQL system includes adjustable flow pumps, compressed air-flow control devices, and spray-assisted air atomizing nozzles. The parameters of the surface grinder and MQL supply device are presented in Table 1. The measuring instruments are TIME3220 surface roughness profiler and S-3400N scanning electron microscope as shown in Figure 3.

Figure 3. Machining setups and measuring apparatus

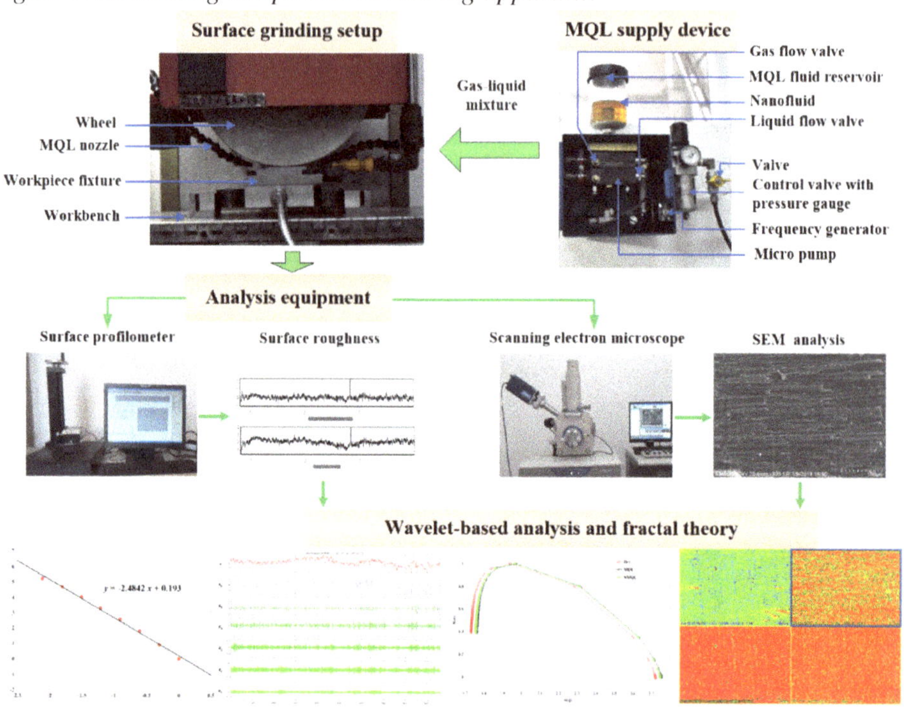

Table 1. Machining conditions

Grinding process parameters	**Unit**	**Value**
Peripheral speed of grinding wheel v_s	m/s	40
Cutting depth a_p	ìm	20
Feed speed v_w	mm/s	30
Gas pressure	bar	6.0
Gas–liquid ratio	--	0.3
MQL flow rate	mL/h	60
MQL nozzle distance	mm	12
MQL nozzle angle	°	15

6.3.2 Methods and Materials

The definition of the lubrication condition is shown in Table 2. CNTs were selected as nanoparticles in NMQL due to excellent heat transfer enhancement performance. However, agglomerating CNT nanoparticles was easy, the lubrication and tribological performances of nanolubricants were affected, and appropriate dispersants were added during the configuration of nanolubricants (Gao et al., 2019). The workpiece material is unidirectional CFRP, and the detailed material parameters are listed in Table 1 of Section 5.2.5.

Table 2. Experimental scheme

Group no.	Lubrication conditions	Contents
1	Dry	/
2	MQL	Pure palm oil
3	NMQL	50 nm CNTs - Pure palm oil

6.4 SURFACE ROUGHNESS

6.4.1 1D Surface Roughness

The *R*a value of the contour arithmetic average deviation containing most information of micro-roughness is preferred to characterize the surface characteristics and behavior in service. The *R*z with a peak height of the reaction profile is selected considering the fiber pull-out phenomenon on the grinding surface of CFRPs. The average width (spacing parameter) *RS*m of common contour elements is selected as the roughness characterization parameters in the horizontal direction. The mean surface roughness *R*a values in the feed and fiber directions were shown in Figure 4, each averaging from five separate measurements and plotted in accordance with the lubrication conditions. The *R*a values of the CFRP grinding surface under MQL and NMQL lubrication in the feed direction are reduced by 18.80% and 17.70%, respectively, compared with that of dry processing. Meanwhile, the *R*a value of the CFRP grinding surface under MQL and NMQL lubrication in the fiber direction is 9.02% and 12.68% lower than that of dry processing, respectively, and the standard deviation under MQL condition is small.

Figure 4. Surface roughness Ra value

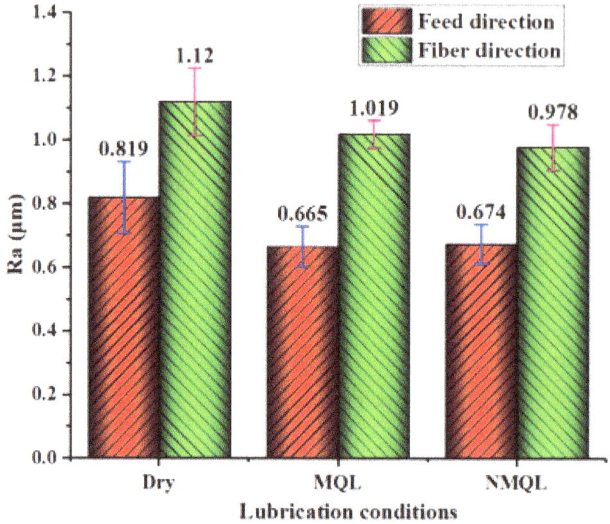

Figure 5 shows the mean surface roughness *R*z values in the feed and fiber directions, each averaging from five separate measurements and plotted in accordance with the lubrication conditions. The *R*z value of the CFRP grinding surface under MQL and NMQL lubrication in the feed direction is 21.68% and 20.78% lower than that of dry processing, respectively. The *R*z value of the CFRP grinding surface under MQL and NMQL lubrication in the fiber direction is 18.72% and 25.06% lower than that of dry processing, respectively.

Figure 5. Surface roughness Rz value

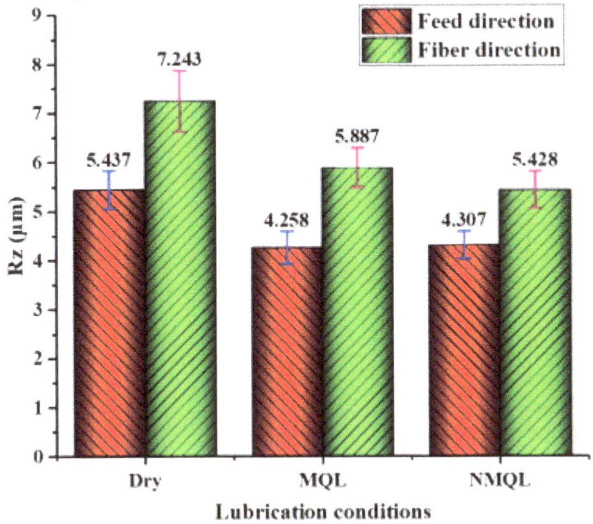

Figure 6 shows the mean surface roughness *RS*m values in the feed and fiber directions, each averaging from five separate measurements and plotted in accordance with the lubrication conditions. The *RS*m value of the CFRP grinding surface under MQL and NMQL lubrication in the feed direction is 30.16% and 25.40% lower than that of dry processing, respectively. Meanwhile, the *RS*m value of the CFRP grinding surface under MQL and NMQL lubrication in the fiber direction decreased by 4.77% and 11.43%, respectively, compared with that of dry processing.

Palm oil contains a substantial amount of saturated fatty acids (palmitic acid accounts for approximately 30%–50% and stearic acid), which have better lubricity than that of unsaturated fatty acids (Abdalla and Patel, 2006). Saturated fatty acids have a significant effect on reducing friction and wear, especially stearic acid, which provides strong protection for oil film by strengthening the intermolecular force. In addition, the number of C atoms of saturated fatty acids in palm oil is 16, which can form a lubricating film with strong adsorption, achieving good lubricating performance (Zhang et al., 2015).

Figure 6. Surface roughness RSm value

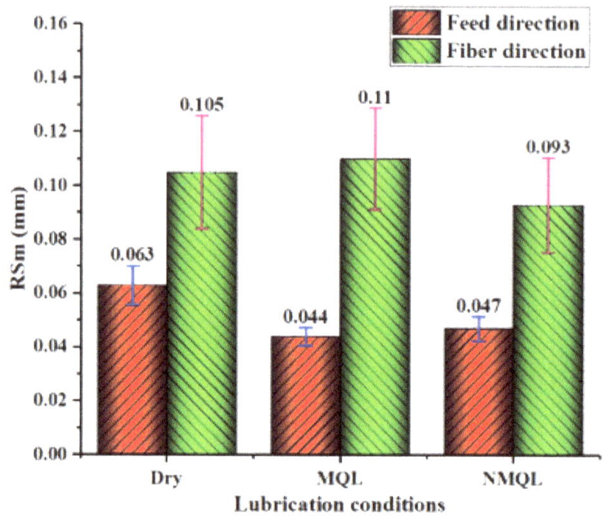

The surface roughness of CFRP is low in the CNT NMQL because the excellent tribological properties of CNT nanoparticles play an effective role in friction reduction and anti-wear at the interface of grinding wheel and CFRP. The grinding effect of the wheel on CFRP is the removal of the fiber by the grains in the micro-scale. In the grinding process, the constraint on the carbon fiber will be reduced when the resin matrix is softened by grinding heat. Some fibers will be separated from the resin matrix under the cutting action of grains, thereby resulting in cracks. This phenomenon further weakens the constraint of the resin matrix and surrounding materials on the fiber. The application of CNT nanoparticles will facilitate embedding into the cracks between the fiber and the resin matrix to strengthen the constraint on the fiber, as shown in Figure 7. By contrast, CNT nanoparticles will be filled into the pits and voids due to fiber fracture under the rolling of the grinding wheel and grains. This phenomenon also reduces surface roughness.

Figure 7. Strong constraint of embedded CNTs on fiber and filling effect

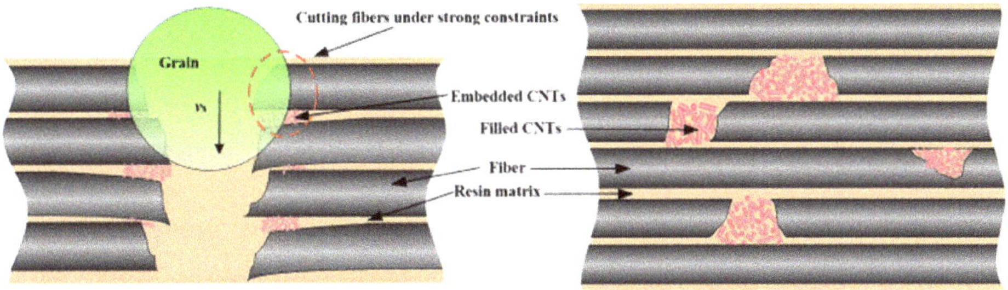

6.4.2 2D Surface Roughness

Different from the characterization of two-dimensional surface topography, the characterization of three-dimensional surface topography is a unified parameter that is defined according to the limited scale surface type and does not use three groups of different parameters to characterize the profile, waviness, and roughness respectively. For example, root mean square deviation is the only parameter in the three-dimensional parameters, and its specific meaning depends on the type of finite scale surface used in filtering, rather than the profile parameter *Pq*, waviness parameter *Wq*, and roughness parameter *Rq*. *Sa* and *Sq* are the parameters that express the amplitude characteristics of surface roughness, and the calculation principle is the same as that of the two-dimensional parameters. However, the range of three-dimensional parameters is a plane area, and the range of points is no longer limited to a certain contour. *Sa* represents the arithmetic mean value of the absolute height values of the points on the measured surface and the reference plane in the sampling area. It can generally describe the height deviation of the surface and is used to describe the adhesion and material removal rate of the composite surface after machining. *Sq* is the standard deviation of surface height distribution, which is more significant than *Sa* in reflecting the distribution of parameter height and characterizing roughness. The results prove that the *Sa* and *Sq* of dry machining are the largest, and the surface quality was the worst, as shown in Figure 8. The *Sq* / *Sa* was also large, reaching 1.282, which indicated that the peak or valley height of the surface was relatively obvious. In other words, the surface limit value was obvious. Compared with the results of dry cutting, *Sa* and *Sq* of MQL were reduced by 2.2% and 3.6% respectively. Compared with the results of dry grinding, *Sa* and *Sq* of CNT biological lubricant were reduced by 8.4% and 7.9%, respectively. The root mean square deviation *Sq* of the workpiece surface in CNT biological lubricant was small, which indicated fewer wave crest features. This condition was due to fewer fiber burrs or protuberances on the surface. The value of each surface roughness is obtained by averaging the five groups of measured values. The standard deviation is calculated according to the formula below

$$\sigma=\sqrt{\left((x_1-x)^2+(x_2-x)^2+\cdots\cdots+(x_n-x)^2\right)/n} \tag{6.27}$$

where σ is the standard deviation, x is the average of five groups of surface roughness, x_1, x_2, • • • x_n are the surface roughness values for all, n is the amount of groups of surface roughness, which is 5.

Figure 8. 3D morphology and surface roughness of grinding under different lubrication conditions.

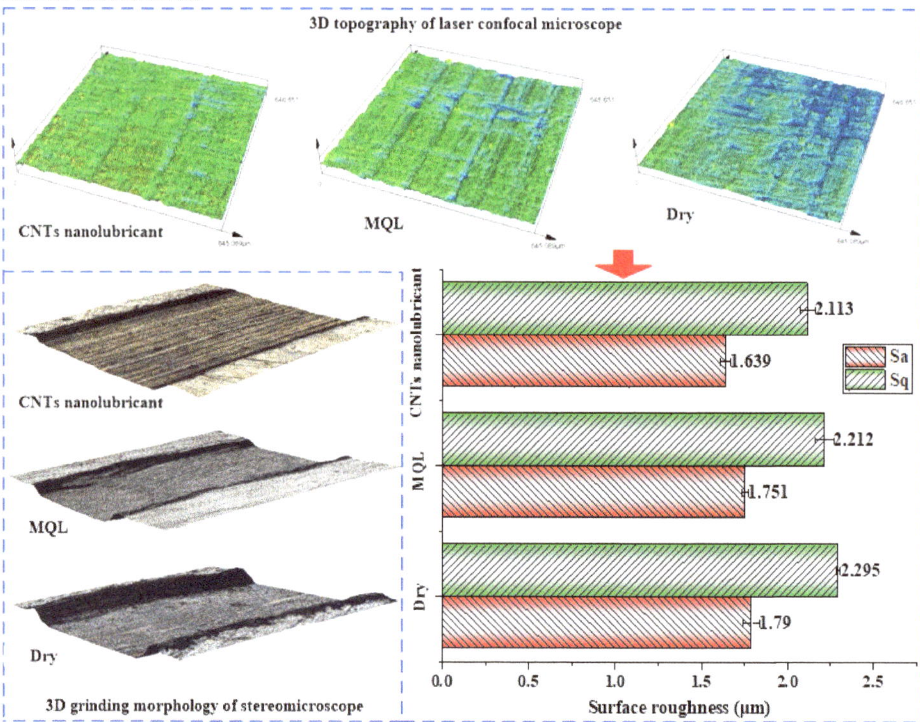

6.5 FRACTAL DIMENSION

6.3.2 2D Fractal Dimension of Surface Roughness

The fractal dimension calculated from the surface roughness data is shown in Figure 9. Compared with dry and MQL grinding, the surface fractal dimension of NMQL grinding is the lowest in the feed and fiber directions. Compared with dry grinding, the two directions are reduced by 2.36% and 1.12%. This finding shows that the amplitude of the surface profile height of grinding CFRP under NMQL condition slightly changes and the surface is relatively flat. The standard deviation of fractal dimension is the lowest under NMQL condition in the fiber direction.

Figure 9. Surface roughness signal one-dimensional fractal dimension

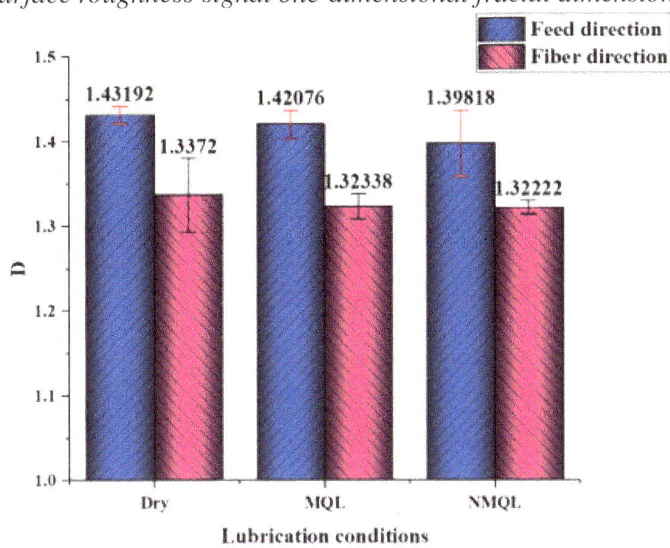

6.3.3 3D Fractal Dimension of SEM Image

The three-dimensional fractal dimension of SEM topography with magnifications of 300 and 500 is calculated. The 3D fractal dimension values of the CFRP grinding surface under different working conditions are shown in Figure 10. For SEM images at 300 times magnification, the fractal dimension of the NMQL grinding surface is 2.319, which is reduced by 0.047 and 0.2085 relative to dry grinding and MQL grinding. For SEM images at 500 times magnification, the fractal dimension of the NMQL grinding surface is 2.2976, reduced by 0.1866 and 0.0961 relative to dry grinding and MQL grinding. NMQL has the smallest 3D fractal dimension at 300 and 500 times, indicating improved fractal features and smooth grinding surfaces.

Figure 10. Three-dimensional fractal dimension of SEM image

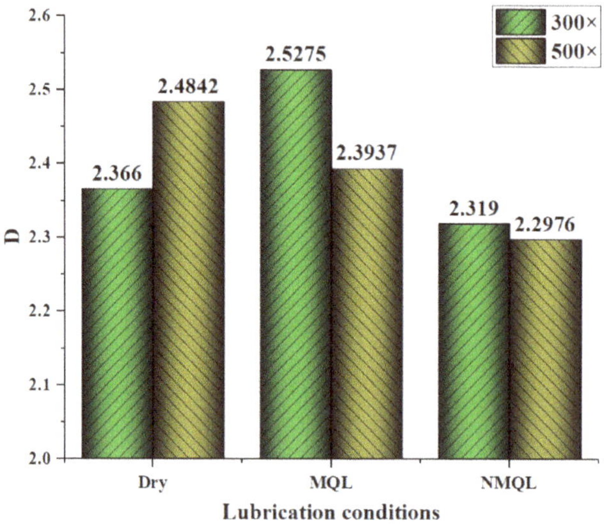

6.6 MULTIFRACTAL CHARACTERISTICS

6.6.1 Feeding Direction

The multifractal spectrum of grinding surface roughness in the feeding direction is shown in Figure 11. The multifractal spectra calculated in the feed direction are all left hooked, and the calculated parameters of the multifractal spectra are all larger than 0. Thus, the low height of the surface roughness profile curve occupies the main position. The corresponding height of the surface topography in CFRP grinding is also low, indicating the presence of numerous valleys on the surface. Figure 12 shows the spectral width and spectral difference in the multifractal spectrum of grinding surface roughness under three lubrication conditions in the feed direction. The multifractal spectrum width $\Delta\alpha$ of MQL and NMQL grinding surface is small in the feed direction. This finding indicates that the height fluctuation of the grinding surface is small and the roughness is low. The spectral difference Δf of the multifractal spectrum under three lubrication modes similar to the spectral width $\Delta\alpha$. When the spectral difference Δf is small, the probability distribution of peaks and valleys on the machined surface is equal. The large spectral difference Δf

of dry grinding indicates the presence of a few peaks and additional valleys on the corresponding surface. This phenomenon is due to the block pull-out of multiple fibers on the surface of CFRP by dry grinding, resulting in pits.

Figure 11. Multifractal spectrum of CFRP grinding surface roughness in the feeding direction

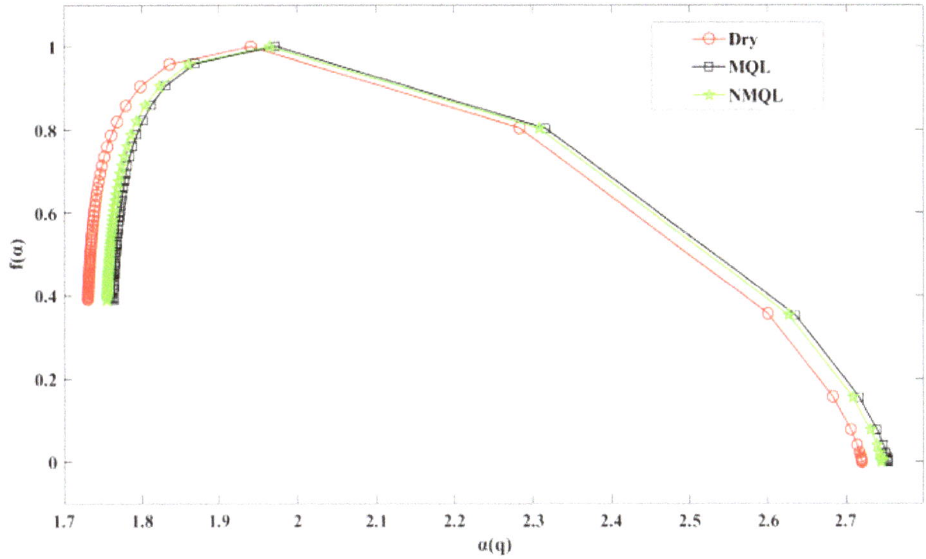

Figure 12. Δα and Δf of the multifractal spectrum in the feeding direction

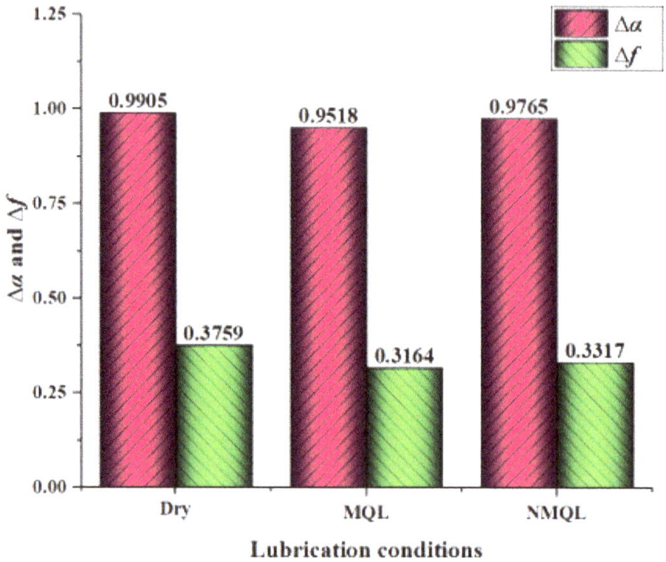

6.6.2 Fiber Direction

The multifractal spectrum of grinding surface roughness in the fiber direction is shown in Figure 13. The multifractal spectra calculated in the fiber direction are all left hooked, and the parameters of the multifractal spectra calculated are all larger than 0. This finding shows that the low height of the surface roughness profile curve occupies the main position. The corresponding height of the surface topography in CFRP grinding is also low, indicating the presence of additional valleys on the surface. Figure 14 shows the spectral width and difference in the multifractal spectrum of grinding surface roughness under three lubrication conditions in the feed direction. The multifractal spectrum width $\Delta\ \alpha$ and spectral difference Δf of MQL and NMQL grinding surfaces in the fiber direction are small, and the reduction is larger than that in the feed direction. This finding indicates that the height fluctuation of the grinding surface in this direction is small and the roughness is low. MQL and NMQL play a crucial role in improving the surface quality in the fiber direction. The contrast results reveal that the multifractal spectrum spectral difference Δf under three lubrication conditions is similar to the spectral width $\Delta\ \alpha$. When the spectral difference Δf is small, the probability distribution of peaks and valleys on the machined surface is equal. The large spectral difference Δf of dry grinding indicates the presence of a few peaks and additional valleys on the CFRP surface. This phenomenon is due to the multifiber pull-out on the CFRP surface in dry grinding, thus resulting in pits.

Figure 13. Multifractal spectrum of CFRP grinding surface roughness in the fiber direction

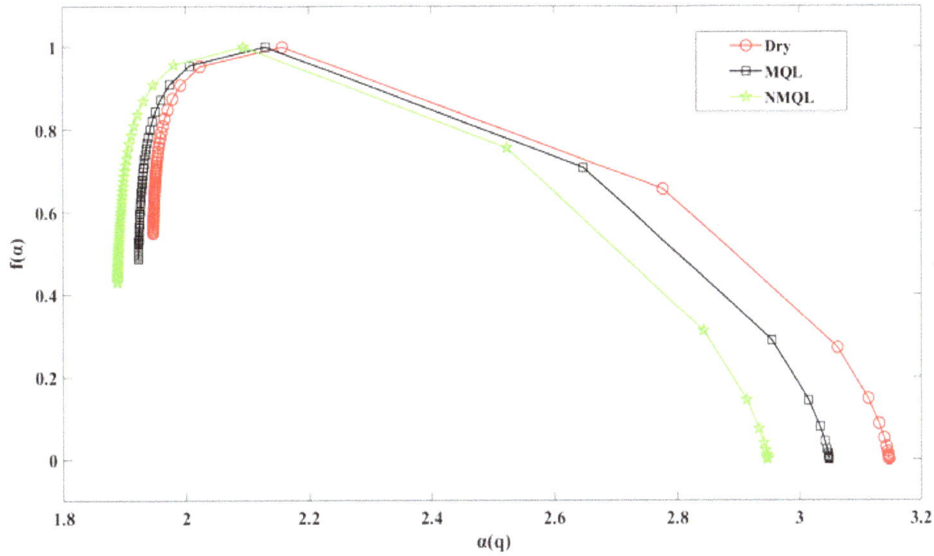

Figure 14. Δα and Δf of the multifractal spectrum in the fiber direction

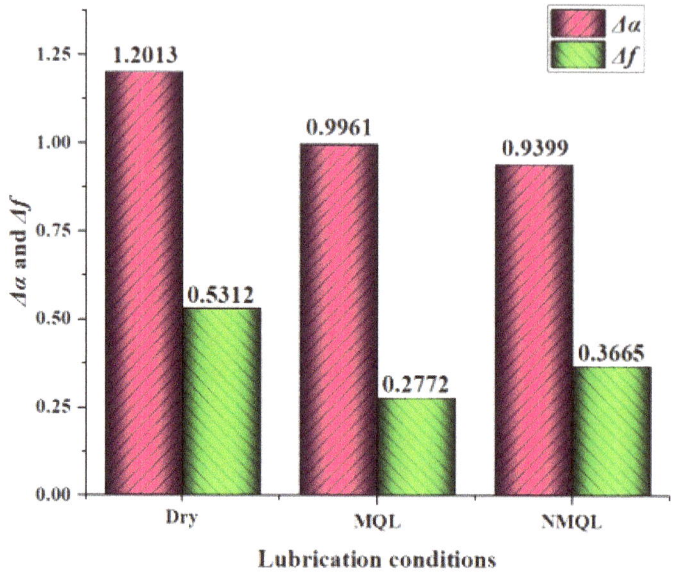

6.7 MICRO-MORPHOLOGY

The surface roughness measurement value could not reflect the real surface topography characteristics of CFRP processing because surface roughness is a scale-related parameter. The measurement value varies with the resolution of the measuring instrument and the change in sampling conditions. Thus, the inherent characteristics of grinding surface topography could not be reflected. Compared with the surface roughness, the machined surface morphology and defects are one of the main focuses in the study of the CFRP surface quality. Figure 15 shows the SEM micro-morphology of the CFRP grinding surface under different conditions. In the grinding edge, edge breakage occurs in dry grinding, while the edges of MQL and NMQL grinding are relatively neat, and edge breakage is absent. Severe damages, such as multifiber pull-out and breakage occurred on the CFRP surface in dry grinding. Serious resin smearing occurred only in dry grinding due to the reduction in strength and properties of a resin with the increase in temperature for the resin matrix sensitivity to temperature. The heat-softened resin adhered to the grinding wheel, extruded onto the CFRP surface of CFRP, or directly smeared on the CFRP. By contrast, the resin attached to the grinding wheel will adhere and carry a large amount of fiber debris, which will adhere to the grinding wheel in the form of a mixture of resin and fiber debris. This phenomenon causes serious blockage of the grinding wheel and further deteriorates the processing conditions. Partially cut-off fibers occur in dry and MQL grinding. By contrast, partially cut-off fibers do not occur in NMQL grinding or are less likely to occur because CNT nanolubricants spread over the surface to form oil films. CNT nanoparticles form a dense protective film at the interface between grain and fibers as shown in Figure 16. This film prevents the grain from removing partially exposed fibers. In addition, single fiber breakage and void phenomena occur in all three processing conditions due to fiber pullout or peeling.

Figure 15. SEM image of CFRP grinding surface under different conditions

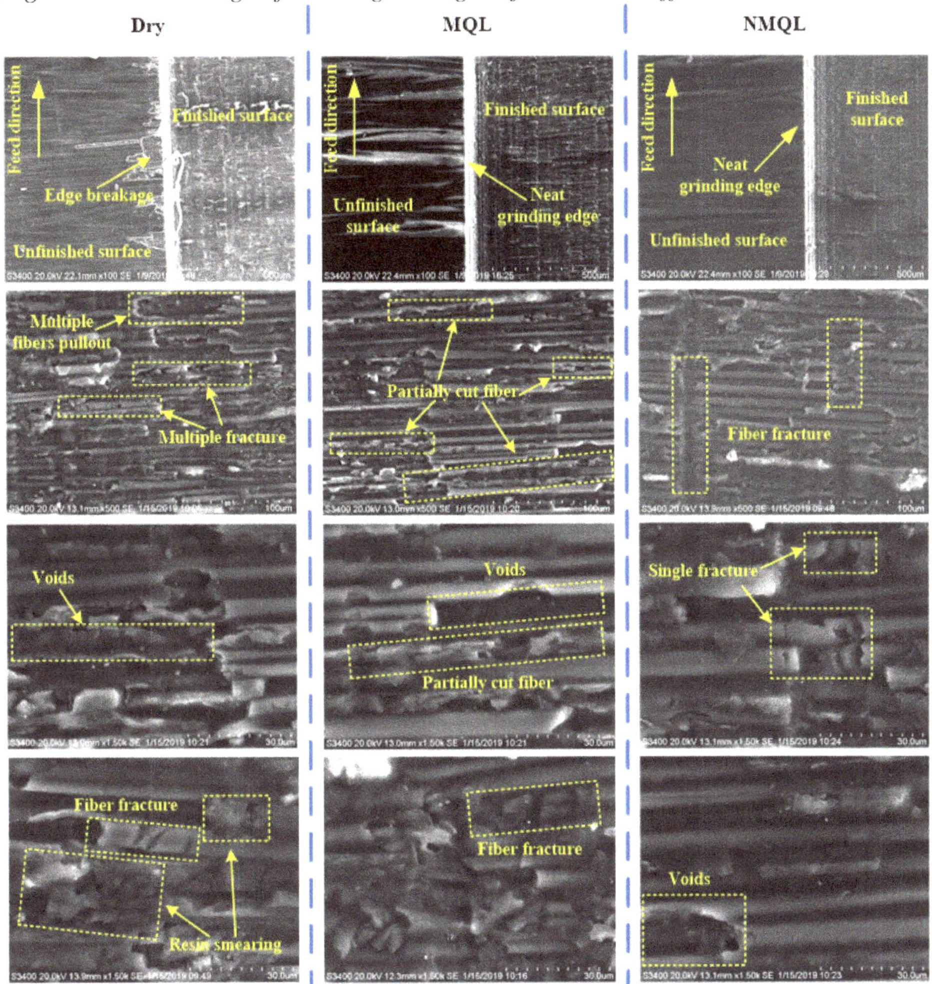

Figure 16. Protective effect of CNT nanolubricants on the grain–fiber interface

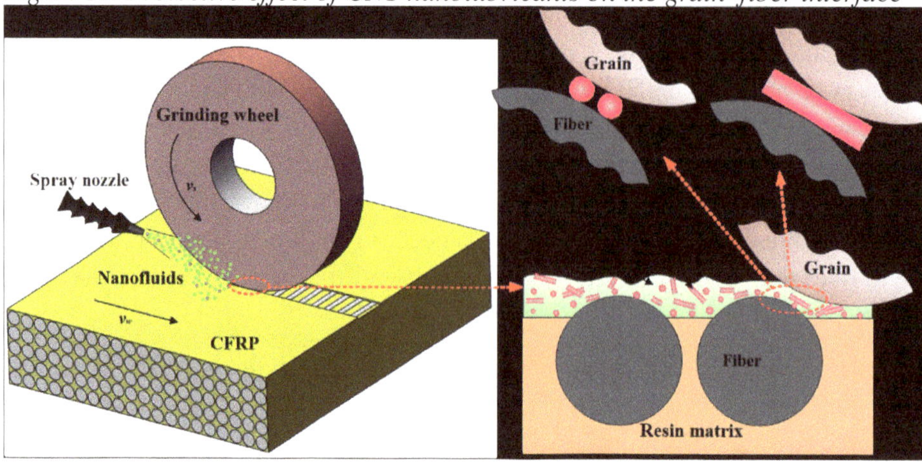

6.8 QUANTITATIVE CHARACTERIZATION OF GRINDING DAMAGE

6.8.1 Energy Distribution of Roughness Signal

The energy of low- and high-frequency detail signals after wavelet decomposition of CFRP grinding surface roughness signal under different lubrication conditions is shown in Figure 17. The energy proportion of the high-frequency detail signal of the CFRP surface in dry grinding (feed direction 0.0097, fiber direction 0.0014) is the highest, followed by MQL grinding. NMQL grinding (feed direction 0.0053, fiber direction 0.0014) is the lowest. Compared with dry grinding, the energy proportion of the high-frequency detail signal of NMQL grinding is reduced by 46.85%. This finding shows that the texture features of machined surface roughness corresponding to NMQL demonstrated the least changes.

Figure 17. Energy distribution of roughness signal by one-dimensional wavelet decomposition

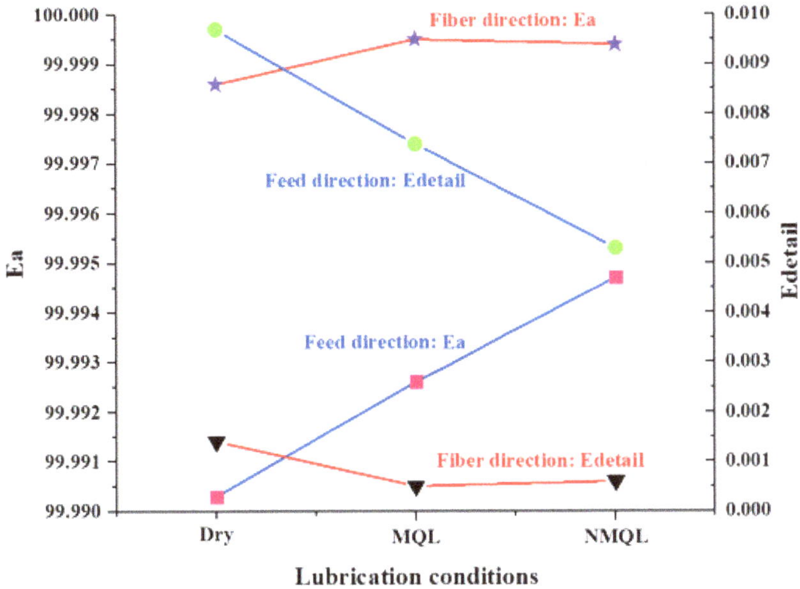

6.8.2 Energy Distribution of SEM Image

Similarly, taking the SEM image of the MQL grinding surface as an example, four sub-images are obtained after wavelet decomposition. Figure 18 shows (a) smooth approximate sub-images, (b) horizontal sub-images, (c) vertical sub-images, and (d) diagonal sub-images.

Figure 18. Wavelet decomposition sub-images of SEM image

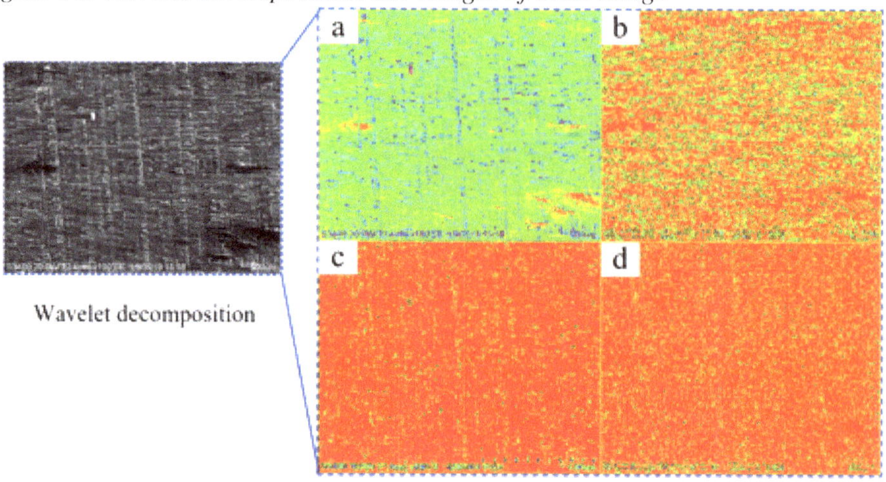

The SEM image is decomposed by wavelet transform, and the energy proportion of each decomposed sub-image is calculated. The variation trend of the singular morphology of the processed surface of CFRP in the decomposition direction is characterized by the energy size of each sub-image. The energy proportion of each decomposed sub-image is calculated by MATLAB programming. The energy of each high-frequency detail sub-image after wavelet decomposition of the CFRP machined surface image is shown in Figure 19.

These changes in surface morphology characteristics are mainly reflected in the high-frequency information part of the SEM image because of fiber pull-out, fiber breakage, processing pits, and other characteristics of the surface morphology of CFRP processing. Therefore, the wavelet decomposition is used to calculate the energy proportion of each detail sub-image after wavelet decomposition to determine the energy distribution of CFRP and the change in defects on the grinding surface in different directions.

Figure 19. Energy distribution of the wavelet decomposed high-frequency sub-images

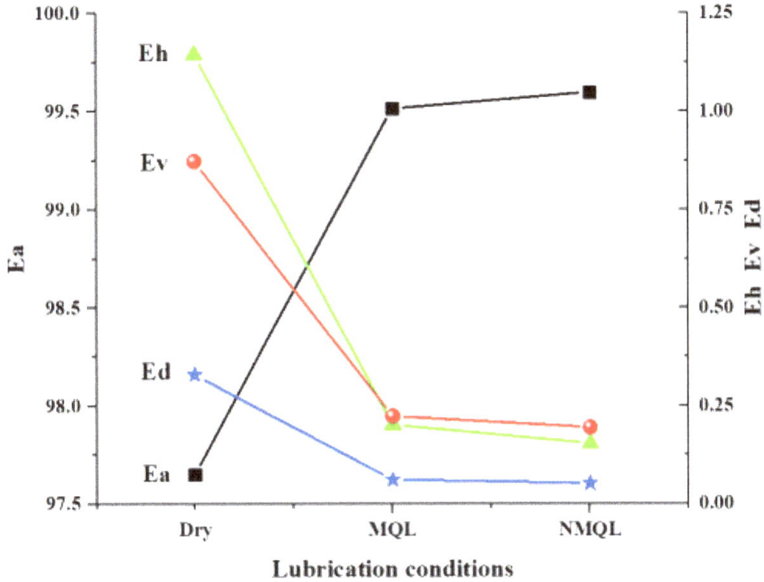

The energy proportion of three detail sub-images in the wavelet decomposition image of the dry grinding surface is the highest. The energy proportion of sub-images (b) (horizontal high frequency), (c) (vertical high frequency), and (d) (diagonal high frequency) is 1.1457, 0.8738, and 0.3297, respectively. Therefore, the defect characteristics of the grinding surface take the largest proportion in the horizontal direction. The texture change information of the surface topography of the processed image is mainly reflected in the horizontal details.

The energy proportion of three detail sub-images in the wavelet decomposition image of the MQL grinding surface is between dry processing and NMQL. The energy proportion of horizontal and vertical detail sub-images is almost the same, that is, the defect change characteristics in the fiber and feed directions are the same. The energy proportion of three detail sub-images in the wavelet decomposition image of the NMQL grinding surface is the lowest. The energy proportion of sub-images (b) (horizontal high frequency), (c) (vertical high frequency), and (d) (diagonal high frequency) is 0.155, 0.1952, and 0.0522, respectively, The proportion of defect change characteristics in the horizontal direction (fiber direction) of NMQL grinding surface is less than that in the vertical direction (feed direction). This finding is also consistent with the energy proportion of one-dimensional roughness signal wavelet analysis.

6.9 SUMMARY

This chapter investigates the grinding performance of CFRP composite materials using NMQL technology, with a focus on characterizing and analyzing the surface integrity and machining defects during grinding. Firstly, the surface roughness *Ra*, *Rz*, and *RSm* were compared. Secondly, the overall fractal characteristics of the machined surface in different dimensions (two-dimensional and three-dimensional) were quantitatively characterized using fractal dimension. Furthermore, the local morphology characteristics of the CFRP machined surface were analyzed using multifractal spectrum, spectral width, and spectral difference. Then, SEM was used to qualitatively analyze the microstructure, details, processing defects, and damage of CFRP surface under different grinding conditions. Furthermore, wavelet analysis and wavelet decomposition energy distribution ratio were used to analyze the high-frequency texture feature changes of machined surface roughness signals and SEM images in different directions, quantitatively characterizing grinding defects.

(1) Compared with dry grinding, the surface roughness *Ra*, *Rz*, and *RSm* reduced by 17.70%, 20.78%, and 25.40%, respectively, using NMQL and by 12.68%, 25.06%, and 11.43%, respectively, in the fiber direction.
(2) In the feed and fiber directions, the 2D fractal dimension of the surface roughness profile curve in NMQL grinding is reduced by 2.36% and 1.12%, respectively. The 3D fractal dimension of the SEM micro-topographic image was reduced by 4.75% (average magnifications of 300 and 500).
(3) In the feed direction, the spectral width and difference of MQL and NMQL grinding CFRP surfaces are slightly lower than those of dry grinding, and MQL is slightly better than NMQL. However, MQL and NMQL have advantages in the fiber direction. Compared with dry grinding, the spectral width and difference in NMQL are reduced by 21.76% and 31%, respectively.
(4) The energy proportion of the high-frequency detail signal of CFRP surface roughness grinded by NMQL is the lowest (feed and fiber directions are 0.0053 and 0.0006, respectively). Compared with dry grinding, the energy proportion of the high-frequency detail signal of the CFRP surface is reduced by 46.85%. This finding shows that the micromorphology features of machined surface roughness corresponded to the least NMQL changes.

(5) The energy proportion of the three detail sub-images in the NMQL grinding surface SEM wavelet decomposition diagram is the lowest. Moreover, the energy proportion of horizontal, vertical, and diagonal high frequencies is 0.155, 0.1952, and 0.0522, respectively. Therefore, the proportion of the damage characteristics in the horizontal direction (fiber direction) is less than that in the vertical direction (feed direction). This finding is also consistent with the energy proportion of the surface roughness wavelet analysis.

REFERENCES

Abdalla, H. S., & Patel, S. (2006). The performance and oxidation stability of sustainable metalworking fluid derived from vegetable extracts. *Proceedings of the Institution of Mechanical Engineers. Part B, Journal of Engineering Manufacture*, *220*(12), 2027–2040. DOI: 10.1243/09544054JEM357

Adibi, H., Esmaeili, H., & Rezaei, S. M. (2018). Study on minimum quantity lubrication (MQL) in grinding of carbon fiber-reinforced SiC matrix composites (CMCs). *International Journal of Advanced Manufacturing Technology*, *95*(9), 3753–3767. DOI: 10.1007/s00170-017-1464-x

Almeida, J. H. S.Jr, Souza, S. D., Botelho, E. C., & Amico, S. C. (2016). Carbon fiber-reinforced epoxy filament-wound composite laminates exposed to hygrothermal conditioning. *Journal of Materials Science*, *51*(9), 4697–4708. DOI: 10.1007/s10853-016-9787-9

Brown, C. A., Charles, P. D., Johnsen, W. A., & Chesters, S. (1993). Fractal analysis of topographic data by the patchwork method. *Wear*, *161*(1-2), 61–67. DOI: 10.1016/0043-1648(93)90453-S

Cheng, X. Q., Yasir, B., & Li, Z. H. (2011). Effects of hygrothermal environmental conditions on compressive strength of CFRP stitched laminates. *Journal of Reinforced Plastics and Composites*, *30*(2), 110–122. DOI: 10.1177/0731684410384894

Davim, J. P., Rubio, J. C., & Abrao, A. M. (2007). A novel approach based on digital image analysis to evaluate the delamination factor after drilling composite laminates. *Composites Science and Technology*, *67*(9), 1939–1945. DOI: 10.1016/j.compscitech.2006.10.009

Debnath, S., Reddy, M. M., & Yi, Q. S. (2014). Environmental friendly cutting fluids and cooling techniques in machining: A review. *Journal of Cleaner Production*, *83*, 33–47. DOI: 10.1016/j.jclepro.2014.07.071

Dogra, M., Sharma, V. S., Dureja, J. S., & Gill, S. S. (2018). Environment-friendly technological advancements to enhance the sustainability in surface grinding-A review. *Journal of Cleaner Production*, *197*, 218–231. DOI: 10.1016/j.jclepro.2018.05.280

Fu, R., Jia, Z., Wang, F., Jin, Y., Sun, D., Yang, L., & Cheng, D. (2018). Drill-exit temperature characteristics in drilling of UD and MD CFRP composites based on infrared thermography. *International Journal of Machine Tools & Manufacture*, *135*, 24–37. DOI: 10.1016/j.ijmachtools.2018.08.002

Gao, T., Li, C., Zhang, Y., Yang, M., Jia, D., Jin, T., Hou, Y., & Li, R. (2019). Dispersing mechanism and tribological performance of vegetable oil-based CNT nanofluids with different surfactants. *Tribology International*, *131*, 51–63. DOI: 10.1016/j.triboint.2018.10.025

Geng, D., Teng, Y., Liu, Y., Shao, Z., Jiang, X., & Zhang, D. (2019). Experimental study on drilling load and hole quality during rotary ultrasonic helical machining of small-diameter CFRP holes. *Journal of Materials Processing Technology*, *270*, 195–205. DOI: 10.1016/j.jmatprotec.2019.03.001

Gu, G., Wang, D., Wu, S., Zhou, S., & Zhang, B. (2024). Research status and prospect of ultrasonic vibration and minimum quantity lubrication processing of nickel-based alloys. *Intelligent and Sustainable Manufacturing*, *1*(1), 10006. DOI: 10.35534/ism.2024.10006

Hadad, M. J., Tawakoli, T., Sadeghi, M. H., & Sadeghi, B. (2012). Temperature and energy partition in minimum quantity lubrication-MQL grinding process. *International Journal of Machine Tools & Manufacture*, *54*, 10–17. DOI: 10.1016/j.ijmachtools.2011.11.010

Iskandar, Y., Tendolkar, A., Attia, M. H., Hendrick, P., Damir, A., & Diakodimitris, C. (2014). Flow visualization and characterization for optimized MQL machining of composites. *CIRP Annals*, *63*(1), 77–80. DOI: 10.1016/j.cirp.2014.03.078

Lawal, S. A., Choudhury, I. A., & Nukman, Y. (2012). Application of vegetable oil-based metalworking fluids in machining ferrous metals—A review. *International Journal of Machine Tools & Manufacture*, *52*(1), 1–12. DOI: 10.1016/j.ijmachtools.2011.09.003

Li, H. N., Xie, K. G., Wu, B., & Zhu, W. Q. (2020). Generation of textured diamond abrasive tools by continuous-wave CO2 laser: Laser parameter effects and optimisation. *Journal of Materials Processing Technology*, *275*, 116279. DOI: 10.1016/j.jmatprotec.2019.116279

Liao, Z., Abdelhafeez, A., Li, H., Yang, Y., Diaz, O. G., & Axinte, D. (2019). State-of-the-art of surface integrity in machining of metal matrix composites. *International Journal of Machine Tools & Manufacture*, *143*, 63–91. DOI: 10.1016/j.ijmachtools.2019.05.006

Mia, M., Gupta, M. K., Singh, G., Królczyk, G., & Pimenov, D. Y. (2018). An approach to cleaner production for machining hardened steel using different cooling-lubrication conditions. *Journal of Cleaner Production*, *187*, 1069–1081. DOI: 10.1016/j.jclepro.2018.03.279

Molaie, M. M., Akbari, J., & Movahhedy, M. R. (2016). Ultrasonic assisted grinding process with minimum quantity lubrication using oil-based nanofluids. *Journal of Cleaner Production*, *129*, 212–222. DOI: 10.1016/j.jclepro.2016.04.080

Qiao, G., Xiong, X., Yang, J., Zhen, D., & Zhang, F. (2025). Cooling and lubrication mechanisms of nanofluid minimum quantity lubrication milling based on interfacial micro-equivalent contact modelling. *Journal of Manufacturing Processes*, *141*, 1551–1563. DOI: 10.1016/j.jmapro.2025.03.080

Rodriguez, R. L., Lopes, J. C., Hildebrandt, R. A., Perez, R. R. V., Diniz, A. E., de Angelo Sanchez, L. E., Rodrigues, A. R., de Mello, H. J., de Aguiar, P. R., & Bianchi, E. C. (2019). Evaluation of grinding process using simultaneously MQL technique and cleaning jet on grinding wheel surface. *Journal of Materials Processing Technology*, *271*, 357–367. DOI: 10.1016/j.jmatprotec.2019.03.019

Santiuste, C., Miguélez, H., & Soldani, X. (2011). Out-of-plane failure mechanisms in LFRP composite cutting. *Composite Structures*, *93*(11), 2706–2713. DOI: 10.1016/j.compstruct.2011.05.034

Sarkar, N., & Chaudhuri, B. B. (1992). An efficient approach to estimate fractal dimension of textural images. *Pattern Recognition*, *25*(9), 1035–1041. DOI: 10.1016/0031-3203(92)90066-R

Sheikh-Ahmad, J., Urban, N., & Cheraghi, H. (2012). Machining damage in edge trimming of CFRP. *Materials and Manufacturing Processes*, *27*(7), 802–808. DOI: 10.1080/10426914.2011.648253

Shi, H., Yuan, S., Li, Z., Song, H., & Qian, J. (2019). Evaluation of surface roughness based on sampling array for rotary ultrasonic machining of carbon fiber reinforced polymer composites. *Measurement*, *138*, 175–181. DOI: 10.1016/j.measurement.2019.02.002

Soo, S. L., Shyha, I. S., Barnett, T., Aspinwall, D. K., & Sim, W. M. (2012). Grinding performance and workpiece integrity when superabrasive edge routing carbon fibre reinforced plastic (CFRP) composites. *CIRP Annals*, *61*(1), 295–298. DOI: 10.1016/j.cirp.2012.03.042

Su, Y. (2019). Effect of the cutting speed on the cutting mechanism in machining CFRP. *Composite Structures*, *220*, 662–676. DOI: 10.1016/j.compstruct.2019.04.052

Sun, X., Wu, Z. Q., & Tian, Y. (2003). *Fractal Principle and Application*. University of Science and Technology of China Press. 53-88.

Voss, R., Henerichs, M., & Kuster, F. (2016). Comparison of conventional drilling and orbital drilling in machining carbon fibre reinforced plastics (CFRP). *CIRP Annals*, *65*(1), 137–140. DOI: 10.1016/j.cirp.2016.04.001

Wang, H., Ning, F., Hu, Y., Li, Y., Wang, X., & Cong, W. (2018). Edge trimming of carbon fiber-reinforced plastic composites using rotary ultrasonic machining: Effects of tool orientations. *International Journal of Advanced Manufacturing Technology*, *98*(5), 1641–1653. DOI: 10.1007/s00170-018-2355-5

Wang, X. M., & Zhang, L. C. (2003). An experimental investigation into the orthogonal cutting of unidirectional fibre reinforced plastics. *International Journal of Machine Tools & Manufacture*, *43*(10), 1015–1022. DOI: 10.1016/S0890-6955(03)00090-7

Wickramasinghe, K. C., Sasahara, H., Abd Rahim, E., & Perera, G. I. P. (2020). Green Metalworking Fluids for sustainable machining applications: A review. *Journal of Cleaner Production*, *257*, 120552. DOI: 10.1016/j.jclepro.2020.120552

Wu, W., Li, C., Yang, M., Zhang, Y., Jia, D., Hou, Y., Li, R., Cao, H., & Han, Z. (2019). Specific energy and G ratio of grinding cemented carbide under different cooling and lubrication conditions. *International Journal of Advanced Manufacturing Technology*, *105*(1), 67–82. DOI: 10.1007/s00170-019-04156-5

Yang, M., & Li, C. (Eds.). (2025). *Heat Transfer Mechanism and Temperature Field Model of MQL Micro-Grinding Biomedical Materials*. IGI Global, 2-97.

Yashiro, T., Ogawa, T., & Sasahara, H. (2013). Temperature measurement of cutting tool and machined surface layer in milling of CFRP. *International Journal of Machine Tools & Manufacture*, *70*, 63–69. DOI: 10.1016/j.ijmachtools.2013.03.009

Zhang, Y., Li, C., Ji, H., Yang, X., Yang, M., Jia, D., Zhang, X., Li, R., & Wang, J. (2017). Analysis of grinding mechanics and improved predictive force model based on material-removal and plastic-stacking mechanisms. *International Journal of Machine Tools & Manufacture*, *122*, 81–97. DOI: 10.1016/j.ijmachtools.2017.06.002

Zhang, Y., Li, C., Jia, D., Zhang, D., & Zhang, X. (2015). Experimental evaluation of MoS_2 nanoparticles in jet MQL grinding with different types of vegetable oil as base oil. *Journal of Cleaner Production*, *87*, 930–940. DOI: 10.1016/j.jclepro.2014.10.027

Zhu, G., Yuan, S., & Chen, B. (2019). Numerical and experimental optimizations of nozzle distance in minimum quantity lubrication (MQL) milling process. *International Journal of Advanced Manufacturing Technology*, *101*(1), 565–578. DOI: 10.1007/s00170-018-2928-3

Chapter 7
Material Removal Mechanical Behavior of CFRP Longitudinal Grinding

ABSTRACT

In the context of 2D multi-angle ultrasonic assisted nanofluid MQL grinding of CFRP, fundamental issues such as the material removal mechanics for different FOA remain unaddressed. This chapter elucidates the material removal mechanism in longitudinal grinding of CFRP with varying FOAs. It analyzes the fracture mechanics of fiber during negative-rake-angle cutting by abrasive grains, specifically investigating the buckling failure and elliptical contact mechanics of 0° FOA. The interlayer shear failure mechanism and the bending-deformation-fracture removal process of 45° FOA are clarified. Furthermore, the chapter reveals the bending deformation and fracture removal mechanics of 90° FOA, along with the matrix cracking caused by longitudinal compression and associated fiber shear fracture. The buckling failure and bulk removal process of fiber bundles in 135° FOA are also examined. The revelation and analysis of these mechanical behaviors establish boundary conditions and provide a theoretical foundation for developing a grinding force model applicable to different FOAs.

7.1 INTRODUCTION

Carbon fiber reinforced polymer (CFRP) composites have become critical materials for weight reduction in the aerospace industry due to their high specific stiffness and strength. Furthermore, CFRP enables integrated design and manufac-

DOI: 10.4018/979-8-3373-9013-0.ch007

turing of structures and properties, facilitating the forming of large and complex components while significantly reducing the number of fasteners required (Gao et al., 2022;Wang et al., 2020). Research has also revealed innovative application prospects for such carbon fiber composites in military marine equipment manufacturing, where their superior vibration damping and noise reduction performance make them ideal for fabricating connection frames for condensate pumps in nuclear submarine reactors. Although CFRP possesses near-net-shape forming capability, extensive secondary machining is still necessary to meet component accuracy and assembly requirements (Chen et al., 2023a; An et al., 2024). Currently, cutting serves as the primary processing method, while grinding is the key process for ensuring surface integrity and dimensional precision (Luna et al., 2020). Precision components such as aircraft engine casings, sealing adjustment plates, and their mating positioning surfaces all require machining via precision grinding methods (Gong et al., 2024; Guo et al., 2023). However, as a highly complex material, CFRP exhibits a multiphase composition of fibers, resin, and interfaces at the micro-level, and a laminated structure with heterogeneous anisotropy at the macro-level (Wang et al., 2022). Its high strength, high modulus, anisotropy, and layered characteristics render it a typical difficult-to-machine material (Wang et al., 2023). Under the severe grinding conditions of high temperature, high pressure, and high speed, the material removal mechanism of CFRP becomes even more complex (Qu et al., 2022; Rao et al., 2023; Wu and Zhang, 2023). Traditional machining techniques are prone to causing damage such as fiber fracture, delamination, and cracking, making it difficult to achieve the required accuracy and surface quality (Chen et al., 2023b). This can ultimately lead to component scrapping, triggering catastrophic accidents and significant economic losses.

The material removal mechanism in grinding significantly influences grinding forces and surface integrity (Qu et al., 2024). The randomness of abrasive grains on the grinding wheel leads to complex contact behavior and cutting mechanics between the grains and the workpiece material. Consequently, the material removal mechanism in grinding is a complex process encompassing ductile material removal, brittle material removal, and frictional behavior (Zhang et al., 2024). Due to the unique structural characteristics of CFRP, including heterogeneity, anisotropy, multiphase multiscale composition, and laminated structure, its material removal mechanism is even more intricate. Under different material removal behaviors, grinding forces manifest in various forms (Zhang et al., 2023). Therefore, a comprehensive analysis of material removal mechanical behavior is crucial. A deep understanding of the material removal mechanism serves as the foundation for establishing an accurate grinding force model and a prerequisite for ensuring its reliability (Wei et al., 2023).

Through single-grit scratching experiments on CFRP, Ning et al. (2019) observed a large ductile removal zone before the onset of continuous brittle fracture and crack formation, with a ductile-to-brittle transition occurring at a distance of 54.32 μm. As the scratch depth increased, fiber fracture, fiber-matrix debonding, and macro-cracking occurred significantly, indicating a brittle removal mode involving delamination. Furthermore, regarding grinding processes, Wang et al. (2018) found that grinding along a FOA of 90° produced more unsuccessfully removed fibers, fiber pull-out, and burr formation. The more severe the carbon fiber fracture, the more debris generated, leading to increased cracking and resulting in higher surface roughness. Hu and Zhang (2003) observed that the geometry of grinding chips in unidirectional CFRP primarily depends on fiber orientation. Chips at 0° FOA mainly consist of pulled-out fractured fibers, while those at 90° FOA are predominantly fine powder. Liu et al. (2017) found that grinding forces do not change with increasing wheel grit size, which is attributed to the corresponding increase in the number of active abrasive grains. Wang et al. (2024) discovered that fiber pull-out occurs when the grinding feed direction aligns with the FOA. High-speed grinding contributes to improving the surface quality of CFRP, whereas excessively small or large abrasive grains may degrade surface quality. Similar to ceramic matrix FRCs, Dong et al. (2023) found that grinding forces are positively correlated with the maximum undeformed chip thickness. When agmax exceeds the grain protrusion height, the squeezing effect of abrasive grains on CFRP material becomes more pronounced, leading to severe tearing damage. Wen et al. (2023) compared the grinding performance of thermoset and thermoplastic matrix FRCs. They found that the surface roughness is lowest at FOA = 90° for both, due to excessive transverse compression in the contact zone, fiber bending fracture along the axis, and smaller fracture cross-sections resulting in less surface damage. For thermoset matrices, surface roughness is worst at FOA = 135°, attributed to damage occurring beneath the grinding surface when the load exceeds the fiber bending strength, and voids forming after broken fiber fragments are removed. For thermoplastic matrices, surface roughness is worst at FOA = 0°, and at FOA = 135°, surface voids are reduced due to the softened matrix coating the workpiece surface. Furthermore, they compared the grinding behavior of UD and MD fiber-reinforced thermoplastic matrix composites (Wen et al., 2024). The variation of grinding forces with grinding parameters was consistent for both MD and UD, but the grinding force for MD was slightly lower than for UD. Simultaneously, chip formation and material removal mechanisms differed between the two composites, indicating that the grinding performance of MD is not simply a superposition of UD.

However, fundamental issues regarding the material removal mechanics of CFRP with different fiber orientations remain unresolved in the context of multi-angle two-dimensional ultrasonic-assisted nanolubricant MQL grinding, a process particularly relevant for machining this typical difficult-to-machine aerospace material. This chapter further elucidates the material removal mechanisms in longitudinal grinding of CFRP with varying fiber orientations. It analyzes the fiber fracture mechanics under negative rake angle cutting by abrasive grains, investigates the buckling failure and elliptical contact mechanics of 0° fibers, clarifies the interlaminar shear failure mechanism and the bending deformation fracture removal process of 45° fibers, reveals the bending deformation and fracture removal mechanics of 90° fibers, including matrix cracking and fiber shear fracture induced by longitudinal compression, and examines the buckling failure and fiber bundle bulk removal process of 135° fibers. Through the elucidation and analysis of these mechanical behaviors, this study provides the boundary conditions and theoretical foundation for establishing grinding force models for different fiber orientations.

While Chapters 4 through 6 primarily examined the transverse grinding configuration, the fiber orientation relative to the cutting direction fundamentally alters the material removal behavior. This chapter pivots to longitudinal grinding of CFRP, investigating the distinct mechanical responses—including fiber fracture, matrix smearing, and interfacial debonding—that arise when the abrasive traverses parallel to the fiber axis. The influence of CFRP anisotropy on the cutting process is primarily reflected in the fiber/resin removal mechanisms, which vary with the FOA, resulting in directional differences during grinding and cutting. Different FOAs lead to distinct deformation states and failure modes in CFRP, thereby generating variations in chip morphology, thermomechanical behavior and its evolution, as well as surface integrity and damage. Accordingly, the FOA is defined as the clockwise angle between the carbon fiber axis and the grinding direction, as illustrated in Figure 1. When FOA = 0° (or 180°), the grinding direction (i.e., the abrasive grain cutting direction) is aligned with the fiber axis, referred to as parallel fiber grinding. When 0° < FOA < 90°, the grinding direction forms an acute angle with the fiber axis, with a typical case being 45°, defined as forward fiber grinding or acute-angle fiber grinding. When FOA = 90°, the grinding direction is perpendicular to the fiber axis, referred to as vertical fiber grinding. When 90° < FOA < 180°, the grinding direction forms an obtuse angle with the fiber axis, exemplified by 135°, defined as reverse fiber grinding or obtuse-angle fiber grinding.

Figure 1. The FOA of carbon fibers in longitudinal grinding of CFRP. (a) Parallel fiber grinding, (b) Along fiber grinding, (c) Vertical fiber grinding, (d) Reverse fiber grinding

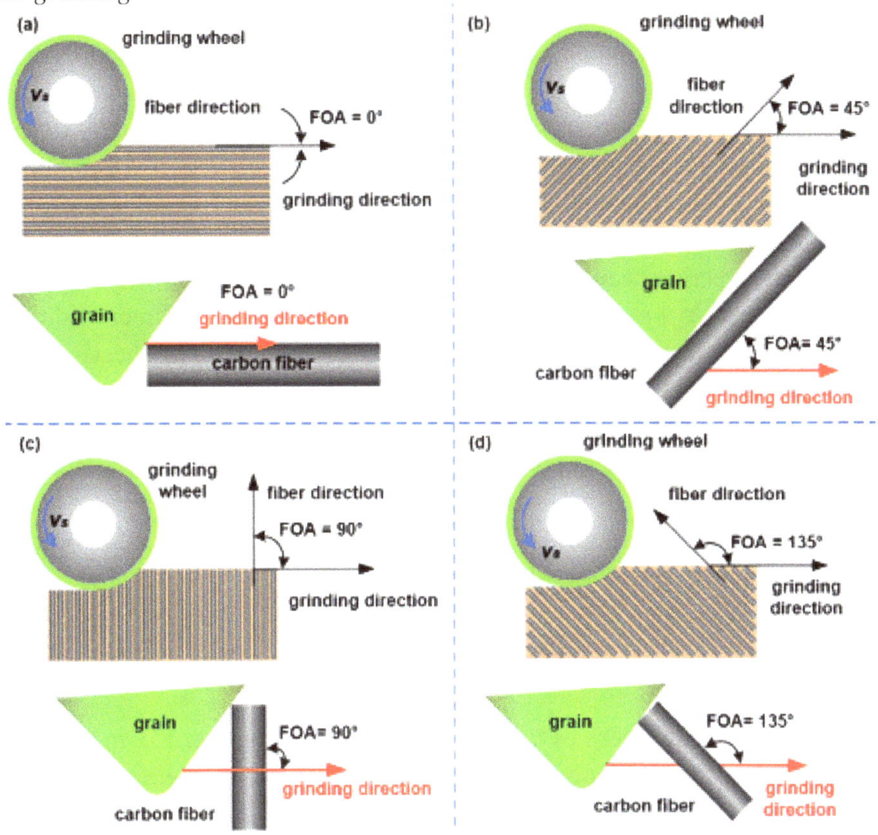

7.2 PARALLEL FIBER GRINDING

When unidirectional composite laminates are subjected to longitudinal compression, carbon fibers can only sustain minimal compressive stress. Even if the matrix remains within its elastic range, it cannot fully prevent the slender fibers from undergoing micro-buckling. Two distinct modes of fiber micro-buckling may occur: extensional (out-of-phase) buckling and shear (in-phase) buckling. In practice, fiber micro-buckling in composites is typically mixed-mode, involving both extensional and shear mechanisms, though one tends to dominate—commonly the shear mode. Longitudinal compressive failure in unidirectional CFRP manifests in multiple forms, including fiber micro-buckling, shear fracture, and transverse cracking (Gao et al.,

2024). Under longitudinal compression, the heterogeneity between resin and fiber induces transverse expansion, leading to cracking at the interface and within the matrix. Additionally, fiber micro-buckling contributes to transverse cracking. When the fiber-matrix bonding is strong and the fiber volume fraction V_f is high, carbon fibers undergo micro-buckling as illustrated in Figure 2, causing periodic tensile-compressive deformation in the resin transverse to the fiber direction. At high V_f, micro-buckling of the form shown in Figure 3 occurs, leading primarily to shear deformation in the resin matrix. In contrast, at low V_f, individual fibers or bundles may buckle at different times and in varying directions. When V_f is sufficiently high, buckling within the width plane is largely suppressed, and out-of-plane buckling in the thickness direction becomes more probable.

The computational model for longitudinal compression of unidirectional CFRP is simplified as a layered structure composed of alternating fiber sheets and matrix sheets bonded together and subjected to axial compression. It is assumed that only the fiber sheets carry compressive load, while the matrix provides transverse support. The combined thickness of one fiber sheet and one matrix sheet is taken as unity. Accordingly, the thicknesses of the fiber and matrix sheets are V_f and V_m, respectively, and the dimension of each sheet along the fiber direction is denoted as l. The buckling shape of the fiber sheet is assumed as:

$$u_y = a_m \cdot \sin\frac{m\pi L}{l}\left(m = 1,2,3,\cdots\right) \tag{7.1}$$

Calculate the strain energies δU_f and δU_m for the fiber and matrix in their deviated states, respectively, along with the work δW done by the external force. Here, δ denotes the variation from the critical state. The critical load is then determined via the energy relationship:

$$\delta W = \delta U_f + \delta U_m \tag{7.2}$$

7.2.1 Tensile-Compressive Buckling Failure

Under bending loads, the flexural strength of unidirectional CFRP is generally higher than its longitudinal compressive strength. This enhancement arises from the presence of the tensile zone, where carbon fibers remain relatively straight, thereby suppressing local buckling in the compressive region. The average stress in the combined compressive-tensile zone remains lower than the maximum compressive and tensile stresses at the outer surfaces of the CFRP workpiece. Plotting the longitudinal compressive strength against carbon fiber volume fraction yields Figure 2. The graph indicates that when the fiber volume fraction is below a critical value, failure occurs by extensional buckling instability. Beyond this critical volume fraction V_f, shear-type buckling instability becomes dominant. The mechanisms governing compressive failure are complex, and current analysis primarily relies on energy methods.

When the fiber spacing is sufficiently large, extensional (out-of-phase) buckling occurs, inducing alternating transverse tensile and compressive strains in the matrix, hence this mode is also referred to as transverse tensile-compressive buckling. As shown in Figure 2, in this failure mode, adjacent fiber sheets buckle in opposite directions, causing the matrix sheets to undergo alternating transverse tension and compression. Assuming the strain in the matrix sheet is independent of spatial coordinates, the strain ε_m is expressed as:

$$\varepsilon_m = \frac{2u_y}{V_m}. \tag{7.3}$$

The stress σ_m is:

$$\sigma_m = \frac{2u_y E_m}{V_m}, \tag{7.4}$$

so,

$$\delta U_m = \frac{1}{2}\int_0^l E_m\left(\frac{2u_y}{V_m}\right)^2 d(V_m L). \tag{7.5}$$

Substituting u_y into the above equation yields:

$$\delta U_m = \frac{E_m a_m^2 l}{V_m}. \tag{7.6}$$

By using the strain energy formula of the buckling rod, the strain energy of the fiber sheet deviating from its state can be obtained as δU_f,

$$\delta U_f = \frac{1}{2}E_f I_f \int_0^l \frac{2^2 u_y^2}{L^4} dL, \tag{7.7}$$

where I_f is the moment of inertia of the fiber sheet. Considering that the moment of inertia of a unit width fiber sheet is $I_f = V_f^3/12$, and substituting u_y into the above equation,

$$\delta U_f = \frac{m^4 \pi^4 E_f V_f^3 a_m^2}{48 l^3}. \tag{7.8}$$

Similarly, by using the formula for the external force work of the buckling rod, the work done by the fiber external force can be obtained:

$$\delta W = \frac{1}{2} P \int_0^1 \left(\frac{dv}{dL}\right)^2 dL = \frac{p a_m^2 m^2 \pi^2}{4l}. \tag{7.9}$$

Therefore,

$$P = \frac{\pi^2 E_f V_f^2}{12 l^2} \left[m^2 + \frac{48 E_m l^4}{\pi^4 E_f V_f^3 V_m m^2} \right]. \tag{7.10}$$

The condition for taking the minimum value of P is:

$$\begin{cases} \frac{dP(m^2)}{dm^2} = 0 \\ \frac{d^2 P(m^2)}{d(m^2)^2} > 0 \end{cases}, \tag{7.11}$$

$m^2 = \sqrt{\frac{48 E_m l^4}{\pi^4 E_f V_f^3 V_m}}$ is substituted into the P calculation formula to obtain the critical buckling load P_{cr} of the fiber sheet:

$$P_{cr} = 2 V_f \sqrt{\frac{E_f V_f E_m}{3 V_m}}. \tag{7.12}$$

Figure 2. Mechanical behavior of buckling removal of tension-compression type fibers

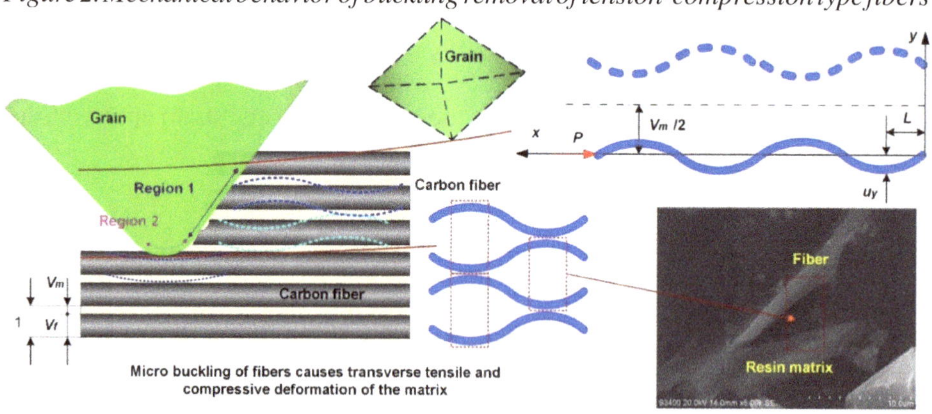

7.2.2 Shear Buckling Failure

Figure 3 shows the matrix exhibiting shear failure, accompanied by fiber fracture. Carbon fibers undergo a mixed-mode failure characterized by sudden bending and buckling around a specific cross-section. This behavior is influenced by the local state of the fibers at that section, such as non-uniform strain, distortion, fracture, and waviness, as well as by the condition of the surrounding resin matrix, including resin-rich or resin-deficient zones, defects, and microcracks. The integrity of the fiber-matrix interface, including its adhesion quality and any debonding, also plays a critical role. Weakened load-bearing support from the interface and matrix, combined with inherent fiber defects and damage, leads to fiber bending and wrinkling, ultimately inducing shear instability. It should be noted that the concept of "instability" here is not entirely equivalent to classical buckling. In carbon fiber composites, compressive strains and stresses are not uniformly distributed among fiber bundles or individual filaments. Those fibers or bundles with poor alignment or subjected to higher compressive stresses may buckle first, imposing lateral loads on adjacent fibers and promoting localized buckling.

When fibers buckle in the same direction, the resulting deformation is dominated by shear strain in the matrix. This buckling mode, often referred to as shear-type micro-buckling, is the most common form of fiber instability and is the dominant mechanism observed in this study. As illustrated in Figure 3, the buckled fiber sheets deform in a mutually aligned manner during shear-type buckling. Given that the shear modulus of the fibers is substantially higher than that of the matrix, the shear deformation within the fiber sheets can be neglected. Detailed finite element numerical modeling and simulations corresponding to this figure are provided in Chapter 9. To express the strain energy of the matrix sheet in shear-type buckling, it is necessary to establish the relationship between the shear strain in the matrix sheet and the buckling morphology of the fiber sheets. As can be observed in Figure 3,

$$\begin{cases} \alpha_1 = \mathrm{du}_y / dL \\ \alpha_2 = \dfrac{V_f}{V_m} \dfrac{\mathrm{du}_y}{dL} \end{cases} . \tag{7.13}$$

Because of the shear strain of the matrix $\gamma_m = \alpha_1 + \alpha_2$,

$$\gamma_{\mathrm{m}} = \left(1 + \frac{V_f}{V_m}\right) \frac{\mathrm{du}_y}{dL} . \tag{7.14}$$

The shear stress is:

$$\tau_{\mathrm{m}} = G_m\left(1+\frac{V_f}{V_m}\right)\frac{du_y}{dL}, \tag{7.15}$$

therefore,

$$\delta U_m = \frac{G_m}{2}\left(1+\frac{V_f}{V_m}\right)^2 V_m a_m^2. \tag{7.16}$$

Substituting u_y in yields:

$$\delta U_m = \frac{G_m m^2\pi^2}{4l}\left(1+\frac{V_f}{V_m}\right)^2 V_m a_m^2. \tag{7.17}$$

Substituting the above formulas into P yields:

$$P = G_m\left(1+\frac{V_f}{V_m}\right)^2 V_m + \frac{m^2\pi^2 E_f V_f^3}{12l^2}. \tag{7.18}$$

The second term can be omitted because the bending half-wave is longer than the fiber diameter. Therefore, the above equation can be transformed into:

$$P_{cr} = \frac{G_m}{1-V_f}. \tag{7.19}$$

The corresponding critical stress and strain of the fibers are:

$$\begin{cases} \sigma_{\mathrm{fcr}} = \dfrac{G_m}{V_f(1-V_f)} \\ \varepsilon_{\mathrm{fcr}} = \dfrac{G_m^2}{E_f V_f(1-V_f)} \end{cases}. \tag{7.20}$$

Figure 3. Mechanical behavior of shear type fiber bending deformation removal

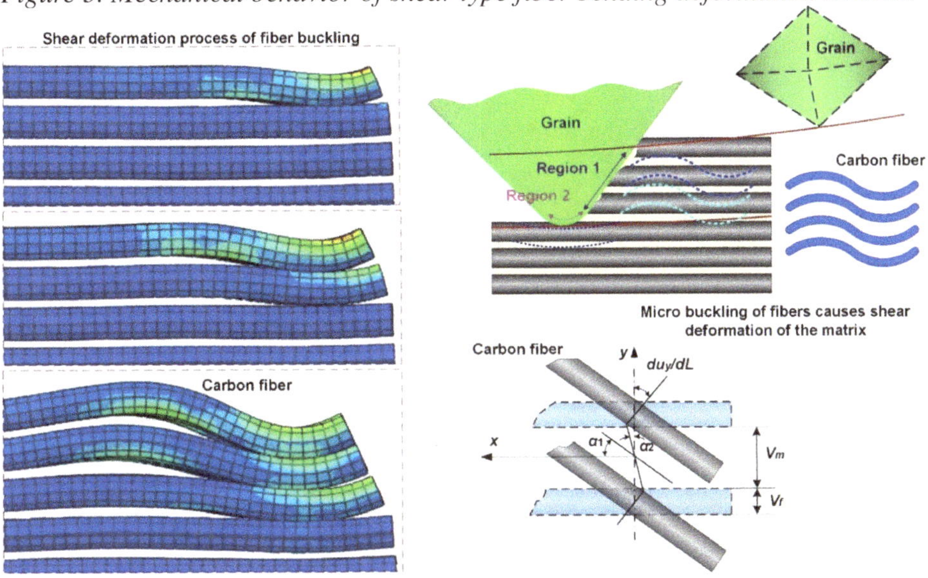

7.2.3 Normal Contact Force in the Elliptical Domain

As shown in Figure 4, the top of the grains forms an elliptical contact with the fibers. The length of the major axis along the x-axis is $2a_e$, while the length of the minor axis along the z-axis is $2b_e$. The eccentricity of the ellipse is calculated as follows:

$$k_e = \sqrt{1 - \left(\frac{b_e}{a_e}\right)^2} \tag{7.21}$$

The semimajor axis is obtained from the arc length formula,

$$a_e = \left(\pi - 2\theta_{gf}\right) \cdot r_g \tag{7.22}$$

where,

$$\theta_{gf} = \arcsin\frac{r_g - h_{gf}}{r_g} \tag{7.23}$$

$$h_{gf} = a_g - n_e \cdot d_f \tag{7.24}$$

The semiminor axis is obtained from the arc length formula,

$$b_e = 2\xi_{gf} \cdot r_g \tag{7.25}$$

According to the cosine theorem,

$$\xi_{gf} = \arccos\frac{r_g^2 + \left(r_f + r_g - h_{gf}\right)^2 - r_f^2}{2 \cdot r_g \cdot \left(r_f + r_g - h_{gf}\right)} \tag{7.26}$$

Figure 4. Elliptical domain contact mechanical behavior of grain tip on fibers in region 2

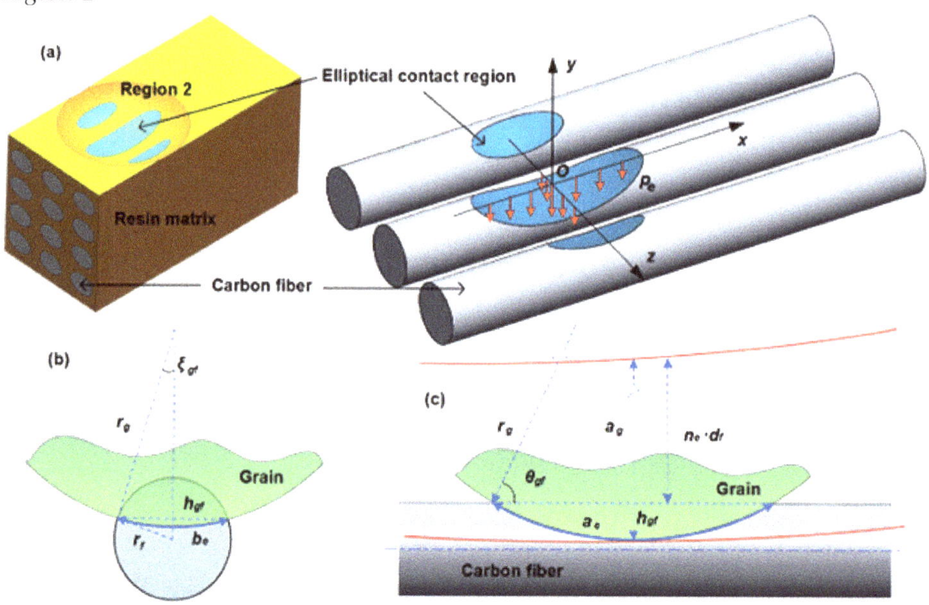

The pressure distribution of the elliptical contact area is as follows according to the theory of the point load of the elastic half space:

$$p_e = p_{max}\left[1 - (x/a_e)^2 - (z/b_e)^2\right]^{1/2} \tag{7.27}$$

According to the theory of maximum elongation line strain, the fracture criterion for fibers is:

$$\sigma_y - \upsilon\left(\sigma_x + \sigma_z\right) = \sigma \tag{7.28}$$

7.3 ALONG-FIBER GRINDING

7.3.1 Interlayer Shear

When a abrasive grain cuts FOA = 45° CFRP, the resin matrix develops initial cracks under shear stress, which subsequently propagate and lead to fracture as the grain advances. The interface phase undergoes shear deformation between the carbon fiber and resin matrix until the shear strength limit is reached, resulting in interfacial debonding. As illustrated in Figure 5, in Region 1, the fibers severed by the abrasive grain remain in contact with adjacent fibers along the cutting direction. Interlaminar shear occurs between the cut fibers and the subsequent fibers, causing the material to slide along the rake face of the grain under further grinding action. The shear force component acting on the abrasive grain in Region 1 is calculated as the product of the interfacial shear strength S_i and the shear area:

$$F_{45\_Sh} = S_i \cdot \frac{1}{2} b_g l_{sh} \tag{7.29}$$

$$l_{sh} = \frac{a_g}{\cos\theta_f} \tag{7.30}$$

where b_g is the grain width and l_{sh} is the crack length.

Figure 5. Interlayer shear and debris friction

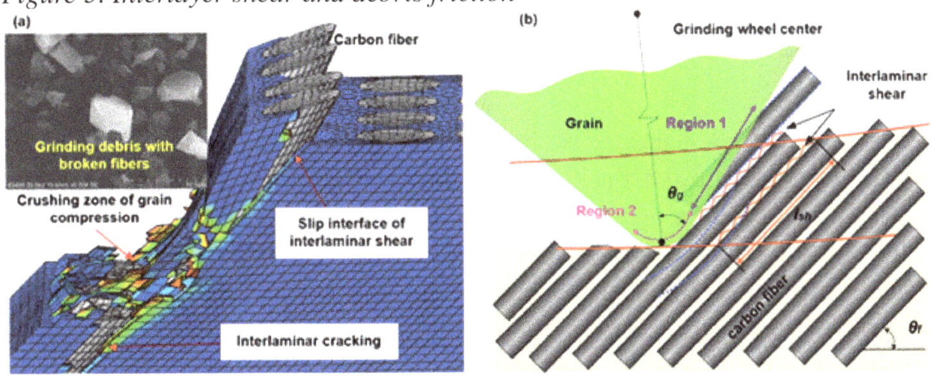

7.3.2 Carbon Fiber Bending Deformation and Fracture Removal

In Region 2, carbon fibers undergo bending deformation and fracture. Starting from point B_{45} at the tip of the abrasive grain in the cutting direction, the fiber is subjected to a force perpendicular to its axis, leading to bending deformation. Simultaneously, the fiber is supported and constrained by the underlying CFRP material yet to be cut, which resists fiber deflection. This reaction force is denoted as p_m. Below point C_{45}, the fiber is also influenced by the interfacial bonding force, represented as p_b. Figure 6 illustrates a schematic of single-grit cutting of an individual carbon fiber. Assuming an elastic correlation between fiber deformation and the mechanical response of the surrounding material, the bending fracture force model for a single carbon fiber is established using the elastic foundation beam theory. To facilitate generalized analysis and calculation, the composite material supporting the fiber can be treated as an equivalent homogeneous material (EHM). Let p_m be the reaction force on the fiber cross-section with deflection w; then $p_m = wk_m$, where k_{mn} is the normal constraint parameter of the surrounding EHM.

$$k_{mn} = \frac{1.23 E_{mf}}{C_{mn}\left(1 - v_{mf}^2\right)} \left[\frac{E_{mf} d_f^4}{C_{mn}\left(1 - v_{mf}^2\right) E_f I_f}\right]^{0.11} \tag{7.31}$$

where C_{mn} is the coefficient of the action mode (C_{mn} =1 under uniform pressure and C_{mn} =1.3 under uniform deflection), and in this study, C_{mn} =1.06. The elastic modulus E_{mf} and Poisson's ratio v_{mf} of the equivalent homogeneous material EHM can be calculated by the mixing rules of composite materials (Feng and Cook, 1983):

$$E_{mf} = \frac{E_m}{1 - v_m^2} \tag{7.32}$$

$$v_{mf} = \frac{v_m}{1 - v_m} \tag{7.33}$$

To determine the actual deflection at point C_{45} as a function of the still-intact bending beam, a critical deflection is introduced. This parameter characterizes the maximum deformation that the matrix or matrix/fiber bond can sustain without failure. As long as this critical value is not exceeded, the matrix/fiber connection between adjacent carbon fibers remains intact. Thus, point C_{45} defines the position where the critical deflection has not yet been reached. As the uncut carbon fiber continues to deform, spalling stress develops at the crack immediately ahead of the cutting edge, leading to an increase in the vertical crack length h_{g_45} and causing C_{45} to shift toward the lower end of the carbon fiber. To consistently describe fiber deformation and calculate the resulting cutting force on the abrasive grain, the deformation function of the carbon fiber must be determined. Based on variations in loading along the negative y-direction, the deflection curve function is divided into three segments, as illustrated in Figure 6. These three independent sections are combined by defining boundary and transition conditions. An x/y coordinate system is established as shown in Figure 6, and a representative volume element is selected to formulate equilibrium equations for force and moment. The abrasive grain engages in negative-rake cutting, with the cutting edge acting on the top of the fiber at point A_{45}. When the carbon fiber in the grinding zone is bent to a certain extent, interfacial cracking occurs below the grinding arc zone (below point B_{45}). The depth of this cracking is denoted by h_{g_45}, and point C_{45} indicates the bottom of the interfacial crack.

Figure 6. Bending deformation and fracture of 45° fiber during grain cutting

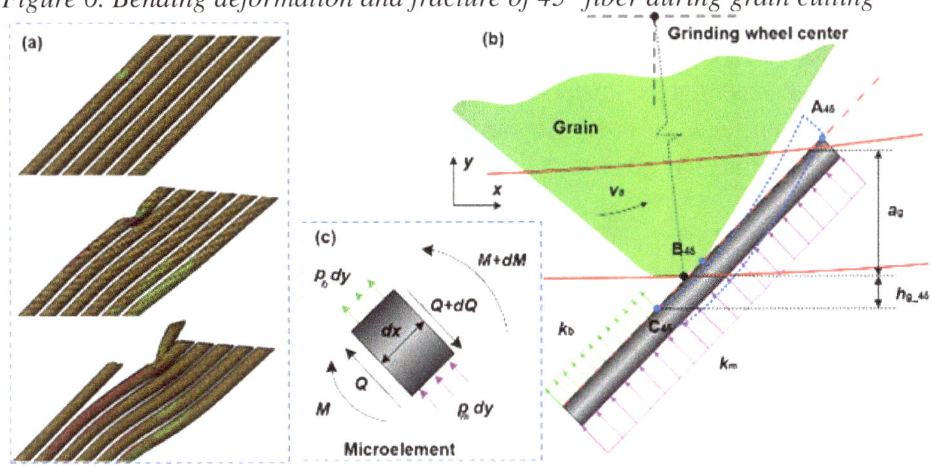

An equilibrium equation is established for the bending moment and force of the representative volume unit dy of the carbon fiber:

$$\begin{cases} Q-(Q+dQ)+\left(k_{mn}+k_b\right)wdy = 0 \\ M-(M+dM)+(Q+dQ)dy+\left(k_{mn}+k_b\right)dy^2 = 0 \end{cases} \tag{7.34}$$

where M is the bending moment and Q is the shear force. The deformation control equation of the carbon fiber elastic foundation beam is:

$$E_{ft}I_f\frac{d^4w}{dy^4}+\left(k_{mn}+k_b\right)w = 0 \tag{7.35}$$

7.4 VERTICAL FIBER GRINDING

7.4.1 Carbon Fiber Bending Fracture

The calculation method for the abrasive cutting removal force of 90° CFRP grinding is similar to that of 45° CFRP grinding. However, there are differences in the cutting interference state and fiber deformation boundary conditions, as shown in Figure 7. Therefore, this section only focuses on the mechanical behavior analysis and calculation of these differences. Based on the cutting action point and cutting state of grains on fibers, it is important to consider the tangential constraint effect of the surrounding equivalent homogeneous material on fiber support, in addition to the normal constraint parameters. This requires the introduction of the tangential constraint parameter, k_{mt}, supported by surrounding materials (Feng and Cook, 1983). The expression for k_{mt} is given:

$$k_{mt} = \frac{E_{mf} d_f}{4C_{mt}\left(1 + v_{mf}\right)} \left[\frac{2\left(1 - v_{mf}^2\right) E_f I_f}{E_{mf} d_f}\right]^{1/3}, \tag{7.36}$$

C_{mt} is the property parameter of the surrounding supporting material, and in this paper, it is taken as 0.7.

Figure 7. Removal of 90° fiber bending fracture by grain cutting

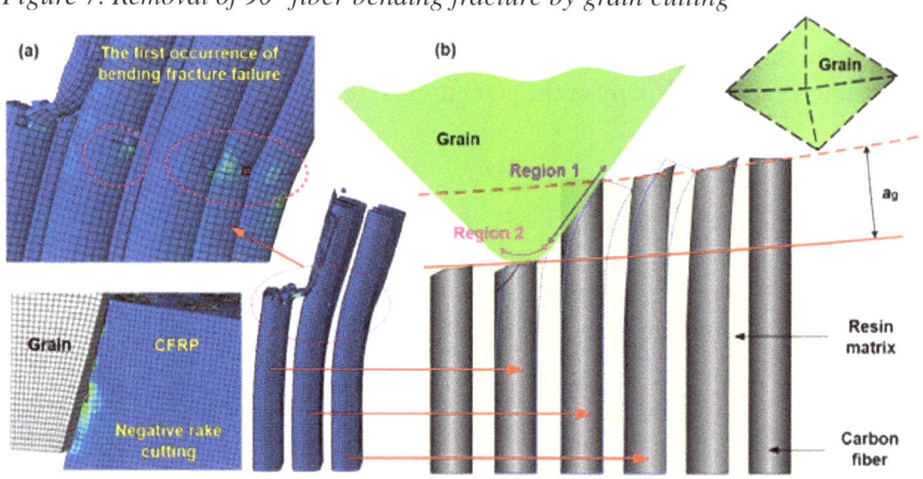

Based on the variation in loading along the negative y-direction, the deflection curve function for 90° fiber cutting deformation is divided into two segments, as shown in Figure 8. A representative volume element is selected to establish the equilibrium equations for force and moment. The abrasive grain performs negative-rake cutting, with its cutting edge acting on the top of the fiber at point A_{90}. When the carbon fiber within the grinding zone is bent to a certain extent, interfacial cracking occurs below the grinding arc region (beneath point B_{90}). The cracking depth is denoted as h_g, and point C_{90} indicates the bottom of the interfacial crack. By formulating and rearranging the equilibrium equations of force and moment for the representative volume element of the carbon fiber, the governing differential equation for the deformation of the carbon fiber modeled as an elastic foundation beam is derived as:

$$E_{ft} I_f \frac{d^4 w}{d y^4} - k_{mt} \frac{d^2 w}{d y^2} + \left(k_{mn} + k_b \right) w = 0 \tag{7.37}$$

Figure 8. Bending fracture force model for grain cutting 90 ° fibers in zone 1

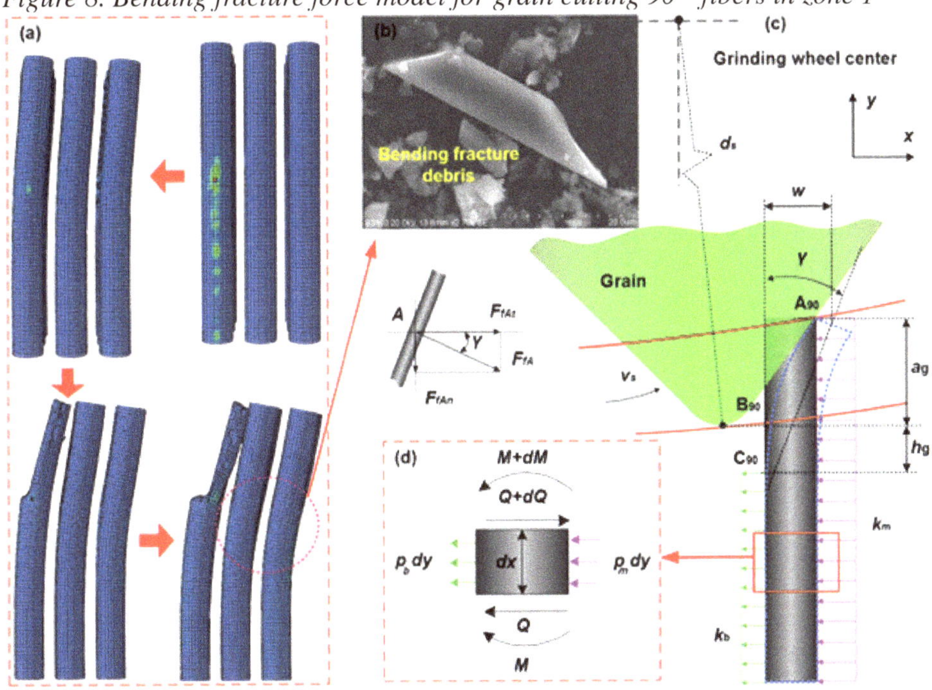

7.4.2 Matrix Cracking and Fiber Shear Fracture Under Longitudinal Compression

The compressive failure of resin-based composites is primarily governed by the failure of the matrix material. During 90° grinding of unidirectional CFRP, Region 2 experiences longitudinal compression, which induces debonding and cracking along the fiber direction, as illustrated in Figure 9. At this point, the transverse tensile strain ε_t in the CFRP reaches its ultimate transverse failure strain ε_{tu}, i.e., ε $\varepsilon_t = \varepsilon_{tu}$. Denoting the longitudinal stress in the CFRP as σ_l, the transverse strain ε_t under uniaxial loading is given by:

$$\varepsilon_t = -\frac{\upsilon_f V_f + \upsilon_m V_m}{E_f V_f + E_m V_m}\sigma_l \tag{7.38}$$

Typically, the transverse failure strain of CFRP composites is lower than the failure strain of the pure matrix. According to the following empirical formula:

$$\varepsilon_{tu} = \left(1 - \sqrt[3]{V_f}\right)\varepsilon_{mu} \tag{7.39}$$

When the external load reaches the level causing cracking failure, $\sigma_l = C_{Lc}$ and $\varepsilon_t = \varepsilon_{tu}$. Substituting these into the two equations above yields the longitudinal compressive limit as:

$$C_{Lc} = \frac{E_f V_f + E_m V_m}{\upsilon_f V_f + \upsilon_m V_m}\left(1 - \sqrt[3]{V_f}\right)\varepsilon_{mu} \tag{7.40}$$

Furthermore, due to the high fiber volume fraction of the CFRP used in this study, shear failure also occurs under longitudinal compression, which depends primarily on the shear strength of the fibers. The ultimate longitudinal compressive stress at shear failure is given by:

$$\mathrm{S_{Lc}} = 2\tau_{\mathrm{fu}}\left[\mathrm{V_f} + \frac{\left(1 - \mathrm{V}_f\right)E_m}{E_f}\right] \tag{7.41}$$

where τ_{fu} denotes the shear strength of the fiber.

Figure 9. Longitudinal compression of fibers by grains during 90° CFRP grinding

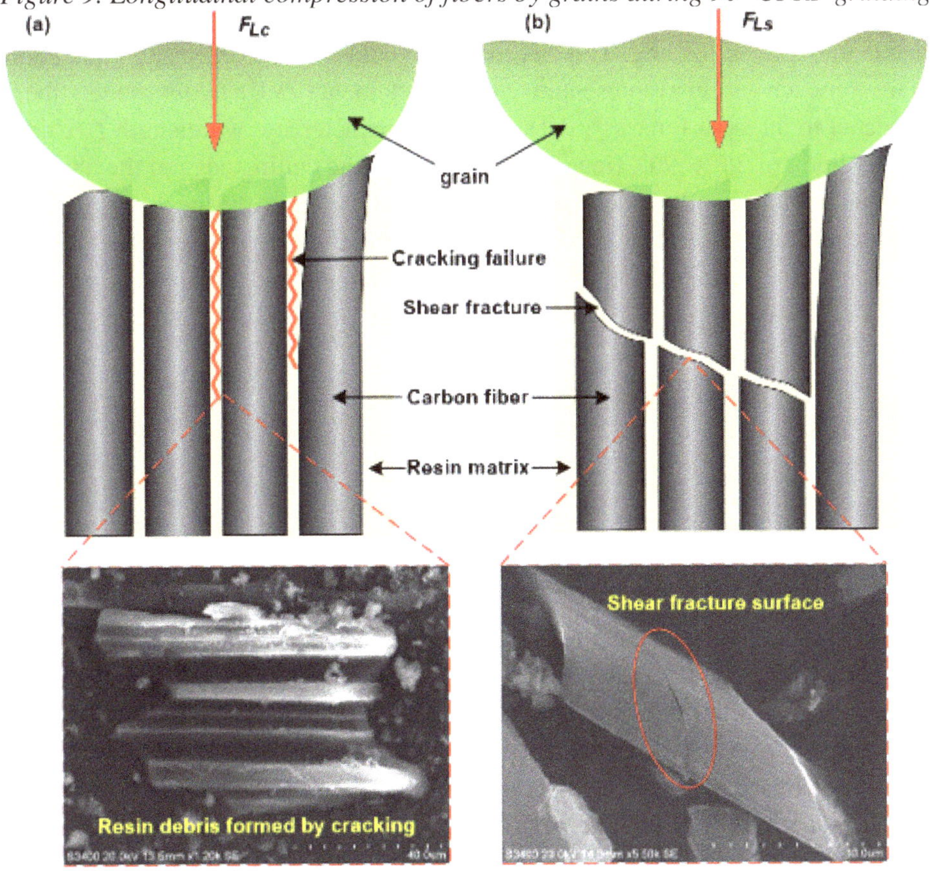

7.5 REVERSE FIBER GRINDING

7.5.1 Buckling Failure of Carbon Fibers

As illustrated in Figure 10, the abrasive grain removal mechanism in 135° carbon fiber-reinforced polymer (CFRP) can be categorized into two regimes based on the $2\theta_g$ of the abrasive grain: $2\theta_g$<90° and $2\theta_g$>90°. When $2\theta_g$ > 90°, it is hypothesized that fibers in contact with the abrasive grain undergo transverse micro-buckling failure due to axial compression. The corresponding strength of the embedded fibers is defined as the micro-buckling strength σ_{mb}. Once this strength is exceeded, fibers fail via transverse micro-buckling, and the detached fibers, along with matrix debris, are expelled along the rake face of the abrasive grain through the cutting motion. Micro-buckling describes a coordinated structural failure in unidirectional CFRP (UD-CFRP), wherein a significant portion of adjacent fibers simultaneously buckle under axial loading. This behavior is modeled using an elastic foundation beam approach within a representative volume element (RVE). In this model, fibers subjected to an axial force P_{mb} deflect over a characteristic wavelength L. Under the condition of $2\theta_g > 90°$, fibers interacting with the cutting edge of the abrasive grain are predominantly under compressive load. Consequently, the dominant fiber fracture mechanism in Region 1 is micro-buckling. By applying the total potential energy method in conjunction with variational principles, and accounting for matrix slip, the critical compressive strength of the CFRP material is derived as:

$$\sigma_{\mathrm{mb}} = G_m\left(\mathrm{V}_f + \frac{E_m}{E_{f1}}(1 - \mathrm{V}_f)\right) \cdot \left[2(1+v_m)\sqrt{\frac{\pi\sqrt{\pi}\,\eta\, r_f}{3\frac{E_m}{E_{f1}}\left(\mathrm{V}_f\frac{E_m}{E_{f1}} + 1 - \mathrm{V}_f\right)\left(1 + \mathrm{V}_f\upsilon_f + \upsilon_m(1 - \mathrm{V}_f)\right)}} + 1 - \xi - \frac{\sin(\pi\xi)}{2\pi}\right] \tag{7.42}$$

The parameter η represents the interfacial fiber matrix bonding performance. For a double-sided embedded fiber, η=2, whereas for unilateral embedding, η=1. The parameter ξ describes the matrix slip ratio related to the wavelength L, $\xi = 2s/L$. The values for ξ and η can be estimated through simulation or experimentation, and this study refers to the experimental research conducted by Naik and Kumar, where ξ= 0.02 and η=1.98 (Naik and Kumar, 1999).

Figure 10. Schematic diagram of carbon fiber buckling failure

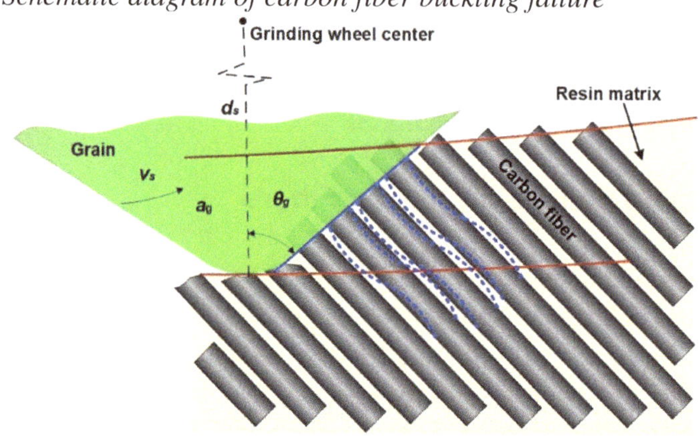

7.5.2 Bending and Fracture of Carbon Fibers

The bending deformation and fracture model of FOA = 135° CFRPs when 2θg < 90° are illustrated in Figure 11. The mechanical analysis and calculation methods used are the same as those for 90° CFRPs and will not be repeated here.

Figure 11. Bending fracture model of 135° carbon fiber

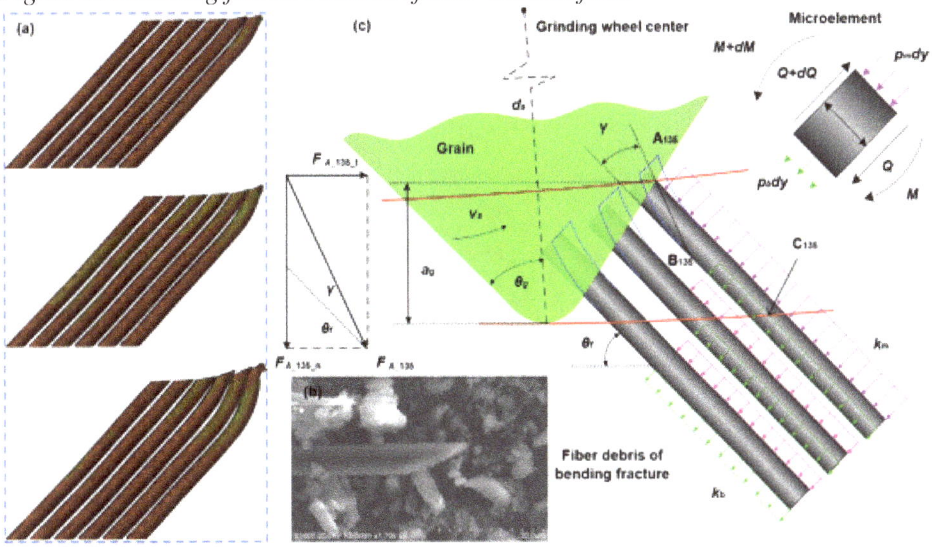

7.5.3 Block Removal of Fiber Bundles

As shown in Figure 12, microscopic observation of grinding chips from 135° CFRP reveals the occurrence of fiber-bundle removal in the form of bulk fracture. The fracture-induced removal of fiber bundles occurs progressively: after a portion of carbon fibers within a bundle fractures, the stress state of the remaining fibers changes accordingly, resulting in a bundle strength that is lower than the strength of individual fibers. Based on probabilistic statistics, the relationship between the strength distribution of the fiber bundle and the strength of individual fibers can be established. The average fracture stress σ_{fb} (i.e., the bundle strength) of a single fiber within the bundle follows a normal distribution,

$$\omega\left(\sigma_{fb}\right) = \frac{1}{\phi_{fb}\sqrt{2\pi}}\exp-\left[\frac{\left(\sigma_{fb}-\overline{\sigma_{fb}}\right)^2}{2\phi_{fb}}\right] \tag{7.43}$$

where $\omega(\sigma_{fb})$ is the density distribution function of the fiber bundle strength, σ_{fb_mv} is the average strength of the fiber bundle,

$$\sigma_{fb\_mv} = \sigma_{f\max}\left[1 - F\left(\sigma_{f\max}\right)\right] \tag{7.44}$$

$F(\sigma)$ is the strength distribution function of a single fiber, and σ_{fmax} is the average fiber stress in the bundle of fibers with the maximum fracture load, obtained by taking the extreme value of the formula $\sigma[1\text{-}F(\sigma)]$,

$$\sigma_{f\max} = \left(L_{mf}\alpha_{135}\beta_{135}\right)^{-\frac{1}{\beta_{135}}} \tag{7.45}$$

Therefore,

$$\overline{\sigma_{fb}} = \left(L_{mf}\alpha_{135}\beta_{135}e\right)^{-\frac{1}{\beta_{135}}} \tag{7.46}$$

The standard deviation is determined using:

$$\phi_{fb} = \sigma_{f\max}\sqrt{F\left(\sigma_{f\max}\right)\left[1 - F\left(\sigma_{f\max}\right)\right]/N_{mf}} \tag{7.47}$$

α_{135} and β_{135} are the scale and shape parameters, respectively, with $\beta_{135} > 1$, indicating the two parameters of the distribution.

Figure 12. Schematic diagram of multi fiber block removal

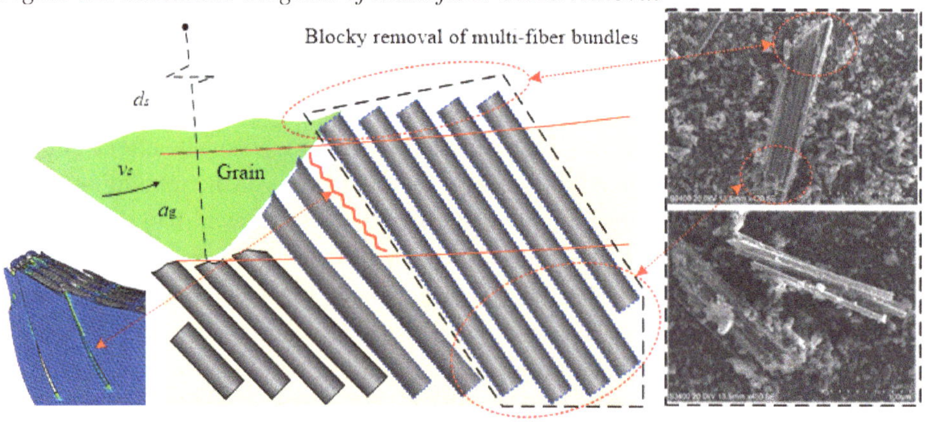

7.6 SUMMARY

This chapter further reveals the material removal mechanisms in longitudinal grinding of CFRP with different fiber orientations, analyzes the fiber fracture mechanics under negative-rake cutting by abrasive grains, investigates the buckling failure and elliptical contact mechanics of 0° fibers, elucidates the interlaminar shear failure mechanism and the bending deformation fracture process of 45° fibers, uncovers the bending deformation and fracture-based removal behavior of 90° fibers, as well as the matrix cracking and fiber shear fracture induced by longitudinal compression, and examines the buckling failure and bulk removal process of fiber bundles in 135° fibers.

(1) The micro-buckling deformation behavior of 0° carbon fibers under negative-rake cutting by abrasive grains is revealed. The material removal mechanics associated with shear-mode (dominant) and extensional-mode buckling failures in 0° carbon fibers are investigated. The shear instability state of carbon fibers is analyzed, and the strain energy of the deviated states of carbon fibers and the matrix is derived based on different fiber buckling morphologies. The critical load is then determined through energy relationships. Additionally, for the elliptical contact mechanics between the abrasive grain tip and horizontal carbon fibers, the fiber fracture criterion is established based on the elliptical contact pressure distribution at the grain/fiber interface and the maximum principal strain theory in material mechanics.
(2) The interlaminar shear-slip behavior between 45° carbon fibers severed by abrasive grains and adjacent fibers along the cutting direction is analyzed. The interlaminar shear mechanics, influenced collectively by interfacial shear strength, shear area, abrasive grain width, and crack length, are revealed. The debonding and cracking along the fiber direction in 90° carbon fibers caused by longitudinal compression from the abrasive grain tip are investigated, and the mechanical behavior of matrix cracking under longitudinal compression as well as the shear fracture mechanism of carbon fibers are elucidated.
(3) The bending deformation and fracture removal mechanics of 45°, 90°, and 135° carbon fibers are studied. The boundary conditions at the contact points between abrasive grains and carbon fibers under different fiber orientations are clarified. Based on the elastic foundation beam theory, the equilibrium equations of force and moment for a representative volume element of carbon fiber are established. Subsequently, the fiber bending fracture deformation process is interpreted by combining the deformation governing equation with cutting simulation results under given cutting contact boundary conditions.

(4) The transverse micro-buckling behavior of 135° carbon fibers under axial compression induced by abrasive grains is analyzed. The coordinated structural failure resulting from simultaneous transverse micro-buckling of adjacent carbon fibers is investigated. The bulk removal mechanics of fiber bundles in 135° CFRP are explored, the mapping relationship between fiber bundle strength distribution and single-fiber strength is clarified, and the influence of fiber bundle fracture behavior on the average fracture stress of individual fibers within the bundle is examined.

REFERENCES

An, Q., Yang, J., Li, J., Liu, G., Chen, M., & Li, C. (2023). A state-of-the-art review on the intelligent tool holders in machining. *Intelligent and Sustainable Manufacturing*, *1*(1), 10002. DOI: 10.35534/ism.2024.10002

Chen, D., Lin, H., Chen, Y., Yan, C., & Xu, J. (2023a). Wear behaviors of PCD and diamond-coated tools during low-frequency vibration-assisted drilling CFRP/Ti stacks. *International Journal of Advanced Manufacturing Technology*, *128*(1), 595–609. DOI: 10.1007/s00170-023-11899-9

Chen, R., Li, S., Zhou, Y., Qiu, X., Li, P., Zhang, H., & Wang, Z. (2023b). Damage formation and evolution mechanisms in drilling CFRP with prefabricated delamination defects: Simulation and experimentation. *Journal of Materials Research and Technology*, *26*, 6994–7011. DOI: 10.1016/j.jmrt.2023.09.065

Dong, Z., Wang, Z., Tian, J., Kang, R., & Bao, Y. (2023). Investigation on tearing damage of CFRP circular cell honeycomb in end-face grinding. *Composite Structures*, *325*, 117616. DOI: 10.1016/j.compstruct.2023.117616

Feng, Z., & Cook, R. D. (1983). Beam elements on two-parameter elastic foundations. *Journal of Engineering Mechanics*, *109*(6), 1390–1402. DOI: 10.1061/(ASCE)0733-9399(1983)109:6(1390)

Gao, T., Xu, P., Wang, W., Zhang, Y., Xu, W., Wang, Y., An, Q., & Li, C. (2024). Force model of ultrasonic empowered minimum quantity lubrication grinding CFRP. *International Journal of Mechanical Sciences*, *280*, 109522. DOI: 10.1016/j.ijmecsci.2024.109522

Gao, T., Zhang, Y., Li, C., Wang, Y., Chen, Y., An, Q., Zhang, S., Li, H., Cao, H., Ali, H. M., Zhou, Z., & Sharma, S. (2022). Fiber-reinforced composites in milling and grinding: Machining bottlenecks and advanced strategies. *Frontiers of Mechanical Engineering*, *17*(2), 24. DOI: 10.1007/s11465-022-0680-8

Gong, P., Zhang, Y., Wang, C., Cui, X., Li, R., Sharma, S., Liu, M., Gao, T., Zhou, Z., Wang, X., Dambatta, Y. S., & Li, C. (2024). Residual stress generation in grinding: Mechanism and modeling. *Journal of Materials Processing Technology*, *324*, 118262. DOI: 10.1016/j.jmatprotec.2023.118262

Guo, Z., Guo, B., Wu, G., Xiang, Y., Meng, Q., Jia, J., Zhao, Q., Li, K., & Zeng, Z. (2023). Three-dimensional topography modelling and grinding performance evaluating of micro-structured CVD diamond grinding wheel. *International Journal of Mechanical Sciences*, *244*, 108079. DOI: 10.1016/j.ijmecsci.2022.108079

Hu, N. S., & Zhang, L. C. (2003). A study on the grindability of multidirectional carbon fibre-reinforced plastics. *Journal of Materials Processing Technology*, *140*(1-3), 152–156. DOI: 10.1016/S0924-0136(03)00704-0

Liu, S., Chen, T., & Wu, C. (2017). Rotary ultrasonic face grinding of carbon fiber reinforced plastic (CFRP): A study on cutting force model. *International Journal of Advanced Manufacturing Technology*, *89*(1), 847–856. DOI: 10.1007/s00170-016-9151-x

Luna, G. G., Axinte, D., & Novovic, D. (2020). Influence of grit geometry and fibre orientation on the abrasive material removal mechanisms of SiC/SiC Ceramic Matrix Composites (CMCs). *International Journal of Machine Tools & Manufacture*, *157*, 103580. DOI: 10.1016/j.ijmachtools.2020.103580

Naik, N. K., & Kumar, R. S. (1999). Compressive strength of unidirectional composites: Evaluation and comparison of prediction models. *Composite Structures*, *46*(3), 299–308. DOI: 10.1016/S0263-8223(99)00098-7

Ning, F., Wang, H., & Cong, W. (2019). Rotary ultrasonic machining of carbon fiber reinforced plastic composites: A study on fiber material removal mechanism through single-grain scratching. *International Journal of Advanced Manufacturing Technology*, *103*(1), 1095–1104. DOI: 10.1007/s00170-019-03433-7

Qu, S., Wei, C., Yang, Y., Yao, P., Chu, D., Gong, Y., Zhao, D., & Zhang, X. (2024). Grinding mechanism and surface quality evaluation strategy of single crystal 4H-SiC. *Tribology International*, *194*, 109515. DOI: 10.1016/j.triboint.2024.109515

Qu, S., Yao, P., Gong, Y., Yang, Y., Chu, D., & Zhu, Q. (2022). Modelling and grinding characteristics of unidirectional C–SiCs. *Ceramics International*, *48*(6), 8314–8324. DOI: 10.1016/j.ceramint.2021.12.036

Rao, X., Zhang, F., & Xu, J. (2023). Wheel wear behavior and its influence on grinding performance in electrical discharge diamond grinding of reaction-bonded silicon carbide. *Journal of Manufacturing Processes*, *85*, 904–914. DOI: 10.1016/j.jmapro.2022.12.024

Wang, C., Li, P., Li, S., Qiu, X., Niu, Q., Li, C., & Ko, T. J. (2023). Study on the mechanism and performance of longitudinal-torsional ultrasonic vibration assisted drilling CFRP/Ti stack. *Journal of Manufacturing Processes*, *92*, 453–465. DOI: 10.1016/j.jmapro.2023.03.005

Wang, D., Jiao, F., & Mao, X. (2020). Mechanics of thrust force on chisel edge in carbon fiber reinforced polymer (CFRP) drilling based on bending failure theory. *International Journal of Mechanical Sciences*, *169*, 105336. DOI: 10.1016/j.ijmecsci.2019.105336

Wang, F., Jin, X., Wang, X., Jia, Z., & Zhang, B. (2022). Numerical investigation on the influences of processing conditions on damage in the CFRTP cutting using a novel Elastic–plastic damage model. *Applied Composite Materials*, *29*(5), 2063–2094. DOI: 10.1007/s10443-022-10041-4

Wang, F., Xuan, S., Chang, Z., Jin, K., Gao, Y., Wang, H., & Song, Q. (2024). Effect of grinding parameters on industrial robot grinding of CFRP and defect formation mechanism. *International Journal of Precision Engineering and Manufacturing-Green Technology*, *11*(2), 427–438. DOI: 10.1007/s40684-023-00561-0

Wang, H., Ning, F., Hu, Y., & Cong, W. (2018). Surface grinding of CFRP composites using rotary ultrasonic machining: A comparison of workpiece machining orientations. *International Journal of Advanced Manufacturing Technology*, *95*(5), 2917–2930. DOI: 10.1007/s00170-017-1401-z

Wei, C., He, C., Chen, G., Sun, Y., & Ren, C. (2023). Material removal mechanism and corresponding models in the grinding process: A critical review. *Journal of Manufacturing Processes*, *103*, 354–392. DOI: 10.1016/j.jmapro.2023.08.045

Wen, Z., Zhao, H., Yan, X., Li, P., & Ding, H. (2024). Grinding performance investigation of multidirectional CF/PEEK composites—A comparative with unidirectional CF/PEEK. *Polymer Composites*, *45*(10), 8889–8904. DOI: 10.1002/pc.28383

Wen, Z., Zhao, H., Yan, X., Peng, Z., & Ding, H. (2023). Grinding performance study of CF/epoxy and CF/PEEK composites. *Polymer Composites*, *44*(8), 4565–4578. DOI: 10.1002/pc.27420

Wu, Z., & Zhang, L. (2023). Analytical grinding force prediction with random abrasive grains of grinding wheels. *International Journal of Mechanical Sciences*, *250*, 108310. DOI: 10.1016/j.ijmecsci.2023.108310

Zhang B, LU S X, Rabiey M, Axinte, D., & Bleicher, F. (2023). Grinding of composite materials. *CIRP Annals-Manufacturing Technology, 72*(2), 645-671.

Zhang, H., Zhao, Z., Li, J., Ye, L., & Liu, Y. (2024). Review on abrasive machining technology of SiC ceramic composites. *Micromachines*, *15*(1), 106. DOI: 10.3390/mi15010106 PMID: 38258225

APPENDIX

Table 1. Abbreviation

CFRP	—	Carbon fiber reinforced polymer	MQL	—	Minimum quantity lubrication
FOA	—	Fiber orientation angle	NMQL	—	Nanolubricant minimum quantity lubrication
UVAM	—	Ultrasonic vibration assisted machining	EHM	—	Equivalent homogeneous material
UVAG	—	Ultrasonic vibration assisted grinding	RVE	—	Representative volume element
θ_f	°	Fiber orientation angle	θ_U	°	Combined vibration angle of two-dimensional ultrasonic vibration
d_s	mm	Wheel diameter	A_U	μm	Amplitude of tangential ultrasonic vibration
v_s	m/s	Circumferential velocity of the grinding wheel	B_U	μm	Amplitude of axial ultrasonic vibration
v_w	mm/s	Feed rate of the workpiece	ϕ_U	°	Phase angle of ultrasonic vibration
a_p	μm	Grinding depth	m_U	—	Number of ultrasonic vibrations
a_g	μm	Thickness of undeformed chips	f_U	kHz	Frequency of ultrasonic vibration
Δt	s	The time required for the grains to complete the cutting process in the grinding area	CNTs	—	Carbon nanotubes
COF	—	Coefficient of friction	SEM	—	Scanning electron microscope
l_{min}	μm	Minimum distance between two fibers	l_{max}	μm	Maximum distance between two fibers
d_f	μm	Diameter of carbon fiber	d_{max}	μm	Maximum particle size of grains
d_{mean}	μm	Average diameter of grains	d_{min}	μm	Minimum particle size of grains
d_{12}	μm	The distance between the centers of two grains	V_g	—	Volume content of grains in the grinding wheel body
S_o	—	Organization number of grinding wheel	L_r	μm	Theoretical grains spacing
N_g	—	Total number of grains in the grinding arc area	η	—	Interface bonding parameter between fibers and matrix
k_{cg}	—	Equivalent coefficient of dynamic stiffness for traditional grinding systems	k_{uvag}	—	Dynamic stiffness equivalent coefficient of ultrasonic assisted grinding system
σ_f	GPa	Breaking strength of carbon fibers	l_c	μm	Critical length of carbon fiber fracture
N_{eg}	—	Total number of grains in the grinding arc area	N_{90-s}	—	Number of grains that cause longitudinal compression shear fracture on 90 ° fibers
N_{90-c}	—	Number of grains that cause longitudinal compression matrix cracking on 90 ° fibers	N_{0-Sh}	—	Number of grains that produce shear type buckling on 0 ° fibers

continued on following page

Table 1. Continued

N_{135-bf}	—	Number of grains that produce blocky removal on 135 ° fibers	N_{0-TC}	—	Number of grains that produce tensile and compressive buckling on 0 ° fibers
λ	µm	Effective spacing of cutting grains	k_{mn}	N/mm²	Normal constraint parameters of surrounding supporting materials
$a_{gmax(n)}$	µm	The maximum undeformed chip thickness of the *nth* dynamic effective grain	k_{mt}	N	Tangential constraint parameters of surrounding supporting materials
h_g	µm	Depth of interface cracking along the fiber direction	k_b	—	Equivalent modulus of fiber/resin interface bonding
C_{mn}	—	Coefficient of action mode	σ_b	MPa	Bond strength of the interface
C_{mt}	—	Property parameters of surrounding supporting materials	γ	°	Bending deformation angle of fibers
n_g	—	Number of consecutive adjacent cutting grains	γ_g	°	The front angle of the grain cutting edge
$2\theta_g$	°	Top cone angle of grains	γ_m	—	Shear strain of the matrix
r_g	µm	The radius of the blunt circle of the grinding edge of the grain	w	µm	Deformation deflection of carbon fibers in the *x*-direction
v_f	—	Poisson's ratio of carbon fiber	E_{fa}	GPa	Axial Young's modulus of carbon fibers
v_m	—	Poisson's ratio of resin matrix	E_{ft}	GPa	Transverse Young's modulus of carbon fibers
v_{mf}	—	Poisson's ratio of equivalent homogeneous material	E_i	Mpa	Elastic modulus of interface phase
v_i	—	Poisson's ratio of interfacial phase	E_m	GPa	Modulus of resin matrix
V_f	—	Volume fraction of carbon fiber	E_{mf}	GPa	Modulus of equivalent homogeneous materials
V_m	—	Volume fraction of resin matrix	t_i	µm	Thickness of interface phase
V_i	—	Interface phase volume fraction	r_i	µm	Radius of the interface phase
I_f	—	Polar moment of inertia of carbon fiber	S_i	GPa	Shear strength of interfacial phase
n	—	Material Constant	T_i	GPa	Tensile strength of interface phase
τ_{fu}	GPa	Shear strength of fiber	C_{Lc}	GPa	Cracking strength of longitudinal compression
P	N	Buckling load of fiber sheets	S_{Lc}	GPa	Longitudinal compressive strength of fibers
P_{cr}	N	Buckling critical load of fiber sheets	τ_m	GPa	Shear stress of matrix
G_f	GPa	Shear modulus of carbon fiber	δU_f	J	Strain energy of carbon fiber in deviating state
G_m	Mpa	Shear modulus of resin matrix	δU_m	J	The Strain energy of the deviant state of the matrix
δW	J	Acting of external forces on fibers	M_1	N·m	Bending moment on carbon fiber
k_1	—	Deviation slope of carbon fiber deformation	M_2	N·m	Bending moment on carbon fiber

continued on following page

Table 1. Continued

k_2	—	Deviation slope of carbon fiber deformation	M_3	N·m	Bending moment on carbon fiber
k_3	—	Deviation slope of carbon fiber deformation	Q_1	N	Force on carbon fiber
ε_m	—	Strain of matrix	Q_2	N	Force on carbon fiber
σ_m	GPa	Stress in the matrix	Q_3	N	Force on carbon fiber
η_d	μm	The distance from any point on the fiber to the neutral axis	σ_{fcr}	GPa	Critical stress of fibers
J_f	—	Inertia moment of fiber sheet	σ_{fb}	GPa	The average fracture stress of fibers in fiber bundles
σ_{fmax}	GPa	Average stress of fibers in the maximum breaking load fiber bundle	σ_{fb_mv}	GPa	Mean value of fiber bundle strength
α_{45}	—	Scaling parameter	L_{mf}	μm	Length of the fiber bundle
β_{45}	—	Shape parameter	N_{mf}	—	Number of fibers in the fiber bundle
μ_f	—	Coefficient of friction	F_{0_Sh}	N	Fiber removal shear deformation force of 0° CFRP
F_{0_e}	N	Elliptical domain contact force of grains on fibers	F_{0_TC}	N	Tensile and compressive deformation force of fiber removal of 0° CFRP
F_{45_Sh}	N	Interlayer shear force of 45 ° CFRP	F_{A_90}	N	Bending fracture force of 90° carbon fiber
F_{B_45}	N	Bending fracture force of 45° carbon fiber	F_{A_135}	N	Bending fracture force of 135° carbon fiber
F_{Lc}	N	Matrix cracking force under longitudinal compressive at 90° CFRP	σ_{mb}	N	Critical compressive strength of CFRP materials considering matrix slip
F_{Ls}	N	Fiber shear fracture force longitudinal compression at 90° CFRP	P_{mb}	N	The average combined force of fiber micro buckling on grains
F_{Ni} $(i=0, 45, 90, 135)$	N	Normal grinding force of different fiber orientations	A_{gmb}	mm^2	The area of fiber micro buckling on grains
K_f	—	Correction factor	F_{mf}	N	Fiber bundle block removal force
F_n	N	Normal grinding force	F_t	N	Tangential grinding force

Chapter 8
Force Model of Ultrasonic Vibration Nanolubricant MQL Longitudinal Grinding CFRP

ABSTRACT

This chapter investigates the material removal mechanism of CFRP with different FOAs and establishes a force model of multiangle 2D ultrasonic vibration nanolubricant MQL longitudinal grinding. Firstly, a homogeneous random distribution model for fibers is developed, wherein the fiber diameter follows a normal distribution, to predict the mechanical properties of the interfacial phase. Then, a model of the diamond grinding wheel is developed, accounting for the normal distribution of both the protrusion height and the size of grains. Furthermore, based on the geometric kinematics of the grain trajectory and intermittent cutting state, the influence of 2D ultrasonic vibration angles on the dynamic stiffness of the grinding system was analyzed. Finally, grinding force models of entire domain of FOAs under different ultrasonic vibration modes and lubrications were interconnected. Experimental assessments demonstrated that the grinding force model has acceptable estimation error and effectively captures the mechanics of CFRPs with different FOAs.

8.1 INTRODUCTION

Machining forces and quality are significantly influenced by cooling and lubrication conditions (Danish et al., 2021). The hygroscopic nature of the resin matrix causes CFRP to absorb moisture and swell, which markedly degrades its mechanical properties (Scida et al., 2013; Nandagopal et al., 2020). This characteristic

DOI: 10.4018/979-8-3373-9013-0.ch008

limits the application of flood cooling and lubrication, making dry processing the predominant method. However, dry processing is associated with deterioration of surface integrity, dust hazards, wheel clogging, and various defects (Haddad et al., 2014). The temperature in dry grinding often exceeds the thermal softening point of the resin, exacerbating thermal deformation and degrading the ground surface (Kodama et al., 2020).

Researchers have confirmed the unique advantages of minimum quantity lubrication (MQL) in composite machining (Senthilkumar et al., 2018; Liu et al., 2024). This innovative method effectively integrates lubrication and cooling functions by atomizing a minimal quantity of lubricant with high-pressure gas into droplets (Wang et al., 2022a; Yang et al., 2024; Gong et al., 2024), which are then delivered at high velocity to the grinding zone (Yang et al., 2023a). The debris-free surface produced by MQL eliminates the need for subsequent cleaning of aerospace components, while the use of biodegradable and non-toxic vegetable oil as the base oil enables clean manufacturing (Zhang et al., 2023). Cococcetta et al. (2013) found that MQL lubricants reduce tool/workpiece friction and minimize tool wear. Qu et al. (2020a) demonstrated that MQL can significantly lower grinding temperature. Esmaeili et al. (2019) determined that directional pressurized delivery of MQL oil mist effectively penetrates the pore network between the grinding wheel and workpiece. However, Pervaiz et al. (2020) reported contrary findings, indicating that externally configured MQL strategies perform unsatisfactorily, likely due to subsurface cutting occurring during MQL-assisted machining of CFRP, which causes a portion of the lubricant to flow into cavities. Iskandar et al. (2014) noted that a spray mode with a low oil flow rate and a high gas flow rate enhances droplet breakup, generating more coherent jets with fewer vortices and facilitating easier penetration into the cutting zone. Divergent perspectives have also emerged in research. Helmy et al. (2018) discovered that compared to flood cooling, MQL spray cooling increased the cutting force because it failed to dissipate heat entirely and remove dust effectively. Nevertheless, MQL offers the advantage of reducing cutting fluid pollution and provides machining results close to the standards of dry machining (Xu et al., 2023). Most importantly, MQL mitigates the issue of moisture absorption and swelling in CFRP, which otherwise leads to decreased mechanical properties.

To enhance the tribological characteristics and heat dissipation capability of MQL technology, a more efficient and cleaner approach known as nanofluid minimum quantity lubrication (NMQL) has been proposed (Li et al., 2020; Duan et al., 2024). This technique involves adding nanoparticles with various physicochemical properties to a base oil, followed by thorough dispersion and mixing to create a nanofluid, also referred to as a nanolubricant (Hu et al., 2023). Subsequently, this nanolubricant is atomized using high-pressure airflow and sprayed into the grinding wedge zone (Chu et al., 2023; Song et al., 2025). NMQL technology effectively improves the heat dissipation and lubrication performance of MQL(Wang et al., 2022b), while the nanoparticles contribute significantly to enhancing lubricity due to their remarkable friction-reducing and anti-wear properties (Saha et al., 2023). Qu et al. (2020b, 2022) successfully determined the optimal parameters for carbon-based NMQL and observed that the bonding state between fibers and the matrix depends on lubrication conditions and grain sharpness.

MQL is currently an essential lubrication-cooling method that can effectively replace dry machining. Moreover, the incorporation of nanoparticles addresses the limitations of MQL in terms of heat transfer and friction reduction performance (Li, 2023). However, due to the presence of a gas barrier layer, a small quantity of nanolubricant struggles to reach the grinding wedge zone effectively, preventing nanolubricant droplets from fully exerting their lubrication and heat dissipation properties. Furthermore, nanoparticles increase the viscosity of the lubricant, thereby enhancing oil film thickness and stability. However, this increased viscosity reduces lubricant fluidity and spreading capability, making it more difficult to penetrate the grinding zone. Ultimately, this results in a decreased effective flow rate, making it challenging to satisfy the dual requirements of sufficient oil film thickness and efficient infiltration for nanolubricants.

Ultrasonic Vibration-Assisted Machining (UVAM) is a technique that employs low-amplitude, high-frequency vibrations applied to the cutting tool or workpiece, utilizing ultrasonic energy to alter the material removal process (Yang et al., 2020; Ding et al., 2023). By superimposing micro-scale amplitude displacement at ultrasonic frequencies onto the tool-tip motion, UVAM modifies the cutting trajectory of the tool tip (Zhao et al., 2023; Liu et al., 2025; Li et al., 2022a). To address the aforementioned technical challenges, ultrasonic vibration-empowered nanolubricant MQL grinding, which incorporates pump suction effects, intermittent cutting, and hydrodynamic lubrication, is anticipated to alleviate the issue of insufficient effective lubricant flow rate. Ni et al. (2019) found that the combination of UVAM and MQL greatly enhances tool wear performance by providing excellent lubrication, which mitigates the occurrence of surface micro-cracks and friction marks. Niu et al. (2023) demonstrated that the separation cutting characteristics of UVAM improve the cooling/lubrication performance of MQL, leading to reduced friction at the tool-chip contact interface and better heat dissipation. Moreover, the coupling effect of UVAM and MQL concurrently decreases the forces and heat generated during the cutting process. Research findings by Das and Pandivelan (2020) showed that, compared with conventional grinding (CG), the combination of UVAG and MQL can reduce the tangential force by 42% and the normal force by 31%. Yang et al. (2023b) applied ultrasonic vibration-assisted grinding (UVAG) and MQL for bone micro-grinding and achieved reductions of 75.1% in the normal grinding force and 82.9% in the tangential grinding force compared to dry grinding. Dambatta et al.

(2024) reported that introducing ultrasonic vibration and SiO_2 nanofluid into MQL can decrease the specific grinding energy, normal and tangential grinding forces, and surface roughness by 65%, 57%, 65%, and 18%, respectively.

However, for special composites with laminated structures, multiphase heterogeneity, and anisotropy, the effect of ultrasonic vibration on improving the performance of MQL nanolubricants differs considerably from that in traditional grinding. The grinding force is an important indicator for investigating the influence of input parameters on grinding outputs (Li et al., 2023). It originates from the plastic and elastic deformation of materials, as well as interactions between grinding debris, specimens, and grains. It serves as the basis for studying grinding processes, designing and selecting appropriate tools, and evaluating material machinability. The grinding force is a critical parameter in the material removal process, directly affecting machining efficiency, surface quality, and tool life. Force models are the most direct and comprehensive mathematical expressions for describing material removal mechanical behavior and play a vital role in the application of grinding across various fields (Gao et al., 2017). Mechanistic modeling of CFRP grinding is key to effectively controlling damage generation during FRC machining. However, ensuring high accuracy and low estimation errors in force models is challenging due to the inhomogeneity, heterogeneity, anisotropy, and low thermal dissipation characteristics of these FRCs. Wang et al. (2019, 2020a, 2020b) established grinding force models for CFRP edge surface grinding and reported that rotary ultrasonic machining can reduce grinding forces and surface roughness at FOAs of 90° and 0°, respectively (Wang et al., 2018). Liang et al. (2019) designed a grinding tool with an ordered distribution of grains and established a force model based on the maximum undeformed chip thickness in UVAG. Liu et al. (2017) developed a mechanical model for ultrasonic vibration-assisted face grinding of CFRPs. The results indicate that as the spindle speed increases, the force reduction rate in the UVAG method decreases more slowly than that in traditional grinding. Ning et al. (2019) modeled the surface grinding feed force for CFRPs by introducing a fracture volume factor and confirmed that ultrasonic vibration significantly reduces debonding and pull-out. Ran et al. (2023) considered the anisotropy and nonuniformity of SiCf/SiC composites, as well as the interaction state between grains and materials. They proposed a multiscale elastic mechanics model and an equivalent fracture toughness applied to UVAG force modeling. The average errors in predicting the normal and tangential grinding forces were 9.29% and 6.16%, respectively. (Li et al., 2022b) proposed a discrete numerical model for 2D ultrasound-assisted end face grinding of SiC to describe dynamic cutting behavior. They also established a grinding force model considering the material removal mechanism with a workpiece profile update procedure, controlling the prediction error within 11.14%. Cao et al. (2024) found that ultrasonic vibration can reduce scraping damage and the scraping load.

Having explored the baseline mechanical behavior of longitudinal grinding in Chapter 7, this chapter introduces the additional complexity of ultrasonic vibration superimposed on the grinding motion. A dedicated force model is developed to capture the synergistic effects of nanolubricant MQL and intermittent ultrasonic impacts, providing a predictive framework for this hybrid longitudinal grinding process. In summary, current research on the material removal mechanical behavior and grinding force models of CFRPs primarily focuses on end face grinding under dry conditions. Additionally, the application of ultrasonic vibration is mostly limited to one-dimensional or non-adjustable two-dimensional modes. However, the fundamental issues related to the material removal mechanical behavior for different FOAs in ultrasonic vibration-empowered nanolubricant MQL grinding remain unresolved. To address these technical challenges and research gaps, this chapter proposes a novel method for multi-angle 2D ultrasonic-empowered nanolubricant MQL grinding of CFRPs. It systematically investigates the material removal mechanical behaviors under different FOAs and innovatively establishes a multi-condition grinding force model. A uniform random distribution model for fibers was developed, and the mechanical properties of the interfacial phase were predicted. A grinding wheel model with a normal distribution of grain protrusion height and grain size was constructed. Based on the geometric kinematics analysis of the grain motion trajectory in multi-angle 2D UVAG, the intermittent grinding states and dynamic stiffness of the UVAG system at different ultrasonic vibration angles were investigated. A material removal mechanism and mechanical model for CFRP grinding at different FOAs, considering the micro-removal behaviors of fibers, resin matrix, and interfaces, was proposed. Finally, a CFRP grinding force model with multi-angle 2D ultrasonic vibration-empowered nanolubricant MQL for different FOAs was established and experimentally validated.

8.2 MATERIAL MODEL

8.2.1 Representative Volume Element

To facilitate the acquisition of mechanical properties of CFRP for supporting the grinding force model, the following assumptions are first made: (1) Both the constituent materials (carbon fiber, resin matrix, and interface) and the CFRP are continuous, with excellent bonding between the matrix and carbon fibers, ensuring uniform strain among the constituents along the fiber direction under load; (2) All constituent materials and their properties are homogeneous; (3) Prior to loading, neither the CFRP workpiece nor its constituents are under stress, and no transverse stresses develop between the resin matrix and carbon fibers after loading; (4) The properties of the CFRP constituents in the composite state remain consistent with those in the uncombined state, and both carbon fibers and resin matrix are isotropic. Based on these assumptions, a representative volume element (RVE) capable of capturing the microstructural characteristics of the constituents and the overall behavior of unidirectional CFRP is selected for analysis. Macroscopically, the stress and strain within the RVE of unidirectional CFRP are uniform, whereas microscopically, they are heterogeneous. The RVE is employed to establish stress/strain relationships that reflect the strength and elastic properties of CFRP.

8.2.2 Material Model of the Interface Phase

The interface phase is a critical component among the three essential elements of CFRP—the resin matrix, carbon fibers, and the interface. It is formed through physicochemical interactions between the resin matrix and carbon fibers, which exhibit distinct morphologies and properties. The formation of the interface phase begins with the infiltration and contact between the resin matrix and carbon fibers, followed by their mutual interactions that lead to the establishment of a stable interfacial layer. As a uniquely important constituent in CFRPs, the interface serves as the medium through which loads sustained by either the resin matrix or carbon fibers are transferred to one another, thereby giving rise to the overall macroscopic mechanical behavior. Consequently, the interfacial phase in CFRP is a distinctive component with material characteristics different from those of carbon fibers and the resin matrix, and it significantly influences the material removal mechanisms and grinding-related mechanical responses of CFRP composites. Based on the representative volume element (RVE), the following interfacial property parameters are predicted. The volume fraction of the interface phase V_i, is given by:

$$V_i = 0.163\,V_f^2 \tag{8.1}$$

So, the interface thickness t_i is,

$$t_i = \left(\sqrt{1 + 0.163\,V_f} - 1\right) r_f \tag{8.2}$$

The elastic modulus of the interface phase E_i is,

$$E_i(r) = \left(E_f - E_m\right)\left(\frac{r_i - r}{r_i - r_f}\right)^n + E_m \tag{8.3}$$

where r_i is the interface radius. Assuming a uniform variation in E_i, it follows that:

$$E_i = \frac{E_f + nE_m}{n+1} \tag{8.4}$$

where n is a material constant, taken as 1.66. Similarly, the Poisson's ratio of the interface is given by:

$$\upsilon_i = \frac{\upsilon_f + n\upsilon_m}{n+1} \tag{8.5}$$

The interfacial phase primarily undergoes shear and tensile failure. The shear strength S_i and tensile strength T_i are considered equal (Deng et al., 1998):

$$T_i = S_i = \sigma_f r_f / l_c \tag{8.6}$$

where l_c represents the critical length of carbon fiber breakage, and σ_f denotes the tensile strength of the carbon fiber. The mechanical property parameters of the unidirectional CFRP workpiece material selected for this study are summarized in Table 1.

Table 1. Material property parameters of the CFRPs

Material	Property parameters	Value
CFRP	Density / g/cm^3	1.575
	Hardness / HRB	68-72
Carbon fiber (70%)	Density / g/cm^3	1.8
	Elastic modulus / GPa	230
	Axial tensile strength / GPa	3.45
	Shear strength / GPa	0.38
	Poisson's ratio	0.3
	Fracture toughness / J/m^2	2
	Elongation at break / %	1.5
Resin matrix (22%)	Density / g/cm^3	1.05
	Elastic modulus / GPa	3.4
	Shear modulus(G_m)/ GPa	1.02
	Axial tensile strength / MPa	85
	Poisson's ratio	0.4
	Fracture toughness / J/m^2	500
	Elongation at break / %	5.5

continued on following page

Table 1. Continued

Material	Property parameters	Value
interface phase (8%)	Thickness / ìm	0.194
	Poisson's ratio	0.36
	Elastic modulus / GPa	88.6
	Tensile strength / MPa	32.8
	Shear strength / MPa	32.8

8.2.3 Uniform Random Arrangement Model for Fibers

The transverse section of the unidirectional CFRP composite was polished, and the arrangement of fibers within the matrix was obtained using laser confocal microscopy or SEM, as shown in Fig. 5-1. In reality, carbon fibers are randomly and uniformly distributed within the CFRP. To prevent fiber interference due to excessively small spacings, the maximum and minimum inter-fiber distances were defined as $l_{max} = d_f$ and $l_{min} = 0$, respectively. Figure 1 illustrates the construction principle and generation process of the uniform random distribution model for carbon fibers. The procedure is as follows:A random coordinate (x_1, y_1), representing the first carbon fiber, is generated.A second random coordinate is generated based on a distance l_1((l_{min} -l_{max}) and an orientation angle θ_{r1} (0-2π) relative to the first fiber. New carbon fibers are iteratively generated by repeating the previous step until no additional fiber positions can be accommodated in the surrounding area. The same algorithm is then applied to the second coordinate point (second carbon fiber), and Steps 2 and 3 are repeated until the square region is entirely filled. Furthermore, cross-sectional observations from laser confocal microscopy indicate that carbon fiber diameters are not identical. Therefore, the fiber diameter is defined to follow a normal distribution with a mean of 7 μm and a standard deviation of 0.3.

Figure 1. Generation principle and simulation results of the carbon fiber random distribution. (a) Microscopic morphology of random distribution of carbon fibers in CFRP. (b) The generation principle of fiber random distribution. (c) Histogram of probability distribution of carbon fiber diameter. (d) Simulation of fiber random distribution

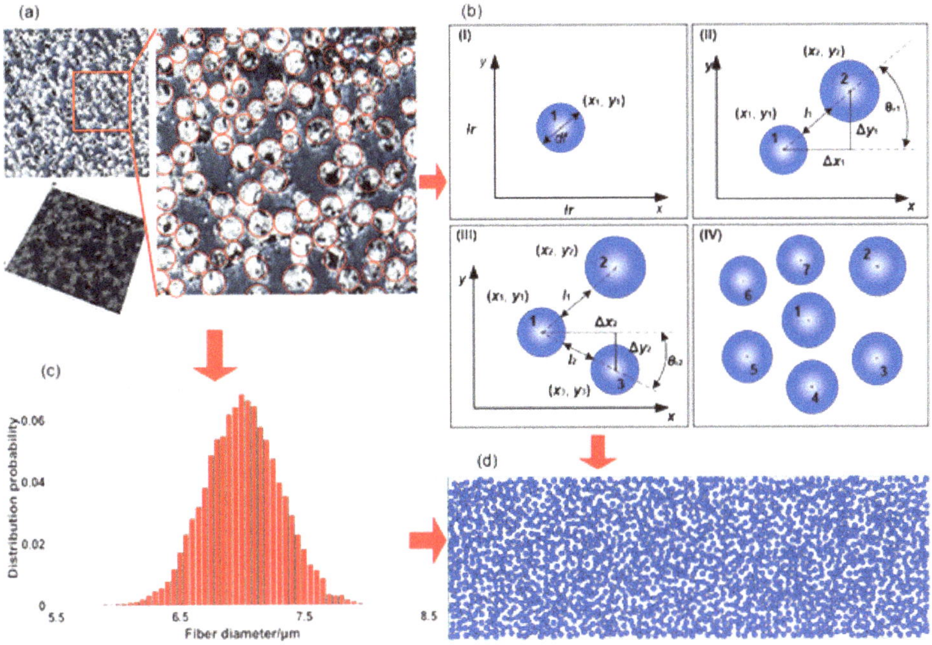

8.3 GRINDING WHEEL MODELING

In practice, a grinding abrasive grain typically features multiple tip cone angles, with the tip cone angles $2\theta_g$ generally ranging between 80° and 145°. If the proportion of acute $2\theta_g$ increases, the probability of the grain cutting with a positive rake angle also rises. Thus, the proportional distribution of different $2\theta_g$ plays a critical role in determining the grinding behavior of the abrasive grain. Both the cutting-edge tip radius r_g and the proportional distribution of the $2\theta_g$ are related to the size of the abrasive grain. Based on statistical probability theory, this study defines the diamond abrasive $2\theta_g$ as a random variable following a normal distribution with a mean of 97° and a standard deviation of 11.

To investigate the different contact and cutting states of abrasive grains on fibers and reveal the geometric interaction morphology between carbon fibers and diamond grains in the grinding zone, this section establishes a diamond grinding wheel model.

This model provides a theoretical basis and data support for the subsequent CFRP grinding force model. Whether a grain actually participates in grinding depends on its protrusion height and spacing. The protrusion heights and distribution of abrasive grains on the wheel surface are non-uniform. Moreover, grains with lower protrusion heights may not effectively engage in material removal, resulting in the number of active grains being less than the total number of surface grains. Here, the abrasive grain size is assumed to follow a normal distribution, and thus the mean grain diameter d_{mean} is given by:

$$d_{mean} = \frac{d_{\max} + d_{\min}}{2} \tag{8.7}$$

where $d_{\min}$ and $d_{\max}$ represent the minimum and maximum abrasive grain sizes, respectively. The specific parameters of the diamond grinding wheel are listed in Table 2.

Table 2. Grinding wheel parameters

Grinding wheel parameters	**Values**
grains materials	Diamond
Binder	Resin
Diameter / mm	300
Width / mm	20
Granularity	150#
Maximum size of grains / ìm	106
Minimum size of grains / ìm	75
Grinding wheel organization number	13

The distribution function and standard deviation of the grain size distribution are derived from the properties of the normal distribution function, as follows:

$$\sigma = \frac{d_{\max} - d_{\min}}{6} \tag{8.8}$$

$$\phi(d) = \frac{1}{\sqrt{2\pi} \cdot \sigma} \cdot e^{\left[-\frac{(d-d_{mean})^2}{2\sigma^2}\right]} \tag{8.9}$$

The distance between the outer surfaces of two abrasive grains is utilized to prevent mutual interference:

$$d_{12} < R_1 + R_2 \tag{8.10}$$

where d_{12} denotes the distance between the centers of these two grains, and R_1 and R_2 represent the radii of grain 1 and grain 2, respectively. The volumetric content of abrasive grains V_g can be obtained using the wheel structure number S_0:

$$V_g = 2 \bullet (31 - S_0) \tag{8.11}$$

Assuming a uniform distribution of abrasive grains, the theoretical inter-grain spacing L_r can be calculated as:

$$L_r = d_{mean} \cdot \left(\sqrt{\frac{\pi}{4V_g}} - 1 \right) \tag{8.12}$$

Subsequently, the number of abrasive grains on the wheel can be determined from the wheel width b and the grinding contact arc length L_g:

$$N_x = \frac{L_g}{d_{mean} + L_r} \tag{8.13}$$

$$N_y = \frac{b}{d_{mean} + L_r} \tag{8.14}$$

Thus, the total number of abrasive grains N is given by:

$$N = N_x \bullet N_y \tag{8.15}$$

Based on the total number of abrasive grains N in the grinding zone and the probability distribution of grain size, the grinding wheel abrasive grain population model matrix $\boldsymbol{G}(d)$ is obtained:

$$G\left(d\right) = \begin{bmatrix} d_{11} & d_{12} & \cdots & d_{1N_y} \\ d_{21} & d_{22} & \cdots & d_{2N_y} \\ \vdots & \vdots & \ddots & \vdots \\ d_{N_x1} & d_{N_x1} & \cdots & d_{N_xN_y} \end{bmatrix} \tag{8.16}$$

The "Grain Vibration Method" is employed to achieve a uniform random protrusion height of abrasive grains within the wheel matrix. The iterative equation for the axial vibration of grains is as follows:

$$z_{gkn} = z_{gkn} + \delta \cdot z_{gkn} + \delta \cdot z_{gkn} + \cdots + \delta \cdot z_{gkn} \tag{8.17}$$

Setting the initial Z-coordinate of the abrasive grains to 0, the array of axial grain positions is obtained as follows:

$$G\begin{pmatrix} \\ z \\ \\ \end{pmatrix} = \frac{G(d)}{2} + G(z_g) = \begin{bmatrix} z_{11} & z_{12} & \cdots & z_{1N_y} \\ z_{21} & z_{22} & \cdots & z_{2N_y} \\ \vdots & \vdots & \ddots & \vdots \\ z_{N_x1} & z_{N_x1} & \cdots & z_{N_xN_y} \end{bmatrix} \tag{8.18}$$

Variations in grinding parameters lead to changes in the grain depth of cut and the number of dynamic active grains. Whether a static active grain actually engages in cutting is determined by its protrusion height and its distance from the preceding active grain. Consequently, the maximum undeformed chip thickness $a_{\text{gmax(n)}}$ for a dynamic active grain can be calculated as:

$$a_{\text{gmax}(n)} = 2 \cdot \lambda_{(n\sim n-1)} \cdot \frac{v_\omega}{v_s} \cdot \sqrt{\frac{a_p}{D}} + \left(a_{p(n)} - a_{p(n-1)}\right) \tag{8.19}$$

Figure 2 illustrates the grinding wheel model, along with the probability distributions of the abrasive grain $2\theta_g$ and the number of active grains.

Figure 2. Simulation results of diamond grinding wheel model

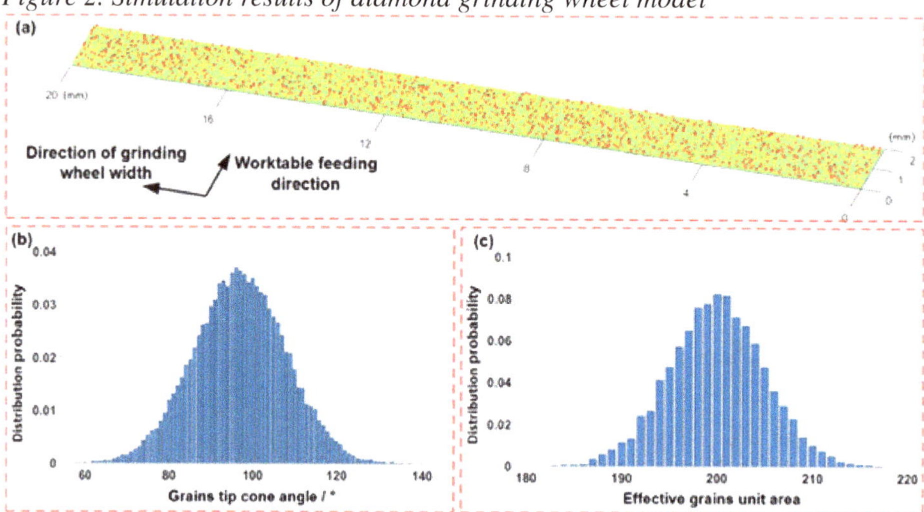

8.4 DYNAMIC STIFFNESS

Due to the workpiece vibration frequency significantly exceeding the natural frequency of the grinding system, the interaction between the grinding wheel and the workpiece manifests as intermittent multiple micro-impacts. Consequently, under identical grinding conditions, the dynamic stiffness of the ultrasonic vibration-assisted grinding system can be substantially increased, by several times, compared to that of conventional grinding. Here, T represents the period of ultrasonic vibration, and T_c denotes the actual duration of cutting engagement by an abrasive grain. Based on the analysis of grain motion trajectories and intermittent cutting characteristics in multi-directional two-dimensional ultrasonic vibration-assisted grinding presented in Section 2.2, the actual cutting time of abrasive grains under different two-dimensional ultrasonic vibration angles can be determined. Let k_{cg} be the equivalent dynamic stiffness coefficient in conventional grinding; then the equivalent dynamic stiffness coefficient in ultrasonic-assisted grinding, k_{uvag}, is given by:

$$k_{uvag} = \left(T/T_c\right)k_{cg} \tag{8.20}$$

8.5 GRINDING FORCE MODEL

8.5.1 Fiber Orientation Mechanics Analysis

Owing to the variation in carbon fiber orientation, the mechanical behavior during material removal and the chip formation mechanisms in unidirectional CFRP grinding exhibit significant differences. Depending on the loading conditions in the contact zone, bending, buckling, compression, or shear may dominate, necessitating separate detailed analyses for each case. The interaction between the CFRP model and the diamond grinding wheel model is further analyzed to elucidate the different contact states and cutting behaviors of abrasive grains with carbon fibers. Building on the analysis of CFRP grinding material removal mechanisms in Chapter 7, mechanical models corresponding to different fiber orientations, as illustrated in Figure 3, must be established individually. The subsequent step involves solving the distinct mechanical models for grinding forces across the four typical fiber orientation ranges represented by Eq. (8.21).

Figure 3. Schematic diagram of CFRP grinding with different fiber orientations

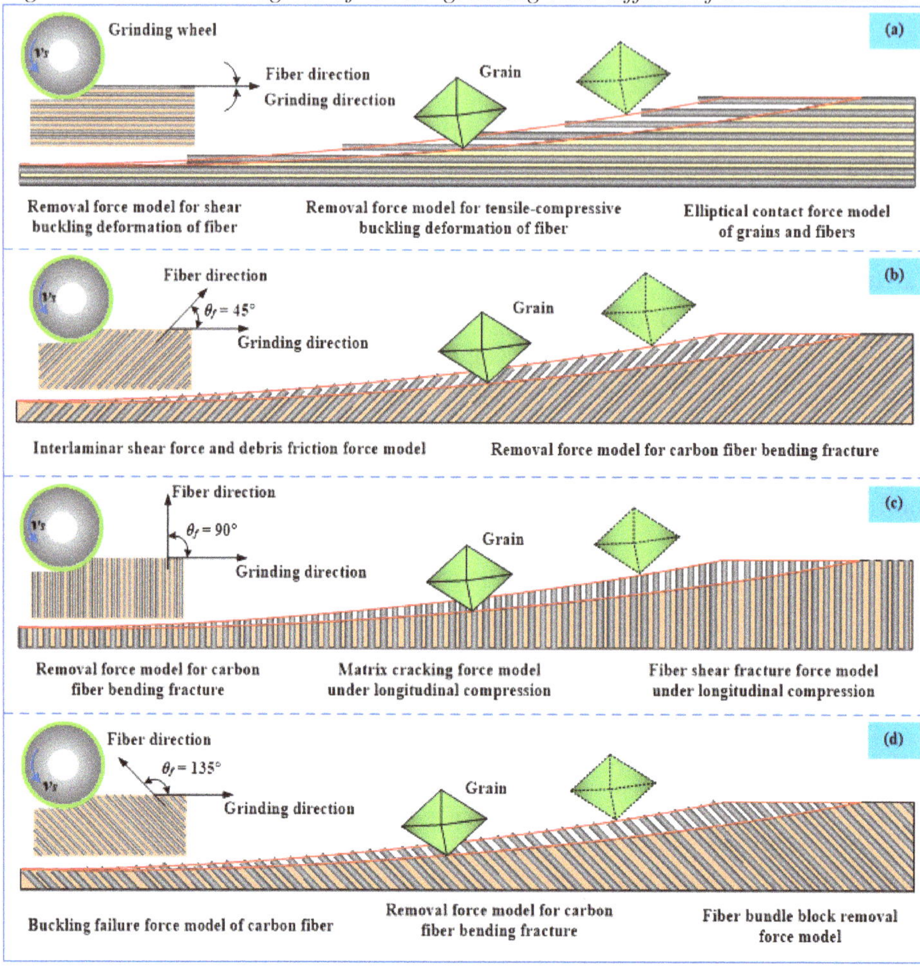

$$\mathrm{F}_{\mathrm{Total}}=\begin{cases}\mathrm{F}_N=\begin{cases}F_{0\_N}, & -10^\circ \le \theta_f \le 10^\circ\\ F_{45\_N}, & 30^\circ \le \theta_f \le 60^\circ\\ F_{90\_N}, & 80^\circ \le \theta_f \le 100^\circ\\ F_{135\_N} & 120^\circ \le \theta_f \le 150^\circ\end{cases}\\ \mathrm{F}_{\mathrm{T}}=\begin{cases}\mu_{\mathrm{f}}\cdot F_{0\_N}+F_{0\_T}, & -10^\circ \le \theta_f \le 10^\circ\\ \mu_{\mathrm{f}}\cdot F_{45\_N}+F_{45\_T}, & 30^\circ \le \theta_f \le 60^\circ\\ \mu_{\mathrm{f}}\cdot F_{90\_N}+F_{90\_T}, & 80^\circ \le \theta_f \le 100^\circ\\ \mu_{\mathrm{f}}\cdot F_{135\_N}+F_{135\_T} & 120^\circ \le \theta_f \le 150^\circ\end{cases}\end{cases} \tag{8.21}$$

8.5.2 Parallel Fiber Grinding Force

For the shear-mode buckling removal of 0° carbon fibers, based on the mechanical behavior analysis and stress-strain relationships in Section 7.2, the shear deformation force F_{0_Sh} for fiber removal in 0° CFRP grinding is given by:

$$\mathrm{F}_{0\_Sh} = \frac{\pi d_f^2 G_m}{4 V_f \left(1 - V_f\right)} \tag{8.22}$$

For the extensional-mode buckling removal of 0° carbon fibers, the critical buckling load P_{cr} is obtained from Eq. (7.20), and the corresponding critical fiber stress is $\sigma_{fcr} = P_{cr}/V_f$. Thus,

$$\sigma_{\mathrm{fcr}} = 2\sqrt{\frac{V_f E_f E_m}{3 V_m}} \tag{8.23}$$

Therefore, the tensile-compressive deformation force F_{0_TC} for fiber removal in 0° CFRP grinding is expressed as:

$$\mathrm{F}_{0\_TC} = \frac{\pi \mathrm{d}_f^2}{2}\sqrt{\frac{V_f E_f E_m}{3 V_m}} \tag{8.24}$$

For the elliptical normal contact force between the abrasive grain tip and 0° CFRP carbon fibers, based on the grain/fiber contact mechanics analysis in Section 7.2.3, the stress along the y-axis is calculated as:

$$\frac{\sigma_x}{p_{\mathrm{emax}}} = \frac{2 b_e}{k_e^2 a_e}\left(\Omega_x + \upsilon \Omega_x'\right) \tag{8.25}$$

$$\frac{\sigma_z}{p_{\mathrm{emax}}} = \frac{2 b_e}{k_e^2 a_e}\left(\Omega_z + \upsilon \Omega_z'\right) \tag{8.26}$$

$$\frac{\sigma_y}{p_{\mathrm{emax}}} = -\frac{b_e}{k_e^2 a_e}\left(\frac{1 - T_e^2}{T_e}\right) \tag{8.27}$$

where,

$$\Omega_x = -\tfrac{1}{2}\left(1 - T_e\right) + \xi\left[F\left(\phi_e, k_e\right) - E\left(\phi_e, k_e\right)\right] \tag{8.28}$$

$$\Omega_x' = 1 - \frac{a_e^2 T_e}{b_e^2} + \xi\left[\frac{a_e^2}{b_e^2} E\left(\phi_e, k_e\right) - F\left(\phi_e, k_e\right)\right] \tag{8.29}$$

$$\Omega_y = \frac{1}{2} + \frac{1}{2T_e} - \frac{a_e^2 T_e}{b_e^2} + \xi\left[\frac{a_e^2}{b_e^2} E\left(\phi_e, k_e\right) - F\left(\phi_e, k_e\right)\right] \tag{8.30}$$

$$\Omega_y' = -1 + T_e + \xi\left[F\left(\phi_e, k_e\right) - E\left(\phi_e, k_e\right)\right] \tag{8.31}$$

$$T_e = \sqrt{\frac{b_e^2 + h_{gf}^2}{a_e^2 + h_{gf}^2}} \tag{8.32}$$

$$\xi = \frac{y_e}{a_e} = \cot \phi_e \tag{8.33}$$

$$\phi_e = \operatorname{arccot}\left(\frac{y_e}{a_e}\right) \tag{8.34}$$

where the elliptic integrals $F(\phi, k)$ and $E(\phi, k)$ are obtained by referring to standard elliptic integral tables. Hence, the elliptical contact force exerted by a single abrasive grain on the fiber is given by:

$$F_{0\_e} = \frac{2\pi a_e b_e p_{emax}}{3} \tag{8.35}$$

8.5.3 Along-Fiber Grinding Force

For the interlaminar shear force at FOA = 45°, its components in the normal and tangential directions are respectively:

$$\mathrm{F}_{45\_Sh\_t} = \mathrm{F}_{45\_Sh} \cdot \cos\theta_f \tag{8.36}$$

$$\mathrm{F}_{45\_Sh\_n} = \mathrm{F}_{45\_Sh} \cdot \sin\theta_f \tag{8.37}$$

For the bending deformation and fracture removal at FOA = 45°, the general solution of the governing equation for the carbon fiber elastic foundation beam (Eq. 7.35) is:

$$w(y) = e^{\beta y}(A\cos\beta y + B\sin\beta y) + e^{-\beta y}(C\cos\beta y + D\sin\beta y) \tag{8.38}$$

Using the following mathematical relation,

$$\begin{cases} \cosh(\beta y) = (e^{\beta y} + e^{-\beta y})/2 \\ \sinh(\beta y) = (e^{\beta y} - e^{-\beta y})/2 \end{cases} \tag{8.39}$$

$$\begin{cases} e^{\beta y} = \cosh(\beta y) + \sinh(\beta y) \\ e^{\beta y} = \cosh(\beta y) - \sinh(\beta y) \end{cases} \tag{8.40}$$

the general solution can be simplified as:

$$w(y) = \cosh\beta y(A\cos\beta y + B\sin\beta y) + \sinh\beta y(C\cos\beta y + D\sin\beta y) \tag{8.41}$$

The deflection of the fiber at different deformation stages is given by,

$$w_1(\beta_m y) = A_1\varphi_1 + B_1(\varphi_2 + \varphi_4) + C_1(\varphi_2 - \varphi_4) + D_1\varphi_3,\ 0 \le y \le a_g \tag{8.42}$$

$$\begin{aligned} & w_2(\beta_m y) \\ &= A_2\varphi_1 + B_2(\varphi_2 + \varphi_4) + C_2(\varphi_2 - \varphi_4) + D_2\varphi_3,\ a_g \le y \le a_g + h_{g\_45} \end{aligned} \tag{8.43}$$

$$\begin{aligned} w_3(\beta_{mb} y) &= A_3(\varphi_1 + \varphi_2 - \varphi_4) + B_3(\varphi_2 + \varphi_3 + \varphi_4) + \\ &+ C_3(\varphi_1 - \varphi_2 + \varphi_4) + D_3(\varphi_2 - \varphi_3 + \varphi_4),\ y \ge a_g + h_{g\_45} \end{aligned} \tag{8.44}$$

where,

$$\begin{cases} \varphi_1(\beta y) = cosh(\beta y) \cdot cos(\beta y) \\ \varphi_2(\beta y) = [\cosh(\beta y) \cdot \sin(\beta y) + \sinh(\beta y) \cdot \cos(\beta y)]/2 \\ \varphi_3(\beta y) = sinh(\beta y) \cdot sin(\beta y) \\ \varphi_4(\beta y) = [\cosh(\beta y) \cdot \sin(\beta y) - \sinh(\beta y) \cdot \cos(\beta y)]/2 \end{cases} \tag{8.45}$$

$$\beta = \begin{cases} \beta_m = \sqrt[4]{\dfrac{k_{mn}}{4E_{ft}I_f}} \\ \beta_{mb} = \sqrt[4]{\dfrac{k_{mn} + k_b}{4E_{ft}I_f}} \end{cases} \tag{8.46}$$

Therefore, the cutting force F_{B_45} exerted by the abrasive grain on the carbon fiber is:

$$\mathrm{F}_{B\_45} = \int_0^{a_g} k_{mn} w_1 dy + \int_{a_g}^{a_g+h_{g\_45}} k_{mn} w_2 dy + \int_{a_g+h_{g\_45}}^{\infty} (\mathrm{k}_{mn} + \mathrm{k}_b)\, w_3 dy \tag{8.47}$$

The slope of the fiber deformation is obtained by the first derivative of the deflection

$$k_1 = \frac{dw_1}{dy} = \beta_m\{-2A_1\varphi_4 + B_1[\varphi_1 + \varphi_3] + C_1[\varphi_1 - \varphi_3] + 2D_1\varphi_2\} \tag{8.48}$$

$$k_2 = \frac{dw_2}{dy} = \beta_m\{-2A_2\varphi_4 + B_2[\varphi_1 + \varphi_3] + C_2[\varphi_1 - \varphi_3] + 2D_2\varphi_2\} \tag{8.49}$$

$$k_3 = \frac{dw_2}{dy} = \beta_{mb}\{A_3[\varphi_1 - \varphi_3 - 2\varphi_4] +$$
$$+B_3[\varphi_1 + 2\varphi_2 + \varphi_3] - C_3[\varphi_1 - \varphi_3 + 2\varphi_4] + 2D_3[\varphi_1 - 2\varphi_2 + \varphi_3]\} \quad (8.50)$$

the bending moment on the micro-segment of the carbon fiber is obtained by the second derivative

$$M_1 = -E_{ft}I_f\frac{d^2w_1}{dy^2} = 2E_{ft}I_f\beta_m^2\{A_1\varphi_3 - B_1[\varphi_2 - \varphi_4] + C_1[\varphi_2 + \varphi_4] + D_1\varphi_1\} \quad (8.51)$$

$$M_2 = -E_{ft}I_f\frac{d^2w_2}{dy^2} = 2E_{ft}I_f\beta_m^2\{A_2\varphi_3 - B_2[\varphi_2 - \varphi_4] + C_2[\varphi_2 + \varphi_4] + D_2\varphi_1\} \quad (8.52)$$

$$M_3 = -E_{ft}I_f\frac{d^2w_2}{dy^2} = 2E_{ft}I_f\beta_{mb}^2\{A_3[\varphi_2 + \varphi_3 + \varphi_4] -$$
$$-B_3[\varphi_1 + \varphi_2 - \varphi_4] - C_3[\varphi_2 - \varphi_3 + \varphi_4] + D_3[\varphi_1 - \varphi_2 + \varphi_4]\} \quad (8.53)$$

and the shear force on the micro-segment is obtained by the third derivative

$$Q_1 = -E_{ft}I_f\frac{d^3w_1}{dy^3} = 2E_{ft}I_f\beta_m^3\{2A_1\varphi_2 - B_1[\varphi_1 - \varphi_3] + C_1[\varphi_1 + \varphi_3] + D_1\varphi_4\} \quad (8.54)$$

$$Q_2 = -E_{ft}I_f\frac{d^3w_1}{dy^3} = 2E_{ft}I_f\beta_m^3\{2A_2\varphi_2 - B_2[\varphi_1 - \varphi_3] + C_2[\varphi_1 + \varphi_3] + D_2\varphi_4\} \quad (8.55)$$

$$Q_3 = -E_{ft}I_f\frac{d^3w_3}{dy^3} = 2E_{ft}I_f\beta_{mb}^3\{A_3[\varphi_1 + 2\varphi_2 + \varphi_3] -$$
$$-B_3[\varphi_1 - \varphi_3 - 2\varphi_4] - C_3[\varphi_1 - 2\varphi_2 + \varphi_3] + D_3[\varphi_1 - \varphi_3 + 2\varphi_4]\} \quad (8.56)$$

The solution is determined by the following boundary conditions: At the top of the carbon fiber (point A_{45}), y=0,

$$M_1 = 0 \quad (8.57)$$

$$Q_1 = 0 \quad (8.58)$$

At the grain cutting point (point B_{45}), $y = a_g$,

$$w_1 = w_2 =$$
$$= A_U\cos(2\pi f_U t) - B_U\cos(2\pi f_U t + \phi)\cdot\cos\theta + v_w t + \frac{d_s}{2}\sin\left(\frac{v_s}{d_s/2}t\right) \quad (8.59)$$

$$k_1 = k_2 \quad (8.60)$$

$$M_1 = M_2 \tag{8.61}$$

At the fiber-matrix debonding point (point C_{45}), $y = a_g + h_{g\_45}$,

$$w_2 = w_3 = \sigma_b / k_b \tag{8.62}$$

$$k_2 = k_3 \tag{8.63}$$

$$M_2 = M_3 \tag{8.64}$$

$$Q_2 = Q_3 \tag{8.65}$$

At the infinite far end of the fiber, $y = +\infty$,

$$w_3 = 0 \tag{8.66}$$

By solving the deflection equation for fiber bending fracture deformation using the above boundary conditions combined with cutting simulation results, w_1, w_2, and w_3 are obtained. Substituting these into Eq. (8.47) yields F_{B_45}. Consequently, the normal and tangential components of the bending fracture force for 45° fibers are respectively:

$$\mathrm{F}_{B\_45\_n} = \mathrm{F}_{B\_45} \cos\theta_f \tag{8.67}$$

$$\mathrm{F}_{B\_45\_t} = \mathrm{F}_{B\_45} \sin\theta_f + \mu_\mathrm{f} \mathrm{F}_{B\_45\_n} \tag{8.68}$$

8.5.4 Vertical Fiber Grinding Force

Regarding the bending-induced fracture removal mechanism in CFRP grinding with 90° fiber orientation, this study addresses the case where $k_{mt} \leq \left(4 k_{mn} E_f I_f\right)^{1/2}$. The general solution is given as follows:

$$w(y) = \mathrm{e}^{\beta y}(A\cos\beta y + B\sin\beta y) + \mathrm{e}^{-\beta y}(C\cos\beta y + D\sin\beta y) \tag{8.69}$$

$$\begin{gathered} w(y) = A\cos(\beta_1 y)\cosh(\beta_2 y) + \\ + B\cos(\beta_1 y)\sinh(\beta_2 y) + C\sin(\beta_1 y)\cosh(\beta_2 y) + D\sin(\beta_1 y)\sinh(\beta_2 y) \end{gathered} \tag{8.70}$$

where,

$$\beta_1 = \sqrt{\beta_{mb}^2 - \frac{k_{mt}}{4E_f I_f}} \tag{8.71}$$

$$\beta_2 = \sqrt{\beta_{mb}^2 + \frac{k_{mt}}{4E_f I_f}} \tag{8.72}$$

Therefore, the cutting force F_{A_90} exerted by the abrasive grain on the carbon fiber can be expressed as:

$$\mathrm{F}_{A\_90} = \int_0^{a_g+h_g} k_{mn} w_1 dy + \int_{a_g+h_{g\_45}}^{\infty} (\mathrm{k}_{mn} + \mathrm{k}_b) w_2 dy \tag{8.73}$$

For different deformation stages of the fiber,

$$w_1(\beta_m y) = A_1\varphi_1 + B_1(\varphi_2 + \varphi_4) + C_1(\varphi_2 - \varphi_4) + D_1\varphi_3,\ 0 \le y \le a_g + h_g \tag{8.74}$$

$$\begin{aligned} w_2(\beta_{mb} y) &= A_2(\varphi_1 + \varphi_2 - \varphi_4) + B_2(\varphi_2 + \varphi_3 + \varphi_4) + \\ &+ C_2(\varphi_1 - \varphi_2 + \varphi_4) + D_2(\varphi_2 - \varphi_3 + \varphi_4),\ y \ge a_g + h_g \end{aligned} \tag{8.75}$$

the slope of fiber deformation is obtained by taking the first derivative of the deflection,

$$k_1 = \frac{dw_1}{dy} = \beta_m\{-2A_1\varphi_4 + B_1[\varphi_1 + \varphi_3] + C_1[\varphi_1 - \varphi_3] + 2D_1\varphi_2\} \tag{8.76}$$

$$\begin{aligned} k_2 &= \frac{dw_2}{dy} = \beta_{mb}\{A_2[\varphi_1 - \varphi_3 - 2\varphi_4] + \\ &+ B_2[\varphi_1 + 2\varphi_2 + \varphi_3] - C_2[\varphi_1 - \varphi_3 + 2\varphi_4] + 2D_2[\varphi_1 - 2\varphi_2 + \varphi_3]\} \end{aligned} \tag{8.77}$$

the bending moment is derived from the second derivative,

$$M_1 = -E_{ft} I_f \frac{d^2 w_1}{dy^2} = 2E_{ft} I_f \beta_m^2 \{A_1\varphi_3 - B_1[\varphi_2 - \varphi_4] + C_1[\varphi_2 + \varphi_4] + D_1\varphi_1\} \tag{8.78}$$

$$\begin{aligned} M_2 &= -E_{ft} I_f \frac{d^2 w_2}{dy^2} = 2E_{ft} I_f \beta_{mb}^2 \{A_2[\varphi_2 + \varphi_3 + \varphi_4] - \\ &- B_2[\varphi_1 + \varphi_2 - \varphi_4] - C_2[\varphi_2 - \varphi_3 + \varphi_4] + D_2[\varphi_1 - \varphi_2 + \varphi_4]\} \end{aligned} \tag{8.79}$$

and the shear force is determined from the third derivative,

$$Q_1 = -E_{ft} I_f \frac{d^3 w_1}{dy^3} = 2E_{ft} I_f \beta_m^3 \{2A_1\varphi_2 - B_1[\varphi_1 - \varphi_3] + C_1[\varphi_1 + \varphi_3] + D_1\varphi_4\} \tag{8.80}$$

$$Q_2 = -E_{ft}I_f\frac{d^3w_2}{dy^3} = 2E_{ft}I_f\beta_{mb}^3\{A_2[\varphi_1 + 2\varphi_2 + \varphi_3] - \\ -B_2[\varphi_1 - \varphi_3 - 2\varphi_4] - C_2[\varphi_1 - 2\varphi_2 + \varphi_3] + D_2[\varphi_1 - \varphi_3 + 2\varphi_4]\} \tag{8.81}$$

By applying the boundary conditions, At point A_{90}, $y = 0$:

$$w_1 = A_U\cos(2\pi f_U t) - B_U\cos(2\pi f_U t + \phi)\cdot\cos\theta + v_w t + \frac{d_s}{2}\sin\left(\frac{v_s}{d_s/2}t\right) \tag{8.82}$$

$$Q_1 = F_{A\_90} \tag{8.83}$$

At point C_{90}, $y = a_g + h_g$:

$$w_1 = w_2 = \frac{\sigma_b}{k_b} \tag{8.8}$$

$$k_1 = k_2 \tag{8.85}$$

$$M_1 = M_2 \tag{8.86}$$

$$Q_1 = Q_2 \tag{8.87}$$

At infinity, $y = +\infty$:

$$w_2 = 0 \tag{8.88}$$

The constants *A*, *B*, *C*, *D*, are solved. Subsequently, the fiber deformation equation w (y) is obtained, from which the maximum deflection can be determined. The normal and tangential components of the bending fracture force for 90° fiber orientation are respectively given as:

$$F_{A\_90\_n} = F_{A\_90}\sin\gamma \tag{8.89}$$

$$F_{A\_90\_t} = F_{A\_90}\cos\gamma + \mu_f F_{A\_90\_n} \tag{8.90}$$

Concerning the longitudinal compressive force exerted by the abrasive grain on 90° CFRP, Equations (7.40) and (7.41) yield the following expression,

$$F_{Lc} = C_{Lc}\cdot A_{Lc} \tag{8.91}$$

$$F_{Ls} = S_{Lc}\cdot A_{Lc} \text{ (8.92)}$$

where A_{Lc} represents the longitudinal compressive contact area, taken as $\pi a_e^2/4$. The resulting normal force components, F_{Lc} and F_{Ls}, correspond to the following tangential force:

$$F_{Lc\_t} = \mu_f \cdot F_{Lc} \tag{8.93}$$

$$F_{Ls\_t} = \mu_f \cdot F_{Ls} \tag{8.94}$$

8.5.5 Reverse Fiber Grinding Force

According to Equation (7.42), the axial linear force increases and fiber fracture occurs at a certain distance from the cutting edge. Thus, the average force exerted on the abrasive grain due to fiber micro-buckling is given by:

$$\overline{P_{mb}} = A_{gmb}\sigma_{mb} = \frac{1}{2}b_g\frac{a_g}{\cos\theta_g}\sigma_{mb} \tag{8.95}$$

where A_{gmb} represents the contact area between the abrasive grain and the workpiece.

$$A_{gmb} = \frac{1}{2}b_g\frac{a_g}{\cos\theta_g} \tag{8.96}$$

To calculate the average force acting on the abrasive grain induced by micro-buckling in Region 1, the following assumptions are made: First, the axial force due to micro-buckling increases linearly along the fiber. Second, micro-buckling failure occurs at a certain distance from the grain/fiber contact point. Consequently, upon fiber fracture, the abrasive grain loses contact with the residual fiber still attached to the workpiece. To determine the average resultant force P_{mb} on the abrasive grain, a correction factor K_f is introduced. When continuous grain/fiber contact is maintained, $K_f = 0.5$; when fracture occurs at a certain distance, $K_f < 0.5$. In this study, K_f is selected as a random number following a normal distribution with a mean of 0.33 and a standard deviation of 0.015.

For the bulk removal force F_{mf} of 135° CFRP fiber bundles, based on the mechanical behavior analysis in Section 7.5.3:

$$F_{mf} = \frac{N_{mf}\pi d_f^2\left(L_{mf}\alpha_{135}\beta_{135}e\right)^{-\frac{1}{\beta_{135}}}}{4} \tag{8.97}$$

where N_{mf} is the number of fibers in the bundle, and L_{mf} is the fiber length within the bundle. According to statistical analysis of SEM micrographs of bulk removal debris from fiber bundles, N_{mf} follows a normal distribution with a mean of 22 and a standard deviation of 3.5; L_{mf} follows a normal distribution with a mean of 163 μm and a standard deviation of 17. Thus, the normal and tangential components of the bulk removal force for the fiber bundle are respectively expressed as:

$$F_{mf\_n} = F_{mf} \cdot \sin\theta_f \tag{8.98}$$

$$F_{mf\_t} = -F_{mf} \cdot \cos\theta_f + \mu_f \cdot F_{mf\_n} \tag{8.99}$$

Finally, the calculation method for the bending deformation and fracture removal force of carbon fibers in 135° CFRP is the same as that for 90° CFRP, and will not be reiterated here. The normal and tangential components of the final fiber bending fracture force generated by the abrasive grain are respectively given as:

$$\mathrm{F}_{A\_135\_n} = \mathrm{F}_{A\_135} \sin\left(\theta_f + \gamma\right) \tag{8.100}$$

$$\mathrm{F}_{A\_135\_t} = \mathrm{F}_{A\_135} \cos\left(\theta_f + \gamma\right) + \mu_{\mathrm{f}}\, \mathrm{F}_{A\_135\_n} \tag{8.101}$$

8.5.6 Interconnecting Mechanical Models

The total grinding force model for CFRP was obtained by applying interpolation to the fiber orientation gaps between the four typical FOAs. This study adopted a conventional linear interpolation approach, as in practical scenarios, the mechanical behavior of material removal and chip formation mechanisms also superimpose at each typical FOA. In other words, these typical FOAs represent inflection points in the variation of grinding force with respect to FOA. Using the linear interpolation method, model predictions were derived for the full range of fiber orientations, i.e., 0°≤ FOA ≤180°. Additionally, the average values from different tests in the friction and wear experiments described in Section 4.5 were used as the coefficient of friction μ_f for different FOAs and incorporated into the force model.

To account for the enhancement effect of ultrasonic vibration on nano-lubricant MQL, an interfacial ultrasonic-enabled lubrication gain coefficient δ_μ was introduced for different two-dimensional ultrasonic vibration directions. This coefficient was determined by comprehensively considering both droplet spreading behavior and hydrodynamic lubrication characteristics under ultrasonic vibration conditions, combined with experimental results from Section 3.7. The values adopted in this study are $\delta_{\mu\text{-}U45}$ =0.751 for 2D ultrasonic vibration at θ_U =45°, and $\delta_{\mu\text{-}U90}$ =0.763 for θ_U =90°. The normal and tangential grinding forces on the wheel were obtained by summing the grinding forces of all dynamic active abrasive grains in the grinding zone:

$$
F_N = \begin{cases} N_{eg} \cdot F_{0\_e}, & \theta_f = 0° \\ N_{eg} \cdot (F_{45\_Sh\_n} + F_{B\_45\_n}), & \theta_f = 45° \\ N_{eg} \cdot F_{A\_90\_n} + N_{90\_s} \cdot F_{Ls} + N_{90\_c} \cdot F_{Lc}, & \theta_f = 90° \\ N_{135\_bf} \cdot F_{A\_90\_n} + N_{eg} \cdot F_{mf\_n} & \theta_f = 135° \end{cases} \tag{8.102}
$$

$$
F_T = \begin{cases} N_{0\_Sh} \cdot F_{0\_Sh} + N_{0\_TC} \cdot F_{0\_TC} + \mu_f \cdot F_{N0}, & \theta_f = 0° \\ N_{eg} \cdot (F_{45\_Sh\_t} + F_{B\_45\_t}) + \mu_f \cdot F_{N45}, & \theta_f = 45° \\ N_{eg} \cdot F_{A\_90\_t} + \mu_f \cdot F_{N90}, & \theta_f = 90° \\ N_{135\_bf} \cdot F_{A\_90\_t} + N_{eg} \cdot F_{mf\_t} + \mu_f \cdot F_{N135} & \theta_f = 135° \end{cases} \tag{8.103}
$$

8.6 EXPERIMENTAL VERIFICATION

8.6.1 Multiangle 2D Ultrasonic Vibration Assisted Grinding Experimental System

Experiments were conducted on a K-36 CNC surface grinding machine equipped with a two-dimensional multi-directional UVAG setup and an MQL supply system. Grinding forces in the normal, tangential, and axial directions were measured and recorded in real-time using a YDM-III99 dynamometer. The average of five measurements was taken as the experimental value of the grinding force, as illustrated in Figure 4. The ground surface morphology and chip formation of CFRP were characterized using an Olympus laser confocal microscope and an S-3400N scanning electron microscope.

Figure 4. Ultrasonic vibration assisted MQL grinding experimental system

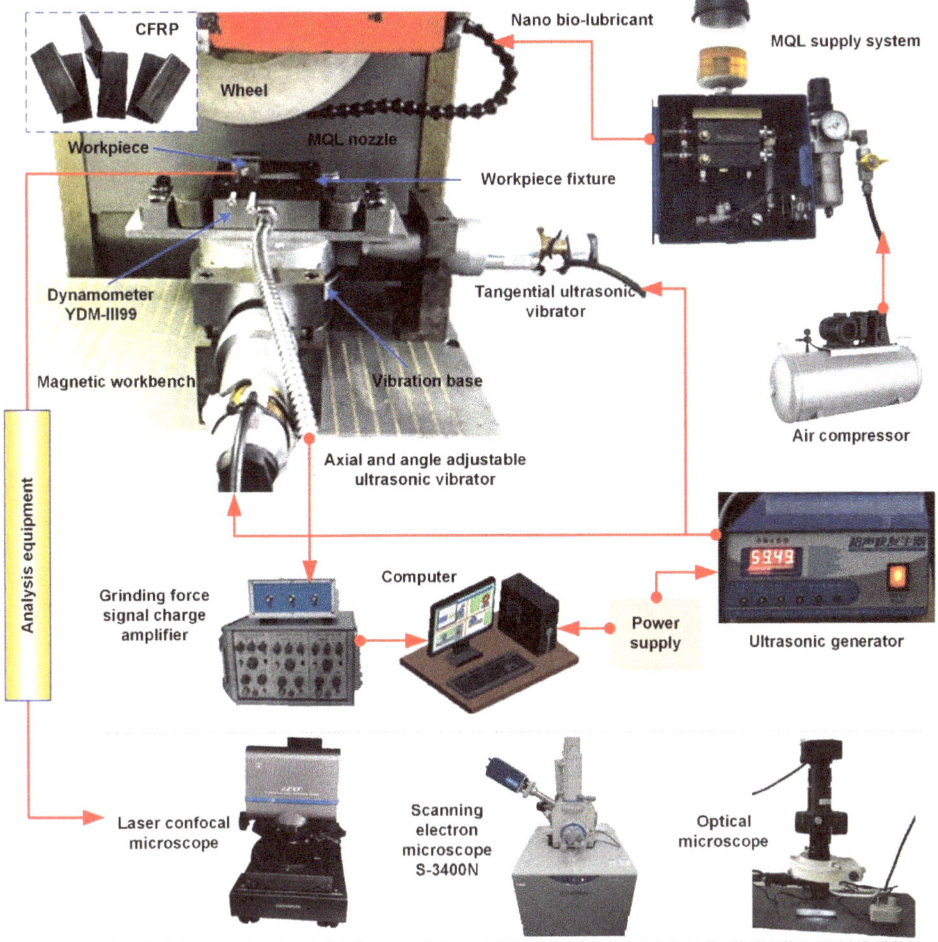

8.6.2 Experimental Materials and Methods

The physical and mechanical properties of the CFRP workpiece and its constituents are provided in Table 1 in Section 8.2.2. The grinding directions for the workpiece were set at longitudinal orientations of 0°, 45°, 90°, and 135°. The parameters of the diamond grinding wheel are listed in Table 2 in Section 8.3. Experiments were conducted under three different conditions: conventional dry grinding, ultrasonic-assisted nano-lubricant MQL with $\theta_U = 45°$, and ultrasonic-assisted nano-lubricant MQL with $\theta_U = 90°$. The specific experimental plan and parameters for multi-directional two-dimensional ultrasonic vibration-assisted nano-lubricant MQL grinding are summarized in Table 4.

Table 3. Experimental parameters of the nanollubricant MQL 2D multiangle UVAG

Parameters		Value
Grinding		
Grinding pattern		Surface grinding
Circumferential linear velocity of grinding wheel v_s (m/s)		30
Feed speed v_w (mm/s)		10
Cutting depth a_p (ìm)		30
FOAs of CFRP grinding		0°, 45°, 90°, 135°
Jet and supply of MQL		
MQL flow rate (ml/h)		50
MQL nozzle distance (mm)		20
MQL nozzle angle (°)		15
MQL gas pressure (MPa		0.3
2D multi-angle ultrasonic vibration		
Direction of application	Tangential	Axial (Angle adjustable)
Combined vibration angle $_{èU}$ (°)	45; 90	
Frequency (kHz)	Theoretical frequency 20, Actual frequency 19.960	Theoretical frequency 20, Actual frequency 19.933
Amplitude (ìm)	10	10

8.7 RESULTS VALIDATION AND ANALYSIS DISCUSSION

A comparison between the predicted results of the grinding force model and experimental values for CFRP with different fiber orientations under various grinding conditions is shown in Figure 5. The maximum error for the normal grinding force F_n is 12.82%, while that for the tangential grinding force F_t is 16.97%. Nevertheless, such errors are reasonable and acceptable, as they arise from multiple influencing factors during CFRP specimen preparation and grinding force measurement. The main reasons are analyzed as follows: (1) Fiber arrangement is neither perfectly straight nor uniformly distributed; (2) The equivalent homogeneous material cannot fully represent the actual material surrounding the fibers; (3) While idealized materials are assumed to be free of inherent defects, actual materials may contain manufacturing flaws such as voids. Comparisons with results from other force models demonstrate that the proposed grinding force prediction model for CFRP effectively captures and validates the material removal mechanics for different fiber orientations and provides reliable grinding force predictions.

As indicated by the experimental results in Figure 5, grinding forces vary significantly with fiber orientation. Under all conditions, the normal grinding force F_n is the lowest at FOA = 0°. In conventional dry grinding and $\theta_U = 45°$ ultrasonic vibration-assisted grinding (UVAG) with MQL, F_n reaches its maximum at FOA = 45°, whereas under $\theta_U = 90°$ UVAG MQL, the maximum occurs at FOA = 90°. For the tangential grinding force F_t, the minimum is consistently observed at FOA = 0°, and the maximum at FOA = 45° across all conditions. UVAG MQL effectively reduces grinding forces, with $\theta_U = 90°$ UVAG MQL exhibiting the most pronounced effect. Compared to conventional grinding, this condition reduces F_n by 16.29%, 18.47%, 5.03%, and 14.53%, and F_t by 25.21%, 33.32%, 24.16%, and 34.35% for FOAs of 0°, 45°, 90°, and 135°, respectively. For $\theta_U = 45°$ UVAG MQL, the reductions in F_n are 5.82%, 12.44%, 7.7%, and 7.94%, and in F_t are 6.53%, 10.81%, 17.82%, and 42.59% for the respective FOAs. Finally, the mechanisms behind grinding force reduction in ultrasonic vibration-assisted grinding are analyzed as follows: (1) Axial ultrasonic vibration promotes fracture of carbon fibers in the grinding zone, facilitating subsequent material removal by abrasive grains. (2) Intermittent cutting induced by tangential ultrasonic vibration reduces wheel-workpiece contact time and enhances the dynamic stiffness of the system. (3) The hydrodynamic lubrication effect generated by ultrasonic vibration provides driving force for the penetration, spreading, and transport of nano-lubricants into the grinding zone, improving the effective flow rate and cooling/lubrication performance. The formation of a nano-lubricant hydrodynamic film further enhances interfacial lubrication conditions. (4) Two-dimensional ultrasonic vibration applied to the CFRP workpiece increases the spreading rate of nano-lubricant droplets on the CFRP surface, improving lubrication and heat exchange efficiency, which contributes to the reduction of grinding forces and grinding temperatures.

Figure 5. Comparison of predicted and experimental values of CFRP grinding force with different FOAs under different machining conditions

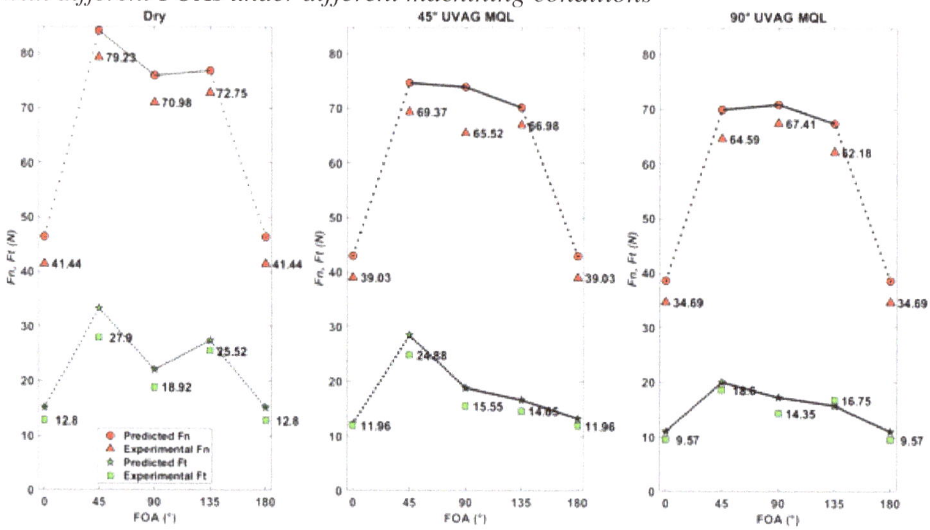

8.8 SUMMARY

This chapter establishes a uniform random distribution model for carbon fibers in CFRP and a diamond grinding wheel model where abrasive grain protrusion heights and diameters follow a normal distribution. It analyzes the geometric interaction and intermittent cutting states between grains and the workpiece under different ultrasonic parameters, investigates the influence of ultrasonic vibration on the dynamic stiffness of the grinding system, and reveals the material removal mechanisms in negative-rake cutting of CFRP with varying fiber orientations. On this basis, an ultrasonic-assisted grinding force model for CFRP under both dry and nano-lubricant MQL conditions is developed and experimentally validated.

(1) The role of the interface phase in transmitting loads between fibers and the matrix is analyzed, clarifying its influence on the material removal mechanism and grinding behavior of CFRP. The mechanical properties of the interface phase are calculated based on a representative volume element (RVE). Using SEM micrographs of polished CFRP cross-sections, a model is established with normally distributed fiber diameters and uniformly randomly distributed fiber positions. A diamond grinding wheel model is developed based on probabilistic statistics, considering normally distributed grain protrusion heights, apex

angles, and sizes. Different grain-fiber interaction states and cutting behaviors are analyzed, leading to the formulation of grinding force models for various fiber orientations.

(2) A parallel fiber grinding force model is established based on shear- and extensional-mode buckling removal of 0° fibers and elliptical contact mechanics between grains and fibers. For 45° forward fiber grinding, the deformation governing equation of carbon fibers modeled as elastic foundation beams is solved, and a force model is developed by analyzing debonding and slip during fiber removal. For 90° vertical fiber grinding, a force model incorporating bending fracture, matrix cracking under longitudinal compression, and fiber shear fracture is established by analyzing geometric boundary conditions of fiber bending. For 135° reverse fiber grinding, a force model is formulated considering micro-buckling removal forces, bulk fiber bundle removal forces, and bending fracture forces. Finally, accounting for interfacial friction under different conditions and introducing an ultrasonic-enabled MQL lubrication gain coefficient and a dynamic stiffness equivalent coefficient, a full-range fiber orientation grinding force model for CFRP is constructed using linear interpolation.

(3) A multi-directional two-dimensional ultrasonic-assisted nano-lubricant MQL grinding experimental system was designed and built. Experimental validation of the grinding force model was conducted for CFRP with different fiber orientations under various conditions. Comparing predicted and experimental grinding forces across all FOAs and conditions, the maximum error for normal grinding force F_n is 12.82%, and for tangential grinding force F_t it is 16.97%.

(4) Grinding forces vary significantly with FOAs. Under all conditions, the normal grinding force F_n is lowest at FOA = 0°. Under conventional dry grinding and θ_U = 45° ultrasonic vibration-assisted grinding (UVAG) with MQL, F_n peaks at FOA = 45°, whereas under θU = 90° UVAG MQL, it peaks at FOA = 90°. UVAG MQL effectively reduces grinding forces, with θ_U = 90° UVAG MQL showing the most significant reduction. Compared to conventional grinding, it reduces F_n by 16.29%, 18.47%, 5.03%, and 14.53%, and F_t by 25.21%, 33.32%, 24.16%, and 34.35% for different FOAs, respectively. For θ_U = 45° UVAG MQL, compared to conventional grinding, F_n is reduced by 5.82%, 12.44%, 7.7%, and 7.94%, and F_t by 6.53%, 10.81%, 17.82%, and 42.59% for different FOAs, respectively.

REFERENCES

Cao, S., Zhang, X., Wu, C., Yang, M., & Xia, K. (2024). Towards understanding the material removal mechanism on the effect of ultrasonic vibration and anisotropy for unidirectional CFRP by scratching. *Journal of Materials Processing Technology*, *324*, 118250. DOI: 10.1016/j.jmatprotec.2023.118250

Chu, A., Li, C., Zhou, Z., Liu, B., Zhang, Y., Yang, M., Gao, T., Liu, M., Zhang, N., Dambatta, Y., & Sharma, S. (2023). Nanofluids minimal quantity lubrication machining: From mechanisms to application. *Lubricants (Basel, Switzerland)*, *11*(10), 422. DOI: 10.3390/lubricants11100422

Dambatta, Y. S., Li, C., Sayuti, M., Sarhan, A. A., Yang, M., Li, B., Chu, A., Liu, M., Zhang, Y., Said, Z., & Zhou, Z. (2024). Grindability evaluation of ultrasonic assisted grinding of silicon nitride ceramic using minimum quantity lubrication based SiO2 nanofluid. *Chinese Journal of Mechanical Engineering*, *37*(1), 25. DOI: 10.1186/s10033-024-01011-x

Danish, M., Gupta, M. K., Rubaiee, S., Ahmed, A., Mahfouz, A., & Jamil, M. (2021). Machinability investigations on CFRP composites: A comparison between sustainable cooling conditions. *International Journal of Advanced Manufacturing Technology*, *114*(11), 3201–3216. DOI: 10.1007/s00170-021-07073-8

Das, S., & Pandivelan, C. (2020). Grinding characteristics during ultrasonic vibration assisted grinding of alumina ceramic in selected dry and MQL conditions. *Materials Research Express*, *7*(8), 085404. DOI: 10.1088/2053-1591/abad14

Deng, S., Ye, L., Mai, Y. W., & Liu, H. Y. (1998). Evaluation of fibre tensile strength and fibre/matrix adhesion using single fibre fragmentation tests. *Composites. Part A, Applied Science and Manufacturing*, *29*(4), 423–434. DOI: 10.1016/S1359-835X(97)00094-8

Ding, W., Huang, Q., Zhao, B., Cao, Y., Tang, M., Deng, M., Liu, G., Zhao, Z., & Chen, Q. (2023). Wear characteristics of white corundum abrasive wheel in ultrasonic vibration-assisted grinding of AISI 9310 steel. *Ceramics International*, *49*(8), 12832–12839. DOI: 10.1016/j.ceramint.2022.12.153

Duan, Z., Wang, S., Wang, Z., Li, C., Li, Y., Song, J., Liu, J., & Liu, X. (2024). Tool wear mechanisms in cold plasma and nano-lubricant multi-energy field coupled micro-milling of Al-Li alloy. *Tribology International*, *192*, 109337. DOI: 10.1016/j.triboint.2024.109337

Esmaeili, H., Adibi, H., & Rezaei, S. M. (2019). An efficient strategy for grinding carbon fiber-reinforced silicon carbide composite using minimum quantity lubricant. *Ceramics International, 45*(8), 10852–10864. DOI: 10.1016/j.ceramint.2019.02.163

Gao, C. Y., Xiao, J. Z., Zhang, L. C., & Ke, Y. L. (2017). On the static and dynamic properties of fiber-reinforced polymer composites: A three-phase constitutive model. *Journal of Thermoplastic Composite Materials, 30*(11), 1560–1577. DOI: 10.1177/0892705716646418

Gong, P., Zhang, Y., Cui, X., Xu, S., Yang, M., Jia, D., & Li, C. (2024). Lubricant transportation mechanism and wear resistance of different arrangement textured turning tools. *Tribology International, 196*, 109704. DOI: 10.1016/j.triboint.2024.109704

Haddad, M., Zitoune, R., Eyma, F., & Castanie, B. (2014). Study of the surface defects and dust generated during trimming of CFRP: Influence of tool geometry, machining parameters and cutting speed range. *Composites. Part A, Applied Science and Manufacturing, 66*, 142–154. DOI: 10.1016/j.compositesa.2014.07.005

Helmy, M. O., El-Hofy, M. H., & El-Hofy, H. (2018). Effect of cutting fluid delivery method on ultrasonic assisted edge trimming of multidirectional CFRP composites at different machining conditions. *Procedia CIRP, 68*, 450–455. DOI: 10.1016/j.procir.2017.12.077

Hu, S., Li, C., Zhou, Z., Liu, B., Zhang, Y., Yang, M., Li, B., Gao, T., Liu, M., Cui, X., Wang, X., Xu, W., Dambatta, Y., Li, R., & Sharma, S. (2023). Nanoparticle-enhanced coolants in machining: Mechanism, application, and prospects. *Frontiers of Mechanical Engineering, 18*(4), 53. DOI: 10.1007/s11465-023-0769-8

Iskandar, Y., Tendolkar, A., Attia, M. H., Hendrick, P., Damir, A., & Diakodimitris, C. (2014). Flow visualization and characterization for optimized MQL machining of composites. *CIRP Annals, 63*(1), 77–80. DOI: 10.1016/j.cirp.2014.03.078

Kodama, H., Okazaki, S., Jiang, Y., Yoden, H., & Ohashi, K. (2020). Thermal influence on surface layer of carbon fiber reinforced plastic (CFRP) in grinding. *Precision Engineering, 65*, 53–63. DOI: 10.1016/j.precisioneng.2020.04.005

Li, C. (2023). *Thermodynamic mechanism of MQL grinding with nano bio-lubricant.* Springer Nature, 43-74.

Li, H., Chen, T., Duan, Z., Zhang, Y., & Li, H. (2022b). A grinding force model in two-dimensional ultrasonic-assisted grinding of silicon carbide. *Journal of Materials Processing Technology, 304*, 117568. DOI: 10.1016/j.jmatprotec.2022.117568

Li, L., Zhang, Y., Cui, X., Said, Z., Sharma, S., Liu, M., Gao, T., Zhou, Z., Wang, X., & Li, C. (2023). Mechanical behavior and modeling of grinding force: A comparative analysis. *Journal of Manufacturing Processes*, *102*, 921–954. DOI: 10.1016/j.jmapro.2023.07.074

Li, M., Yu, T., Zhang, R., Yang, L., Ma, Z., Li, B., Wang, X., Wang, W., & Zhao, J. (2020). Experimental evaluation of an eco-friendly grinding process combining minimum quantity lubrication and graphene-enhanced plant-oil-based cutting fluid. *Journal of Cleaner Production*, *244*, 118747. DOI: 10.1016/j.jclepro.2019.118747

Li, Z., Zhu, L., Yang, Z., Ma, J., & Cao, W. (2022a). Investigation of tool-workpiece contact rate and milling force in elliptical ultrasonic vibration-assisted milling. *International Journal of Advanced Manufacturing Technology*, *118*(1), 585–601. DOI: 10.1007/s00170-021-07946-y

Liang, Y., Chen, Y., Chen, B., Fan, B., Yan, C., & Fu, Y. (2019). Feasibility of ultrasonic vibration assisted grinding for carbon fiber reinforced polymer with monolayer brazed grinding tools. *International Journal of Precision Engineering and Manufacturing*, *20*(7), 1083–1094. DOI: 10.1007/s12541-019-00135-8

Liu, C., Zhang, Y., Zhu, L., Li, Q., Shu, X., Qin, S., Wang, D., & Shi, W. (2025). A Review of Ultrasonic Vibration-Assisted Grinding for Advanced Materials. *Intelligent and Sustainable Manufacturing*, *2*(1), 10001. DOI: 10.70322/ism.2025.10001

Liu, D., Ni, C., Wang, Y., & Zhu, L. (2024). Review of serrated chip characteristics and formation mechanism from conventional to additively manufactured titanium alloys. *Journal of Alloys and Compounds*, *970*, 172573. DOI: 10.1016/j.jallcom.2023.172573

Liu, S., Chen, T., & Wu, C. (2017). Rotary ultrasonic face grinding of carbon fiber reinforced plastic (CFRP): A study on cutting force model. *International Journal of Advanced Manufacturing Technology*, *89*(1), 847–856. DOI: 10.1007/s00170-016-9151-x

Nandagopal, R. A., Boay, C. G., & Narasimalu, S. (2020). An empirical model to predict the strength degradation of the hygrothermal aged CFRP material. *Composite Structures*, *236*, 111876. DOI: 10.1016/j.compstruct.2020.111876

Ni, C., Zhu, L., & Yang, Z. (2019). Comparative investigation of tool wear mechanism and corresponding machined surface characterization in feed-direction ultrasonic vibration assisted milling of Ti–6Al–4V from dynamic view. *Wear*, *436*, 203006. DOI: 10.1016/j.wear.2019.203006

Ning, F., Wang, H., & Cong, W. (2019). Rotary ultrasonic machining of carbon fiber reinforced plastic composites: A study on fiber material removal mechanism through single-grain scratching. *International Journal of Advanced Manufacturing Technology*, *103*(1), 1095–1104. DOI: 10.1007/s00170-019-03433-7

Niu, Q., Rong, J., Jing, L., Gao, H., Tang, S., Qiu, X., Liu, L., Wang, X., & Dai, F. (2023). Study on force-thermal characteristics and cutting performance of titanium alloy milled by ultrasonic vibration and minimum quantity lubrication. *Journal of Manufacturing Processes*, *95*, 115–130. DOI: 10.1016/j.jmapro.2023.04.002

Pervaiz, S., Kannan, S., Huo, D., & Mamidala, R. (2020). Ecofriendly inclined drilling of carbon fiber-reinforced polymers (CFRP). *International Journal of Advanced Manufacturing Technology*, *111*(7), 2127–2153. DOI: 10.1007/s00170-020-06203-y

Qu, S., Gong, Y., Yang, Y., Sun, Y., Wen, X., & Qi, Y. (2020a). Investigating minimum quantity lubrication in unidirectional Cf/SiC composite grinding. *Ceramics International*, *46*(3), 3582–3591. DOI: 10.1016/j.ceramint.2019.10.076

Qu, S., Gong, Y., Yang, Y., Wang, W., Liang, C., & Han, B. (2020b). An investigation of carbon nanofluid minimum quantity lubrication for grinding unidirectional carbon fibre-reinforced ceramic matrix composites. *Journal of Cleaner Production*, *249*, 119353. DOI: 10.1016/j.jclepro.2019.119353

Qu, S., Yao, P., Gong, Y., Chu, D., Yang, Y., Li, C., Wang, Z., Zhang, X., & Hou, Y. (2022). Environmentally friendly grinding of C/SiCs using carbon nanofluid minimum quantity lubrication technology. *Journal of Cleaner Production*, *366*, 132898. DOI: 10.1016/j.jclepro.2022.132898

Ran, Y., Kang, R., Dong, Z., Jin, Z., & Bao, Y. (2023). Ultrasonic assisted grinding force model considering anisotropy of SiCf/SiC composites. *International Journal of Mechanical Sciences*, *250*, 108311. DOI: 10.1016/j.ijmecsci.2023.108311

Saha, S., Deb, S., & Bandyopadhyay, P. P. (2023). Tool wear induced burr formation and concomitant reduction in MQL wetting capability in micro-milling. *International Journal of Mechanical Sciences*, *245*, 108095. DOI: 10.1016/j.ijmecsci.2022.108095

Scida, D., Assarar, M., Poilâne, C., & Ayad, R. (2013). Influence of hygrothermal ageing on the damage mechanisms of flax-fibre reinforced epoxy composite. *Composites. Part B, Engineering*, *48*, 51–58. DOI: 10.1016/j.compositesb.2012.12.010

Senthilkumar, M., Prabukarthi, A., & Krishnaraj, V. (2018). Machining of CFRP/Ti6Al4V stacks under minimal quantity lubricating condition. *Journal of Mechanical Science and Technology*, *32*(8), 3787–3796. DOI: 10.1007/s12206-018-0731-6

Song, Y. X., Li, C. H., Zhou, Z. M., Liu, B., Sharma, S., Dambatta, Y. S., Zhang, Y., Yang, M., Gao, T., Liu, M., Cui, X., Wang, X., Xu, W., Li, R., & Wang, D. Z. (2025). Nanobiolubricant grinding: A comprehensive review. *Advances in Manufacturing*, *13*(1), 1–42. DOI: 10.1007/s40436-023-00477-7

Wang, H., Hu, Y., Cong, W., & Hu, Z. (2019). A mechanistic model on feeding-directional cutting force in surface grinding of CFRP composites using rotary ultrasonic machining with horizontal ultrasonic vibration. *International Journal of Mechanical Sciences*, *155*, 450–460. DOI: 10.1016/j.ijmecsci.2019.03.009

Wang, H., Ning, F., Hu, Y., & Cong, W. (2018). Surface grinding of CFRP composites using rotary ultrasonic machining: A comparison of workpiece machining orientations. *International Journal of Advanced Manufacturing Technology*, *95*(5), 2917–2930. DOI: 10.1007/s00170-017-1401-z

Wang, H., Pei, Z. J., & Cong, W. (2020a). A feeding-directional cutting force model for end surface grinding of CFRP composites using rotary ultrasonic machining with elliptical ultrasonic vibration. *International Journal of Machine Tools & Manufacture*, *152*, 103540. DOI: 10.1016/j.ijmachtools.2020.103540

Wang, H., Pei, Z. J., & Cong, W. (2020b). A mechanistic cutting force model based on ductile and brittle fracture material removal modes for edge surface grinding of CFRP composites using rotary ultrasonic machining. *International Journal of Mechanical Sciences*, *176*, 105551. DOI: 10.1016/j.ijmecsci.2020.105551

Wang, X., Li, C., Zhang, Y., Ali, H. M., Sharma, S., Li, R., Yang, M., Said, Z., & Liu, X. (2022b). Tribology of enhanced turning using biolubricants: A comparative assessment. *Tribology International*, *174*, 107766. DOI: 10.1016/j.triboint.2022.107766

Wang, Z., Liang, Y., Li, H., & Yu, T. (2022a). Milling force and tool wear mechanisms on milling TC21 titanium alloy under different lubrication conditions. *International Journal of Advanced Manufacturing Technology*, *123*(1), 169–185. DOI: 10.1007/s00170-022-10108-3

Xu, W., Li, C., Cui, X., Zhang, Y., Yang, M., Gao, T., Liu, M., Wang, X., Zhou, Z., Sharma, S., & Dambatta, Y. S. (2023). Atomization mechanism and machinability evaluation with electrically charged nanolubricant grinding of GH4169. *Journal of Manufacturing Processes*, *106*, 480–493. DOI: 10.1016/j.jmapro.2023.10.037

Yang, M., Kong, M., Li, C., Long, Y., Zhang, Y., Sharma, S., Li, R., Gao, T., Liu, M., Cui, X., Wang, X., Ma, X., & Yang, Y. (2023a). Temperature field model in surface grinding: A comparative assessment. *International Journal of Extreme Manufacturing*, *5*(4), 042011. DOI: 10.1088/2631-7990/acf4d4

Yang, M., Ma, H., Hao, J., Li, Z., Li, R., Zhou, Z., Gao, T., Liu, M., Cui, X., Wang, X., Zhang, Y., Dambatta, Y., Long, Y., & Li, C. (2024). Droplet size distribution model of needle electrode electrostatic atomization and milling nickel-based alloy performance evaluation. *Journal of Manufacturing Processes*, *119*, 682–698. DOI: 10.1016/j.jmapro.2024.04.011

Yang, Y., Yang, M., Li, C., Li, R., Said, Z., Ali, H. M., & Sharma, S. (2023b). Machinability of ultrasonic vibration-assisted micro-grinding in biological bone using nanolubricant. *Frontiers of Mechanical Engineering*, *18*(1), 1. DOI: 10.1007/s11465-022-0717-z

Yang, Z., Zhu, L., Zhang, G., Ni, C., & Lin, B. (2020). Review of ultrasonic vibration-assisted machining in advanced materials. *International Journal of Machine Tools & Manufacture*, *156*, 103594. DOI: 10.1016/j.ijmachtools.2020.103594

Zhang, X., Li, C., Zhou, Z., Liu, B., Zhang, Y., Yang, M., Gao, T., Liu, M., Zhang, N., Said, Z., Sharma, S., & Ali, H. M. (2023). Vegetable oil-based nanolubricants in machining: From physicochemical properties to application. *Chinese Journal of Mechanical Engineering*, *36*(1), 76. DOI: 10.1186/s10033-023-00895-5

Zhao, B., Huang, Q., Cao, Y., Ding, W., Xu, J., & Liu, G. (2023). Thermal analysis of ultrasonic vibration-assisted grinding with moment-triangle heat sources. *International Journal of Heat and Mass Transfer*, *216*, 124552. DOI: 10.1016/j.ijheatmasstransfer.2023.124552

Chapter 9
Numerical Analysis and Experimental Quantitative Assessment of Damage Mechanism:
Formation, Evolution, and Distribution

ABSTRACT

In this chapter, Firstly, The instantaneous dynamic generation, conduction, accumulation of grinding heat and its thermal effects were revealed. Furthermore, a 3D microscale three phase finite element numerical model was established to determine the removal and damage mechanisms of different FOAs using single grain grinding. Finally, the grinding experiments were conducted under different ultrasonic and lubrication conditions. The results showed the damage formation and evolution mechanisms with different FOAs are various which complement the debris generation mechanisms. The damage is mainly distributed along the grinding direction, with the lowest damage at 0° FOA, and the highest at 45°. The 2D ultrasonic assisted nanolubricant MQL with 90° coupled vibration angle (θU) has an optimal damage reduction effect with a maximum of 27.56% compared to conventional grinding. However, 45° θU only showed excellent damage suppression on 135° FOA of 40.38%. This paper provides theoretical support for the damage suppression strategy of CFRP grinding.

DOI: 10.4018/979-8-3373-9013-0.ch009

9.1 INTRODUCTION

The widespread application of carbon fiber-reinforced polymers (CFRP) is highly dependent on the advancement of their manufacturing and machining technologies (Wen et al., 2023). Characterized by high specific strength and specific stiffness, CFRP composites exhibit multidirectional lamination and heterogeneous anisotropy at the macroscopic level, while comprising a multiphase system of resin matrix, carbon fibers, and interfaces at the microscopic scale. The complex interactions between fibers and matrix during material removal in cutting and grinding processes classify CFRP as a typical difficult-to-machine aerospace material. During cutting and grinding operations, CFRP is prone to carbon fiber fracture or matrix cracking, leading to quality defects such as burrs and cracks on machined surfaces. These defects severely compromise component machining accuracy and surface quality. Moreover, conventional tools and processes often induce severe damage such as delamination, making high-quality processing extremely challenging. Additional issues include short tool life, frequent tool changes, difficulties in achieving high-precision digital machining, low efficiency, and high scrap rates—ultimately resulting in significant economic losses and potential catastrophic failures.

Consequently, investigating the formation and evolution mechanisms of grinding-induced damage in CFRP represents a critical prerequisite for achieving high-quality machining. Dong et al. (2023) discovered that tearing damage during the initial grinding engagement stage correlates with weak stiffness and radial grinding forces, with both radial grinding force and maximum tear depth varying with changes in maximum undeformed chip thickness. Tian et al. (2023) demonstrated that milling CFRP circular-celled honeycomb structures produces greater tearing than grinding, while employing chamfered or rounded grinding wheels enhances machining zone stiffness and suppresses damage. Wang et al. (2024) found that either excessively small or large abrasive grains can degrade surface quality, and fiber pull-out may occur when the feed direction aligns with the FOA. Researchers have explored various assisted grinding techniques; notably, Yan et al. (2024) proposed vortex tube cooling-assisted grinding, which reduces grinding zone temperature, inhibits resin smearing, improves surface quality, and minimizes wheel clogging.

Furthermore, UVAG has emerged as an advanced CFRP machining method and represents the most extensively researched and prominent direction (Ran et al., 2023; Gu et al., 2023; Liu et al., 2025). Grinding constitutes a multi-point cutting process characterized by multiple interference effects due to irregular grain geometries and inconsistent grinding depths (Yao et al., 2023; Xu et al., 2024). Therefore, single-grit grinding studies provide an essential methodology for accurately capturing complex material removal behaviors (Zhang et al., 2024). Cao et al. (2024) confirmed through scratch tests that ultrasonic vibration suppresses machining damage and reduces scratch loads. Lin and Zhou (2024) observed that the intermittent impact cutting on fiber surfaces and instantaneous velocity enhancement induced by ultrasonic vibration reduce interface debonding and limit the localized propagation of matrix damage. Similarly, Liang et al. (2021) found that UVAG generates lower grinding forces and shorter fiber fragments compared to conventional slot grinding, with maximum fiber chip lengths at 0° and 90° orientations reduced by approximately 78% and 66%, respectively. Ding et al. (2024) demonstrated that grinding forces are highest perpendicular to the fiber direction and lowest along the longitudinal fiber direction. The most severe surface damage occurs during longitudinal fiber grinding, where extensive fiber pull-out from the matrix is observed. Compared with conventional grinding (CG), UVAG reduces surface roughness by 19.1%. Zhao et al. (2023) attributed the intermittent cutting behavior between grinding wheel and workpiece to a significant 24.2%–51.8% reduction in grinding temperature and a 2.8-fold increase in material removal rate, thereby preventing sudden burning during high-speed grinding.

Cooling and lubrication serve as crucial means for reducing grinding forces and heat generation, minimizing friction and wear, and improving surface integrity (Yang and Li, 2025; Gong et al., 2024). However, CFRP's mechanical properties degrade due to moisture absorption and swelling characteristics, limiting the application of traditional flood lubrication. Researchers have confirmed the unique advantages of MQL in composite machining (Yang et al., 2023, 2024a). As an innovative method integrating lubrication and cooling, MQL has proven effective for various difficult-to-machine materials (Xu et al., 2022; Yang et al., 2024b; Duan et al., 2024). Qu et al. (2022) successfully identified optimal parameters for carbon-based NMQL, finding that the bonding state between fibers and matrix depends on lubrication conditions and grain sharpness. Other scholars have verified that ultrasonic vibration enhances the lubricant penetration and transport performance of MQL (Namlu et al., 2024). Niu et al. (2023) demonstrated that the separation cutting characteristics of ultrasonic vibration improve MQL cooling/lubrication performance, reducing tool-chip interface friction, enhancing heat dissipation, and decreasing mechanical and thermal loads. Naskar et al. (2021) proposed a hybrid grinding method combining UVAG with MQL, discovering that higher lubricant fluidity reduces resistance to vibration transmission to the grinding zone, thereby increasing effective amplitude with improved fluid lubricity. Duman et al. (2024) found that simultaneous application of ultrasonic vibration and MQL prevents surface defects, reduces average surface roughness by 24%–35%, lowers machining costs by approximately 20%–30%, and significantly contributes to carbon emission reduction. Namlu et al. (2024) confirmed that the combination of ultrasonic vibration with hybrid nanofluid MQL offers substantial advantages over conventional machining methods, reducing cutting forces (up to 37.6%) and surface roughness (up to 37.4%), while producing highly uniform surface textures characterized by minimal surface defects and significantly suppressed burr formation.

The preceding chapters have characterized the forces and surface morphologies arising from the proposed hybrid process. This chapter delves deeper into the underlying damage mechanisms, employing numerical analysis alongside experimental quantitative assessment to trace the formation, evolution, and spatial distribution of grinding-induced damage in CFRP. The aim is to connect the observed mechanical signatures to the specific modes of fiber–matrix failure. Current research consistently identifies the FOA as a critical factor influencing grinding forces, damage initiation, and propagation in CFRP. However, the formation and evolution mechanisms of grinding damage under high-speed, high-pressure, and high-temperature boundary conditions across different FOAs remain unclear. Furthermore, the inherent random characteristics of abrasive grains on grinding wheel surfaces compound the complexity of damage mechanisms in heterogeneous, anisotropic CFRP grinding. This study employs finite element numerical analysis combined with grinding experiments to investigate the thermomechanical coupling damage mechanisms in CFRP grinding. We reveal the instantaneous dynamic generation mechanism of grinding heat for different FOAs and analyze the influence of heat conduction and accumulation on the thermal expansion coefficient. A 3D microscale finite element model is established to determine material removal and damage mechanisms during negative rake angle cutting of CFRP with varying FOAs. Finally, we propose a novel multi-angle 2D ultrasonic nanofluid MQL grinding method and conduct grinding experiments under different conditions. Comparative analysis of damage distribution characteristics and quantitative characterization aims to provide theoretical guidance and data support for damage suppression strategies in precision grinding of CFRP components.

9.2 INSTANTANEOUS DYNAMIC HEAT GENERATION MECHANISM

Grinding heat plays a crucial role in determining grinding performance and product quality. Once the grinding temperature exceeds the glass transition temperature of the resin matrix, degradation occurs, resulting in thermal damage and the formation of a heat-sensitive coating on the machined surface. The FOA significantly influences the generation, conduction, and accumulation of heat during the grinding of CFRP. The majority of the energy expended during material deformation is converted into heat during CFRP grinding. However, the dissipation of this heat into the environment is hindered by the low thermal conductivity of CFRP, leading to thermal damage and degradation of material properties. As illustrated in Figure 1, the heat sources can be categorized into three components: material removal heat generation, friction heat generated between the grains and debris, and friction heat generated between the grains and the CFRP workpiece surface.

Figure 1. Heat generation in grain grinding of CFRP

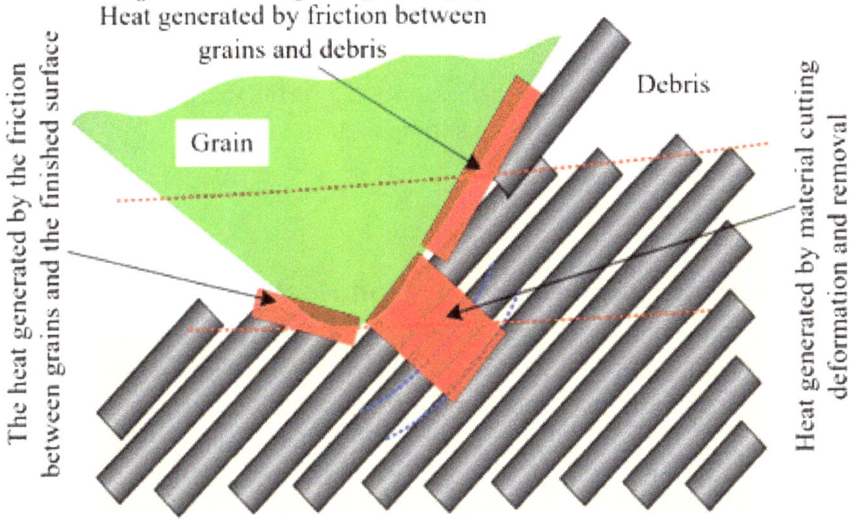

9.2.1 Heat Generation From Material Cutting Removal and Friction Between Grains and Debris

The dissipation of plastic energy is caused by grinding force of single grain, its heat flux density is,

$$q_p = \eta_p \sigma \dot{\varepsilon}_p \tag{9.1}$$

Among them, η_p represents the factor that converts plastic energy dissipation into heat, $\dot{\varepsilon}_p$ is the effective plastic strain rate. The heat flux density generated by the grains friction on CFRP and chips is,

$$q_f = \eta_f \, \eta_s \, \tau \, \dot{\gamma}_f \tag{9.2}$$

Among them, η_f is the proportion of frictional energy converted into heat, η_s represents the proportion factor of frictional heat transfer to debris, τ is the frictional stress between grains and representative CFRP units, and $\dot{\gamma}_f$ is the slip ratio between grains and debris.

By applying an energy balance principle, the total heat generated through plastic deformation and frictional energy equals the temperature rise in the CFRP minus the heat transferred to the surrounding environment. Thus, the heat flux balance equation for CFRP can be formulated as:

$$q_p + q_f = C\frac{\partial T}{\partial t} - \lambda_q \frac{\partial^2 T}{\partial x^2} \tag{9.3}$$

Among them, C is the specific heat capacity, λ_q is the thermal conductivity, so the temperature rise of the workpiece at time $t+\Delta t$ is,

$$\Delta T_{pf} = \left(\frac{q_p + q_f}{C} + \frac{q}{C}\frac{\partial^2 T}{\partial x^2}\right)\Delta t \tag{9.4}$$

9.2.2 Instantaneous Dynamic Friction Heat Generation

A predictive model for frictional heat has been established, whereby the heat transfer per unit area per unit time is proportional to the temperature gradient.

$$q_H = \frac{\Delta Q_H}{A_H \Delta t} = -\lambda_H \frac{dT_H}{d\delta_H} \tag{9.5}$$

Where q_H is the heat flux density, $\Delta Q/\Delta t$ is the heat flux rate, A_H is the contact surface area, λ_H is the thermal conductivity, and $dT_H/d\delta_H$ is the temperature gradient.

The friction between grains and CFRP surface significantly affects heat generation. The derivation of the temperature rise prediction model for CFRP grinding is as follows (Shaw and Cookson, 2005),

$$\Delta T = \frac{0.754 R_h q_{hf}}{k_\omega}\sqrt{\frac{a_\omega L_\omega}{V_s}} \tag{9.6}$$

Among them, R_h is the percentage of heat transferred to the workpiece, q_{hf} is the heat flux density in interaction area between the grain and the workpiece, k_w is the material's thermal conductivity, L_w is the length of the interaction area, and a_w is the material's thermal diffusion coefficient,

$$a_\omega = \frac{k_\omega}{\rho c_\omega} \tag{9.7}$$

where c_ω and ρ are the specific heat capacity and density of CFRP. Assuming that all frictional work was converted into heat, therefore,

$$q_{\mathrm{hf}} = \frac{\mu F_n v_s}{S_{gh}} \tag{9.8}$$

In the formula, S_{gh} is the frictional contact area (calculated by approximating the grains as a cone),

$$S_{gh} = \frac{\pi}{2}\left(\frac{a_g}{\cos\theta_g}\right)^2 \tag{9.9}$$

Finally, the local temperature of friction between grains and CFRP interface can be obtained,

$$T_H = T_0 + \Delta T \tag{9.10}$$

where T_0 is room temperature.

9.3 HEAT IMPACT ON THERMAL EXPANSION

The plastic deformation of fibers can be disregarded, as the elongation of fibers is minimal, often resulting in brittle fracture. Consequently, the sole heat source for the fibers is frictional energy. The effect of thermal expansion on the overall strain during grinding can be described as follows (Merino-Pérez et al., 2015),

$$\varepsilon_H = \varepsilon_{H\_0} - \varepsilon_{H\_T} \tag{9.11}$$

The thermal strain in the longitudinal and transverse direction is,

$$\varepsilon^{\mathrm{l}}_{H\_T} = \alpha_l \Delta T \tag{9.12}$$

$$\varepsilon^{\mathrm{t}}_{H\_T} = \alpha_{\mathrm{t}} \Delta T \tag{9.13}$$

In the equation, ε_{H_0} is the initial strain (due to heating), ε_{H_T} is the thermal strain, α_l and α_t are the longitudinal and transverse thermal expansion coefficients of CFRP, and ΔT is the temperature increment.

Under no external stress, subjected to a uniform temperature change ΔT, the longitudinal and transverse thermal strains are $\varepsilon^{\mathrm{l}}_{\mathrm{T}}$ and $\varepsilon^{t}_{\mathrm{T}}$. Figure 2 shows the free expansion deformation of resin matrix region and fiber region under the action of ΔT. Due to the absence of external stress, so,

$$\sigma_{fl} V_f + \sigma_{ml} V_m = 0 \tag{9.14}$$

Although there are differences in the thermal expansion coefficients of the fibers and matrix—as well as in their free expansion along the longitudinal direction—this expansion does not occur due to the strong bond between the fibers and matrix. Therefore, the longitudinal strain ε_T^l remains the same, and the micro-stress generated in the fibers and matrix compensates for the displacement differences caused by the varying expansions. Thus, we have:

$$\varepsilon_T^l = \varepsilon_{fl} = \varepsilon_{ml} \tag{9.15}$$

When the temperature changes ΔT, the longitudinal stress-strain relationship between fibers and matrix,

$$\begin{cases} \varepsilon_{ml} = \dfrac{\sigma_{ml}}{E_m} + \alpha_m \Delta T \\ \varepsilon_{fl} = \dfrac{\sigma_{fl}}{E_{fl}} + \alpha_{fl} \Delta T \end{cases} \tag{9.16}$$

Among them, α_m and α_{fl} are the thermal expansion coefficients of the matrix and the longitudinal thermal expansion coefficient of the fibers (α_m is taken as 72×10^{-6}/K, α_{ft} is taken as -1×10^{-6}/K), defined by the effective thermal expansion coefficient,

$$\alpha_l = \frac{\varepsilon_T^l}{\Delta T} \tag{9.17}$$

So,

$$\alpha_l = \frac{\alpha_{fl} E_{fl} V_f + \alpha_m E_m V_m}{E_{fl} V_f + E_m V_m} \tag{9.18}$$

The transverse thermal strain is,

$$\varepsilon_T^t = \varepsilon_{ft} V_f + \varepsilon_{mt} V_m \tag{9.19}$$

When the temperature changes ΔT, the transverse stress-strain relationship between fibers and matrix,

$$\begin{cases} \varepsilon_{mt} = -\upsilon_f \dfrac{\sigma_{mt}}{E_m} + \alpha_m \Delta T \\ \varepsilon_{ft} = -\upsilon_m \dfrac{\sigma_{ft}}{E_{ft}} + \alpha_{ft} \Delta T \end{cases} \tag{9.20}$$

Among them, α_{ft} is the transverse thermal expansion coefficient of the fiber (α_{ft} is taken as 28×10^{-6}/K). According to the definition of transverse thermal expansion coefficient,

$$\alpha_t = \frac{\varepsilon_T^t}{\Delta T} \tag{9.21}$$

So,

$$\alpha_t = \alpha_f V_f (1 + \upsilon_f) + \alpha_m V_m (1 + \upsilon_m) - (\upsilon_m V_m + \upsilon_f V_f) \alpha_l \tag{9.22}$$

Figure 2. Schematic diagram for calculating the thermal expansion coefficient of CFRP

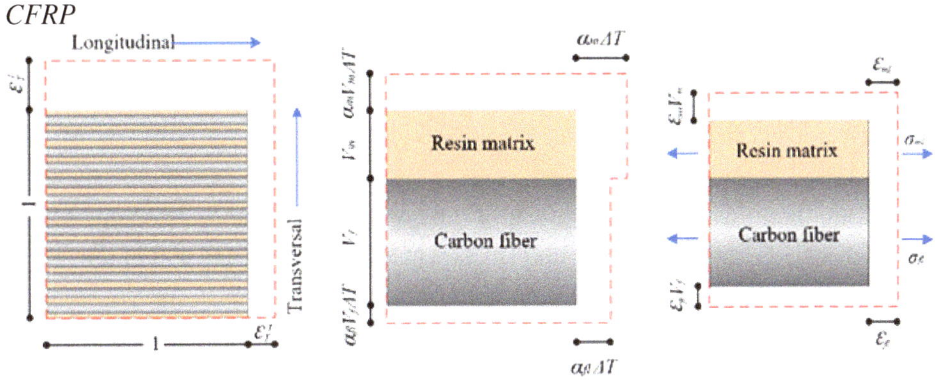

9.4 MATERIAL MODELING

9.4.1 Continuum Damage Model for Carbon Fibers

This study defines carbon fiber as a transversely isotropic linear elastic material, as it does not undergo plastic deformation prior to material failure. A user-developed subroutine, VUMAT, was created to define the constitutive model for the fiber. When the stress exceeds the ultimate strength, the carbon fiber will fail, resulting in the removal of the element from the finite element model. The unit type for the carbon fiber and resin matrix is C3D8R (8-node reduced integration element), equipped with automatic hourglass control and universal contact type. The material property parameters of the carbon fiber can be found in Chapter 8.

The stress-strain response of a carbon fiber unit under longitudinal uniaxial tension and compression is illustrated in Figure 3. When the unit stress σ_{11} reaches the tensile strength X_t, the value of the damage variable d^f_{t1} signifies a completely damaged state. All other stress components in the fibers undergo a similar process. While modeling compression damage as a progressive process may align more closely with real-world behavior, it would extend computation time significantly. Therefore, it is assumed that compression failure occurs more rapidly. The dissipated energy associated with the fiber damage corresponds to the shaded area in Figure 3. The increment of elastic strain energy U_{start} is defined as:

$$U_{start} = U_{end} + \Delta U/\rho_f \tag{9.23}$$

$$\Delta U = \sum_{i,j=1}^{3} \Delta \varepsilon_{ij} \sigma_{ij} \tag{9.24}$$

$$\sigma_{ij} = \left(\sigma_{ij}^{end} + \sigma_{ij}^{start}\right)/2 \tag{9.25}$$

Among them, $\Delta \varepsilon_{ij}$ is the increment of the normal and shear components of the strain tensor, σ_{ij}^{start} and σ_{ij}^{end} are the unit stresses at the beginning and end of the increment, respectively. ρ_f is the mass density of the fiber. As shown in the shaded area in Figure 3, ΔU represents the increment of elastic strain energy density.

Figure 3. Stress-strain response of carbon fibers under longitudinal tension and compression

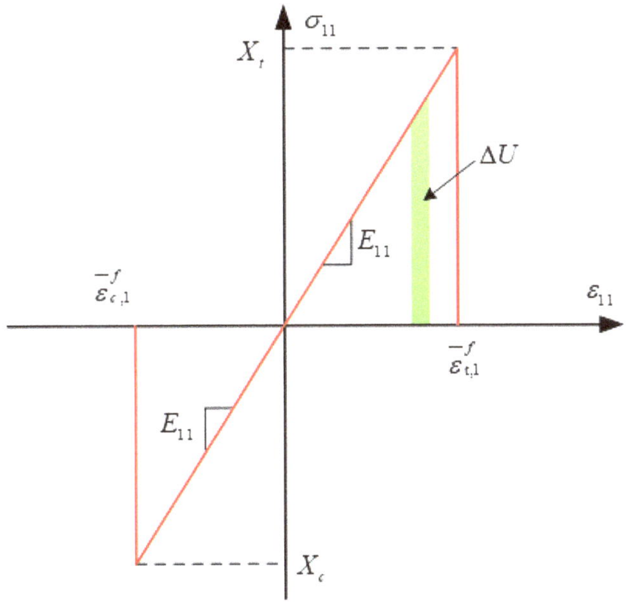

9.4.2 Constitutive Model for Matrix Resins

In terms of material strength modeling, CFRP exhibits unique characteristics inherent to materials with complex internal structures, which are heavily influenced by the stress state. A method for formalizing the types of stress states using triaxial parameters of stress tensors was proposed (Lomakin & Fedulov, 2013; Alexandrov et al., 2011; Fedulov et al., 2017).

$$\xi=\sigma/\sigma_0 \tag{9.26}$$

Where, $\sigma=\sigma_{ii}/3$, $\sigma_0=\sqrt{\frac{3}{2}S_{ij}S_{ij}}$, $S_{ij} = \sigma_{ij} - \delta_{ij}\sigma$, $\delta_{ij} = 0(i \neq j)$, $\delta_{ii} = 1$.

This parameter characterizes the relationship between shear stress intensity and static pressure. It facilitates the identification of loading types and the subsequent adjustment of material responses accordingly. A key advantage of triaxial parameters is that material properties vary smoothly and continuously as the stress state changes. Table 1 presents the values of stress triaxiality corresponding to various stress states.

The plasticity criteria applicable to stress state-related materials take a general form,

$$f(\xi) \bullet \sigma_0 = k_0 \tag{9.27}$$

among them, k_0 is a material constant representing the size of the yield limit of the material, such as the yield limit $f(\xi) = 1$ under Von Mises plastic conditions. There is a significant linear correlation between yield strength and triaxiality parameters. According to reference (Merino-Pérez et al., 2015), $f(\xi) = 1 + C\xi$, where C is equivalent to $\tan(\psi)$, where ψ is the material friction angle or expansion angle.

Table 1. Examples of values of the triaxiality parameter with corresponding stress states

Triaxial stress parameters ξ	**Stress state**
$-\infty$	Triaxial compression
-2/3	Biaxial compression
-1/3	Uniaxial compression
0	Pure shear
1/3	Uniaxial stretching
2/3	Biaxial stretching
$+\infty$	Triaxial stretching

The next step in plastic modeling requires selecting the hardening law that affects the yield limit parameters. The plastic criterion that includes hardening is,

$$\left(1 + C\xi\right)\sigma_0 = k\left(\varepsilon_{eq}^{pl}\right) \tag{9.28}$$

Among them, ε_{eq}^{pl} is the equivalent plastic strain, expressed as,

$$\varepsilon_{eq}^{pl} = \int \frac{\sigma_{ij} d\varepsilon_{ij}^{pl}}{\sigma_t^{pl}} \tag{9.29}$$

Among them, σ_t^{pl} is the tensile stress of the uniaxial tension curve. Material property parameters of the resin matrix is shown in Table 2.

Table 2. Material property parameters of the resin matrix

Property parameters	**Value**
Volume fraction / %	22
Density / g/cm³	1.05
Elastic modulus / GPa	3.4
Shear modulus / GPa	1.02
Axial tensile strength / MPa	85
Poisson's ratio	0.4
Fracture toughness / J/m²	500
Elongation at break / %	5.5

Once the material plastic flow model is established, it is essential to determine the criteria for damage accumulation and failure. Establishing a failure criterion that results in less damage accumulation under compressive conditions and more under tensile loads is crucial. The damage parameters proposed in references (Fedulov et al., 2017; Hooputra et al., 2004) are expressed as:

$$D^{pl} = \int \frac{d\varepsilon_{eq}^{pl}}{\varepsilon_D^{pl}(\xi)} \tag{9.30}$$

Where $\varepsilon_D^{pl}(\xi)$ is the experimentally determined dependency relationship of a given material, defined as a piecewise linear function. D^{pl} is the failure criterion, and the parameters depend on the triaxial stress parameters. Therefore, the model can be adjusted to reflect increased damage accumulation under tensile loads and reduced accumulation under compressive loads. Prior to the material meeting its failure criteria, the damage parameter D^{pl} does not influence the modeling; at this juncture ($D^{pl} = 1$), the material stiffness becomes zero. The aforementioned plastic model exhibits a linear relationship with its parameters and failure conditions (Formula (9.30)) and can be integrated into the material model library of Abaqus software.

9.4.3 Zero-Thickness Cohesive Elements of Interface

Modeling the interface phase between fiber and resin was accomplished using zero thickness cohesive elements (COH3D8). The behavior of these elements was defined based on the traction-separation law to accurately simulate the initiation and propagation of interface cracking. Figure 4 illustrates the constitutive relationship under pure tensile or shear loads. The AB segment represents the linear elastic stage of the cohesive element, and damage initiates upon reaching point B. The BC segment signifies the damage evolution stage, culminating in complete failure at point

C, at which point the cohesive element is deleted. A bilinear model is employed to characterize the adhesive contact properties, while a linear elastic model describes the elastic response of the interface prior to damage initiation.

$$\begin{Bmatrix} t_n \\ t_s \\ t_t \end{Bmatrix} = \begin{bmatrix} K_{nn} & 0 & 0 \\ 0 & K_{ss} & 0 \\ 0 & 0 & K_{tt} \end{bmatrix} \begin{Bmatrix} \varepsilon_n \\ \varepsilon_s \\ \varepsilon_t \end{Bmatrix} \tag{9.31}$$

Among them, t_n is the normal stress component, t_s is the shear stress component along the fiber direction, and t_t is the shear stress component perpendicular to the fiber direction. K_{nn} is the normal elastic modulus, K_{ss} is the shear elastic modulus along the fiber direction, and K_{tt} is the shear elastic modulus perpendicular to the fiber direction. ε_n is the normal strain component, ε_s is the shear strain component along the fiber direction, and ε_t is the shear strain component perpendicular to the fiber direction.

Figure 4. Traction separation response of zero thickness cohesive element under normal-shear loads

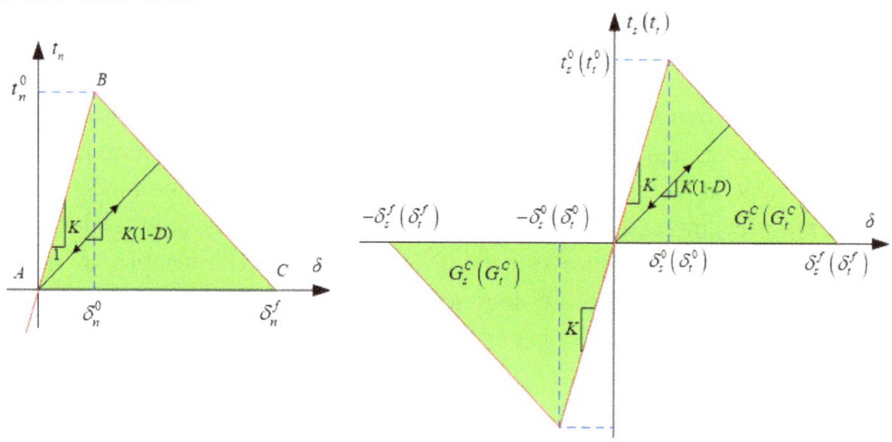

To model interface damage, the secondary stress criterion was adopted; damage onset occurs when the contact stress satisfies the conditions outlined in (Brewer and Lagace, 1988),

$$\left\{\frac{\langle t_n \rangle}{t_n^0}\right\}^2 + \left\{\frac{t_s}{t_s^0}\right\}^2 + \left\{\frac{t_t}{t_t^0}\right\}^2 = 1 \tag{9.32}$$

Among them t_n^0, t_s^0, and t_t^0 represent the cohesive strength in the normal and two tangential directions, respectively, that is, the maximum stress reached at the onset of damage in each direction. $\langle t_n \rangle = (t_n + |t_n|)/2$, indicating that pure normal compressive stress will not cause damage. The interface material parameters of numerical analysis model is shown in Table 3.

Table 3. Material property parameters of the CFRPs interface

Property parameters	**Value**
Poisson's ratio	0.36
Elastic modulus / GPa	88.6
Tensile strength / MPa	32.8
Shear strength / MPa	32.8
Cohesive stiffness / $\times 10^5$ MPa/mm	6.4
Normal cohesive strength / MPa	43
Tangential cohesive strength / MPa	50
Normal fracture energy / kJ/m^2	1.7
Fracture energy in shear direction / kJ/m^2	2.0
B-K mixed mode interaction index	1.09

Two methodologies exist for determining whether the cohesive element has completely failed: the maximum displacement method and the fracture energy method. In this study, the fracture energy method is utilized as the final failure criterion, coupled with the Benzeggagh-Kenane (B-K) criterion for energy (Benzeggagh and Kenane, 1996). The evolutionary behavior of mixed-mode progressive damage is defined following its occurrence, characterized by a total fracture energy,

$$G^C = G_n^C + \left(G_s^C - G_n^C\right)\left(\frac{G_s + G_t}{G_n + G_s + G_t}\right)^{\eta_C} \tag{9.33}$$

Among them, G_n^C and G_s^C are the fracture energies in the normal and shear directions. G_n, G_s and G_t are the work done by the traction stress in the normal and shear directions, respectively. η_C is the B-K mixed mode interaction index.

9.5 3D MESOSCOPIC FEM OF GRINDING MECHANICAL-THERMAL COUPLING

On the basis of the geometric structure of CFRP, each component is individually defined and modeled for microscopic numerical analysis, followed by the establishment of contact relationships and mutual constraints among the components. To enhance the efficiency of numerical analysis, a modeling investigation is conducted on a small section of the domain concerning the negative rake angle cutting of abrasive particles. In order to perform numerical analysis of the material removal process during the negative rake cutting of CFRP, the material constitutive model includes initial response, damage initiation criteria, and damage evolution mechanisms following material failure.

9.5.1 Geometric Modeling and Boundary Conditions

To emphasize the negative rake angle cutting of grains while taking into account the characteristics of diamond grains, the geometric shape of the grain cutting surface is depicted in Figure 5, where the rake angle of the cutting edge is set at -30°. The numerical model disregards the effects of abrasive wear and deformation, treating the grains as analytical rigid bodies to enhance computational efficiency. The grains are permitted to move solely along the negative y-axis at a cutting speed of 30 m/s with a cutting depth of 20 μm. The physical and mechanical properties of the grains are outlined in Table 4. The CFRP workpiece is constrained by fixing the bottom and right degrees of freedom. The unidirectional CFRP negative rake cutting process is simulated with FOAs of 0°, 45°, 90°, and 135°. When determining the grid partitioning size, a balance must be struck between model accuracy and computational speed; while finer grids yield high computational accuracy, they also result in slower calculation speeds. Given the three-phase modeling of CFRP workpieces, a mesh local subdivision method cannot be employed. Additionally, the radial size of the interface cohesive force model is extremely small, and excessive mesh partitioning could lead to partitioning failure; therefore, the mesh used for cohesive forces is finer than that of the adjacent materials.

Table 4. Physical and mechanical properties of diamond grains

Property parameters	Value
Density / $\times 10^3$ kg/m^3	4.12
Hardness / GPa	50
Young's modulus / GPa	776

continued on following page

Table 4. Continued

Property parameters	Value
Poisson's ratio	0.07
Transverse fracture strength / GPa	1.2

Figure 5. Numerical analysis diagram of CFRP cutting with negative rake angle of diamond grains

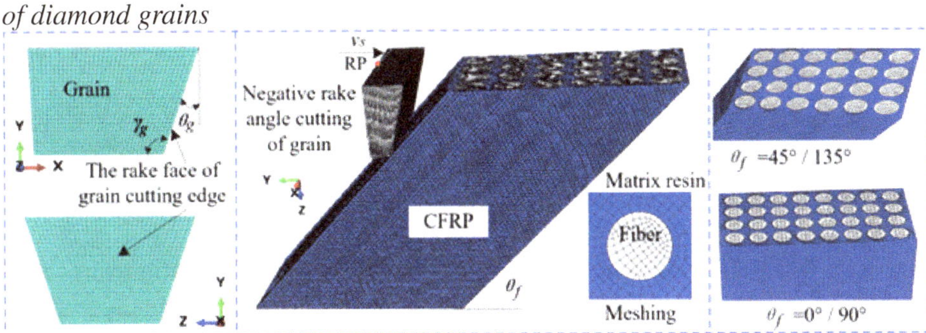

9.5.2 Contact and Friction

To simulate the mechanical interaction between grains and CFRPs, the "penalty contact" formula, which incorporates Coulomb friction, has been utilized within the ABAQUS platform. This approach establishes penalty contact among all components of the workpiece to prevent infiltration between units. The normal contact behavior at the contact surface is characterized as hard contact. By default, infiltration is disregarded until a unit is deleted. The friction coefficient between the grain and the workpiece is derived from friction and wear tests from Chapter 4. For the friction coefficient between fiber and matrix, a value of 0.5 has been established (Schön, 2000). The microdamage mechanism of carbon fiber negative rake cutting has been examined through a discrete continuous medium model. Continuous damage mechanics effectively describes the failure of fibers and the gradual softening of the matrix. The numerical analysis damage criteria for CFRP are presented in Table 5 (Liang et al., 2019; Hashin and Rotem, 1973).

Table 5. CFRP damage criteria for negative rake cutting

Failure mode	**Failure criteria**
Longitudinal tensile damage of fibers ($\sigma_{11} \geq 0$)	$d^{f}_{t1} = \sigma_{11}/X_t$
Longitudinal compression damage of fibers ($\sigma_{11} < 0$)	
Transverse tensile damage of fibers ($\sigma_{22} \geq 0$ or $\sigma_{33} \geq 0$)	,
Transverse compression damage of fibers ($\sigma_{22} < 0$ or $\sigma_{33} < 0$)	,
In-plane shear damage of fibers	
External shear damage of fibers	

9.6 EXPERIMENTAL DETAILS

Experiments were conducted using the multiangle 2D UVAG MQL grinding system, as illustrated in Figure 6. The amplitude of ultrasonic vibration was set at 10 μm, with a frequency of 20 kHz. The experiments were performed under three distinct conditions: traditional dry grinding, 45° UVAG with nanolubricants in a MQL setup, and 90° UVAG with nanolubricants in an MQL setup. The grinding directions of the workpiece were oriented at 0°, 45°, 90°, and 135° in the longitudinal direction, respectively. A resin-bonded diamond grinding wheel with a diameter of 300 mm, a width of 20 mm, an organization number of 13, and a granularity of 150 mesh was utilized for the experiments. The physical and mechanical properties of grains are shown in Table 6.

Figure 6. Multiangle 2D ultrasonic assisted MQL grinding experimental system

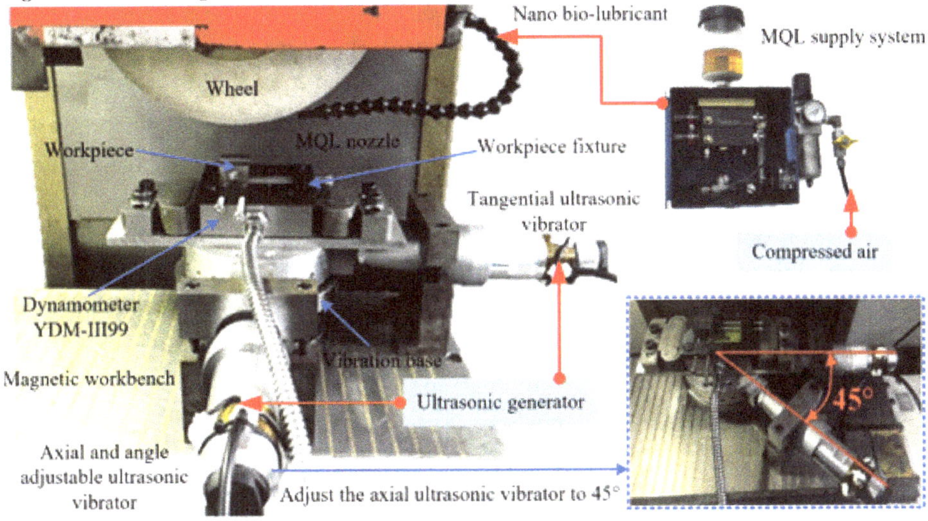

Table 6. Physical and mechanical properties of diamond grains

Property parameters	**Values**
Density / $\times 10^3$ kg/m^3	4.12
Hardness / GPa	50
Young's modulus / GPa	776
Poisson's ratio	0.07
Transverse fracture strength / GPa	1.2

9.7 NUMERICAL AND EXPERIMENTAL RESULT ANALYSIS

9.7.1 Surface Damage Mechanism and Evolution

The numerical analysis results and micromorphology of CFRP grinding surface damage at varying FOAs are presented in Figure 7, illustrating the mechanisms behind damage formation. At an FOA of 0°, the primary material removal mechanism involves the debonding and cracking of fibers/resin along the fiber direction due to fiber buckling. This leads to fiber slippage and removal, resulting in elongated rod-shaped debris from the stripped carbon fibers. This removal process can cause significant damage, including fiber loss and fracture. At an FOA of 45°, the debris removal process entails fiber bending and fracturing driven by the action of abrasive grains, as well as interlayer shear separation along the fiber direction. This can lead to damage such as fiber fragmentation and resin smearing on the surface. When the FOA is adjusted to 90°, the material removal mechanism shifts to fiber bending and fracture, along with cracking of the resin matrix and shear fracture of fibers due to the longitudinal compression exerted by the grains. Consequently, this results in the formation of surface microcracks and subsurface damage, along with notable resin smearing and adhered fiber debris on the surface. At an FOA of 135°, carbon fibers initially experience micro buckling as a result of reverse fiber grinding by the grains. This is followed by interlayer tensile fractures that propagate into the subsurface layer, and subsequent compression shear separation and bending fracture of the carbon fibers due to grain pushing. The result of a 135° FOA is the removal of a significant quantity of block-shaped fiber bundles, leading to potential damages such as fiber bundle pull-out and surface voids.

Figure 7. Grinding surface damage numerical analysis results and micromorphology of different FOAs

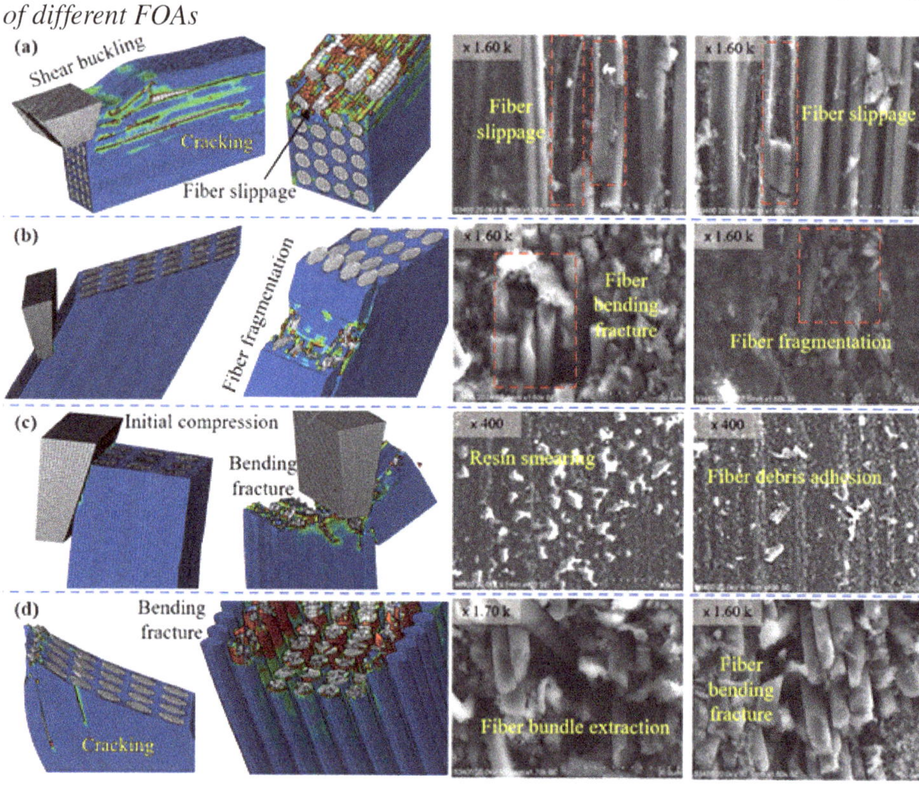

9.7.2 Mechanism of Debris Formation

Observing and characterizing the microstructure of debris is essential for conducting a comprehensive analysis of material removal and damage formation mechanisms. As illustrated in Figure 8 (a) - (d), the CFRP debris produced during grinding at a 0° FOA primarily consists of relatively long fiber segments along with a minor proportion of fragmented powdery fibers. The harsh conditions associated with traditional dry grinding result in high thermal loads that soften the resin matrix, leading to fiber adhesion to the resin matrix, as depicted in Figure 8 (c). Furthermore, Figure 6 (d) reveals that the resin matrix has peeled away from the fibers due to shear forces generated by the bending of the fibers, which occurs as the grains cut parallel to the fiber axis. In the case of the CFRP debris from the 45° FOA grinding direction, shown in Figure 8 (e)-(h), the debris mainly comprises powdered fibers and short fibers, along with some longer fibers. Notably, the cross-sections of both long and short fiber debris predominantly exhibit bending fracture surfaces, as shown in Figure 8(f). The formation of powdery fiber debris can be attributed to the squeezing effect in the region where grains directly contact the fibers. After direct cutting, the fibers that act as back support undergo bending, fracture, and subsequent removal. As shown in Figure 8 (i)-(l), the microstructure of CFRP debris in the 90° FOA grinding direction consists of a mixture of long fibers and powdery debris. The formation of powdery fiber debris and bent, broken fibers is similar to that in the 45° FOA, but the proportion of powdery fiber debris is lower in the 90° FOA. This difference arises because, in the 45° FOA, the fibers are supported by the back and undergo bending deformation, causing the grains cut at the negative rake angle to slip along the fiber surface, increasing direct contact and compression. This results in a larger amount of powdery fiber debris. In contrast, the 90° FOA CFRP provides stronger back support, facilitating fiber removal with a lower grinding force and reducing the area of direct contact and compression with the grains. This observation is further supported by the comparison of grinding forces, where the grinding force at 45° FOA exceeds that at 90° FOA. Lastly, the micromorphology of CFRP debris resulting from grinding at a 135° FOA is displayed in Figure 8 (m)-(p). This debris comprises a combination of block-shaped fiber bundles, powdery fiber debris, and long fibers. In Figure 8 (p), it is evident that the surfaces subjected to the cutting action of grains are heavily smeared with resin, and the cross-sections exhibit a sloping surface at a certain angle relative to the fiber direction. Conversely, the fiber bundles at the opposing end display an irregular cross-section, attributable to bending and breaking at their ends, as shown in Figure 8 (o).

Figure 8. Comparison of micromorphology of CFRP grinding chips with different FOAs

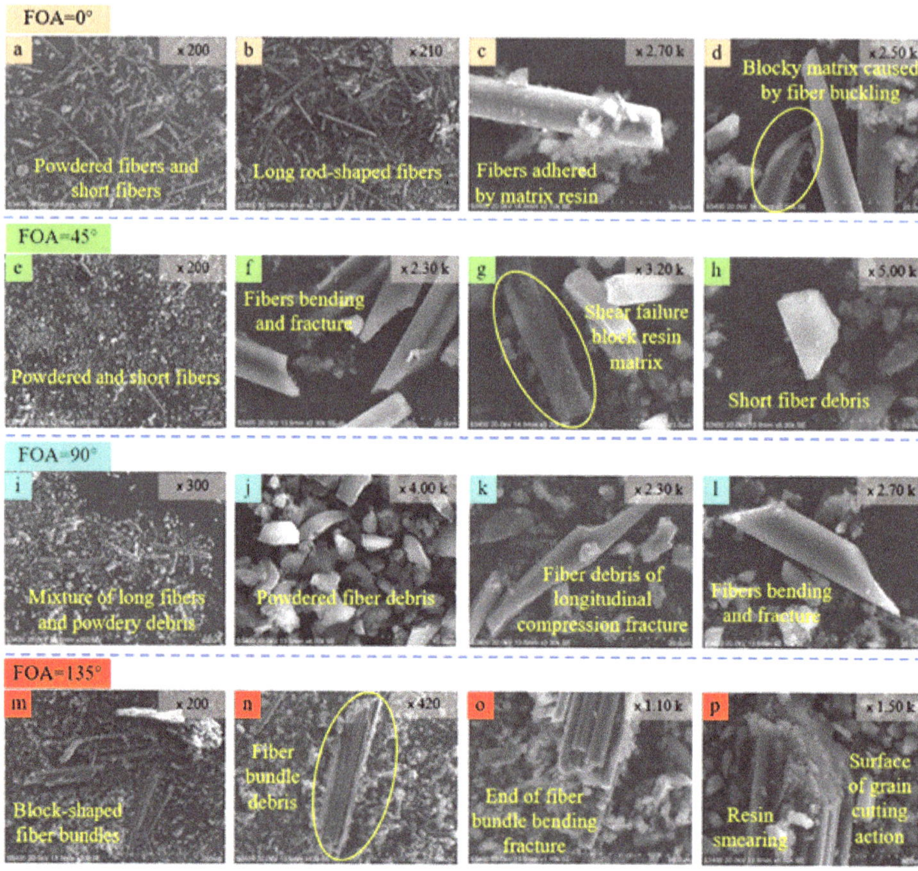

9.7.3 Damage Distribution

The grinding of CFRP materials can lead to various mechanical and thermal damages, including carbon fiber pull-out, fiber loss, and cracking. Understanding the distribution characteristics of damage resulting from CFRP grinding is essential for analyzing damage mechanisms and developing effective suppression strategies. Notably, there are different patterns in the variation of surface roughness profiles when comparing CFRP grinding surfaces subjected to different FOAs in both parallel and perpendicular grinding directions. The high-frequency components of the surface roughness profile provide a quantitative measure of damage characteristics. To further investigate, wavelet decomposition and analysis were performed on the surface roughness signals of CFRP grinding across varying FOAs. Specifically, a

5-layer wavelet decomposition using the sym2 wavelet was applied to the surface roughness signals of CFRP, resulting in the energy ratios of the low-frequency detail signal (Ea) and high-frequency detail signal (Ed), as illustrated in Figure 9 (a). The Ed in the vertical grinding direction was consistently lower than that in the parallel grinding direction across all FOAs. This suggests that damage characteristics predominantly accumulate along the parallel grinding direction. Additionally, the Ed for CFRP with a 0° FOA is comparatively lower than that for other FOAs, with minimal variation observed between parallel and vertical grinding directions, indicating optimal surface roughness texture features. Conversely, the Ed for CFRP with a 45° FOA is notably higher, peaking at 5.59% in the parallel grinding direction.

The energy distribution ratios of the two-dimensional (2D) wavelet decomposition detail subimages for CFRP grinding surfaces at different FOAs are depicted in Figure 9(b). In this context, Ev, Eh, and Edi correspond to the energy distribution ratios of the wavelet decomposition detail subimages oriented horizontally, vertically, and diagonally, respectively. The figure demonstrates that the energy proportion of high-frequency detail signals on surfaces perpendicular to the grinding direction is consistently lower than that found in the parallel grinding direction across all FOAs. There is a significant discrepancy in the energy distribution ratios of the 2D wavelet decomposition detail subimages for the 0° CFRP grinding surface, particularly between the vertical and parallel grinding directions. The proportion of high-frequency detail signals in the parallel grinding direction is notably higher—by 181.63%—compared to that in the perpendicular direction, aligning with the findings from the 1D wavelet decomposition. For the ground surfaces with FOAs of 45°, 90°, and 135°, the energy distribution ratio from the 2D wavelet decomposition detail subimages demonstrates less sensitivity to fiber orientation, with the diagonal high-frequency Edi remaining largely unaffected by the FOA.

Figure 9. (a) The influence of FOA on the 1D wavelet decomposition energy distribution ratio; (b) 2D wavelet decomposition energy distribution ratio of CERP grinding surfaces with different FOAs

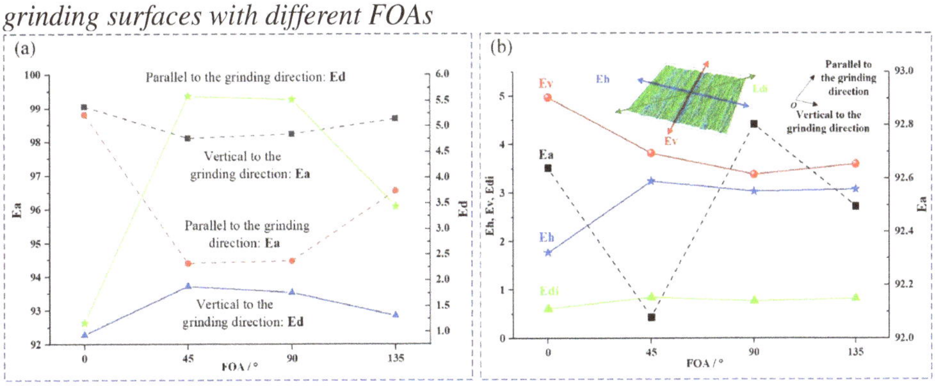

9.7.4 Quantitative Damage Comparison

A comparison of CFRP surface damage was conducted between traditional dry grinding and 2D ultrasonic-assisted nanolubricant MQL grinding conditions. The results, shown in Figure 10, illustrate the 2D wavelet decomposition energy distribution ratio of CFRP grinding surfaces with various FOAs under different working conditions. It is evident that Ev exhibits the highest proportion of high-frequency detail signal energy, indicating a significant extent of damage is distributed parallel to the grinding direction, followed by the vertical grinding direction, while the diagonal direction shows the least damage distribution. The application of ultrasonic-assisted nanolubricant MQL appears to reduce the energy distribution ratio of high-frequency detail signals in the 2D wavelet decomposition of CFRP grinding surfaces. It was observed that at $\theta_U = 90°$, there is a notable reduction in the energy ratio of high-frequency detail signals, with decreases of 7.42%, 27.56%, 5.03%, and 2.57% across the four FOAs compared to dry grinding. Furthermore, the 2D ultrasonic-assisted nanolubricant MQL at $\theta_U = 45°$ produced a significant reduction of 40.38% in FOA = 135° CFRP. The mechanism behind the damage suppression effect of ultrasonic-empowered nanolubricants can be categorized into several aspects. Firstly, the palm oil-based nanolubricant creates a stable lubricating oil film between the grinding wheel and the CFRP interface, which significantly enhances anti-wear and friction reduction properties, consequently lowering the grinding force. Secondly, CNT nanolubricants possess excellent heat transfer capabilities that effectively facilitate grinding heat dissipation, thereby inhibiting resin matrix softening and the occurrence of smearing. Preventing matrix softening is also advantageous in maintaining strong constraints on the removal of fibers and reducing the formation of interfacial cracking defects. Under weakened constraint conditions, multiple carbon fibers exhibit a piling effect, akin to the cutting action of grain on multiple carbon fibers, leading to pull-out fracture modes for the fibers in these conditions. Additionally, the material experiences significant compression deformations and energy absorption, resulting in increased grinding forces. Lastly, the incorporation of nanoparticles in nanolubricants plays a critical role in filling gaps and surface pits created by fiber loss, further contributing to the reduction of surface damage.

Figure 10. Energy distribution ratio of 2D wavelet decomposition on 2D ultrasonic nanolubricant MQL grinding surface

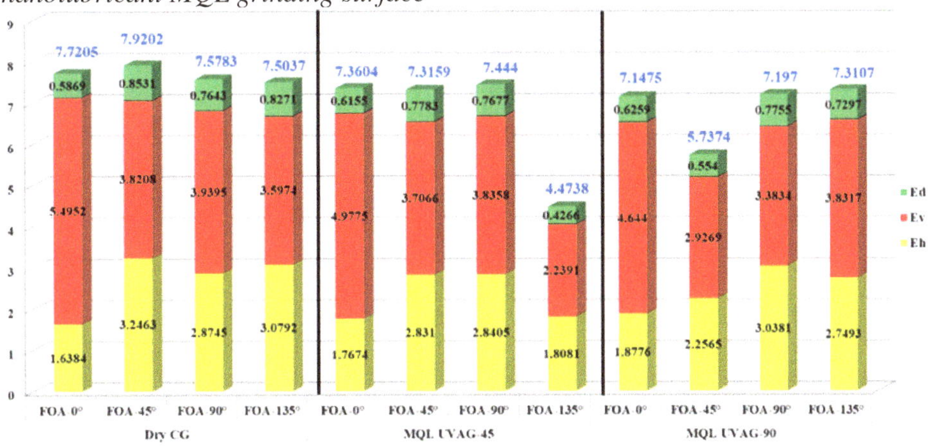

9.8 SUMMARY

The grinding damage of CFRP with FOA=0° is fiber loss and fiber fracture, and the resulting fiber debris is mostly long fibers with resin attached. The damage at FOA=45° is fiber breakage and resin coating, with most of the debris being powdery fibers and short fibers. The damage with FOA=90° is surface microcracks and subsurface damage, resulting in a mixture of long fibers and powdery debris. The damage with FOA=135° is characterized by fiber block pull-out and surface voids, and the debris is a mixture of block fiber bundles, powdery fiber debris, and long fibers.

The energy distribution (Ed) in the vertical grinding direction is lower than in the parallel grinding direction, with damage predominantly aligned along the parallel grinding direction. Among the various FOAs, the Ed of surface roughness signals is lowest at 0°, indicating minimal changes in surface roughness texture features, while it is highest at 45°, particularly in the parallel grinding direction. Aside from an FOA of 0°, the impact of FOA on the energy distribution ratio from wavelet decomposition in 2D micromorphology images remains relatively modest.

The parallel direction indicates the highest damage distribution, followed by the vertical grinding direction, while the diagonal direction shows the least damage. Notably, the application of nanolubricant MQL 2D UVAG with $\theta_U = 90°$ significantly reduces the energy ratio of high-frequency detail signals, achieving a maximum decrease of 27.56% compared to traditional dry grinding. Furthermore, the same technique applied with $\theta_U = 45°$ results in a remarkable reduction of 40.38% in damage to the FOA of 135°.

The interaction between the grinding wheel and the material during dynamic grinding presents complexity. Establishing clear relationships between UVAG parameters and material removal modes remains challenging. This complexity hinders the ability to effectively judge and control material removal methods. Therefore, developing a universal grinding force model for CFRP under varying material removal modes presents a significant research challenge and a critical area of focus for future investigations.

REFERENCES

Alexandrov, S., Jeng, Y. R., & Lomakin, E. (2011). Effect of pressure-dependency of the yield criterion on the development of plastic zones and the distribution of residual stresses in thin annular disks. *Journal of Applied Mechanics*, *78*(3), 031012. DOI: 10.1115/1.4003361

Benzeggagh, M. L., & Kenane, M. J. C. S. (1996). Measurement of mixed-mode delamination fracture toughness of unidirectional glass/epoxy composites with mixed-mode bending apparatus. *Composites Science and Technology*, *56*(4), 439–449. DOI: 10.1016/0266-3538(96)00005-X

Brewer, J. C., & Lagace, P. A. (1988). Quadratic stress criterion for initiation of delamination. *Journal of Composite Materials*, *22*(12), 1141–1155. DOI: 10.1177/002199838802201205

Cao, S., Zhang, X., Wu, C., Yang, M., & Xia, K. (2024). Towards understanding the material removal mechanism on the effect of ultrasonic vibration and anisotropy for unidirectional CFRP by scratching. *Journal of Materials Processing Technology*, *324*, 118250. DOI: 10.1016/j.jmatprotec.2023.118250

Ding, W., Liu, G., Chen, T., Feng, S., Zhao, Z., Zhao, B., & Cui, H. (2024). Effects of fiber orientations on removal behaviors in ultrasonic vibration-assisted grinding continuous fiber reinforced Ti matrix composite. *International Journal of Advanced Manufacturing Technology*, *133*(9), 4223–4234. DOI: 10.1007/s00170-024-14009-5

Dong, Z., Wang, Z., Tian, J., Kang, R., & Bao, Y. (2023). Investigation on tearing damage of CFRP circular cell honeycomb in end-face grinding. *Composite Structures*, *325*, 117616. DOI: 10.1016/j.compstruct.2023.117616

Duan, Z., Wang, S., Li, C., Wang, Z., Bian, P., Sun, J., Song, J., & Liu, X. (2024). Cold plasma and different nano-lubricants multi-energy field coupling-assisted micro-milling of Al-Li alloy 2195-T8 and flow rate optimization. *Journal of Manufacturing Processes*, *127*, 218–237. DOI: 10.1016/j.jmapro.2024.07.146

Duman, E., Yapan, Y. F., Salvi, H., Sofuoğlu, M. A., Khanna, N., & Uysal, A. (2024). Investigation of ultrasonic vibration assisted orthogonal turning under dry and minimum quantity lubrication conditions and performing sustainability analyses. *Journal of Cleaner Production*, *434*, 140187. DOI: 10.1016/j.jclepro.2023.140187

Fedulov, B. N., Safonov, A. A., Kantor, M. M., & Lomov, S. V. (2017). Modelling of thermoplastic polymer failure in fiber reinforced composites. *Composite Structures*, *163*, 293–301. DOI: 10.1016/j.compstruct.2016.11.091

Gong, P., Zhang, Y., Cui, X., Xu, S., Yang, M., Jia, D., & Li, C. (2024). Lubricant transportation mechanism and wear resistance of different arrangement textured turning tools. *Tribology International*, *196*, 109704. DOI: 10.1016/j.triboint.2024.109704

Gu, G., Wu, S., Wang, D., Zhou, S., Zhu, L., An, Q., Guo, H., & Li, C. (2023). A review of the research on the variation of tool's motion trajectory and its influence on the formation mechanism of surface quality in ultrasonic vibration machining. *Journal of Manufacturing Processes*, *107*, 294–319. DOI: 10.1016/j.jmapro.2023.10.025

Hashin, Z., & Rotem, A. (1973). A fatigue failure criterion for fiber reinforced materials. *Journal of Composite Materials*, *7*(4), 448–464. DOI: 10.1177/002199837300700404

Hooputra, H., Gese, H., Dell, H., & Werner, H. (2004). A comprehensive failure model for crashworthiness simulation of aluminium extrusions. *International Journal of Crashworthiness*, *9*(5), 449–464. DOI: 10.1533/ijcr.2004.0289

Liang, Y., Chen, Y., Chen, B., Fan, B., Yan, C., & Fu, Y. (2019). Feasibility of ultrasonic vibration assisted grinding for carbon fiber reinforced polymer with monolayer brazed grinding tools. *International Journal of Precision Engineering and Manufacturing*, *20*(7), 1083–1094. DOI: 10.1007/s12541-019-00135-8

Liang, Y., Chen, Y., Zhu, Y., Ji, J., & Ding, W. (2021). Investigations on tool clogging and surface integrity in ultrasonic vibration-assisted slot grinding of unidirectional CFRP. *International Journal of Advanced Manufacturing Technology*, *112*(5), 1557–1570. DOI: 10.1007/s00170-020-06364-w

Lin, H., & Zhou, M. (2024). Fracture mechanism of ultrasound-assisted scratching 2D-SiCf/SiC composite fibers with different fiber orientations. *Journal of Materials Research and Technology*, *30*, 9048–9060. DOI: 10.1016/j.jmrt.2024.05.225

Liu, C., Zhang, Y., Zhu, L., Li, Q., Shu, X., Qin, S., Wang, D., & Shi, W. (2025). A Review of Ultrasonic Vibration-Assisted Grinding for Advanced Materials. *Intelligent and Sustainable Manufacturing*, *2*(1), 10001. DOI: 10.70322/ism.2025.10001

Lomakin, E. V., & Fedulov, B. N. (2013). Plane strain extension of a strip made of a material with stress state type dependent properties and weakened by cuts with circular base. *Mechanics of Solids*, *48*(4), 424–430. DOI: 10.3103/S0025654413040109

Merino-Pérez, J. L., Royer, R., Ayvar-Soberanis, S., Merson, E., & Hodzic, A. (2015). On the temperatures developed in CFRP drilling using uncoated WC-Co tools Part I: Workpiece constituents, cutting speed and heat dissipation. *Composite Structures*, *123*, 161–168. DOI: 10.1016/j.compstruct.2014.12.033

Namlu, R. H., Lotfi, B., & Kılıç, S. E. (2024). Enhancing machining efficiency of Ti-6Al-4V through multi-axial ultrasonic vibration-assisted machining and hybrid nanofluid minimum quantity lubrication. *Journal of Manufacturing Processes, 119*, 348–371. DOI: 10.1016/j.jmapro.2024.03.073

Naskar, A., Choudhary, A., & Paul, S. (2021). Surface generation in ultrasonic-assisted high-speed superabrasive grinding under minimum quantity cooling lubrication with various fluids. *Tribology International, 156*, 106815. DOI: 10.1016/j.triboint.2020.106815

Niu, Q., Rong, J., Jing, L., Gao, H., Tang, S., Qiu, X., Liu, L., Wang, X., & Dai, F. (2023). Study on force-thermal characteristics and cutting performance of titanium alloy milled by ultrasonic vibration and minimum quantity lubrication. *Journal of Manufacturing Processes, 95*, 115–130. DOI: 10.1016/j.jmapro.2023.04.002

Qu, S., Yao, P., Gong, Y., Chu, D., Yang, Y., Li, C., Wang, Z., Zhang, X., & Hou, Y. (2022). Environmentally friendly grinding of C/SiCs using carbon nanofluid minimum quantity lubrication technology. *Journal of Cleaner Production, 366*, 132898. DOI: 10.1016/j.jclepro.2022.132898

Ran, Y., Kang, R., Dong, Z., Jin, Z., & Bao, Y. (2023). Ultrasonic assisted grinding force model considering anisotropy of SiCf/SiC composites. *International Journal of Mechanical Sciences, 250*, 108311. DOI: 10.1016/j.ijmecsci.2023.108311

Schön, J. (2000). Coefficient of friction of composite delamination surfaces. *Wear, 237*(1), 77–89. DOI: 10.1016/S0043-1648(99)00315-4

Shaw, M. C., & Cookson, J. O. (2005). Metal cutting principles: Vol. 2. *No. 3*. Oxford university press.

Tian, J., Kang, R., Dong, Z., Liu, Z., Guo, D., & Bao, Y. (2023). Multi-scale machining damages of CFRP circular cell honeycomb during end face machining. *Journal of Manufacturing Processes, 86*, 282–293. DOI: 10.1016/j.jmapro.2023.01.006

Wang, F., Xuan, S., Chang, Z., Jin, K., Gao, Y., Wang, H., & Song, Q. (2024). Effect of grinding parameters on industrial robot grinding of CFRP and defect formation mechanism. *International Journal of Precision Engineering and Manufacturing-Green Technology, 11*(2), 427–438. DOI: 10.1007/s40684-023-00561-0

Wen, Z., Zhao, H., Yan, X., Peng, Z., & Ding, H. (2023). Grinding performance study of CF/epoxy and CF/PEEK composites. *Polymer Composites, 44*(8), 4565–4578. DOI: 10.1002/pc.27420

Xu, W., Gao, T., Liu, D., Xu, P., An, Q., Wang, Y., Ding, W., & Li, C. (2024). Progress in grinding mechanical modeling of fiber reinforced SiC ceramic matrix composites. *Acta Mater Compos Sin*, *41*(9), 4628–4653.

Xu, W., Li, C., Zhang, Y., Ali, H. M., Sharma, S., Li, R., Yang, M., Gao, T., Liu, M., Wang, X., Said, Z., Liu, X., & Zhou, Z. (2022). Electrostatic atomization minimum quantity lubrication machining: From mechanism to application. *International Journal of Extreme Manufacturing*, *4*(4), 042003. DOI: 10.1088/2631-7990/ac9652

Yan, X., Zhao, H., Wen, Z., Wang, Y., Li, P., & Ding, H. (2024). Investigation on grinding temperature characteristics of CF/PEEK and material removal mechanism under temperature treatments. *Composites Science and Technology*, *252*, 110621. DOI: 10.1016/j.compscitech.2024.110621

Yang, M., & Li, C. (Eds.). (2025). *Heat Transfer Mechanism and Temperature Field Model of MQL Micro-Grinding Biomedical Materials*. IGI Global, 2-97.

Yang, M., Ma, H., Hao, J., Li, Z., Li, R., Zhou, Z., Gao, T., Liu, M., Cui, X., Wang, X., Zhang, Y., Dambatta, Y. S., Long, Y., & Li, C. (2024a). Droplet size distribution model of needle electrode electrostatic atomization and milling nickel-based alloy performance evaluation. *Journal of Manufacturing Processes*, *119*, 682–698. DOI: 10.1016/j.jmapro.2024.04.011

Yang, M., Ma, H., Li, Z., Hao, J., Liu, M., Cui, X., Zhang, Y., Zhou, Z., Long, Y., & Li, C. (2024b). Force model in electrostatic atomization minimum quantity lubrication milling GH4169 and performance evaluation. *Frontiers of Mechanical Engineering*, *19*(4), 28. DOI: 10.1007/s11465-024-0800-8

Yang, Y., Yang, M., Li, C., Li, R., Said, Z., Ali, H. M., & Sharma, S. (2023). Machinability of ultrasonic vibration-assisted micro-grinding in biological bone using nanolubricant. *Frontiers of Mechanical Engineering*, *18*(1), 1. DOI: 10.1007/s11465-022-0717-z

Yao, L., Liu, Z., Song, Q., Wang, B., & Cai, Y. (2023). Prediction modelling of cutting force in rotary ultrasonic end grinding 2.5 D woven SiO2f/SiO2 ceramic matrix composite. *Composite Structures*, *304*, 116448. DOI: 10.1016/j.compstruct.2022.116448

Zhang, Y., Hu, Z., Chen, Y., Yu, Y., Jin, J., Peng, Q., & Xu, X. (2024). Insights into scratching force in axial ultrasonic vibration-assisted single grain scratching. *Journal of Manufacturing Processes*, *112*, 150–160. DOI: 10.1016/j.jmapro.2024.01.005

Zhao, B., Wu, B., Yue, Y., Ding, W., Xu, J., & Guo, G. (2023). Developing a novel radial ultrasonic vibration-assisted grinding device and evaluating its performance in machining PTMCs. *Chinese Journal of Aeronautics*, *36*(7), 244–256. DOI: 10.1016/j.cja.2023.01.008

Chapter 10

Surface Quality Evaluation of Longitudinal Grinding With Multiangle Ultrasonic Vibration Minimum Quantity Lubrication

ABSTRACT

The aeronautical industry constantly strives for efficient technologies to facilitate high surface integrity in CFRP structural components. This chapter conducts a multi-dimensional systematic evaluation of grinding surface integrity, ranging from global characteristics to local features in. Firstly, line and areal roughness are employed to characterize surface topography, functional performance, fluctuation intensity, and height distribution, with peak heights and valley depths specifically reflecting fiber pull-out damage. Secondly, 2D and 3D fractal dimensions, derived from line roughness profiles and surface microtopography, are used to characterize the overall morphological attributes. Subsequently, multifractal spectra including spectral width and difference, are applied to quantitatively analyze local morphological variations. Finally, SEM and laser confocal microscopy are utilized to observe 2D and 3D surface topography, enabling qualitative analysis of grinding-induced damage such as fiber (bundle) pull-out, fiber fragmentation, resin smearing, and fiber-matrix debonding.

DOI: 10.4018/979-8-3373-9013-0.ch010

10.1 INTRODUCTION

Owing to the anisotropic thermomechanical behavior during CFRP grinding, the resultant surface integrity exhibits significant distinctions from that observed in isotropic materials. The material's high susceptibility to machining-induced damage poses considerable challenges to surface quality assurance—a characteristic manifestation of the difficult machinability of CFRPs (Tang et al., 2013; Qiu et al., 2020). Xiao et al. (2022) identified the coupling effect between fiber orientation and cutting speed as the most influential factor affecting surface roughness, with a contribution rate of 4.8%. The interaction between fiber orientation and cutting edge radius exerted the most pronounced effect on subsurface damage depth, accounting for 5% of the observed variation. Their findings further emphasized the necessity of considering the synergistic effect of cutting speed and rake angle to improve the surface integrity of CFRP composites. Liu et al. (2023) observed a sharp deterioration in surface quality during bending fracture, suggesting that such fracture modes should be avoided whenever possible. Moreover, increasing the radial depth of cut was found to exacerbate damage propagation under bending fracture, warranting the recommendation of smaller cutting depths. The lack of support for the top CFRP layer renders it prone to defects during milling: burr formation occurs at FOA < 90°, whereas delamination arises at FOA > 90°. During transitions in fracture mechanisms, both burrs and delamination may coexist. The type of top-layer fiber defect is independent of the laminate stacking sequence and is determined solely by the FOA of the top layer. Tian et al. (2025) described the burr formation process as involving fiber–matrix debonding, which reduces constraints on the fibers. Insufficiently constrained fibers cannot be effectively removed by the grinding wheel and are consequently transformed into burrs. Dong et al. (2023) identified tearing as an intralaminar damage mechanism, occurring predominantly during the engagement stage in CFRP circular cell structures, where fiber–resin interfacial cracking develops under bending stress. Furthermore, grinding damage originates from thermomechanical coupling effects, specifically resin thermal softening (above the glass transition temperature Tg) and a damage chain described as "thermal softening–interface debonding–fiber fracture." The material's low stiffness amplifies thermal softening–induced damage. A thermal control–based optimization strategy was consequently proposed (wheel speed < 1.5 m/s, 120° < FOA < 140°, workpiece feed rate > 0.2 mm/r) to minimize resin softening, interfacial peeling, and tearing (Dong et al., 2025). Handa & Sooraj (2023) demonstrated that a progressive intermittent cutting strategy induced at lower eccentricity is markedly more effective for grinding heterogeneous fiber–matrix systems compared to conventional surface grinding. Liang et al. (2019) designed single-layer brazed diamond tools with well-defined grain distribution. They found that tools with an inter-grain spacing of 1.2

mm yielded the highest grinding force but improved surface roughness, attributable to the increased number of active grains and overlapping interaction zones.

Lubrication and cooling mediums represent key factors influencing response parameters (Yang et al., 2024a; Zhang et al., 2025), rendering MQL an indispensable cooling-lubrication technique with irreplaceable advantages (Yang et al., 2024b; Kong et al., 2025). Numerous researchers have validated the unique benefits of MQL in machining FRCs (Gao et al., 2024). As a green and sustainable cooling-lubrication technology (Duan et al., 2025), MQL atomizes a minimal amount of lubricant with high-pressure gas into fine droplets, which are transported at high velocity through a nozzle into the grinding zone (Wang et al., 2025a; Xu et al., 2023). Owing to its volatile nature, MQL leaves no residue on machined surfaces, eliminating the need for secondary cleaning of aerospace components after processing. The lubricant forms a thin film in the grinding zone, while the high-pressure air stream primarily facilitates chip removal and cooling (Jia et al., 2023; Wang et al., 2025b). To achieve green and clean manufacturing at the source, biodegradable and non-toxic vegetable oils are generally selected as base oils, significantly reducing environmental and human health impacts (Wang et al., 2024). Li et al. (2024) investigated the grinding of carbon fiber-reinforced thermoplastics (CFRTPs) using oil-on-water (OoW) MQL. They found that OoW-MQL with an oil-to-water ratio of 1:1 outperformed other lubrication conditions, achieving superior surface quality along with the lowest grinding forces and temperatures.

Ultrasonic Vibration-Assisted Machining (UVAM) introduces high-frequency, micro-scale amplitude displacement to the tool-tip motion, altering the cutting trajectory and enabling the instantaneous depth of cut to be significantly smaller than the fiber diameter. This mechanism contributes to improved surface integrity (Yang et al., 2020; Gao et al., 2020). The overlapping cutting paths of abrasive grains under horizontal ultrasonic vibration result in fewer defects. The reduced instantaneous depth of cut and cutting volume in UVAM help minimize fiber fracture, cracking, and matrix damage during CFRP machining (Zhu et al., 2025). Zhang et al. (2024) demonstrated that Rotary Longitudinal-Torsional Ultrasonic Machining (RLTUM) significantly reduces grinding forces, grinding temperature, and exit-hole damage, while improving the surface smoothness of CFRP hole walls. By employing torsional vibration to achieve intermittent cutting in the circumferential direction, RLTUM facilitates automatic chip removal, suppresses chip adhesion to the tool, and reduces heat generation and tool wear. Cao et al. (2024) reported that the introduction of ultrasonic vibration markedly reduces mechanical damage during scratching. This is attributed to the intermittent contact induced by ultrasonic vibration in ultrasonic vibration-assisted scratching, which effectively suppresses the propagation of surface cracks and the occurrence of interfacial defects. Several scholars have begun to integrate ultrasonic vibration with MQL, confirming that synergistic enhancement effects can be achieved (Gu et al., 2024).

With the damage mechanisms elucidated in Chapter 9, this chapter returns to the practical outcome of surface quality, now under the most comprehensive process variant—multiangle ultrasonic vibration MQL longitudinal grinding. By evaluating the surface quality across a range of ultrasonic incidence angles, the study identifies the optimal configuration that minimizes defects and maximizes the integrity of machined CFRP surfaces. It is crucial to systematically compare, characterize, and quantitatively evaluate the surface integrity of CFRP ground with MQL nanolubricant empowered by multi parameter collaborative coupling ultrasonic. This article provides a multidimensional systematic evaluation of the surface integrity of CFRP grinding from overall to local aspects, including surface roughness, fractal dimension, multifractal spectrum, and surface microstructure. Firstly, the surface features, performance, fluctuation degree, unevenness, as well as the peak height and valley depth of the contour are compared and characterized by the line roughness and 3D surface roughness to reflect the fiber pull-out damage on the CFRP grinding surface. Secondly, the overall morphology characteristics of the CFRP grinding surface are characterized by comparing the two-dimensional and three-dimensional fractal dimensions based on the line roughness profile curve and surface microstructure. Then, the local morphology characteristics are quantitatively analyzed and discussed through multifractal spectra, their spectral widths, and spectral differences. Finally, SEM and confocal laser microscopy were used to observe the 2D and 3D surface morphology, respectively, to qualitatively analyze and characterize surface damages such as fiber (bundle) extraction, fiber fragmentation, resin coating, and fiber resin debonding caused by grinding.

10.2 EXPERIMENT AND MATERIALS

As shown in Figure 1, experiments were conducted on a K-36 CNC surface grinder equipped with a two-dimensional multi angle UVAG device and an MQL supply system. The workpiece grinding directions were 0°, 45°, 90°, and 135° in the longitudinal direction, respectively. And the microstructure and debris morphology of CFRP grinding surface were observed and characterized using surface roughness profilometer, Olympus laser confocal microscope, and S-3400N scanning electron microscope, respectively. The experiments were conducted under three working conditions: traditional dry type, $\theta_U = 45°$ ultrasonic nanolubricant MQL, and $\theta_U = 90°$ ultrasonic empowered nano lubricant MQL. The specific experimental plan and parameters for multi angle two-dimensional ultrasonic vibration assisted nano lubricant MQL grinding, as well as the physical and mechanical performance parameters of the workpiece CFRP and its components, and diamond grinding wheel parameters, are detailed in Chapter 8.

Figure 1. Multiangle 2D ultrasonic vibration-empowered nanolubricant MQL grinding experimental system and Surface integrity analysis device

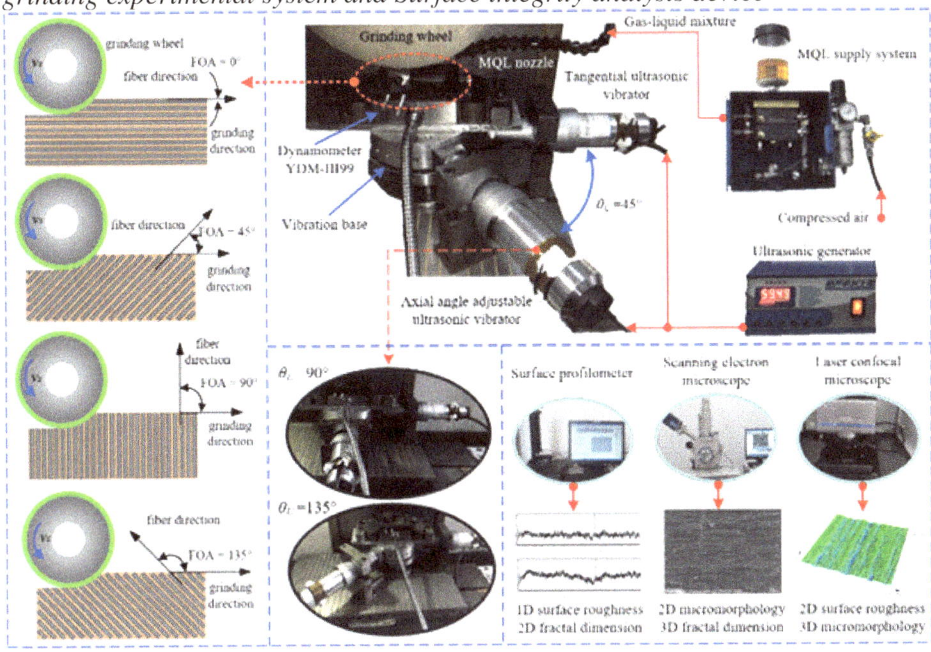

10.3 SURFACE ROUGHNESS

10.3.1 1D Surface Roughness

Different measurement directions for surface roughness can result in variations in the height, size, and spacing of peaks and valleys in the machining profile, due to the directional nature of texture marks in cutting and grinding processes. This leads to different surface roughness measurement results within the same measurement length. For the surface roughness of isotropic homogeneous materials during cutting and grinding, the measurement direction should be selected perpendicular to the direction of the processed texture traces. However, for inhomogeneous anisotropic CFRP composite materials, the selection of surface roughness measurement direction may not be determined as perpendicular to the grinding texture direction due to the directional influence of reinforcing phase carbon fibers, as well as the formation of grinding force thermal damage such as carbon fiber pull-out, resin coating, cracks, carbon fiber fracture, and pits. Therefore, this article conducts measurements in both vertical and parallel grinding directions to comprehensively evaluate the line roughness of CFRP grinding surfaces.

Firstly, the line roughness of CFRP grinding surfaces with different FOAs under different working conditions was compared and characterized. The arithmetic mean deviation *Ra* value of the contour is preferred, which contains most of the micro roughness information and can characterize the basic surface features and performance. As shown in Figure 2 (a), for different grinding conditions, the *Ra* of 45° UVAG MQL is basically the lowest (except for 90° UVAG MQL with FOA=0° and parallel grinding direction). Compared to traditional dry grinding, the reduction degree ranges from 18.08% to 35.73% (with the maximum reduction in the vertical grinding direction at FOA=90°). In addition, under almost all conditions, the surface roughness *Ra* in the parallel grinding direction is smaller than that in the vertical grinding direction. In order to compare the differences in surface roughness *Ra* values between different grinding conditions more clearly and directly, the maximum value of surface roughness *Ra* values in the vertical and parallel grinding directions is taken for comparison. As shown in Figure 2(b), it can be seen that for different FOAs, the surface roughness *Ra* value of CFRP grinding with FOA=0° is the highest, while the *Ra* values of the other three FOAs are not significantly different. Compared with homogeneous isotropic materials, this difference also better demonstrates the difficulty of cutting and grinding non-homogeneous anisotropic CFRP and the complexity of theoretical experimental research. A comprehensive and systematic comparative evaluation is needed in multiple dimensions, directions, and scales.

Figure 2. Surface roughness Ra under different grinding conditions

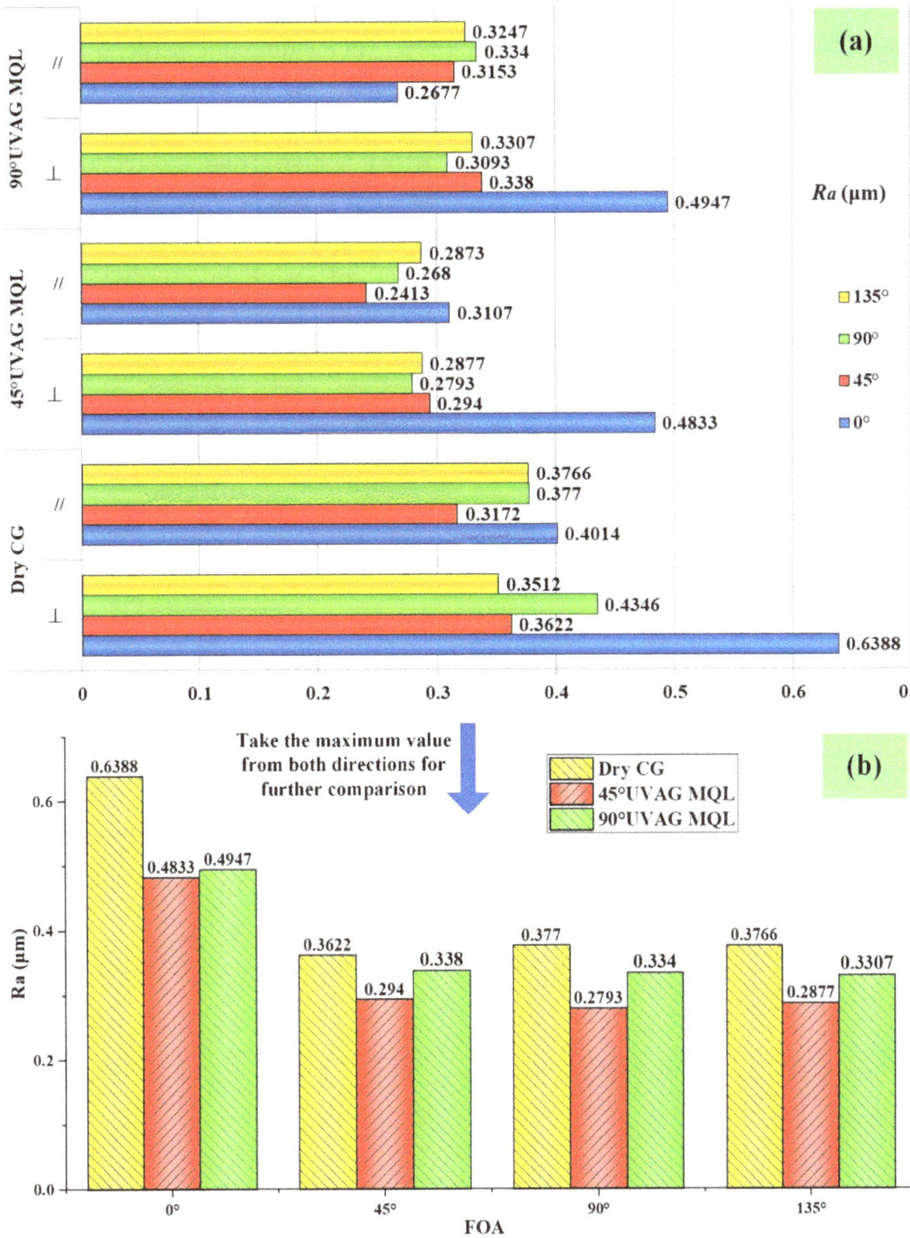

In addition, the average width (spacing parameter) *Rsm* of the selected contour units was used to characterize and compare the roughness in the horizontal direction. As shown in Figure 3, the grinding surface roughness *Rsm* of CFRP with different FOAs under traditional dry and different ultrasonic empowered nanolubricant MQL conditions is presented. As shown in Figure 3(a), the 45°UVAG MQL has a more significant effect on reducing the *Rsm* of FOA=45° and 90°, with a maximum reduction of 20.93% compared to dry grinding (FOA=135° in the direction perpendicular to the grinding direction). However, there is no clear pattern in the comparison between parallel and vertical grinding directions. As shown in Figure 3(b), the maximum surface roughness *Rsm* in the parallel and perpendicular grinding directions was compared. For different FOAs, the FOA=0° had the highest surface roughness *Rsm*, while the one with FOA=45° had the lowest. In addition, 45° UVAG MQL is more conducive to reducing the *Rsm* of FOA=45° and 90°, while 90° UVAG MQL is more conducive to reducing the *Rsm* of FOA=0° and 135°.

Figure 3. Surface roughness Rsm under different grinding conditions

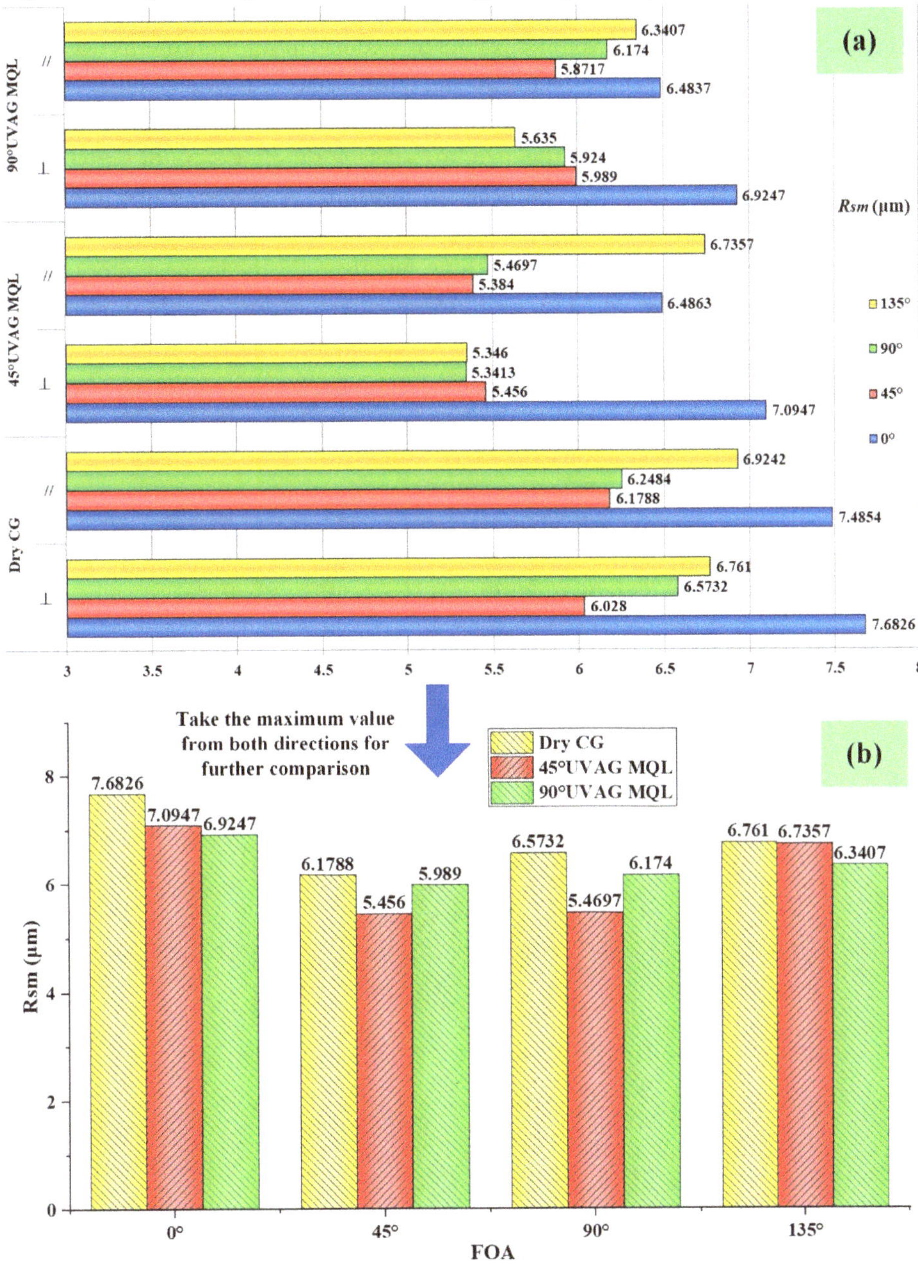

10.3.2 2D Surface Roughness

Furthermore, the surface roughness under different grinding conditions was compared and characterized. Similarly, the *Sa* value was preferred, which contains most of the micro roughness information and characterizes the surface features and usage behavior. As shown in Figure 4, the grinding surface roughness *Sa* of CFRP with different FOAs under different conditions of ultrasonic empowered nanolubricant MQL is presented. For different FOAs, the *Sa* of the CFRP grinding surface with FOA=45° is the lowest in all three conditions, and the *Sa* is the highest in FOA=0°. Moreover, the variation law with FOA is the same under the same working conditions, which is 0°>90°>135°>45°. For the same FOA, ultrasonic empowered nanolubricant MQL has a significant effect on reducing surface roughness *Sa*, among which the 45° UVAG nanolubricant MQL has the best effect on reducing *Sa*. Compared to traditional dry grinding conditions, the FOA decreases by 13.15%, 20.19%, 21.83%, and 22.48% for FOA of 0°, 45°, 90°, and 135°, respectively.

Figure 4. Surface roughness Sa under different grinding conditions

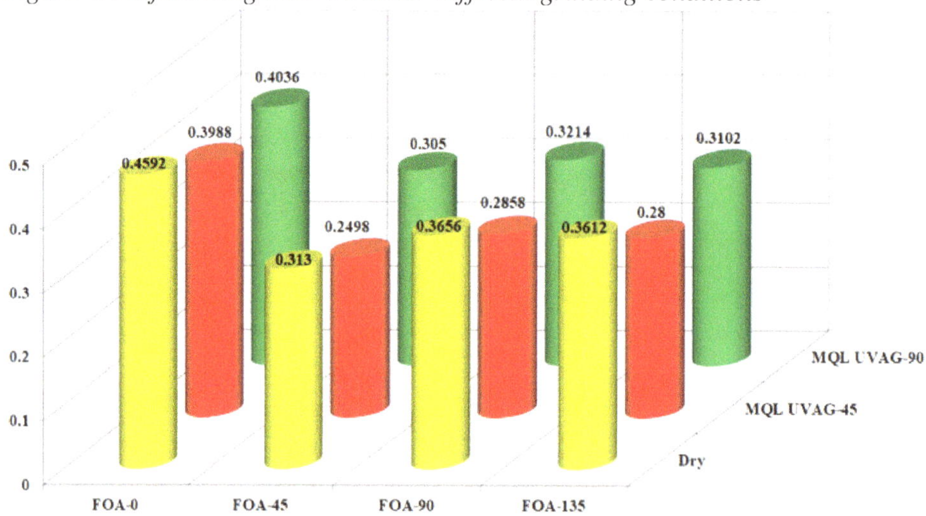

Surface roughness *Sq* is an indicator for evaluating surface roughness, used to describe the degree of small fluctuations and unevenness of a surface. A larger *Sq* indicates a rougher and more undulating surface. On the contrary, when *Sq* is smaller, it indicates that the surface is smoother and the roughness is smaller. As shown in Figure 5, the grinding surface roughness *Sq* of CFRP with different FOAs under different ultrasonic empowered nanolubricant MQL conditions. The impact of FOA on *Sq* follows the same pattern as *Sa*. The surface fluctuation of 0° CFRP grinding is the largest, and the nanolubricant MQL UVAG with θ_U =45° also has the best effect on reducing *Sa*. Compared to traditional dry grinding conditions, the Sa of different FOAs decreased by 13.46%, 17.56%, 23.81%, and 24.81%, respectively. Ultrasonic empowerment of MQL effectively improves the fluctuation and unevenness of the grinding surface.

Figure 5. Surface roughness Sq under different grinding conditions

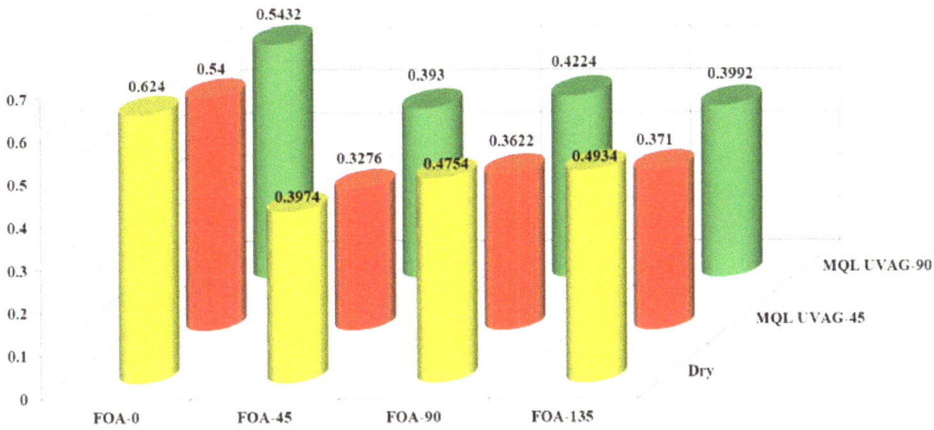

Finally, the peak height of the reaction profile *Sz* was compared and evaluated to characterize fiber pull-out damage. It is the sum of the maximum contour peak height and the maximum contour valley depth within the sampling area. Figure 6 shows the surface roughness *Sz* of CFRP with different FOAs under different conditions of ultrasonic empowered nanolubricant MQL. For different FOAs, the *Sz* values of 135° CFRP are much higher than those of other FOAs, which also confirms that this orientation produces severe fiber bundle block pull-out, resulting in more pits on the surface. The application of 2D ultrasonic empowered nano lubricant MQL reduced the *Sz* value, effectively reducing fiber bundle pull-out damage, especially the reduction effect of θ_U =45° 2D ultrasonic vibration was the most significant, with a 47.26% decrease compared to dry orientation at 135°.

Figure 6. Surface roughness Sz under different grinding conditions

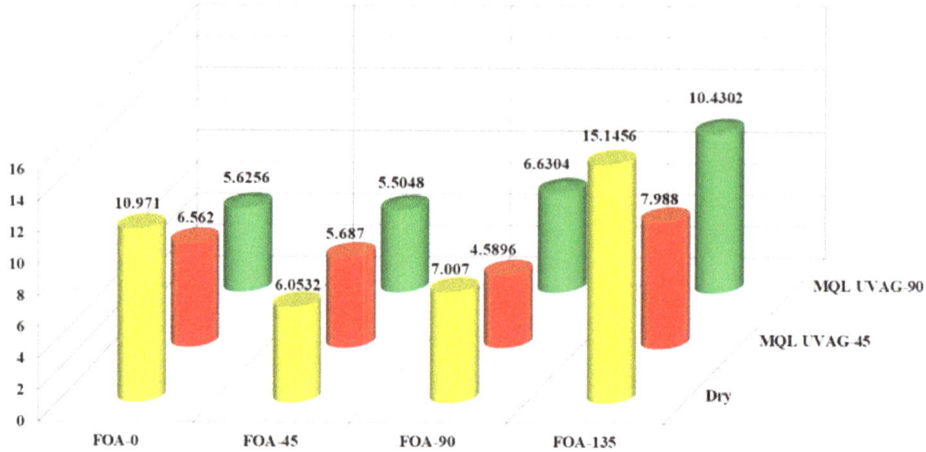

It is worth noting that there is no correlation between the effects of different FOAs on surface roughness and grinding force, as can be seen from the comparative analysis of grinding force in Chapter 8. However, for traditional isotropic homogeneous materials, the positive correlation law generally follows that the higher the grinding force, the higher the surface roughness value. Higher grinding force will cause higher grinding temperature rise, and higher force thermal load will lead to harsh grinding behavior, ultimately resulting in high surface roughness and deteriorating surface integrity. The difference in the grinding results of CFRP also proves the difficulty of cutting and grinding heterogeneous anisotropic fiber composite materials and the complexity of theoretical experimental research.

10.4 FRACTAL DIMENSION

As a scale-free parameter, fractal dimension is a measure of the randomness and irregularity of complex shapes, which can reflect the effectiveness of space occupied by complex shapes and quantitatively characterize fractal characteristics. The fractal dimension believes that under certain conditions, local morphology may have similarities with the overall structure, morphology, etc. The fractal dimension can characterize the specific characteristics and proportion of complex surface structures to the entire surface. The larger the fractal dimension, the more complex and refined the structure. In this study, the box counting dimension method was used to compare and evaluate the overall morphology characteristics of CFRP grinding surfaces with different FOAs. It should be noted that the calculation method for the percentage reduction of fractal dimension in this study is to only calculate the difference compared to the current dimension. Define D_f^i (i=2, 3) as the fractal dimension, and $D_{f\_dry}^i$ and $D_{f\_uv}^i$ as the fractal dimensions under ultrasonic empowered MQL and dry grinding conditions, respectively. The percentage reduction in fractal dimension is $\left(D_{f\_dry}^i - D_{f\_uv}^i\right)/D_{f\_dry}^i - i$.

10.4.1 2D Fractal Dimension

The 2D fractal dimension of CFRP in the vertical grinding direction under different working conditions is shown in Figure 7. The fractal dimension of dry traditional grinding is the highest, and the number of complex surface structures is more numerous and finer. When FOA=0°, 45°, and 135°, the fractal dimension of the 45° UVAG MQL grinding condition is the lowest, and the reduction degree of FOA=135° CFRP compared to dry grinding is the largest, at 16.61%, indicating that 45° UVAG MQL is more suitable for reducing the fractal dimension of the ground surface of FOA=135° CFRP. When FOA=90 °, the fractal dimension of the 90° UVAG MQL grinding condition is the lowest, indicating that its surface has fewer complex structures. This indicates that 90° UVAG MQL is more suitable for reducing the fractal dimension of FOA=90° CFRP grinding surfaces. For different FOAs, the fractal dimension of FOA=0° is the lowest compared to the other three FOAs, which is different from the overall pattern of surface roughness. This indicates that although the roughness and fluctuation of the surface profile of CFRP with FOA=0° grinding are relatively large, the complexity of its overall roughness profile characteristics is lower than that of other fiber orientations, and it can also be said that the self similarity of the profile structure is higher.

Figure 7. 2D fractal dimension perpendicular to the grinding direction

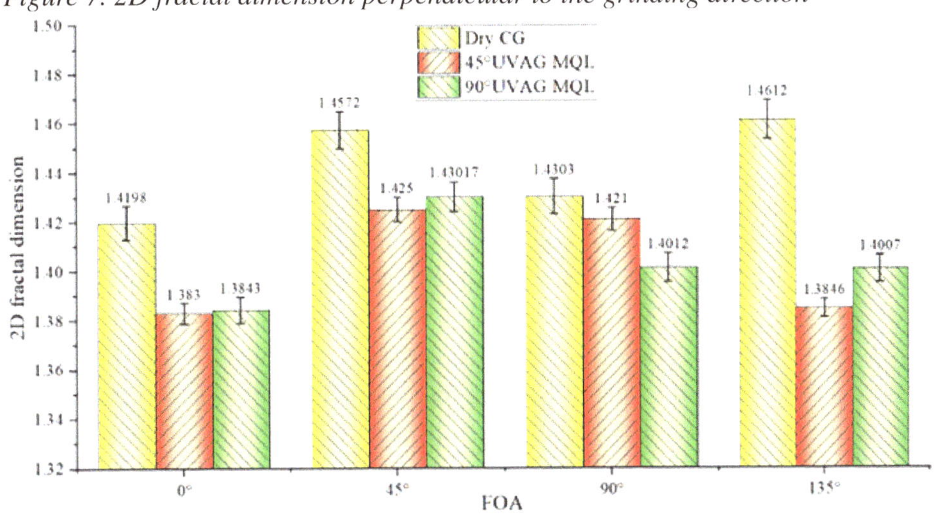

As shown in Figure 8, the 2D fractal dimension of CFRP in parallel grinding direction under different conditions is reduced by UVAG MQL. When FOA=0°, 45°, and 90°, the fractal dimension of the 90° UVAG MQL condition is the lowest, and FOA=45° has the greatest reduction compared to dry grinding, at 20.85%. When FOA=135°, the fractal dimension of the 45° UVAG MQL condition is the lowest. For different FOAs, the fractal dimension with FOA=0° is the lowest when it is the same as in the vertical grinding direction. The reason for the difference in the overall pattern of surface roughness is explained as follows: the automatic repair effect of nanoparticles in nanolubricants fills the damage structural characteristics such as fiber pull-out, fiber fragmentation, and pit voids, reducing the diversity of surface contour structure in CFRP grinding, resulting in a decrease in fractal dimension.

Figure 8. 2D fractal dimension parallel to the grinding direction

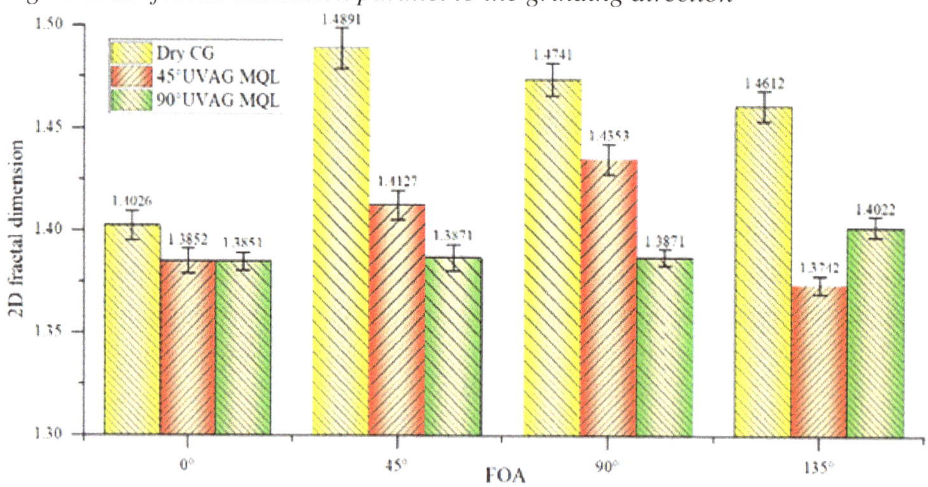

10.4.2 3D Fractal Dimension

The 3D fractal dimension of the microstructure of the grinding surface was obtained using MATLAB. The comparison results are shown in Figure 9. The 3D fractal dimension of traditional dry grinding is the highest under different FOAs. However, different directions of ultrasonic vibration application have varying degrees of reduction in the three-dimensional fractal dimension of CFRP with different FOAs, indicating that the direction of ultrasonic vibration application has adaptability to FOA. From Figure 9, it can be seen that 45° UVAG MQL has the best effect on reducing the 3D fractal dimension of FOA=45° and 135° CFRP, with the best adaptability. Compared with traditional dry grinding, it reduces 17.73% and

10.32%, respectively. However, 90° UVAG MQL has the best effect on reducing the 3D fractal dimension of FOA=0° and 90° CFRP, with the best adaptability, reducing by 8.38% and 29.22% respectively compared to traditional dry grinding.

Figure 9. 3D fractal dimension of CFRP grinding surface with different FOAs under different conditions

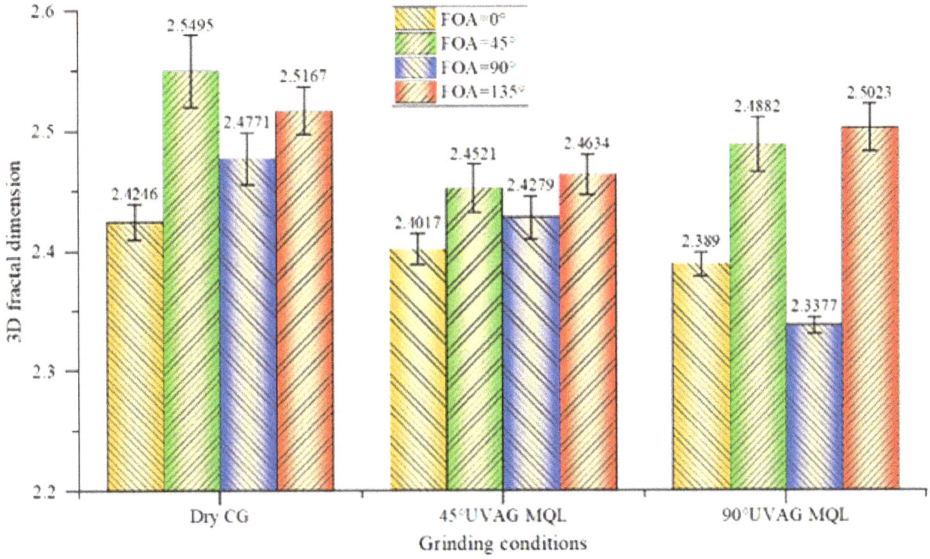

10.5 SURFACE MICROMORPHOLOGY

10.5.1 2D Surface Micromorphology

The SEM microstructure comparison of CFRP grinding surfaces with different FOAs under different ultrasonic vibration and cooling lubrication conditions is shown in Figure 10. Both 45° UVAG MQL and 90° UVAG MQL significantly improved the surface quality and morphology of CFRP grinding, reducing grinding damage. For CFRP with FOA=0°, dry grinding resulted in significant fiber slippage causing voids and fiber loss, as well as continuous fiber fragmentation. Under the 45° UVAG MQL condition, a small amount of carbon fiber fragmentation and loss were also caused, while under the 90° UVAG MQL condition, more carbon fiber fractures occurred. For CFRP with FOA=45°, dry grinding resulted in severe carbon fiber pull-out, while in the other two grinding conditions (45° UVAG MQL and 90° UVAG MQL), only a small amount of residual carbon fiber debris adhered.

For CFRP with FOA=90°, dry grinding resulted in significant resin coating and block shaped debris adhesion. In the other two grinding conditions, except for slight debris adhesion, there were no other obvious grinding damages. For CFRP with FOA=135°, dry grinding caused severe cracking. The 45° UVAG MQL caused voids due to the extraction of carbon fibers. 90° UVAG MQL formed significant adhesion of carbon fiber debris.

Figure 10. Comparison of SEM micromorphology of CFRP grinding surface with different FOAs under different conditions

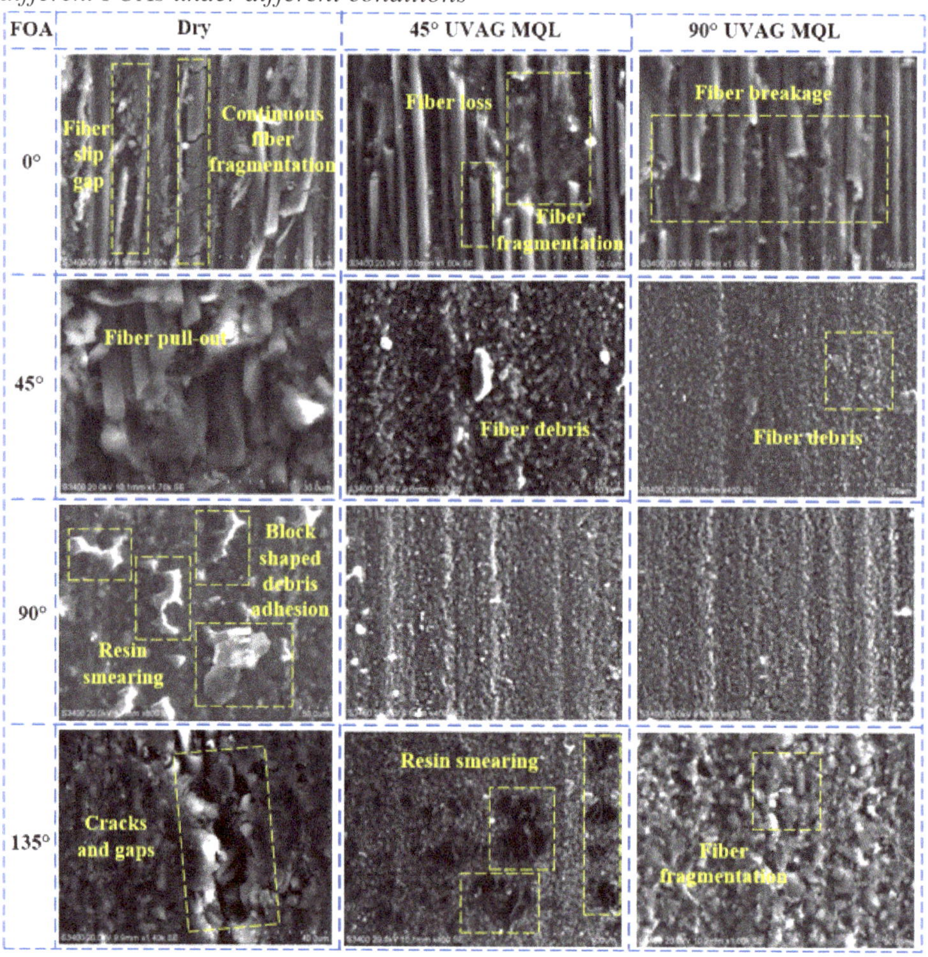

10.5.2 3D Surface Micromorphology

The 3D microstructure comparison of CFRP grinding surfaces with different FOAs under different ultrasonic vibration and cooling lubrication conditions is shown in Figure 11. Among them, the surface morphology of CFRP subjected to dry grinding deteriorated severely, especially for CFRP with FOA=0°, which resulted in significant and severe surface damage such as fiber fracture, fiber pull-out, fiber loss, and fiber slip pits. The CFRP grinding surfaces with FOA=45°, 90°, and 135° also experienced fiber breakage and resin coating damage. Comparatively, the 3D surface micromorphology under 45° UVAG MQL and 90° UVAG MQL grinding conditions is smoother. The FOA=0° CFRP surface under 90° UVAG MQL conditions is optimal. However, under 45° UVAG MQL conditions, fiber fracture and fiber loss damage also occurred on the CFRP surface of FOA=0°. There was no significant difference in the 3D surface microstructure of other FOAs (FOA=45°, 90°, and 135°) under 45° UVAG MQL and 90° UVAG MQL grinding conditions, and no obvious surface damage was found.

Figure 11. Comparison of 3D microscopic morphology of CFRP grinding surface with different FOAs under different conditions

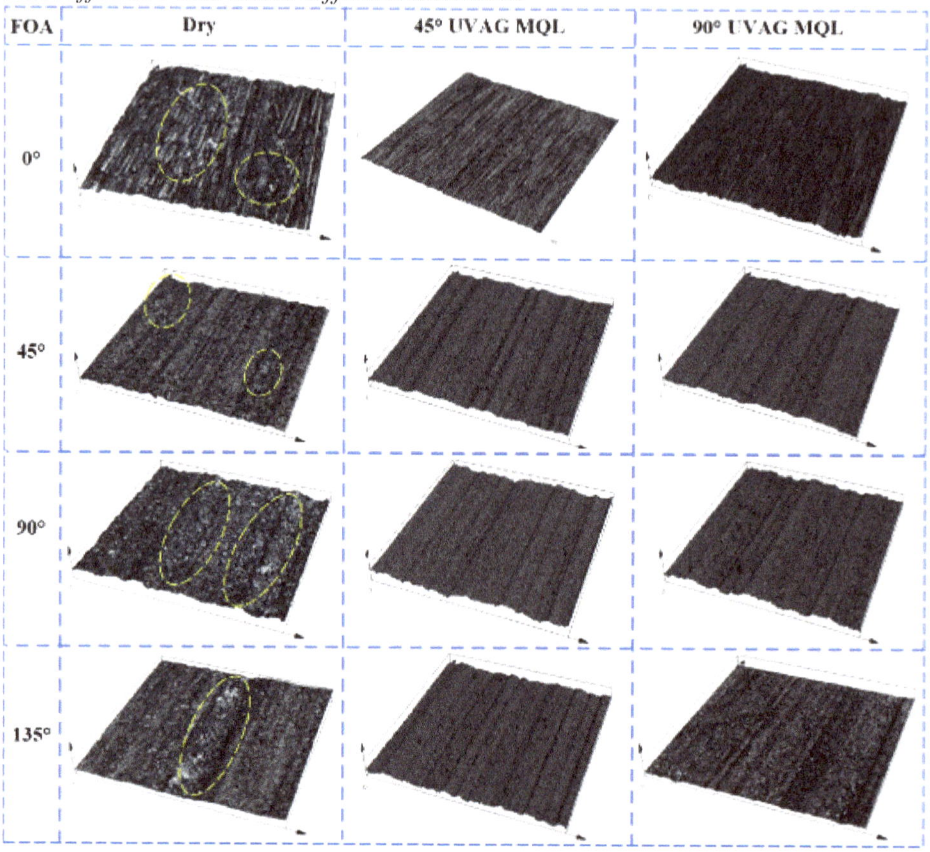

10.6 MULTIFRACTAL SPECTRUM

The overall morphology characteristics of CFRP grinding surfaces have been quantitatively compared and characterized by fractal dimension, but it is not sufficient to fully compare and explain the advantages and disadvantages of the morphology characteristics of CFRP grinding surfaces with different FOAs under different grinding conditions. This section further calculates the multifractal spectrum, its spectral width, and spectral difference to analyze and discuss the local morphology characteristics of CFRP grinding surface. If the multifractal spectrum of the surface roughness of CFRP grinding shows a right hook shape, it indicates that the contour of the surface roughness of CFRP grinding is dominated by high height values, and the morphology height is high. The probability of the valley on the grinding surface being occupied is greater than that of the peak. On the contrary, if the shape is a left hook, it indicates that the roughness profile curve is smaller and the height value has a higher probability. The height of the ground surface morphology is lower, and more valleys than peaks are generated on the machined surface.

10.6.1 Vertical Grinding Direction

As shown in Figure 12, the multifractal spectra of the surface roughness profile of CFRP with different FOAs in the vertical grinding direction under different conditions are presented (Figure 12 a), along with its spectral width α (q) (Figure 12 b) and spectral difference f (q) (Figure 12 c). It can be seen that the multifractal spectra all show a left hook, with α $(q)>0$ and f $(q)>0$, indicating that the surface roughness profile has a smaller proportion of height values and a lower morphology height. This indicates that all grinding surfaces have experienced varying degrees of carbon fiber extraction, resulting in damage such as surface pits and voids. For all FOAs, the spectral width of the multifractal spectrum of the CFRP grinding surface under θ_U =45° UVAG MQL condition is relatively small, with a maximum reduction percentage of 27.14% compared to dry traditional grinding (FOA=90°). This indicates that θ_U =45° UVAG MQL effectively reduces the unevenness and roughness of the grinding surface height. The comparison results of f (q) of multi-fractal spectra under different ultrasonic vibration and cooling lubrication conditions are basically the same as $\alpha(q)$, and the maximum reduction of θ_U =45° UVAG MQL compared to dry traditional grinding is 14.54% (FOA=45°). The smaller the spectral difference, the closer the proportion of peaks and valleys on the surface of CFRP grinding. The spectral difference of traditional dry grinding under all FOAs is relatively large, indicating that the probability of generating peaks on the grinding surface is lower than that of valleys. This also proves the occurrence of carbon fiber pull-out phenomenon on the surface of CFRP produced by traditional dry grinding, and even the block shaped pull-out of carbon fiber bundles can lead to fiber loss, pits, voids and other damages.

Figure 12. Multifractal spectra, spectral width and difference perpendicular to the grinding direction under different grinding conditions

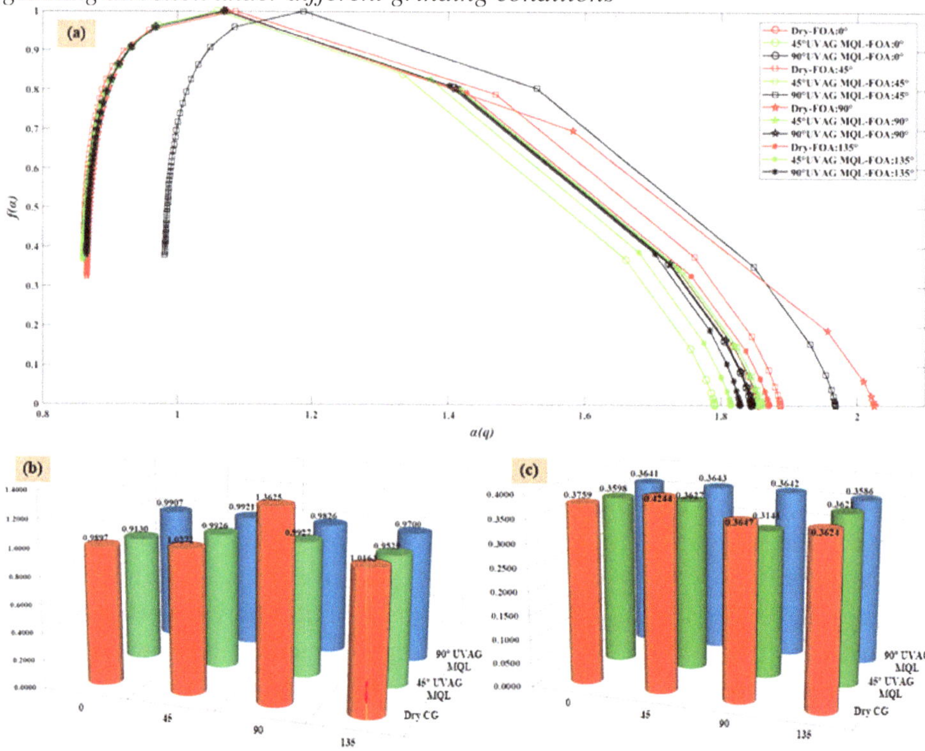

10.6.2 Parallel Grinding Direction

As shown in Figure 13, the multifractal spectra (Figure 13 (a)), spectral width (Figure 13 (b)), and spectral difference (Figure 13 (c)) of the surface roughness profile of CFRP with different FOAs in the parallel grinding direction under different working conditions are presented. The multifractal spectra in the parallel grinding direction are also in the shape of a left hook, with α (q) and f (q)>0, indicating that the surface roughness profile height value is low and there are many shape valleys. This proves that all grinding surfaces have experienced varying degrees of fiber pull-out, resulting in surface pits and other damages. For the spectral width and spectral difference, the multifractal spectrum of the CFRP grinding surface under 45° UVAG MQL condition is relatively small, with a maximum reduction of 3.92% compared to dry traditional grinding (FOA=90°). This indicates that 45° UVAG MQL effectively reduces the unevenness and roughness of the grinding surface height. The comparison results of spectral differences and spectral widths of multifractal spectra under different grinding conditions are basically the same. Compared with dry traditional grinding, the maximum reduction of θ_U=45° UVAG MQL is 10.72% (FOA=90°). The probability distribution of peaks and valleys on the grinding surface under ultrasonic vibration and cooling lubrication conditions is closer. The spectral differences of traditional dry grinding under all FOAs are relatively large, indicating that the probability of valleys appearing on the grinding surface is higher than that of peaks. This also proves that traditional dry grinding of CFRP surfaces results in carbon fiber pull-out, and even block shaped pull-out of carbon fiber bundles, leading to the formation of many fiber defects and pits and other grinding force thermal damage.

Figure 13. Multifractal spectra, spectral width and spectral difference parallel to the grinding direction under different grinding conditions

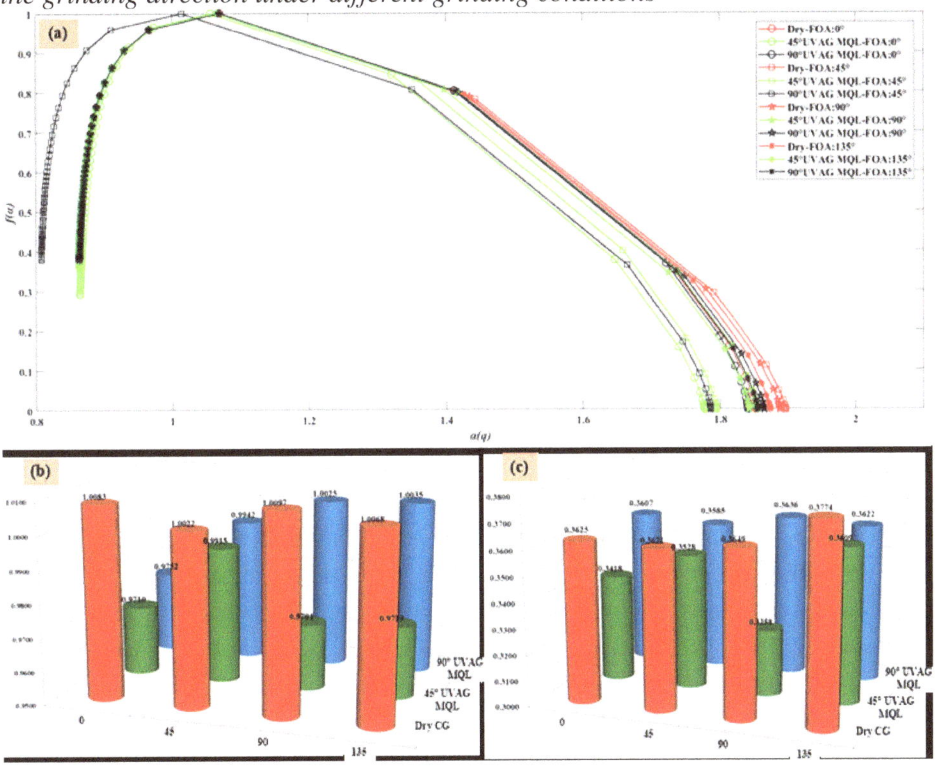

10.7 SUMMARY

This paper compares the line roughness and surface roughness, which reflect basic surface characteristics such as performance, fluctuation, and unevenness. The overall and local morphological characteristics of the CFRP grinding surface were quantitatively characterized. 2D and 3D surface morphology were observed using SEM and confocal laser microscopy, respectively. Qualitative analysis was conducted on grinding surface damage characterized by fiber (bundle) pull-out, fiber fragmentation, resin coating, and fiber resin debonding.

(1) The *Ra* value of 45° UVAG MQL is reduced by a maximum of 35.73% compared to dry traditional grinding (FOA=90° CFRP vertical grinding direction). The *Ra* value in the parallel grinding direction is generally lower than that in the vertical grinding direction, and the surface roughness *Ra* value of CFRP with FOA=0° is the highest. The *Ra* values of the other three FOAs are not significantly different. The 45° UVAG MQL has a significant effect on reducing the *Rsm* values of FOA=45° and 90°, with a maximum reduction of 20.93% compared to traditional dry grinding (vertical grinding direction with FOA=135°). 45° UVAG MQL is more conducive to reducing *Rsm* values at FOA=45° and 90°, while 90° UVAG MQL has a more significant effect on reducing *Rsm* values at FOA=0° and 135°.
(2) The variation law of surface roughness *Sa* with FOA is 0°>90°>135°>45°. The effect of θ_U =45° UVAG MQL on reducing *Sa* is the best. Compared with traditional dry grinding conditions, it reduces *Sa* by 13.15%, 20.19%, 21.83%, and 22.48% for FOA of 0°, 45°, 90°, and 135°, respectively. For surface roughness *Sq*, the influence of FOA on *Sq* is consistent with *Sa*, and θ_U =45° UVAG MQL has the best effect on reducing *Sq*. Compared with dry traditional grinding, it reduces the surface roughness and unevenness of CFRP grinding by 13.46%, 17.56%, 23.81%, and 24.81% when the FOA is 0°, 45°, 90°, and 135°, respectively. For the surface roughness *Sz*, FOA=135° is much higher than other FOAs, indicating severe fiber bundle block pull-out, resulting in more pits on the surface. Compared to dry traditional grinding, UVAG MQL with θ_U =45 ° decreased by 47.26% (FOA=135°). The reduction in surface roughness Sz, particularly the significant decrease at 135° FOA under 45° UVAG MQL, is directly indicative of suppressed fiber bundle pull-out, a defect known to severely degrade the fatigue life and load-bearing capacity of CFRP aerospace fastening holes.
(3) Ultrasonic vibration empowered nanolubricant MQL reduces the fractal dimension of CFRP grinding surface, which differs from the overall pattern of surface roughness. For the 2D fractal dimension, in the vertical grinding direction, 45°

UVAG MQL has a better effect on reducing CFRP with FOA=0°, 45°, and 135°, while 90° UVAG MQL is more effective for FOA=90°. In the direction of parallel grinding, 90° UVAG MQL has a better effect on reducing the 2D fractal dimension of CFRP grinding surfaces with FOA=0°, 45°, and 90°, while 45° UVAG MQL is more suitable for FOA=135°. The fractal dimension of FOA=0° is the lowest among all FOAs. For 3D fractal dimension, the direction of ultrasonic vibration application also has adaptability to FOA. 45° UVAG MQL has the best adaptability to FOA=45° and 135° CFRP, with a reduction of 17.73% and 10.32% compared to dry grinding, respectively. The 90° UVAG MQL has the best adaptability to FOA=0° and 90°, with a reduction of 8.38% and 29.22% compared to dry grinding, respectively.

(4) Under all conditions and fiber orientations, the multifractal spectra in both vertical and parallel grinding directions are left hook shaped, with positive spectral widths and differences. The surface morphology of CFRP grinding is low in height and has many valleys, indicating the occurrence of fiber pull-out causing surface pits. Compared with dry traditional grinding, UVAG MQL with θ_U =45° in the vertical grinding direction reduces the maximum spectral width and spectral difference by 27.14% (FOA=90°) and 14.54% (FOA=45°), respectively. This effectively reduces the unevenness and roughness of the ground surface height, and the probability distribution of peaks and valleys on the surface becomes closer to equality. Traditional dry grinding of CFRP surfaces results in fiber pull-out, and even fiber bundle block pull-out, leading to the formation of numerous pits and other damages. The spectral width and spectral difference of the multifractal spectrum under the condition of θ_U =45° UVAG MQL in the parallel grinding direction are the smallest, with a maximum reduction of 3.92% (FOA=90°) and 10.72% (FOA=90°) compared to dry traditional grinding, respectively.

(5) The UVAG nanolubricant MQL significantly improved the surface morphology of CFRP grinding and reduced grinding damage. The new method significantly reduces the voids and fiber loss caused by fiber slip during grinding of 0° CFRP, as well as continuous fiber fragmentation, reduces fiber pull-out during 45° CFRP grinding, reduces resin coating and block debris adhesion caused by 90° CFRP grinding, and avoids severe cracking caused by 135° CFRP grinding.

REFERENCES

Cao, S., Zhang, X., Wu, C., Yang, M., & Xia, K. (2024). Towards understanding the material removal mechanism on the effect of ultrasonic vibration and anisotropy for unidirectional CFRP by scratching. *Journal of Materials Processing Technology, 324*, 118250. DOI: 10.1016/j.jmatprotec.2023.118250

Dong, Z., Wang, Z., Tian, J., Kang, R., & Bao, Y. (2023). Investigation on tearing damage of CFRP circular cell honeycomb in end-face grinding. *Composite Structures, 325*, 117616. DOI: 10.1016/j.compstruct.2023.117616

Duan, Z., Wang, S., Li, C., Wang, Z., Bian, P., Song, J., & Liu, X. (2025). Tribological and micro-milling performance of surfactant-free microwave plasma-modified Al2O3 nanoparticles biodegradable lubricants. *Journal of Cleaner Production, 493*, 144969. DOI: 10.1016/j.jclepro.2025.144969

Gao, T., Xu, P., Wang, W., Zhang, Y., Xu, W., Wang, Y., An, Q. L., & Li, C. (2024). Force model of ultrasonic empowered minimum quantity lubrication grinding CFRP. *International Journal of Mechanical Sciences, 280*, 109522. DOI: 10.1016/j.ijmecsci.2024.109522

Gao, T., Zhang, X., Li, C., Zhang, Y., Yang, M., Jia, D., Ji, H. J., Zhao, Y. J., Li, R. Z., Yao, P., & Zhu, L. (2020). Surface morphology evaluation of multi-angle 2D ultrasonic vibration integrated with nanofluid minimum quantity lubrication grinding. *Journal of Manufacturing Processes, 51*, 44–61. DOI: 10.1016/j.jmapro.2020.01.024

Gu, G., Wang, D., Wu, S., Zhou, S., & Zhang, B. (2024). Research status and prospect of ultrasonic vibration and minimum quantity lubrication processing of nickel-based alloys. *Intelligent and Sustainable Manufacturing, 1*(1), 10006. DOI: 10.35534/ism.2024.10006

Handa, D., & Sooraj, V. S. (2023). Some constructive observations on the effect of wheel eccentricity during grinding of CFRP. *Journal of Manufacturing Processes, 97*, 62–75. DOI: 10.1016/j.jmapro.2023.04.066

Jia, D., Li, C., Liu, J., Zhang, Y., Yang, M., Gao, T., Said, Z., & Sharma, S. (2023). Prediction model of volume average diameter and analysis of atomization characteristics in electrostatic atomization minimum quantity lubrication. *Friction, 11*(11), 2107–2131. DOI: 10.1007/s40544-022-0734-2

Kong, X., Li, Z., Chen, X., Yang, M., Ma, X., Cui, X., Liu, M. Z., Li, B. K., Dambatta, Y. S., Song, K. P., Wang, G., & Li, C. (2025). Convective heat transfer mechanism of bionic desert beetle micro-grinding tool/bone interface and micro-grinding experiment. *International Journal of Heat and Mass Transfer*, *253*, 127611. DOI: 10.1016/j.ijheatmasstransfer.2025.127611

Li, P., Zhao, H., Yan, X., Wen, Z., Wang, R., & Ding, H. (2024). An oil-on-water minimum quantity lubrication technology for high-quality grinding of carbon fiber reinforced thermoplastic composites. *Polymer Composites*, *45*(15), 14077–14096. DOI: 10.1002/pc.28755

Liang, Y., Chen, Y., Chen, B., Fan, B., Yan, C., & Fu, Y. (2019). Feasibility of ultrasonic vibration assisted grinding for carbon fiber reinforced polymer with monolayer brazed grinding tools. *International Journal of Precision Engineering and Manufacturing*, *20*(7), 1083–1094. DOI: 10.1007/s12541-019-00135-8

Liu, C., Ren, J., Shi, K., & Zhang, Y. (2023). Investigation of fracture mechanism evolution model for UD-CFRP and MD-CFRP during the milling process. *Composite Structures*, *306*, 116585. DOI: 10.1016/j.compstruct.2022.116585

Tian, J., Bao, Y., Chen, K., Dong, Z., Kang, R., Guo, D., & Sun, J. (2025). Prediction and analysis of grinding burr of CFRP circular tube with a rounded corner grinding wheel. *Journal of Manufacturing Processes*, *133*, 249–259. DOI: 10.1016/j.jmapro.2024.11.059

Wang, W., Zhou, Z., Wang, Q., Gao, B., & Mao, C. (2025b). Muti-energy field-assisted grinding of hard and brittle materials: Tools, equipment and mechanisms. *Intelligent and Sustainable Manufacturing*, *2*(2), 10022. DOI: 10.70322/ism.2025.10022

Wang, X., Song, Y., Li, C., Zhang, Y., Ali, H. M., Sharma, S., Li, R. Z., Yang, M., Gao, T., Liu, M. Z., Cui, X., Said, Z., & Zhou, Z. (2024). Nanofluids application in machining: A comprehensive review. *International Journal of Advanced Manufacturing Technology*, *131*(5), 3113–3164. DOI: 10.1007/s00170-022-10767-2

Wang, X., Yang, M., Gao, T., Dong, L., Dambatta, Y. S., Liu, X., Yang, Y. Y., An, Q. L., Zhang, Y. B., & Li, C. (2025a). Lubricant transport mechanism and dynamics model for nepenthes-shaped biomimetic microtexture. *Chinese Journal of Mechanical Engineering*, *38*(1), 36. DOI: 10.1186/s10033-025-01197-8

Xiao, J., Wang, G., Su, H., Huang, P., & Chen, Z. (2022). A study of surface integrity in carbon fiber-reinforced polymer composites cutting with the coupling effect of the multiple machining parameters. *Journal of Composite Materials*, *56*(15), 2385–2397. DOI: 10.1177/00219983221099081

Xu, W., Li, C., Cui, X., Zhang, Y., Yang, M., Gao, T., Liu, M. Z., Wang, X. M., Zhou, Z. M., Sharma, S., & Dambatta, Y. S. (2023). Atomization mechanism and machinability evaluation with electrically charged nanolubricant grinding of GH4169. *Journal of Manufacturing Processes*, *106*, 480–493. DOI: 10.1016/j.jmapro.2023.10.037

Yang, M., Hao, J., Wu, W., Li, Z., Ma, Y., Zhou, Z., Gao, T., Liu, M. Z., Cui, X., Zhang, Y. B., Li, B. K., Ma, X., Dambatta, Y. S., & Li, C. (2024a). Critical cutting thickness model considering subsurface damage of zirconia grinding and friction–wear performance evaluation applied in simulated oral environment. *Tribology International*, *198*, 109881. DOI: 10.1016/j.triboint.2024.109881

Yang, M., Ma, H., Hao, J., Li, Z., Li, R., Zhou, Z., Gao, T., Liu, M. Z., Cui, X., Wang, X. M., Zhang, Y. B., Dambatta, Y. S., Long, Y. Z., & Li, C. (2024b). Droplet size distribution model of needle electrode electrostatic atomization and milling nickel-based alloy performance evaluation. *Journal of Manufacturing Processes*, *119*, 682–698. DOI: 10.1016/j.jmapro.2024.04.011

Yang, Z., Zhu, L., Zhang, G., Ni, C., & Lin, B. (2020). Review of ultrasonic vibration-assisted machining in advanced materials. *International Journal of Machine Tools & Manufacture*, *156*, 103594. DOI: 10.1016/j.ijmachtools.2020.103594

Zhang, Y. B., Li, L. Y., Cui, X., An, Q. L., Xu, P. M., Wang, W., Jia, D. Z., Liu, M. Z., Dambatta, Y. S., & Li, C. H. (2025). Lubricant activity enhanced technologies for sustainable machining: Mechanisms and processability. *Chinese Journal of Aeronautics*, *38*(6), 103203. DOI: 10.1016/j.cja.2024.08.034

Zhang, Z. Q., Jiao, F., Li, Y. X., Wang, X., Niu, Y., & Tong, J. L. (2024). Experimental study on rotary longitudinal-torsional ultrasonic machining of unidirectional CFRP. *Chinese Journal of Aeronautics*, *37*(11), 517–534. DOI: 10.1016/j.cja.2024.02.012

Zhu, F., Zhang, H., Hu, K., Duan, L., Zhang, K., & Xu, X. (2025). A comparative investigation on surface quality of CFRP laminates between rotary drilling and rotary ultrasonic-assisted drilling. *Composite Structures*, *358*, 118964. DOI: 10.1016/j.compstruct.2025.118964

Chapter 11
Research Hotspots and Development Trends in High-Performance Machining of FRC

ABSTRACT

This chapter summarizes the principal conclusions and innovations presented in this monograph. Primarily, it provides a focused analysis of the current research hotspots in the machining of FRCs and offers an outlook on future developmental trends. Research on the machining of FRCs, particularly CFRPs, is evolving toward a system characterized by clarified mechanisms, low-damage processes, intelligent control, and integrated systems. Key research hotspots reflect this evolution. First, understanding multi-scale material removal and damage formation is of paramount importance. Second, advanced tooling technologies and low-damage assisted processes are essential for improving machining quality and efficiency. Furthermore, challenging machining technologies for emerging materials and structures present new frontiers. Future breakthroughs will necessitate deep interdisciplinary integration to establish a complete, predictable, controllable, and optimizable precision machining technology system that links material microstructure to component service performance.

11.1 CONCLUSIONS

Addressing fundamental scientific questions regarding the material removal behavior and thermo-mechanical damage mechanisms in ultrasonic-assisted nanofluid MQL grinding of CFRPs, this study investigated the infiltration and transport

DOI: 10.4018/979-8-3373-9013-0.ch011

behavior of nano-lubricants within the ultrasonically activated wheel/workpiece interface, revealed the multi-phase heterogeneous cross-scale fracture mechanisms and associated macro-micro mechanical behavior, explored the tribological behavior of interfacial microstructures with different fiber orientations, and uncovered the mechanisms of thermo-mechanical damage in CFRP grinding. A grinding force model for CFRPs under various processing conditions was established and experimentally validated. A multi-dimensional, cross-scale, qualitative and quantitative damage characterization and surface integrity evaluation was conducted for multi-parameter coordinated ultrasonic-assisted nanofluid MQL grinding of CFRPs. The main theoretical, experimental, and applied conclusions are as follows:

Tangential ultrasonic vibration induces multiple fine polishing effects, reducing surface roughness, while axial ultrasonic vibration produces groove-widening effects that eliminate residual material peaks caused by grit spacing, thereby improving surface integrity. By adjusting the depth of cut (a) and wheel speed (v), multiple vibrational cutting cycles can be completed within the grinding zone. The differential cutting action of subsequent abrasive grains in 2D UVAG enhances lubrication effectiveness, and unilateral extrusion cutting mitigates workpiece tearing by abrasive grains. The differential cutting effect can be improved by appropriately increasing ultrasonic frequency (f), decreasing v, and increasing a. When the 2D ultrasonic angle (θ) is 45°, the effect is strongest in the direction of grinding zone penetration, and ultrasonic vibration most effectively promotes micro-droplet infiltration.

The droplet spreading rate on CFRP surfaces with different FOAs follows the order: 0° > 90° > 135° > 45°. On 0° FOA surfaces, the stabilized droplet aspect ratio reached 1.87, indicating the most significant anisotropic wetting. The contact angles of CNT nano-lubricants were smaller along the grinding direction than perpendicular to it. For FOAs from 0° to 135°, the perpendicular direction contact angles were higher than the parallel direction by 20.09%, 20.98%, 16.02%, and 5.95%, respectively. Among different FOAs, the 0° CFRP surface exhibited the smallest contact angle, while the 135° surface showed the largest. Under the combined influence of different θ and FOAs, the spreading behavior of nano-lubricant droplets on ground CFRP surfaces was studied, and the effects of ultrasonic direction on spreading rate and stabilized droplet aspect ratio were compared. The application of 2D ultrasonic vibration reduced the stabilized droplet aspect ratio—by 28.34% for 0° CFRP and 18.49% for 45° CFRP at $\theta = 90°$. Ultrasonic vibration also accelerated droplet spreading, allowing droplets to reach a stable state more rapidly. Nano-lubricants significantly improved tribological performance, reducing the friction coefficient by 17.09%, 34.87%, 19.59%, and 27.13% for the four FOAs compared to dry conditions. The wear volume under nano-lubrication followed the order: 45° > 0° > 135° > 90°, with the best anti-wear effect observed at 90° FOA, showing a 48.29% improvement over dry grinding.

The micro-buckling deformation behavior of 0° carbon fibers under negative rake angle cutting was revealed. The material removal mechanisms for 0° fibers were dominated by shear buckling and compression-tension buckling. The elliptical contact mechanics between the abrasive grain tip and horizontal carbon fibers were analyzed. For 45° fibers, the interlayer shear-slip behavior between cut fibers and adjacent fibers along the cutting direction was clarified. For 90° fibers, longitudinal compression by the abrasive grain caused debonding and cracking along the fiber direction, revealing the underlying matrix cracking and fiber shear fracture mechanisms. The bending deformation and fracture removal behavior of 45°, 90°, and 135° carbon fibers were investigated. The boundary conditions at the contact point between the abrasive grain and carbon fiber for different FOAs were established, along with force and moment equilibrium equations for carbon fiber micro-elements, elucidating the fiber bending fracture deformation based on the grain/fiber contact boundary conditions. The transverse micro-buckling behavior of axially compressed carbon fibers and associated coordinated structural failure were analyzed, and the blocky removal mechanics of fiber bundles were explored.

A homogeneous random distribution model for carbon fibers in CFRP and a diamond grinding wheel model with normally distributed grit protrusion heights and diameters were established. The geometric interference and intermittent cutting state under different ultrasonic parameters were analyzed, and the influence of ultrasonic vibration on the dynamic stiffness of the grinding system was investigated. The material removal mechanics of negative rake angle cutting for different FOAs were revealed. Based on this, an ultrasonic-assisted grinding force model for different FOAs under dry and nano-lubricant MQL conditions was developed and experimentally validated. The maximum errors between predicted and experimental grinding forces across all FOAs and conditions were 12.82% for normal grinding force (F) and 16.97% for tangential grinding force (F). UVAG with MQL effectively reduced grinding forces, with θ = 90° UVAG MQL showing the most significant reduction. Compared to conventional grinding, F was reduced by 16.29%, 18.47%, 5.03%, and 14.53%, and F by 25.21%, 33.32%, 24.16%, and 34.35% for different FOAs, respectively.

FOA significantly influences CFRP grinding damage and chip formation mechanisms. For FOA = 0°, damage mainly involves interfacial debonding and cracking due to fiber buckling, leading to fiber missing and fracture on the surface; chips are mostly long fibers with resin adhesion. For FOA = 45°, abrasive grains cause fiber bending fracture and interlayer shear separation, resulting in fiber fragmentation and resin smearing; chips are predominantly powdered/short fibers with some long fibers. For FOA = 90°, damage mechanisms include matrix cracking and fiber shear fracture, causing surface micro-cracks and sub-surface damage; chips are mixtures of long and powdered fibers, with less powder than at 45°. For FOA = 135°, fiber micro-buckling induces interlayer tensile fracture propagating to the sub-surface, followed by bending fracture and extrusion-shear separation, ultimately causing blocky fiber bundle removal, fiber pull-out, and surface voids; chips consist of blocky fiber bundles, powdered fibers, and long fibers.

CFRP grinding damage primarily distributes along the grinding direction. The 0° FOA surface showed the lowest high-frequency detail signal energy ratio, while the 45° FOA showed the highest, especially parallel to the grinding direction. For two-dimensional surface signals, damage distribution was most prominent parallel to the grinding direction, followed by perpendicular, and least along the diagonal. The $\theta = 90°$ 2D ultrasonic-assisted nano-lubricant MQL most significantly reduced the high-frequency detail signal energy ratio, achieving reductions of 7.42%, 27.56%, 5.03%, and 2.57% for the four FOAs compared to dry conventional grinding. Additionally, $\theta = 45°$ 2D ultrasonic-assisted nano-lubricant MQL was most effective for FOA = 135°, with a reduction of 40.38%.

UVAG nano-lubricant MQL effectively reduced the surface roughness and fractal dimension of ground CFRP, though the extent of reduction varied with FOA. No universally significant trend was observed for line roughness and 2D fractal dimension, indicating the limitations of these 2D parameters for evaluating surface integrity in heterogeneous anisotropic CFRP grinding. Overall, 45° UVAG MQL was more effective in reducing 3D areal roughness. Furthermore, 45° UVAG MQL better reduced the 3D fractal dimension for FOA = 45° and 135°, while 90° UVAG MQL was optimal for FOA = 0° and 90°. Multifractal spectra, including spectrum width and difference, indicated that 45° UVAG MQL more effectively reduced surface height variation and roughness in both perpendicular and parallel grinding directions, minimizing fiber pull-out and pit damage. Finally, no absolute correlation was found between surface roughness and grinding force across different FOAs, highlighting the challenging machinability and complexity of theoretical and experimental research on heterogeneous anisotropic fiber composites.

11.2 INNOVATIONS

To address fundamental scientific issues concerning the material removal behavior and thermo-mechanical damage mechanisms in ultrasonic-assisted nanofluid MQL grinding of CFRPs, this study conducted systematic research on the material removal mechanics and surface integrity evaluation in multi-directional two-dimensional ultrasonic-assisted nanofluid MQL grinding of CFRPs. The main innovations are summarized as follows:

The kinematic behavior of abrasive grains in multi-directional 2D ultrasonic-assisted grinding was analyzed, and the synergistic effects of ultrasonic parameters, grinding parameters, and wheel specifications on cutting characteristics were investigated. This study is the first to reveal the mechanisms of nanofluid infiltration and transport within the ultrasonically activated wheel-workpiece interface, and established a dynamic pressure lubricating film thickness model for the grinding zone under ultrasonic-enhanced nanofluid MQL conditions.

The spreading characteristics of CNT nanofluid droplets under different ultrasonic vibration conditions were investigated. This research pioneeringly uncovered the mechanical behavior of droplet contact lines during spreading on ground CFRP surfaces with different fiber orientations. The anisotropic spreading mechanism and capillary action of droplets on 0° fiber-oriented CFRP surfaces were analyzed, along with the correlation between surface roughness and spreading rate. The tribological behavior of nanofluids on micro-structured CFRP surfaces with varying fiber directions was also examined.

The multi-phase heterogeneous cross-scale fracture mechanisms and macro-micro mechanical behavior in CFRP grinding were revealed, and the material removal mechanisms for different fiber orientations were analyzed. Based on a stochastic carbon fiber distribution model and a diamond grinding wheel model, a mechanical model for multi-directional 2D ultrasonic-assisted nanofluid MQL grinding of CFRPs under various processing conditions was established for the first time and validated through grinding experiments.

The transient thermal generation mechanism in negative rake angle grinding of CFRPs was analyzed, and the influence of heat conduction and accumulation on the thermal expansion coefficient of CFRPs was investigated. Through coupled thermo-mechanical 3D meso-scale finite element simulations, the mechanisms of thermo-mechanical damage in CFRP grinding were revealed. Quantitative and qualitative analysis of grinding-induced damage was further conducted by combining experimental results. Finally, a multi-dimensional cross-scale surface integrity evaluation was performed for multi-parameter coordinated ultrasonic-assisted MQL grinding of CFRPs.

11.3 RESEARCH HOTSPOTS

While the present experimental and modeling work is devoted exclusively to CFRP, a brief consideration of other fiber-reinforced composites (FRCs) is necessary to delineate the transferability and limitations of the findings. For glass fiber-reinforced polymers (GFRP), the isotropic, less brittle nature of the fibers tends to moderate the strong fiber-orientation dependence observed in CFRP machining, promote more ductile material removal, and reduce the propensity for abrupt brittle fracture; consequently, quantitative predictions of cutting forces, surface integrity, and delamination cannot be directly transposed without recalibrating damage laws and failure criteria. For ceramic matrix composites (CMC), the extreme hardness and thermal stability of both fibers and matrix shift the dominant material-removal mechanism toward microfracture, grain pull-out, and severe tool wear, thereby making the thermally induced matrix degradation typical of CFRP far less relevant while instead requiring accounting for thermal shock and highly abrasive tool–workpiece interactions. In both cases, the multiphase progressive-damage modeling framework established here can, in principle, be extended, but the constituent material models and governing parameters must be fundamentally revised to capture the distinct fiber–matrix mechanics. Such adaptations would constitute a natural and valuable extension of the current work.

Multi-scale material removal mechanisms and damage formation theory. Current research has moved beyond the sole focus on macroscopic cutting forces, delving deeper into the meso- and micro-scale physical nature of material removal. Hotspots include revealing the competitive mechanisms of fiber fracture, matrix cracking, and interfacial debonding in anisotropic laminates under tool action. By establishing finite element models coupled with composite failure criteria and combining in situ observations via high-speed imaging and scanning electron microscopy, researchers systematically analyze the evolution of cutting forces and chip formation modes under different FOAs. Particularly critical is the prediction of delamination damage—a core failure mode in drilling and edge milling—whose initiation and propagation have become key research topics. Current efforts are directed toward developing theoretical models that account for fluctuations in axial force, tool wear state, and interlaminar toughness, aiming to accurately predict delamination factors (e.g., critical thrust force) and provide a theoretical foundation for process optimization.

Advanced tooling technologies and wear behavior studies. Given the high abrasiveness and difficult machinability of composites, the development of specialized tooling technologies is a direct approach to improving machining quality and efficiency. Research hotspots in this domain encompass innovations in macro-geometry, applications of super-abrasives, and surface modification techniques. In geometric design, emphasis is placed on multi-facet angles, step structures, and secondary edge modifications for drills, as well as eccentric relief angles and wavy edge profiles for milling cutters. The core objective is to guide clean shear fracture of fibers through mechanical design, thereby suppressing burr formation and delamination. In terms of materials, the wear mechanisms and life prediction of polycrystalline diamond (PCD) tools and diamond-coated tools are focal points. By analyzing failure modes such as abrasive wear and coating spallation, and integrating cutting forces and acoustic emission signals to build tool health monitoring models, researchers aim to enable intelligent decision-making for tool replacement.

Low-damage assisted machining process innovations. To proactively mitigate inherent damage in conventional machining, a range of assisted energy-field processes has emerged as a frontier exploration. Ultrasonic vibration-assisted machining imparts high-frequency, micro-amplitude vibrations to the tool in axial or radial directions, periodically altering the tool–workpiece contact state. This effectively reduces average cutting forces and heat, demonstrating excellent burr suppression and surface quality enhancement in drilling and contour machining. Cryogenic-assisted machining primarily utilizes liquid nitrogen or supercritical CO_2 jets for active cooling and lubrication of the cutting zone. By inhibiting the thermal softening of the resin matrix and enhancing local material brittleness, it improves chip morphology and reduces thermally induced damage. Additionally, auxiliary methods such as rigid support and media filling continue to receive attention for controlling machining deformation and exit tearing in thin-walled components.

Process modeling, simulation, and intelligent method applications. With advances in computational power, multiscale simulation and data-driven methods are profoundly transforming process development paradigms. At the mechanistic modeling level, multiscale finite element methods combining mesoscale representative volume element models with macroscale continuum mechanics models are widely used to simulate the entire process from fiber/matrix interface failure to macroscopic workpiece deformation. Concurrently, the discrete element method provides a unique perspective for studying discontinuous chip formation and evacuation behavior. At the data-driven level, machine learning algorithms—such as support vector machines, random forests, and deep learning—are being employed to construct complex nonlinear surrogate models linking process parameters with machining quality indicators. This enables rapid prediction of surface roughness, cutting forces, and damage extent, as well as multi-objective process parameter optimization.

Integrated machining quality inspection and closed-loop control strategies. Achieving online perception and closed-loop control of machining quality is a core requirement of intelligent manufacturing. Research hotspots in this area focus on online monitoring technologies based on multi-source information fusion. By integrating high-frequency dynamometers, acoustic emission sensors, and industrial vision systems, real-time acquisition of force, acoustic, and image signals during machining is achieved. Signal processing and pattern recognition algorithms—such as wavelet transforms and deep learning, are then utilized for early identification and classification of defects like delamination and burrs. Further research aims to integrate these monitoring systems with CNC systems to form adaptive machining closed loops. This involves dynamically adjusting parameters such as spindle speed and feed rate based on real-time detection of workpiece condition or tool wear information, thereby enabling autonomous process optimization and ensuring quality consistency.

Challenging machining technologies for emerging materials and structures. The expanding application of composites continuously gives rise to new machining challenges. "One-shot drilling" of fiber-metal laminates and multi-material stacked structures represents a typical difficulty, requiring solutions to issues such as interfacial damage in dissimilar materials, abrupt tool load changes, and poor chip evacuation. The high hardness and brittleness of ceramic matrix composites impose extreme demands on ultra-precision grinding technology. For high-performance thermoplastic composites, thermal management during machining is critical, necessitating precise control of cutting heat to avoid excessive matrix melting and fiber reorientation. For extreme environments such as aerospace and deep-sea applications, research focuses on integrated coating preparation and machining technologies for composite components with radiation resistance and corrosion resistance. Additionally, through multi-material 3D printing or tailored layups, metamaterial structures with properties such as negative Poisson's ratio and variable stiffness are manufactured. The study of machining mechanisms, development of dedicated processes, and design of specialized tools for these special materials constitute highly challenging frontier directions.

Environmental and economic assessment of dry or minimum lubrication machining technologies. In response to the manufacturing industry's sustainability requirements, the environmental and economic evaluation of dry or minimum quantity lubrication (MQL) machining technologies has become a research hotspot. The core objective is to maximize the reduction of chemical coolant usage while ensuring machining quality. On another front, digital twin technology—as a key enabler bridging the physical and information spaces—is being introduced into the composite machining domain. Combining 3D printing (continuous fiber additive manufacturing) with precision machining enables the integrated forming and high-precision machining of complex structures. Machine learning techniques, such as neural networks and reinforcement learning, are employed to optimize process parameters, reduce trial-and-error costs, and achieve automatic defect identification and classification. By constructing virtual mappings that integrate machine tool dynamics, tool wear models, material variability, and real-time sensor data, digital twin systems can achieve full-factor simulation of machining processes, performance prediction, and proactive optimization of process parameters. This represents the developmental trend of future intelligent machining systems.

To translate the identified sustainability trends into quantifiable assessments, future work should adopt established metrics such as Specific Energy Consumption (SEC), Carbon Footprint Analysis (CFA), and Life Cycle Assessment (LCA). Applying these frameworks would enable a rigorous evaluation of the hybrid process's environmental benefits—particularly the reductions in grinding fluid usage, energy demand, and tool wear observed in this study. In parallel, economic viability could be assessed through metrics like total machining cost per component, tool life extension, and waste disposal savings. Integrating such sustainability and economic indicators would not only strengthen the technological relevance of the proposed process but also align with the growing emphasis on sustainable manufacturing within the broader composites community.

Research on the machining of fiber-reinforced composites is evolving deeply toward "clarified mechanisms, low-damage processes, intelligent control, and integrated systems." Future breakthroughs will rely on the deep interdisciplinary integration of materials science, mechanics, mechanical engineering, and information science. The ultimate goal is to establish a complete precision machining technology system—from material microstructure to component service performance—that is "predictable, controllable, and optimizable."

11.4 DEVELOPMENT TRENDS AND PROSPECTS

Carbon fiber-reinforced polymer (CFRP) composites serve as the performance benchmark and technological vanguard within the fiber-reinforced composite (FRC) family, enjoying the broadest application spectrum. They define the upper limit of mechanical performance for advanced composites and, through their highly mature industrial chain and cutting-edge applications, continuously propel technological innovation and industrial upgrading across the entire field. To meet the growing demands of high-end equipment for FRC components with superior performance, structural integrity, and precision, future research in machining technologies must achieve synergistic breakthroughs in both depth and breadth, establishing a complete technological chain from fundamental theory to engineering application.

At the level of fundamental theory and simulation, current research predominantly focuses on macroscopic phenomenological models. Future efforts must advance towards mechanism studies involving multi-scale and multi-physical field coupling. Emphasis should be placed on developing cross-scale theoretical models spanning from micro-scale fiber/matrix interface debonding, meso-scale fiber fracture and pull-out, to macro-scale material removal. Furthermore, it is crucial to deeply elucidate material response mechanisms under thermo-mechanical-chemical coupled fields (e.g., resin pyrolysis or interface degradation induced by high grinding temperatures). For instance, integrating molecular dynamics simulations with continuum damage mechanics can enable precise prediction of subsurface damage initiation and propagation behavior under varying FOAs. Concurrently, introducing machine learning methods to mine vast amounts of process experimental and simulation data will facilitate the establishment of high-fidelity mappings among process parameters, damage evolution, and final performance. This will lay the foundation for reverse process design based on the paradigm of "theoretical prediction first, followed by process validation."

Regarding core process technology innovation, future development pathways will concentrate on "hybridization with energy-field assistance" and "integration of the process chain." On one hand, there is a need to vigorously develop active hybrid processes combining energy fields—such as ultrasonic vibration, laser, and cryogenic cooling (liquid nitrogen/cold air)—with precision grinding/cutting. Specifically, ultrasonic-assisted grinding can reduce transverse bending stress on fibers via high-frequency impact, facilitating cleaner cutting of brittle carbon fibers. Laser preheating-assisted cutting can locally soften thermoplastic resin matrices, enabling cooperative fiber deformation to reduce burrs and delamination. Cryogenic grinding effectively suppresses heat accumulation in the machining zone, protecting resin properties and minimizing tool adhesion. On the other hand, breakthroughs are required in integrated machining technologies encompassing "precision shaping – surface modification – functionalization creation." An example is completing both precision grinding and subsequent laser surface texturing of composite components in a single setup to simultaneously meet dimensional accuracy, fatigue performance, and specialized functional requirements (e.g., hydrophobicity, anti-icing).

At the level of equipment and tooling systems, realizing the aforementioned high-performance processes relies on the support of specialized and intelligent manufacturing systems. Future equipment development must shift from "general-purpose platforms" to "process-customized" solutions: developing dedicated composite machining centers integrated with multi-axis ultrasonic vibration systems, in-process cooling/lubrication units, and multi-source information sensing (force, temperature, acoustic emission, vision). Key functional modules include adaptive process parameter control systems based on digital twins and online non-destructive evaluation units for subsurface damage. Simultaneously, innovation in grinding tools is paramount. It is essential to develop specialized grinding wheels tailored for different systems, such as carbon fiber and ceramic matrix composites—for example, utilizing graphene-coated diamond abrasive grains to enhance heat dissipation or designing bionic ordered abrasive edges to achieve gentler "shear-dominated" material removal. Integrating in-situ laser dressing and contour detection technologies is key to enabling intelligent tool condition management and ensuring machining consistency.

In terms of technological system and standard establishment, the ultimate goal is to build a fully controllable manufacturing chain for FRCs, from "digital models" to "qualified products." This requires promoting deep interdisciplinary integration among materials science, mechanics, information science, and manufacturing engineering, and constructing an intelligent manufacturing ecosystem encompassing databases for material machinability, process knowledge bases, intelligent decision-making models, and digital quality traceability systems. Moreover, the industry urgently needs to establish a unified standard system for evaluating the machining quality of composites. This includes, in particular, standards for the classification, detection, and allowable limits of subsurface micro-damage (e.g., micro-cracks, fiber pull-out), as well as rapid assessment methods for long-term service performance. Such standards are vital to regulate technological development and support the scalable, high-reliability application of composites in critical sectors like aerospace, new energy vehicles, and wind turbine blades.

Research on the integrated manufacturing covering design, fabrication, machining, and performance evaluation of FRCs to meet specific performance requirements remains limited. The multiphase composition and laminate structure of FRCs afford them excellent designability. Notably, their damping characteristics surpass those of traditional materials, and vibration transmission properties vary across different FRC structures. Vibration transmission paths within FRCs can be actively controlled through design. Therefore, designing and manufacturing FRCs for vibration damping and noise reduction represents a promising future direction. This requires systematically considering the dual performance requirements of vibration damping effectiveness and machining quality during the manufacturing process.

In ultrasonic vibration-assisted machining, the high-frequency impact of the tool exerts multiple effects on the material. Simultaneously, due to the inhomogeneity and anisotropy of composites, specific material parameters—such as the proportion, morphology, arrangement, and bonding interface of the matrix and reinforcement—directly influence the material removal mode. Consequently, the interaction between tool and material becomes more complex during dynamic machining. This makes it difficult to establish the relationship between process parameters and material removal modes in ultrasonic-assisted machining, or to define clear criteria for identifying removal modes, thereby hindering the judgment and control of material removal behavior. Furthermore, the mechanisms of mechanical and thermal stress transfer under ultrasonic assistance for different types of FRCs (with different matrices or fibers) exhibiting heterogeneity and anisotropy remain unclear. This complicates the integration of cutting force models under different material removal modes to establish a universal model.

Currently, the relationships among material removal mechanisms, deformation behavior, and surface formation in FRCs are not fully understood. It is necessary to further reveal deformation characteristics, chip formation mechanisms, the formation mechanisms of machinability and their influencing factors, based on the micro- and nano-mechanical properties of composites. Building upon the micro/nano composition and mechanical properties of composites, research should investigate the formation mechanisms of subsurface damage, including cracks and residual stresses, and the evolution of material functional performance, service life, and reliability. A comprehensive evaluation methodology for surface/subsurface integrity in precision grinding of composites needs to be established. It is essential to elucidate the influence of thermo-mechanical coupling on the mechanical performance of internal interfaces within composites. Furthermore, micromechanical constitutive models and cross-scale material removal models for composites should be developed. Clarifying the mechanical behavior of material removal and the formation process of machined surfaces is necessary to provide a systematic and complete theoretical basis for composite machining research. Ultimately, the goal is to achieve service-performance-oriented control over the surface/subsurface integrity of composites. Additionally, for carbon fiber-reinforced thermoplastic composites, the material removal mechanisms in cutting and grinding are not yet clear. Further investigation is required into the co-removal mechanisms of high-strength fibers and viscous/tough matrices, along with in-depth analysis of the variations in cutting forces/heat and damage formation mechanisms in high-toughness, strongly adhesive thermoplastic composites. Tool development specifically for the high-quality and high-efficiency machining of thermoplastic composites needs refinement. Finally, systematic research on the development and integrated application validation of property-controlled machining processes is also required.

Compilation of References

Abdalla, H. S., & Patel, S. (2006). The performance and oxidation stability of sustainable metalworking fluid derived from vegetable extracts. *Proceedings of the Institution of Mechanical Engineers. Part B, Journal of Engineering Manufacture*, *220*(12), 2027–2040. DOI: 10.1243/09544054JEM357

Abdullah, A., Farhadi, A., & Pak, A. (2012). Ultrasonic-assisted dry creep-feed up-grinding of superalloy Inconel738LC. *Experimental Mechanics*, *52*(7), 843–853. DOI: 10.1007/s11340-011-9557-1

Adibi, H., Esmaeili, H., & Rezaei, S. M. (2018). Study on minimum quantity lubrication (MQL) in grinding of carbon fiber-reinforced SiC matrix composites (CMCs). *International Journal of Advanced Manufacturing Technology*, *95*(9), 3753–3767. DOI: 10.1007/s00170-017-1464-x

Alargova, R. G., Kochijashky, I. I., Sierra, M. L., Kwetkat, K., & Zana, R. (2001). Mixed micellization of dimeric (gemini) surfactants and conventional surfactants: II. CMC and micelle aggregation numbers for various mixtures. *Journal of Colloid and Interface Science*, *235*(1), 119–129. DOI: 10.1006/jcis.2000.7311 PMID: 11237450

Alexandrov, S., Jeng, Y. R., & Lomakin, E. (2011). Effect of pressure-dependency of the yield criterion on the development of plastic zones and the distribution of residual stresses in thin annular disks. *Journal of Applied Mechanics*, *78*(3), 031012. DOI: 10.1115/1.4003361

Ali, S., Kurniawan, R., Xu, M., Ko, T. J., & Gautam, R. K. S. (2025). Effect of High-Speed Longitudinal Ultrasonic Vibration-Assisted Milling on Tribological Characteristics of a Plane Surface. *Journal of Materials Engineering and Performance*, *34*(1), 870–884. DOI: 10.1007/s11665-024-09737-4

Almeida, J. H. S.Jr, Souza, S. D., Botelho, E. C., & Amico, S. C. (2016). Carbon fiber-reinforced epoxy filament-wound composite laminates exposed to hygrothermal conditioning. *Journal of Materials Science*, *51*(9), 4697–4708. DOI: 10.1007/s10853-016-9787-9

Alves, S. M., Barros, B. S., Trajano, M. F., Ribeiro, K. S. B., & Moura, E. J. T. I. (2013). Tribological behavior of vegetable oil-based lubricants with nanoparticles of oxides in boundary lubrication conditions. *Tribology International*, *65*, 28–36. DOI: 10.1016/j.triboint.2013.03.027

Amin, M., Yuan, S., Israr, A., Zhen, L., & Qi, W. (2018b). Development of cutting force prediction model for vibration-assisted slot milling of carbon fiber reinforced polymers. *International Journal of Advanced Manufacturing Technology*, *94*(9), 3863–3874. DOI: 10.1007/s00170-017-1087-2

Amin, M., Yuan, S., Khan, M. Z., Wu, Q., & Zhu, G. (2017). Development of a generalized cutting force prediction model for carbon fiber reinforced polymers based on rotary ultrasonic face milling. *International Journal of Advanced Manufacturing Technology*, *93*(5), 2655–2666. DOI: 10.1007/s00170-017-0469-9

Amin, M., Yuan, S., Khan, M. Z., Zhang, C., & Wu, Q. (2018a). Development of cutting force prediction model for carbon fiber reinforced polymers based on rotary ultrasonic slot milling. *Machining Science and Technology*, *22*(3), 402–426. DOI: 10.1080/10910344.2017.1365895

Anand, P. S. P., Arunachalam, N., & Vijayaraghavan, L. (2018). Study on grinding of pre-sintered zirconia using diamond wheel. *Materials and Manufacturing Processes*, *33*(6), 634–643. DOI: 10.1080/10426914.2017.1364761

An, Q., Yang, J., Li, J., Liu, G., Chen, M., & Li, C. (2023). A state-of-the-art review on the intelligent tool holders in machining. *Intelligent and Sustainable Manufacturing*, *1*(1), 10002. DOI: 10.35534/ism.2024.10002

Asmael, M., Safaei, B., Zeeshan, Q., Zargar, O., & Nuhu, A. A. (2021). Ultrasonic machining of carbon fiber–reinforced plastic composites: A review. *International Journal of Advanced Manufacturing Technology*, *113*(11), 3079–3120. DOI: 10.1007/s00170-021-06722-2

Attwood, D., & Florence, A. T. (1983). *Surfactant Systems: Their Chemistry, Pharmacy and Biology*. Chapman and Hall. DOI: 10.1007/978-94-009-5775-6

Azarhoushang, B., & Tawakoli, T. (2011). Development of a novel ultrasonic unit for grinding of ceramic matrix composites. *International Journal of Advanced Manufacturing Technology*, *57*(9), 945–955. DOI: 10.1007/s00170-011-3347-x

Babbar, A., Sharma, A., Jain, V., & Jain, A. K. (2019). Rotary ultrasonic milling of C/SiC composites fabricated using chemical vapor infiltration and needling technique. *Materials Research Express*, *6*(8), 085607. DOI: 10.1088/2053-1591/ab1bf7

Bai, X., Li, C., Dong, L., & Yin, Q. (2019). Experimental evaluation of the lubrication performances of different nanofluids for minimum quantity lubrication (MQL) in milling Ti-6Al-4V. *International Journal of Advanced Manufacturing Technology*, *101*(9), 2621–2632. DOI: 10.1007/s00170-018-3100-9

Behera, B. C., Setti, D., Ghosh, S., & Rao, P. V. (2017). Spreadability studies of metal working fluids on tool surface and its impact on minimum amount cooling and lubrication turning. *Journal of Materials Processing Technology*, *244*, 1–16. DOI: 10.1016/j.jmatprotec.2017.01.016

Benzeggagh, M. L., & Kenane, M. J. C. S. (1996). Measurement of mixed-mode delamination fracture toughness of unidirectional glass/epoxy composites with mixed-mode bending apparatus. *Composites Science and Technology*, *56*(4), 439–449. DOI: 10.1016/0266-3538(96)00005-X

Bie, W., Chen, F., Wang, X., Chen, S., Fu, Z., & Zhao, B. (2023). Longitudinal-torsional coupled rotary ultrasonic machining end surface grinding of SiCf/SiC composites: A mechanical model of cutting force. *International Journal of Advanced Manufacturing Technology*, *129*(3), 1227–1248. DOI: 10.1007/s00170-023-12360-7

Bin, L. I. N., Haoji, W. A. N. G., Jinhua, W. E. I., & Tianyi, S. U. I. (2021). Diamond wheel grinding characteristics of 3D orthogonal quartz fiber reinforced silica ceramic matrix composite. *Chinese Journal of Aeronautics*, *34*(5), 404–414. DOI: 10.1016/j.cja.2020.12.026

Biot, M. A. (1937). Bending of an infinite beam on an elastic foundation. *Journal of Applied Mechanics*, *4*(1), A1–A7. DOI: 10.1115/1.4008739

Birleanu, C., Pustan, M., Pop, G., Cioaza, M., Popa, F., Lazarescu, L., & Contiu, G. (2022). Experimental investigation of the tribological behaviors of carbon fiber reinforced polymer composites under boundary lubrication. *Polymers*, *14*(18), 3716. DOI: 10.3390/polym14183716 PMID: 36145857

Brewer, J. C., & Lagace, P. A. (1988). Quadratic stress criterion for initiation of delamination. *Journal of Composite Materials*, *22*(12), 1141–1155. DOI: 10.1177/002199838802201205

Brown, C. A., Charles, P. D., Johnsen, W. A., & Chesters, S. (1993). Fractal analysis of topographic data by the patchwork method. *Wear*, *161*(1-2), 61–67. DOI: 10.1016/0043-1648(93)90453-S

Bu, P., Ruan, W., & Liu, J. (2023). Effect of interface and matrix strength on transverse tensile strength of unidirectional CFRP composite. *Modelling and Simulation in Materials Science and Engineering*, *31*(6), 065005. DOI: 10.1088/1361-651X/acde93

Buscaglia, G. C., & Ausas, R. F. (2011). Variational formulations for surface tension, capillarity and wetting. *Computer Methods in Applied Mechanics and Engineering*, *200*(45-46), 3011–3025. DOI: 10.1016/j.cma.2011.06.002

Cao, J., Wu, Y., Lu, D., Fujimoto, M., & Nomura, M. (2014a). Material removal behavior in ultrasonic-assisted scratching of SiC ceramics with a single diamond tool. *International Journal of Machine Tools & Manufacture*, *79*, 49–61. DOI: 10.1016/j.ijmachtools.2014.02.002

Cao, S., Zhang, X., Wu, C., Yang, M., & Xia, K. (2024). Towards understanding the material removal mechanism on the effect of ultrasonic vibration and anisotropy for unidirectional CFRP by scratching. *Journal of Materials Processing Technology*, *324*, 118250. DOI: 10.1016/j.jmatprotec.2023.118250

Cao, X., Lin, B., Wang, Y., & Wang, S. (2014b). Influence of diamond wheel grinding process on surface micro-topography and properties of SiO2/SiO2 composite. *Applied Surface Science*, *292*, 181–189. DOI: 10.1016/j.apsusc.2013.11.109

Cao, X., Lin, B., & Zhang, X. (2013). A study on grinding surface waviness of woven ceramic matrix composites. *Applied Surface Science*, *270*, 503–512. DOI: 10.1016/j.apsusc.2013.01.069

Cao, X., Lin, B., & Zhang, X. (2015). Investigations on grinding process of woven ceramic matrix composite based on reinforced fiber orientations. *Composites. Part B, Engineering*, *71*, 184–192. DOI: 10.1016/j.compositesb.2014.11.029

Cao, Y., Zhao, B., Ding, W., Wu, J., Jia, X., Zhang, J., & Das, R. (2024). Intermittent cutting behavior and grinding force model in ultrasonic vibration-assisted grinding K4002 nickel-based superalloy. *International Journal of Advanced Manufacturing Technology*, *131*(5), 3085–3102. DOI: 10.1007/s00170-024-13053-5

Carlson, B. C. (1977). Elliptic integrals of the first kind. *SIAM Journal on Mathematical Analysis*, *8*(2), 231–242. DOI: 10.1137/0508016

Carlson, B. C. (1987). A table of elliptic integrals of the second kind. *Mathematics of Computation*, *49*(180), 595–606. DOI: 10.1090/S0025-5718-1987-0906192-1

Chen, D., Lin, H., Chen, Y., Yan, C., & Xu, J. (2023a). Wear behaviors of PCD and diamond-coated tools during low-frequency vibration-assisted drilling CFRP/Ti stacks. *International Journal of Advanced Manufacturing Technology*, *128*(1), 595–609. DOI: 10.1007/s00170-023-11899-9

Cheng, X. Q., Yasir, B., & Li, Z. H. (2011). Effects of hygrothermal environmental conditions on compressive strength of CFRP stitched laminates. *Journal of Reinforced Plastics and Composites*, *30*(2), 110–122. DOI: 10.1177/0731684410384894

Chen, H., & Tang, J. (2015). A model for prediction of surface roughness in ultrasonic-assisted grinding. *International Journal of Advanced Manufacturing Technology*, *77*(1-4), 643–651. DOI: 10.1007/s00170-014-6482-3

Chen, J., An, Q., Ming, W., & Chen, M. (2020a). Investigation on machined surface quality in ultrasonic-assisted grinding of Cf/SiC composites based on fracture mechanism of carbon fibers. *International Journal of Advanced Manufacturing Technology*, *109*(5), 1583–1599. DOI: 10.1007/s00170-020-05739-3

Chen, J., Ming, W., An, Q., & Chen, M. (2020b). Mechanism and feasibility of ultrasonic-assisted milling to improve the machined surface quality of 2D Cf/SiC composites. *Ceramics International*, *46*(10), 15122–15136. DOI: 10.1016/j.ceramint.2020.03.047

Chen, J., Shen, J., Huang, H., & Xu, X. (2010). Grinding characteristics in high speed grinding of engineering ceramics with brazed diamond wheels. *Journal of Materials Processing Technology*, *210*(6-7), 899–906. DOI: 10.1016/j.jmatprotec.2010.02.002

Chen, R., Li, S., Zhou, Y., Qiu, X., Li, P., Zhang, H., & Wang, Z. (2023b). Damage formation and evolution mechanisms in drilling CFRP with prefabricated delamination defects: Simulation and experimentation. *Journal of Materials Research and Technology*, *26*, 6994–7011. DOI: 10.1016/j.jmrt.2023.09.065

Chen, T., Li, H., Ye, M., Xiang, C., & Tian, S. (2019). Experimental study on effects of structural characteristics of C/E composite laminates on grinding temperature. *Composites. Part B, Engineering*, *157*, 100–108. DOI: 10.1016/j.compositesb.2018.08.120

Chockalingam, P., & Kuang, K. C. (2015). Effects of abrasive types on grinding of chopped strand mat glass fiber-reinforced polymer composite laminates. *Machining Science and Technology*, *19*(2), 313–324. DOI: 10.1080/10910344.2015.1018534

Chockalingam, P., Kuang, K. C., & Vijayaram, T. R. (2013). Effects of grinding process parameters and coolants on the grindability of GFRP laminates. *Materials and Manufacturing Processes*, *28*(10), 1071–1076. DOI: 10.1080/10426914.2013.792411

Choi, C., Jung, M., Choi, Y., Lee, J., & Oh, J. (2011). Tribological properties of lubricating oil-based nanofluids with metal/carbon nanoparticles. *Journal of Nanoscience and Nanotechnology*, *11*(1), 368–371. DOI: 10.1166/jnn.2011.3227 PMID: 21446457

Choudhary, A., Chakladar, N. D., & Paul, S. (2021). Identification and estimation of defects in high-speed ground C/SiC ceramic matrix composites. *Composite Structures*, *261*, 113274. DOI: 10.1016/j.compstruct.2020.113274

Chu, A., Li, C., Zhou, Z., Liu, B., Zhang, Y., Yang, M., Gao, T., Liu, M., Zhang, N., Dambatta, Y., & Sharma, S. (2023). Nanofluids minimal quantity lubrication machining: From mechanisms to application. *Lubricants (Basel, Switzerland)*, *11*(10), 422. DOI: 10.3390/lubricants11100422

Clint, J. H. (1975). Micellization of mixed nonionic surface active agents. Journal of the Chemical Society, Faraday Transactions 1. *Physical Chemistry in Condensed Phases*, *71*, 1327–1334.

Cococcetta, N. M., Pearl, D., Jahan, M. P., & Ma, J. (2020). Investigating surface finish, burr formation, and tool wear during machining of 3D printed carbon fiber reinforced polymer composite. *Journal of Manufacturing Processes*, *56*, 1304–1316. DOI: 10.1016/j.jmapro.2020.04.025

Cococcetta, N., Jahan, M. P., Schoop, J., Ma, J., Pearl, D., & Hassan, M. (2021). Post-processing of 3D printed thermoplastic CFRP composites using cryogenic machining. *Journal of Manufacturing Processes*, *68*, 332–346. DOI: 10.1016/j.jmapro.2021.05.054

Cong, W. L., Pei, Z. J., Mohanty, N., Van Vleet, E., & Treadwell, C. (2011). Vibration amplitude in rotary ultrasonic machining: a novel measurement method and effects of process variables.

Cui, X., & Li, C. H. (2025). *Magnetic Traction Nanolubricant Grinding: Force Model and Mechanisms*. Nova Science Publishers, 33-44. DOI: 10.52305/PVGY4472

Cui, X., Li, C. H., Ding, W. F., Chen, Y., Mao, C., Xu, X. F., Liu, B., Wang, D. Z., Li, H. N., Zhang, Y. B., Said, Z., Debnath, S., Jamil, M., Ali, H. M., & Sharma, S. (2022). Minimum quantity lubrication machining of aeronautical materials using carbon group nanolubricant: From mechanisms to application. *Chinese Journal of Aeronautics*, *35*(11), 85–112. DOI: 10.1016/j.cja.2021.08.011

Cui, X., Li, C., Zhang, Y., Jia, D., Zhao, Y., Li, R., & Cao, H. (2019). Tribological properties under the grinding wheel and workpiece interface by using graphene nanofluid lubricant. *International Journal of Advanced Manufacturing Technology, 104*(9), 3943–3958. DOI: 10.1007/s00170-019-04129-8

Dambatta, Y. S., Li, C., Sayuti, M., Sarhan, A. A., Yang, M., Li, B., Chu, A., Liu, M., Zhang, Y., Said, Z., & Zhou, Z. (2024). Grindability evaluation of ultrasonic assisted grinding of silicon nitride ceramic using minimum quantity lubrication based SiO2 nanofluid. *Chinese Journal of Mechanical Engineering, 37*(1), 25. DOI: 10.1186/s10033-024-01011-x

Danish, M., Gupta, M. K., Rubaiee, S., Ahmed, A., Mahfouz, A., & Jamil, M. (2021). Machinability investigations on CFRP composites: A comparison between sustainable cooling conditions. *International Journal of Advanced Manufacturing Technology, 114*(11), 3201–3216. DOI: 10.1007/s00170-021-07073-8

Das, S., & Pandivelan, C. (2020). Grinding characteristics during ultrasonic vibration assisted grinding of alumina ceramic in selected dry and MQL conditions. *Materials Research Express, 7*(8), 085404. DOI: 10.1088/2053-1591/abad14

Davim, J. P., Rubio, J. C., & Abrao, A. M. (2007). A novel approach based on digital image analysis to evaluate the delamination factor after drilling composite laminates. *Composites Science and Technology, 67*(9), 1939–1945. DOI: 10.1016/j.compscitech.2006.10.009

Debnath, S., Reddy, M. M., & Yi, Q. S. (2014). Environmental friendly cutting fluids and cooling techniques in machining: A review. *Journal of Cleaner Production, 83*, 33–47. DOI: 10.1016/j.jclepro.2014.07.071

Deng, S., Ye, L., Mai, Y. W., & Liu, H. Y. (1998). Evaluation of fibre tensile strength and fibre/matrix adhesion using single fibre fragmentation tests. *Composites. Part A, Applied Science and Manufacturing, 29*(4), 423–434. DOI: 10.1016/S1359-835X(97)00094-8

Diaz, O. G., & Axinte, D. A. (2017). Towards understanding the cutting and fracture mechanism in ceramic matrix composites. *International Journal of Machine Tools & Manufacture, 118*, 12–25. DOI: 10.1016/j.ijmachtools.2017.03.008

Ding, K., Fu, Y., Su, H., Cui, F., Li, Q., Lei, W., & Xu, H. (2017b). Study on surface/subsurface breakage in ultrasonic assisted grinding of C/SiC composites. *International Journal of Advanced Manufacturing Technology, 91*(9), 3095–3105. DOI: 10.1007/s00170-017-0012-z

Ding, K., Fu, Y., Su, H., Xu, H., Cui, F., & Li, Q. (2017a). Experimental studies on matching performance of grinding and vibration parameters in ultrasonic assisted grinding of SiC ceramics. *International Journal of Advanced Manufacturing Technology*, *88*(9), 2527–2535. DOI: 10.1007/s00170-016-8977-6

Ding, W., Huang, Q., Zhao, B., Cao, Y., Tang, M., Deng, M., Liu, G., Zhao, Z., & Chen, Q. (2023). Wear characteristics of white corundum abrasive wheel in ultrasonic vibration-assisted grinding of AISI 9310 steel. *Ceramics International*, *49*(8), 12832–12839. DOI: 10.1016/j.ceramint.2022.12.153

Ding, W., Liu, G., Chen, T., Feng, S., Zhao, Z., Zhao, B., & Cui, H. (2024). Effects of fiber orientations on removal behaviors in ultrasonic vibration-assisted grinding continuous fiber reinforced Ti matrix composite. *International Journal of Advanced Manufacturing Technology*, *133*(9), 4223–4234. DOI: 10.1007/s00170-024-14009-5

Ding, X., Li, P., Lin, S. C. S., Stratton, Z. S., Nama, N., Guo, F., Slotcavage, D., Mao, X., Shi, J., Costanzo, F., & Huang, T. J. (2013). Surface acoustic wave microfluidics. *Lab on a Chip*, *13*(18), 3626–3649. DOI: 10.1039/c3lc50361e PMID: 23900527

Dogra, M., Sharma, V. S., Dureja, J. S., & Gill, S. S. (2018). Environment-friendly technological advancements to enhance the sustainability in surface grinding-A review. *Journal of Cleaner Production*, *197*, 218–231. DOI: 10.1016/j.jclepro.2018.05.280

Dong, L., Li, C., Bai, X., Zhai, M., Qi, Q., Yin, Q., Lv, X., & Li, L. (2019). Analysis of the cooling performance of Ti–6Al–4V in minimum quantity lubricant milling with different nanoparticles. *International Journal of Advanced Manufacturing Technology*, *103*(5), 2197–2206. DOI: 10.1007/s00170-019-03466-y

Dong, Z., Wang, Z., Tian, J., Kang, R., & Bao, Y. (2023). Investigation on tearing damage of CFRP circular cell honeycomb in end-face grinding. *Composite Structures*, *325*, 117616. DOI: 10.1016/j.compstruct.2023.117616

Dong, Z., Zhang, H., Kang, R., Ran, Y., & Bao, Y. (2022). Mechanical modeling of ultrasonic vibration helical grinding of SiCf/SiC composites. *International Journal of Mechanical Sciences*, *234*, 107701. DOI: 10.1016/j.ijmecsci.2022.107701

Duan, W. H., Wang, Q., & Collins, F. (2011). Dispersion of carbon nanotubes with SDS surfactants: A study from a binding energy perspective. *Chemical Science (Cambridge)*, *2*(7), 1407–1413. DOI: 10.1039/c0sc00616e

Duan, Z. J., Li, C. H., Zhang, Y. B., Dong, L., Bai, X. F., Yang, M., Jia, D. Z., Li, R. Z., Cao, H. J., & Xu, X. F. (2021). Milling surface roughness for 7050 aluminum alloy cavity influenced by nozzle position of nanofluid minimum quantity lubrication. *Chinese Journal of Aeronautics*, *34*(6), 33–53. DOI: 10.1016/j.cja.2020.04.029

Duan, Z., Wang, S., Li, C., Wang, Z., Bian, P., Song, J., & Liu, X. (2025). Tribological and micro-milling performance of surfactant-free microwave plasma-modified Al2O3 nanoparticles biodegradable lubricants. *Journal of Cleaner Production, 493*, 144969. DOI: 10.1016/j.jclepro.2025.144969

Duan, Z., Wang, S., Li, C., Wang, Z., Bian, P., Sun, J., Song, J., & Liu, X. (2024). Cold plasma and different nano-lubricants multi-energy field coupling-assisted micro-milling of Al-Li alloy 2195-T8 and flow rate optimization. *Journal of Manufacturing Processes, 127*, 218–237. DOI: 10.1016/j.jmapro.2024.07.146

Duan, Z., Wang, S., Wang, Z., Li, C., Li, Y., Song, J., Liu, J., & Liu, X. (2024). Tool wear mechanisms in cold plasma and nano-lubricant multi-energy field coupled micro-milling of Al-Li alloy. *Tribology International, 192*, 109337. DOI: 10.1016/j.triboint.2024.109337

Du, J., Ming, W., Ma, J., He, W., Cao, Y., Li, X., & Liu, K. (2018). New observations of the fiber orientations effect on machinability in grinding of C/SiC ceramic matrix composite. *Ceramics International, 44*(12), 13916–13928. DOI: 10.1016/j.ceramint.2018.04.240

Duman, E., Yapan, Y. F., Salvi, H., Sofuoğlu, M. A., Khanna, N., & Uysal, A. (2024). Investigation of ultrasonic vibration assisted orthogonal turning under dry and minimum quantity lubrication conditions and performing sustainability analyses. *Journal of Cleaner Production, 434*, 140187. DOI: 10.1016/j.jclepro.2023.140187

El-Hofy, M. H., & El-Hofy, H. (2019). Laser beam machining of carbon fiber reinforced composites: A review. *International Journal of Advanced Manufacturing Technology, 101*(9), 2965–2975. DOI: 10.1007/s00170-018-2978-6

Esmaeili, H., Adibi, H., & Rezaei, S. M. (2019). An efficient strategy for grinding carbon fiber-reinforced silicon carbide composite using minimum quantity lubricant. *Ceramics International, 45*(8), 10852–10864. DOI: 10.1016/j.ceramint.2019.02.163

Estellé, P., Halelfadl, S., & Maré, T. (2014). Lignin as dispersant for water-based carbon nanotubes nanofluids: Impact on viscosity and thermal conductivity. *International Communications in Heat and Mass Transfer, 57*, 8–12. DOI: 10.1016/j.icheatmasstransfer.2014.07.012

Fan, B. P., Chen, Y., Chen, B. B., Liang, Y. H., & Sun, L. (2019). Finite element analysis on heat distribution ratio during grinding CFRP. *Diamond and Abrasives Engineering, 39*(1), 66–71.

Fedulov, B. N., Safonov, A. A., Kantor, M. M., & Lomov, S. V. (2017). Modelling of thermoplastic polymer failure in fiber reinforced composites. *Composite Structures*, *163*, 293–301. DOI: 10.1016/j.compstruct.2016.11.091

Feng, Z., & Cook, R. D. (1983). Beam elements on two-parameter elastic foundations. *Journal of Engineering Mechanics*, *109*(6), 1390–1402. DOI: 10.1061/(ASCE)0733-9399(1983)109:6(1390)

Fergani, O., Yousfi, M., Ding, Z., Welo, T., & Liang, S. Y. (2017). A physics-based approach to relate grinding process parameters to tribological behavior of ground surfaces. *International Journal of Advanced Manufacturing Technology*, *91*(9), 4151–4161. DOI: 10.1007/s00170-017-0111-x

Fernando, P. K. S. C., Zhang, M., & Pei, Z. (2019). Rotary ultrasonic machining: Effects of tool natural frequency on ultrasonic vibration amplitude. *Machining Science and Technology*, *23*(4), 595–611. DOI: 10.1080/10910344.2019.1575404

Friend, J., & Yeo, L. Y. (2011). Microscale acoustofluidics: Microfluidics driven via acoustics and ultrasonics. *Reviews of Modern Physics*, *83*(2), 647–704. DOI: 10.1103/RevModPhys.83.647

Fu, R., Jia, Z., Wang, F., Jin, Y., Sun, D., Yang, L., & Cheng, D. (2018). Drill-exit temperature characteristics in drilling of UD and MD CFRP composites based on infrared thermography. *International Journal of Machine Tools & Manufacture*, *135*, 24–37. DOI: 10.1016/j.ijmachtools.2018.08.002

Gao, C. Y., Xiao, J. Z., Zhang, L. C., & Ke, Y. L. (2017). On the static and dynamic properties of fiber-reinforced polymer composites: A three-phase constitutive model. *Journal of Thermoplastic Composite Materials*, *30*(11), 1560–1577. DOI: 10.1177/0892705716646418

Gao, T., Li, C., Jia, D., Zhang, Y., Yang, M., Wang, X., Cao, H., Li, R., Ali, H. M., & Xu, X. (2020b). Surface morphology assessment of CFRP transverse grinding using CNT nanofluid minimum quantity lubrication. *Journal of Cleaner Production*, *277*, 123328. DOI: 10.1016/j.jclepro.2020.123328

Gao, T., Li, C., Yang, M., Zhang, Y., Jia, D., Ding, W., Debnath, S., Yu, T., Said, Z., & Wang, J. (2021a). Mechanics analysis and predictive force models for the single-diamond grain grinding of carbon fiber reinforced polymers using CNT nano-lubricant. *Journal of Materials Processing Technology*, *290*, 116976. DOI: 10.1016/j.jmatprotec.2020.116976

Gao, T., Li, C., Zhang, Y., Yang, M., Jia, D., Jin, T., Hou, Y., & Li, R. (2019). Dispersing mechanism and tribological performance of vegetable oil-based CNT nanofluids with different surfactants. *Tribology International*, *131*, 51–63. DOI: 10.1016/j.triboint.2018.10.025

Gao, T., Liu, J., Sun, X., Zhang, Y., Yang, M., Liu, M., Xu, W., An, Q., Wang, D., Xu, P., & Li, C. (2025). Enhanced permeation mechanism and tribological assessment of ultrasonic vibration nanolubricants grinding CFRP. *Tribology International*, *204*, 110494. DOI: 10.1016/j.triboint.2024.110494

Gao, T., Xu, P., Wang, W., Zhang, Y., Xu, W., Wang, Y., An, Q., & Li, C. (2024). Force model of ultrasonic empowered minimum quantity lubrication grinding CFRP. *International Journal of Mechanical Sciences*, *280*, 109522. DOI: 10.1016/j.ijmecsci.2024.109522

Gao, T., Zhang, X., Li, C., Zhang, Y., Yang, M., Jia, D., Ji, H., Zhao, Y., Li, R., Yao, P., & Zhu, L. (2020a). Surface morphology evaluation of multi-angle 2D ultrasonic vibration integrated with nanofluid minimum quantity lubrication grinding. *Journal of Manufacturing Processes*, *51*, 44–61. DOI: 10.1016/j.jmapro.2020.01.024

Gao, T., Zhang, Y., Li, C., Wang, Y., An, Q., Liu, B., Said, Z., & Sharma, S. (2021b). Grindability of carbon fiber reinforced polymer using CNT biological lubricant. *Scientific Reports*, *11*(1), 22535. DOI: 10.1038/s41598-021-02071-y PMID: 34795390

Gao, T., Zhang, Y., Li, C., Wang, Y., Chen, Y., An, Q., Zhang, S., Li, H., Cao, H., Ali, H. M., Zhou, Z., & Sharma, S. (2022). Fiber-reinforced composites in milling and grinding: Machining bottlenecks and advanced strategies. *Frontiers of Mechanical Engineering*, *17*(2), 24. DOI: 10.1007/s11465-022-0680-8

Geier, N., Davim, J. P., & Szalay, T. (2019). Advanced cutting tools and technologies for drilling carbon fibre reinforced polymer (CFRP) composites: A review. *Composites. Part A, Applied Science and Manufacturing*, *125*, 105552. DOI: 10.1016/j.compositesa.2019.105552

Ge, J., Fu, G., Almeida, J. H. S.Jr, Jin, Y., & Sun, D. (2025). Thermal effect in CFRP machining: Temperature field characteristics, heat generation mechanism and thermal damage management. *Composite Structures*, *356*, 118845. DOI: 10.1016/j.compstruct.2025.118845

Geng, D., Teng, Y., Liu, Y., Shao, Z., Jiang, X., & Zhang, D. (2019). Experimental study on drilling load and hole quality during rotary ultrasonic helical machining of small-diameter CFRP holes. *Journal of Materials Processing Technology*, *270*, 195–205. DOI: 10.1016/j.jmatprotec.2019.03.001

Geng, D., Zhang, D., Xu, Y., He, F., Liu, D., & Duan, Z. (2015). Rotary ultrasonic elliptical machining for side milling of CFRP: Tool performance and surface integrity. *Ultrasonics*, *59*, 128–137. DOI: 10.1016/j.ultras.2015.02.006 PMID: 25708349

Ge, X., Xia, Y., & Cao, Z. (2015). Tribological properties and insulation effect of nanometer TiO2 and nanometer SiO2 as additives in grease. *Tribology International*, *92*, 454–461. DOI: 10.1016/j.triboint.2015.07.031

Giasin, K., Ayvar-Soberanis, S., & Hodzic, A. (2016). Evaluation of cryogenic cooling and minimum quantity lubrication effects on machining GLARE laminates using design of experiments. *Journal of Cleaner Production*, *135*, 533–548. DOI: 10.1016/j.jclepro.2016.06.098

Gong, P., Zhang, Y., Cui, X., Xu, S., Yang, M., Jia, D., & Li, C. (2024). Lubricant transportation mechanism and wear resistance of different arrangement textured turning tools. *Tribology International*, *196*, 109704. DOI: 10.1016/j.triboint.2024.109704

Gong, P., Zhang, Y., Wang, C., Cui, X., Li, R., Sharma, S., Liu, M., Gao, T., Zhou, Z., Wang, X., Dambatta, Y. S., & Li, C. (2024). Residual stress generation in grinding: Mechanism and modeling. *Journal of Materials Processing Technology*, *324*, 118262. DOI: 10.1016/j.jmatprotec.2023.118262

Gu, G., Wang, D., Wu, S., Zhou, S., & Zhang, B. (2024). Research status and prospect of ultrasonic vibration and minimum quantity lubrication processing of nickel-based alloys. *Intelligent and Sustainable Manufacturing*, *1*(1), 10006. DOI: 10.35534/ism.2024.10006

Gu, G., Wu, S., Wang, D., Zhou, S., Zhu, L., An, Q., Guo, H., & Li, C. (2023). A review of the research on the variation of tool's motion trajectory and its influence on the formation mechanism of surface quality in ultrasonic vibration machining. *Journal of Manufacturing Processes*, *107*, 294–319. DOI: 10.1016/j.jmapro.2023.10.025

Guo, C., Zhu, X., Liu, J., & Zhang, D. (2018). Cavitation bubble dynamics induced by hydrodynamic pressure oil film in ultrasonic vibration honing. *Journal of Tribology*, *140*(4), 041707. DOI: 10.1115/1.4039409

Guo, J., Barber, G. C., Schall, D. J., Zou, Q., & Jacob, S. B. (2017). Tribological properties of ZnO and WS2 nanofluids using different surfactants. *Wear*, *382*, 8–14. DOI: 10.1016/j.wear.2017.04.003

Guo, M., Tao, J., Wu, C., Luo, C., & Lin, Z. (2023). High-speed grinding fracture mechanism of Cf/SiC composite considering interfacial strength and anisotropy. *Ceramics International*, *49*(2), 2600–2612. DOI: 10.1016/j.ceramint.2022.09.240

Guo, S., Li, C., Zhang, Y., Wang, Y., Li, B., Yang, M., Zhang, X., & Liu, G. (2017). Experimental evaluation of the lubrication performance of mixtures of castor oil with other vegetable oils in MQL grinding of nickel-based alloy. *Journal of Cleaner Production*, *140*, 1060–1076. DOI: 10.1016/j.jclepro.2016.10.073

Guo, Z., Guo, B., Wu, G., Xiang, Y., Meng, Q., Jia, J., Zhao, Q., Li, K., & Zeng, Z. (2023). Three-dimensional topography modelling and grinding performance evaluating of micro-structured CVD diamond grinding wheel. *International Journal of Mechanical Sciences*, *244*, 108079. DOI: 10.1016/j.ijmecsci.2022.108079

Hadad, M. J., Tawakoli, T., Sadeghi, M. H., & Sadeghi, B. (2012). Temperature and energy partition in minimum quantity lubrication-MQL grinding process. *International Journal of Machine Tools & Manufacture*, *54*, 10–17. DOI: 10.1016/j.ijmachtools.2011.11.010

Haddad, M., Zitoune, R., Eyma, F., & Castanie, B. (2014). Study of the surface defects and dust generated during trimming of CFRP: Influence of tool geometry, machining parameters and cutting speed range. *Composites. Part A, Applied Science and Manufacturing*, *66*, 142–154. DOI: 10.1016/j.compositesa.2014.07.005

Hanasaki, S., Fujiwara, J., & Tashiro, T. (2003). 218 Study on Surface Grinding of Unidirectional Carbon Fiber Reinforced Plastics in Dry Method. In *Proceedings of International Conference on Leading Edge Manufacturing in 21st century: LEM21 2003* (pp. 297-302). The Japan Society of Mechanical Engineers. DOI: 10.1299/jsmelem.2003.297

Handa, D., & Sooraj, V. S. (2023). Some constructive observations on the effect of wheel eccentricity during grinding of CFRP. *Journal of Manufacturing Processes*, *97*, 62–75. DOI: 10.1016/j.jmapro.2023.04.066

Hashin, Z., & Rotem, A. (1973). A fatigue failure criterion for fiber reinforced materials. *Journal of Composite Materials*, *7*(4), 448–464. DOI: 10.1177/002199837300700404

Helmy, M. O., El-Hofy, M. H., & El-Hofy, H. (2018). Effect of cutting fluid delivery method on ultrasonic assisted edge trimming of multidirectional CFRP composites at different machining conditions. *Procedia CIRP*, *68*, 450–455. DOI: 10.1016/j.procir.2017.12.077

Hoang, Q. N., Lee, S., Lee, S., & Park, H. (2025). Characterization of CFRP interface properties with varying fiber surface roughness using AFM measurements and finite element modeling. *Composites. Part B, Engineering*, •••, 112990.

Holland, P. M., & Rubingh, D. N. (1992). Mixed surfactant systems: An overview. *ACS Symposium Series*, 501, 2-30. DOI: 10.1021/bk-1992-0501.ch001

Hooputra, H., Gese, H., Dell, H., & Werner, H. (2004). A comprehensive failure model for crashworthiness simulation of aluminium extrusions. *International Journal of Crashworthiness*, *9*(5), 449–464. DOI: 10.1533/ijcr.2004.0289

Hosokawa, A., Hirose, N., Ueda, T., & Furumoto, T. (2014). High-quality machining of CFRP with high helix end mill. *CIRP Annals*, *63*(1), 89–92. DOI: 10.1016/j.cirp.2014.03.084

Huang, B., Wang, W. H., Xiong, Y. F., Wu, X. F., Liu, J. T., Liu, C., & Wang, D. H. (2023). Investigation of force modeling in ultrasonic vibration-assisted drilling SiCf/SiC ceramic matrix composites. *Journal of Manufacturing Processes*, *96*, 21–30. DOI: 10.1016/j.jmapro.2023.04.040

Hu, N. S., & Zhang, L. C. (2003). A study on the grindability of multidirectional carbon fibre-reinforced plastics. *Journal of Materials Processing Technology*, *140*(1-3), 152–156. DOI: 10.1016/S0924-0136(03)00704-0

Hu, S., Li, C., Zhou, Z., Liu, B., Zhang, Y., Yang, M., Li, B., Gao, T., Liu, M., Cui, X., Wang, X., Xu, W., Dambatta, Y., Li, R., & Sharma, S. (2023). Nanoparticle-enhanced coolants in machining: Mechanism, application, and prospects. *Frontiers of Mechanical Engineering*, *18*(4), 53. DOI: 10.1007/s11465-023-0769-8

Hussien, A. A., Qasem, I., Kataraki, P. S., Al-Kouz, W. G., & Janvekar, A. A. (2021). Studying the performance of cutting carbon fibre-reinforced plastic using an abrasive water jet technique. *Strojniški vestnik, 67*(4), 125-141.

Hu, Z. S., Dong, J. X., & Chen, G. X. (1998). Study on antiwear and reducing friction additive of nanometer ferric oxide. *Tribology International*, *31*(7), 355–360. DOI: 10.1016/S0301-679X(98)00042-5

Hwang, Y., Park, H. S., Lee, J. K., & Jung, W. H. (2006). Thermal conductivity and lubrication characteristics of nanofluids. *Current Applied Physics*, *6*, 67–71. DOI: 10.1016/j.cap.2006.01.014

Inoue, H., & Kawaguchi, I. (1990). Study on the grinding mechanism of glass fiber reinforced plastics. *Journal of Engineering Materials and Technology*, *112*(3), 341–345. DOI: 10.1115/1.2903335

Iskandar, Y., Tendolkar, A., Attia, M. H., Hendrick, P., Damir, A., & Diakodimitris, C. (2014). Flow visualization and characterization for optimized MQL machining of composites. *CIRP Annals*, *63*(1), 77–80. DOI: 10.1016/j.cirp.2014.03.078

Islam, S., Yuan, S., & Li, Z. (2020). A cutting force prediction model, experimental studies, and optimization of cutting parameters for rotary ultrasonic face milling of C/SiC composites. *Applied Composite Materials, 27*(4), 407–431. DOI: 10.1007/s10443-020-09815-5

James, S., & Nejadian, S. M. (2020). Experimental study on high-speed saw cutting of hybrid composite stacks using nanoparticle-enhanced minimum quantity lubrication. *International Journal of Advanced Manufacturing Technology, 110*(11), 3077–3090. DOI: 10.1007/s00170-020-06008-z

Jebaratnam, J. M., & Hassan, M. H. (2025). Comprehensive review of drilling strategies for CFRP/Ti stacks in aircraft manufacturing. *Machining Science and Technology, 29*(3), 334–438. DOI: 10.1080/10910344.2025.2475484

Jia, D., Li, C., Liu, J., Zhang, Y., Yang, M., Gao, T., Said, Z., & Sharma, S. (2023). Prediction model of volume average diameter and analysis of atomization characteristics in electrostatic atomization minimum quantity lubrication. *Friction, 11*(11), 2107–2131. DOI: 10.1007/s40544-022-0734-2

Jia, D., Li, C., Zhang, Y., Yang, M., Zhang, X., Li, R., & Ji, H. (2019). Experimental evaluation of surface topographies of NMQL grinding ZrO2 ceramics combining multiangle ultrasonic vibration. *International Journal of Advanced Manufacturing Technology, 100*(1-4), 457–473. DOI: 10.1007/s00170-018-2718-y

Jia, D., Li, C., Zhang, Y., Zhang, D., & Zhang, X. (2016). Experimental research on the influence of the jet parameters of minimum quantity lubrication on the lubricating property of Ni-based alloy grinding. *International Journal of Advanced Manufacturing Technology, 82*(1), 617–630. DOI: 10.1007/s00170-015-7381-y

Jia, D., Zhang, Y., Li, C., Yang, M., Gao, T., Said, Z., & Sharma, S. (2022). Lubrication-enhanced mechanisms of titanium alloy grinding using lecithin biolubricant. *Tribology International, 169*, 107461. DOI: 10.1016/j.triboint.2022.107461

Jiang, T. S., Soo-Gun, O. H., & Slattery, J. C. (1979). Correlation for dynamic contact angle. *Journal of Colloid and Interface Science, 69*(1), 74–77. DOI: 10.1016/0021-9797(79)90081-X

Jia, Z., Fu, R., Niu, B., Qian, B., Bai, Y., & Wang, F. (2016). Novel drill structure for damage reduction in drilling CFRP composites. *International Journal of Machine Tools & Manufacture, 110*, 55–65. DOI: 10.1016/j.ijmachtools.2016.08.006

Jia, Z., Fu, R., Wang, F., Qian, B., & He, C. (2018). Temperature effects in end milling carbon fiber reinforced polymer composites. *Polymer Composites, 39*(2), 437–447. DOI: 10.1002/pc.23954

Johnson, K. L. (1987). *Contact mechanics*. Cambridge university press.

Kodama, H., Okazaki, S., Jiang, Y., Yoden, H., & Ohashi, K. (2020). Thermal influence on surface layer of carbon fiber reinforced plastic (CFRP) in grinding. *Precision Engineering*, *65*, 53–63. DOI: 10.1016/j.precisioneng.2020.04.005

Kong, X., Li, C., Li, Z., Yang, M., Cui, X., Liu, M., Li, B., Zhang, Y., Ma, X., & Li, C. (2025). Biological bone and replacement materials in grinding: Force model and processing capability. *Intelligent and Sustainable Manufacturing*, *2*(1), 10003. DOI: 10.70322/ism.2025.10003

Kong, X., Li, Z., Chen, X., Yang, M., Ma, X., Cui, X., Liu, M. Z., Li, B. K., Dambatta, Y. S., Song, K. P., Wang, G., & Li, C. (2025). Convective heat transfer mechanism of bionic desert beetle micro-grinding tool/bone interface and micro-grinding experiment. *International Journal of Heat and Mass Transfer*, *253*, 127611. DOI: 10.1016/j.ijheatmasstransfer.2025.127611

Larbi, S., Bensaada, R., Bilek, A., & Djebali, S. (2015). Hygrothermal ageing effect on mechanical properties of FRP laminates. In. AIP Conference Proceedings: Vol. 1653. *No. 1* (p. 020066). AIP Publishing LLC. DOI: 10.1063/1.4914257

Lawal, S. A., Choudhury, I. A., & Nukman, Y. (2012). Application of vegetable oil-based metalworking fluids in machining ferrous metals—A review. *International Journal of Machine Tools & Manufacture*, *52*(1), 1–12. DOI: 10.1016/j.ijmachtools.2011.09.003

Leus, M., Gutowski, P., & Bachtiak-Radka, E. (2023). The effect of contact compliance of sliding pair on friction force reduction at longitudinal tangential vibrations. *Tribology International*, *187*, 108701. DOI: 10.1016/j.triboint.2023.108701

Li, C. (2023). *Thermodynamic mechanism of MQL grinding with nano bio-lubricant*. Springer Nature, 43-74.

Li, C., & Ali, H. M. (Eds.). (2019). *Enhanced heat transfer mechanism of nanofluid MQL cooling grinding*. IGI Global, 298-307.

Li, C., & Ali, H. M. (Eds.). (2019). *Enhanced heat transfer mechanism of nanofluid MQL cooling grinding*. IGI Global, 43-60.

Liang, X., & Wu, D. (2019). Tribological properties of carbon-fibre-reinforced plastic against tungsten carbide under dry condition. *Tribology International*, *134*, 118–128. DOI: 10.1016/j.triboint.2019.01.043

Liang, Y., Chen, Y., Chen, B., Fan, B., Yan, C., & Fu, Y. (2019). Feasibility of ultrasonic vibration assisted grinding for carbon fiber reinforced polymer with monolayer brazed grinding tools. *International Journal of Precision Engineering and Manufacturing*, *20*(7), 1083–1094. DOI: 10.1007/s12541-019-00135-8

Liang, Y., Chen, Y., Zhu, Y., Ji, J., & Ding, W. (2021). Investigations on tool clogging and surface integrity in ultrasonic vibration-assisted slot grinding of unidirectional CFRP. *International Journal of Advanced Manufacturing Technology*, *112*(5), 1557–1570. DOI: 10.1007/s00170-020-06364-w

Liao, Z., Abdelhafeez, A., Li, H., Yang, Y., Diaz, O. G., & Axinte, D. (2019). State-of-the-art of surface integrity in machining of metal matrix composites. *International Journal of Machine Tools & Manufacture*, *143*, 63–91. DOI: 10.1016/j.ijmachtools.2019.05.006

Li, B., Li, C., Zhang, Y., Wang, Y., Jia, D., Yang, M., Zhang, N., Wu, Q., Han, Z., & Sun, K. (2017). Heat transfer performance of MQL grinding with different nanofluids for Ni-based alloys using vegetable oil. *Journal of Cleaner Production*, *154*, 1–11. DOI: 10.1016/j.jclepro.2017.03.213

Li, H. N., Xie, K. G., Wu, B., & Zhu, W. Q. (2020). Generation of textured diamond abrasive tools by continuous-wave CO2 laser: Laser parameter effects and optimisation. *Journal of Materials Processing Technology*, *275*, 116279. DOI: 10.1016/j.jmatprotec.2019.116279

Li, H., Chen, T., Duan, Z., Zhang, Y., & Li, H. (2022b). A grinding force model in two-dimensional ultrasonic-assisted grinding of silicon carbide. *Journal of Materials Processing Technology*, *304*, 117568. DOI: 10.1016/j.jmatprotec.2022.117568

Li, H., Lin, B., Wan, S., Wang, Y., & Zhang, X. (2016). An experimental investigation on ultrasonic vibration-assisted grinding of SiO2f/SiO2 composites. *Materials and Manufacturing Processes*, *31*(7), 887–895. DOI: 10.1080/10426914.2015.1090586

Li, J., Zhang, K., Liu, P., & Li, Y. (2018). Tribological Behavior and Microstructure of Carbon Fiber–Reinforced Polymer against Ti6Al4V Alloy in Fretting Contact. *Tribology Transactions*, *61*(2), 256–268. DOI: 10.1080/10402004.2017.1313471

Li, K. M., Hu, Y. M., Yang, Z. Y., & Chen, M. Y. (2012). Experimental study on vibration-assisted grinding. *Journal of Manufacturing Science and Engineering*, *134*(4), 041009. DOI: 10.1115/1.4007102

Li, L., Zhang, Y., Cui, X., Said, Z., Sharma, S., Liu, M., Gao, T., Zhou, Z., Wang, X., & Li, C. (2023). Mechanical behavior and modeling of grinding force: A comparative analysis. *Journal of Manufacturing Processes*, *102*, 921–954. DOI: 10.1016/j.jmapro.2023.07.074

Li, M., Yu, T., Zhang, R., Yang, L., Ma, Z., Li, B., Wang, X. Z., Wang, W., & Zhao, J. (2020). Experimental evaluation of an eco-friendly grinding process combining minimum quantity lubrication and graphene-enhanced plant-oil-based cutting fluid. *Journal of Cleaner Production*, *244*, 118747. DOI: 10.1016/j.jclepro.2019.118747

Lin, H., & Zhou, M. (2024). Fracture mechanism of ultrasound-assisted scratching 2D-SiCf/SiC composite fibers with different fiber orientations. *Journal of Materials Research and Technology*, *30*, 9048–9060. DOI: 10.1016/j.jmrt.2024.05.225

Li, P., Zhao, H., Yan, X., Wen, Z., Wang, R., & Ding, H. (2024). An oil-on-water minimum quantity lubrication technology for high-quality grinding of carbon fiber reinforced thermoplastic composites. *Polymer Composites*, *45*(15), 14077–14096. DOI: 10.1002/pc.28755

Liu, M., & Li, C. (2024). Thermodynamic mechanism of cryogenic air minimum quantity lubrication grinding. IGI Global, 136-176. DOI: 10.4018/979-8-3693-3378-5

Liu, C., Ren, J., Shi, K., & Zhang, Y. (2023). Investigation of fracture mechanism evolution model for UD-CFRP and MD-CFRP during the milling process. *Composite Structures*, *306*, 116585. DOI: 10.1016/j.compstruct.2022.116585

Liu, C., Zhang, Y., Zhu, L., Li, Q., Shu, X., Qin, S., Wang, D., & Shi, W. (2025). A Review of Ultrasonic Vibration-Assisted Grinding for Advanced Materials. *Intelligent and Sustainable Manufacturing*, *2*(1), 10001. DOI: 10.70322/ism.2025.10001

Liu, D., Ni, C., Wang, Y., & Zhu, L. (2024). Review of serrated chip characteristics and formation mechanism from conventional to additively manufactured titanium alloys. *Journal of Alloys and Compounds*, *970*, 172573. DOI: 10.1016/j.jallcom.2023.172573

Liu, H., Lin, J., Sun, Y., & Zhang, J. (2019b). Micro model of carbon fiber/cyanate ester composites and analysis of machining damage mechanism. *Chinese Journal of Mechanical Engineering*, *32*(1), 52. DOI: 10.1186/s10033-019-0364-4

Liu, M., Zhang, Y., Yang, M., Gao, T., Cui, X., Wang, X., & Sharma, S. (2023). Analysis of grinding mechanics and improved grinding force model based on randomized grain geometric characteristics. *Chinese Journal of Aeronautics*, *36*(7), 160–193. DOI: 10.1016/j.cja.2022.11.005

Liu, Q., Huang, G., Cui, C., Tong, Z., & Xu, X. (2019a). Investigation of grinding mechanism of a 2D Cf/C–SiC composite by single-grain scratching. *Ceramics International, 45*(10), 13422–13430. DOI: 10.1016/j.ceramint.2019.04.041

Liu, Q., Huang, G., Fang, C., Cui, C., & Xu, X. (2017c). Experimental investigations on grinding characteristics and removal mechanisms of 2D–Cf/C-SiC composites based on reinforced fiber orientations. *Ceramics International, 43*(17), 15266–15274. DOI: 10.1016/j.ceramint.2017.08.064

Liu, Q., Huang, G., Xu, X., Fang, C., & Cui, C. (2017a). A study on the surface grinding of 2D C/SiC composites. *International Journal of Advanced Manufacturing Technology, 93*(5), 1595–1603. DOI: 10.1007/s00170-017-0626-1

Liu, Q., Huang, G., Xu, X., Fang, C., & Cui, C. (2018). Influence of grinding fiber angles on grinding of the 2D–Cf/C–SiC composites. *Ceramics International, 44*(11), 12774–12782. DOI: 10.1016/j.ceramint.2018.04.083

Liu, S., Chen, T., & Wu, C. (2017b). Rotary ultrasonic face grinding of carbon fiber reinforced plastic (CFRP): A study on cutting force model. *International Journal of Advanced Manufacturing Technology, 89*(1), 847–856. DOI: 10.1007/s00170-016-9151-x

Liu, W., Ni, H., Wang, P., & Chen, H. (2020). Investigation on the tribological performance of micro-dimples textured surface combined with longitudinal or transverse vibration under hydrodynamic lubrication. *International Journal of Mechanical Sciences, 174*, 105474. DOI: 10.1016/j.ijmecsci.2020.105474

Liu, Y., Liu, Z. B., Wang, X. B., & Huang, T. (2021). Experimental study on cutting force and surface quality in ultrasonic vibration-assisted milling of C/SiC composites. *International Journal of Advanced Manufacturing Technology, 112*(7), 2003–2014.

Li, Y., Ge, X., Wang, H., Hu, Y., Ning, F., Cong, W., & Ren, C. (2019a). Study of material removal mechanisms in grinding of C/SiC composites via single-abrasive scratch tests. *Ceramics International, 45*(4), 4729–4738. DOI: 10.1016/j.ceramint.2018.11.165

Li, Y., Ren, C., Wang, H., Hu, Y., Ning, F., Wang, X., & Cong, W. (2019b). Edge surface grinding of CFRP composites using rotary ultrasonic machining: Comparison of two machining methods. *International Journal of Advanced Manufacturing Technology, 100*(9), 3237–3248. DOI: 10.1007/s00170-018-2901-1

Li, Z., Zhu, L., Yang, Z., Ma, J., & Cao, W. (2022a). Investigation of tool-workpiece contact rate and milling force in elliptical ultrasonic vibration-assisted milling. *International Journal of Advanced Manufacturing Technology*, *118*(1), 585–601. DOI: 10.1007/s00170-021-07946-y

Lomakin, E. V., & Fedulov, B. N. (2013). Plane strain extension of a strip made of a material with stress state type dependent properties and weakened by cuts with circular base. *Mechanics of Solids*, *48*(4), 424–430. DOI: 10.3103/S0025654413040109

Long, Y., Huang, D., Tao, N., Chen, G., He, Z., & Zhang, H. (2025). A review on laser drilling of CFRP laminates: Fundamental mechanisms, defect formation, strategies for high-quality processing. *Optics & Laser Technology*, *192*, 113842. DOI: 10.1016/j.optlastec.2025.113842

Luna, G. G., Axinte, D., & Novovic, D. (2020). Influence of grit geometry and fibre orientation on the abrasive material removal mechanisms of SiC/SiC Ceramic Matrix Composites (CMCs). *International Journal of Machine Tools & Manufacture*, *157*, 103580. DOI: 10.1016/j.ijmachtools.2020.103580

Lu, S., & Somasundaran, P. (2008). Tunable synergism/antagonism in a mixed nonionic/anionic surfactant layer at the solid/liquid interface. *Langmuir*, *24*(8), 3874–3879. DOI: 10.1021/la703233d PMID: 18324849

Ma, G., Kang, R., Dong, Z., Yin, S., Bao, Y., & Guo, D. (2020). Hole quality in longitudinal–torsional coupled ultrasonic vibration assisted drilling of carbon fiber reinforced plastics. *Frontiers of Mechanical Engineering*, *15*(4), 538–546. DOI: 10.1007/s11465-020-0598-y

Mahrous, E., Jarauta, A., Chan, T., Ryzhakov, P., Weber, A. Z., Roy, R. V., & Secanell, M. (2020). A particle finite element-based model for droplet spreading analysis. *Physics of Fluids*, *32*(4), 042106. DOI: 10.1063/5.0006033

Malkin, S., & Guo, C. (2007). Thermal analysis of grinding. *CIRP Annals*, *56*(2), 760–782. DOI: 10.1016/j.cirp.2007.10.005

Manservisi, S., & Scardovelli, R. (2009). A variational approach to the contact angle dynamics of spreading droplets. *Computers & Fluids*, *38*(2), 406–424. DOI: 10.1016/j.compfluid.2008.05.001

Mao, C., Zou, H., Zhou, X., Huang, Y., Gan, H., & Zhou, Z. (2014). Analysis of suspension stability for nanofluid applied in minimum quantity lubricant grinding. *International Journal of Advanced Manufacturing Technology*, *71*(9), 2073–2081. DOI: 10.1007/s00170-014-5642-9

Mei, J., Diaz, O. G., & Axinte, D. A. (2017). An approach on capturing the influence of the stochasticity of fibre distributions for modelling the variability of cutting forces in composite materials. *Composites. Part B, Engineering*, *125*, 27–38. DOI: 10.1016/j.compositesb.2017.05.056

Meng, Q. Y., Guo, B., Zhao, Q. L., Li, H. N., Jackson, M. J., Linke, B. S., & Luo, X. (2023). Mark, j., Barara, S., Luo, X.C. Modelling of grinding mechanics: A review [J]. *Chinese Journal of Aeronautics*, *36*(7), 25–39. DOI: 10.1016/j.cja.2022.10.006

Merino-Pérez, J. L., Royer, R., Ayvar-Soberanis, S., Merson, E., & Hodzic, A. (2015). On the temperatures developed in CFRP drilling using uncoated WC-Co tools Part I: Workpiece constituents, cutting speed and heat dissipation. *Composite Structures*, *123*, 161–168. DOI: 10.1016/j.compstruct.2014.12.033

Mia, M., Gupta, M. K., Singh, G., Królczyk, G., & Pimenov, D. Y. (2018). An approach to cleaner production for machining hardened steel using different cooling-lubrication conditions. *Journal of Cleaner Production*, *187*, 1069–1081. DOI: 10.1016/j.jclepro.2018.03.279

Molaie, M. M., Akbari, J., & Movahhedy, M. R. (2016). Ultrasonic assisted grinding process with minimum quantity lubrication using oil-based nanofluids. *Journal of Cleaner Production*, *129*, 212–222. DOI: 10.1016/j.jclepro.2016.04.080

Moore, V. C., Strano, M. S., Haroz, E. H., Hauge, R. H., Smalley, R. E., Schmidt, J., & Talmon, Y. (2003). Individually suspended single-walled carbon nanotubes in various surfactants. *Nano Letters*, *3*(10), 1379–1382. DOI: 10.1021/nl034524j

Morozov, M., & Manor, O. (2018). Vibration-driven mass transfer and dynamic wetting. *Current Opinion in Colloid & Interface Science*, *36*, 37–45. DOI: 10.1016/j.cocis.2017.12.002

Mughal, K., Mughal, M. P., Farooq, M. U., Anwar, S., & Ammarullah, M. I. (2023). Using nano-fluids minimum quantity lubrication (NF-MQL) to improve tool wear characteristics for efficient machining of CFRP/Ti6Al4V aeronautical structural composite. *Processes (Basel, Switzerland)*, *11*(5), 1540. DOI: 10.3390/pr11051540

Naik, N. K., & Kumar, R. S. (1999). Compressive strength of unidirectional composites: Evaluation and comparison of prediction models. *Composite Structures*, *46*(3), 299–308. DOI: 10.1016/S0263-8223(99)00098-7

Namlu, R. H., Lotfi, B., & Kılıç, S. E. (2024). Enhancing machining efficiency of Ti-6Al-4V through multi-axial ultrasonic vibration-assisted machining and hybrid nanofluid minimum quantity lubrication. *Journal of Manufacturing Processes*, *119*, 348–371. DOI: 10.1016/j.jmapro.2024.03.073

Nandagopal, R. A., Boay, C. G., & Narasimalu, S. (2020). An empirical model to predict the strength degradation of the hygrothermal aged CFRP material. *Composite Structures*, *236*, 111876. DOI: 10.1016/j.compstruct.2020.111876

Naskar, A., Choudhary, A., & Paul, S. (2021). Surface generation in ultrasonic-assisted high-speed superabrasive grinding under minimum quantity cooling lubrication with various fluids. *Tribology International*, *156*, 106815. DOI: 10.1016/j.triboint.2020.106815

Nayak, D., Bhatnagar, N., & Mahajan, P. (2005). Machining studies of UD-FRP composites part 2: Finite element analysis. *Machining Science and Technology*, *9*(4), 503–528. DOI: 10.1080/10910340500398183

Ni, C., Zhu, L., Liu, C., & Yang, Z. (2018). Analytical modeling of tool-workpiece contact rate and experimental study in ultrasonic vibration-assisted milling of Ti–6Al–4V. *International Journal of Mechanical Sciences*, *142*, 97–111. DOI: 10.1016/j.ijmecsci.2018.04.037

Ni, C., Zhu, L., & Yang, Z. (2019). Comparative investigation of tool wear mechanism and corresponding machined surface characterization in feed-direction ultrasonic vibration assisted milling of Ti–6Al–4V from dynamic view. *Wear*, *436*, 203006. DOI: 10.1016/j.wear.2019.203006

Ning, F., Cong, W., Wang, H., Hu, Y., Hu, Z., & Pei, Z. (2017). Surface grinding of CFRP composites with rotary ultrasonic machining: A mechanistic model on cutting force in the feed direction. *International Journal of Advanced Manufacturing Technology*, *92*(1), 1217–1229. DOI: 10.1007/s00170-017-0149-9

Ning, F., Wang, H., & Cong, W. (2019). Rotary ultrasonic machining of carbon fiber reinforced plastic composites: A study on fiber material removal mechanism through single-grain scratching. *International Journal of Advanced Manufacturing Technology*, *103*(1), 1095–1104. DOI: 10.1007/s00170-019-03433-7

Ning, Z., Liu, R., Elhajjar, R. F., & Wang, F. (2017). Micro-modeling of thermal properties in carbon fibers reinforced polymer composites with fiber breaks or delamination. *Composites. Part B, Engineering*, *114*, 247–255. DOI: 10.1016/j.compositesb.2017.01.036

Niu, Q., Rong, J., Jing, L., Gao, H., Tang, S., Qiu, X., Liu, L., Wang, X., & Dai, F. (2023). Study on force-thermal characteristics and cutting performance of titanium alloy milled by ultrasonic vibration and minimum quantity lubrication. *Journal of Manufacturing Processes*, *95*, 115–130. DOI: 10.1016/j.jmapro.2023.04.002

Pereira, A. C., Monteiro, S. N., da Silva, L. R. R., Binali, R., Kuntoğlu, M., Machado, A. R., & Pimenov, D. Y. (2025). Machinability and surface integrity of glass fiber reinforced plastic composite: A review. *Journal of Materials Research and Technology*, *35*, 6446–6467. DOI: 10.1016/j.jmrt.2025.02.215

Pervaiz, S., Kannan, S., Huo, D., & Mamidala, R. (2020). Ecofriendly inclined drilling of carbon fiber-reinforced polymers (CFRP). *International Journal of Advanced Manufacturing Technology*, *111*(7), 2127–2153. DOI: 10.1007/s00170-020-06203-y

Qian, M., Xiao, J., Wang, G., Huang, P., Chen, Z., & Han, G. (2020). Evaluation of heat generation using a microscopic cutting model with thermo-mechanical coupling for carbon fiber reinforced polymer composites. *Journal of Reinforced Plastics and Composites*, *39*(21-22), 793–804. DOI: 10.1177/0731684420931589

Qiao, G., Cheng, Z., Zheng, W., Yi, S., & Zhang, F. (2022). Grinding force model for longitudinal-torsional ultrasonic-assisted face grinding of ceramic matrix composites. *International Journal of Advanced Manufacturing Technology*, *120*(11), 7721–7733. DOI: 10.1007/s00170-022-09174-4

Qiao, G., Xiong, X., Yang, J., Zhen, D., & Zhang, F. (2025). Cooling and lubrication mechanisms of nanofluid minimum quantity lubrication milling based on interfacial micro-equivalent contact modelling. *Journal of Manufacturing Processes*, *141*, 1551–1563. DOI: 10.1016/j.jmapro.2025.03.080

Qiu, X., Li, P., Li, C., Niu, Q., Chen, A., Ouyang, P., & Ko, T. J. (2018). Study on chisel edge drilling behavior and step drill structure on delamination in drilling CFRP. *Composite Structures*, *203*, 404–413. DOI: 10.1016/j.compstruct.2018.07.007

Qu, M., Yao, Y., He, J., Ma, X., Liu, S., Feng, J., & Hou, L. (2016). Tribological performance of functionalized ionic liquid and Cu microparticles as lubricating additives in sunflower seed oil. *Tribology International*, *104*, 166–174. DOI: 10.1016/j.triboint.2016.08.035

Qu, S., Gong, Y., Yang, Y., Sun, Y., Wen, X., & Qi, Y. (2020c). Investigating minimum quantity lubrication in unidirectional Cf/SiC composite grinding. *Ceramics International*, *46*(3), 3582–3591. DOI: 10.1016/j.ceramint.2019.10.076

Qu, S., Gong, Y., Yang, Y., Wang, W., Liang, C., & Han, B. (2020b). An investigation of carbon nanofluid minimum quantity lubrication for grinding unidirectional carbon fibre-reinforced ceramic matrix composites. *Journal of Cleaner Production*, *249*, 119353. DOI: 10.1016/j.jclepro.2019.119353

Qu, S., Gong, Y., Yang, Y., Xu, Y., Wang, W., Xin, B., & Pang, S. (2020a). Mechanical model and removal mechanism of unidirectional carbon fibre-reinforced ceramic composites. *International Journal of Mechanical Sciences*, *173*, 105465. DOI: 10.1016/j.ijmecsci.2020.105465

Qu, S., Wei, C., Yang, Y., Yao, P., Chu, D., Gong, Y., Zhao, D., & Zhang, X. (2024). Grinding mechanism and surface quality evaluation strategy of single crystal 4H-SiC. *Tribology International*, *194*, 109515. DOI: 10.1016/j.triboint.2024.109515

Qu, S., Yang, Y., Yao, P., Li, L., Sun, Y., & Chu, D. (2025). Fiber reinforced ceramic matrix composites: From the controlled fabrication to precision machining. *International Journal of Extreme Manufacturing*, *7*(6), 062004. DOI: 10.1088/2631-7990/adeee1

Qu, S., Yao, P., Gong, Y., Chu, D., Yang, Y., Li, C., Wang, Z., Zhang, X., & Hou, Y. (2022). Environmentally friendly grinding of C/SiCs using carbon nanofluid minimum quantity lubrication technology. *Journal of Cleaner Production*, *366*, 132898. DOI: 10.1016/j.jclepro.2022.132898

Qu, S., Yao, P., Gong, Y., Yang, Y., Chu, D., & Zhu, Q. (2022). Modelling and grinding characteristics of unidirectional C–SiCs. *Ceramics International*, *48*(6), 8314–8324. DOI: 10.1016/j.ceramint.2021.12.036

Rabiei, F., Rahimi, A. R., Hadad, M. J., & Saberi, A. (2017). Experimental evaluation of coolant-lubricant properties of nanofluids in ultrasonic assistant MQL grinding. *International Journal of Advanced Manufacturing Technology*, *93*(9-12), 3935–3953. DOI: 10.1007/s00170-017-0774-3

Rahim, E. A., & Sasahara, H. (2011). A study of the effect of palm oil as MQL lubricant on high speed drilling of titanium alloys. *Tribology International*, *44*(3), 309–317. DOI: 10.1016/j.triboint.2010.10.032

Rahimzadeh, A., & Eslamian, M. (2017). Experimental study on the evaporation of sessile droplets excited by vertical and horizontal ultrasonic vibration. *International Journal of Heat and Mass Transfer*, *114*, 786–795. DOI: 10.1016/j.ijheatmasstransfer.2017.06.099

Ran, Y., Kang, R., Dong, Z., Jin, Z., & Bao, Y. (2023). Ultrasonic assisted grinding force model considering anisotropy of SiCf/SiC composites. *International Journal of Mechanical Sciences*, *250*, 108311. DOI: 10.1016/j.ijmecsci.2023.108311

Ran, Y., Kang, R., Sun, J., Dong, Z., Wang, H., & Bao, Y. (2024). Insight into crack propagation induced by fiber orientations during single grain scratching of SiCf/SiC composites using FEM. *Composites. Part A, Applied Science and Manufacturing*, *177*, 107928. DOI: 10.1016/j.compositesa.2023.107928

Rao, X., Zhang, F., & Xu, J. (2023). Wheel wear behavior and its influence on grinding performance in electrical discharge diamond grinding of reaction-bonded silicon carbide. *Journal of Manufacturing Processes*, *85*, 904–914. DOI: 10.1016/j.jmapro.2022.12.024

Rastogi, R., Kaushal, R., Tripathi, S. K., Sharma, A. L., Kaur, I., & Bharadwaj, L. M. (2008). Comparative study of carbon nanotube dispersion using surfactants. *Journal of Colloid and Interface Science*, *328*(2), 421–428. DOI: 10.1016/j.jcis.2008.09.015 PMID: 18848704

Reifsnider, K. L., & Case, S. W. (2002). *Damage tolerance and durability of material systems.*

Ren, W., & e, W. (2007). Boundary conditions for the moving contact line problem. *Physics of Fluids*, *19*(2), 022101. DOI: 10.1063/1.2646754

Rodriguez, R. L., Lopes, J. C., Hildebrandt, R. A., Perez, R. R. V., Diniz, A. E., de Angelo Sanchez, L. E., Rodrigues, A. R., de Mello, H. J., de Aguiar, P. R., & Bianchi, E. C. (2019). Evaluation of grinding process using simultaneously MQL technique and cleaning jet on grinding wheel surface. *Journal of Materials Processing Technology*, *271*, 357–367. DOI: 10.1016/j.jmatprotec.2019.03.019

Rodriguez, R. L., Lopes, J. C., Mancini, S. D., de Ângelo Sanchez, L. E., de Almeida Varasquim, F. M. F., Volpato, R. S., de Mello, H. J., de Aguiar, P. R., & Bianchi, E. C. (2019). Contribution for minimization the usage of cutting fluids in CFRP grinding. *International Journal of Advanced Manufacturing Technology*, *103*(1), 487–497. DOI: 10.1007/s00170-019-03529-0

Saha, S., Deb, S., & Bandyopadhyay, P. P. (2023). Tool wear induced burr formation and concomitant reduction in MQL wetting capability in micro-milling. *International Journal of Mechanical Sciences*, *245*, 108095. DOI: 10.1016/j.ijmecsci.2022.108095

Said, Z., Arora, S., Farooq, S., Sundar, L. S., Li, C., & Allouhi, A. (2022). Recent advances on improved optical, thermal, and radiative characteristics of plasmonic nanofluids: Academic insights and perspectives. *Solar Energy Materials and Solar Cells*, *236*, 111504. DOI: 10.1016/j.solmat.2021.111504

Said, Z., Sharma, P., Sundar, L. S., Afzal, A., & Li, C. (2021). Synthesis, stability, thermophysical properties and AI approach for predictive modelling of Fe3O4 coated MWCNT hybrid nanofluids. *Journal of Molecular Liquids, 340*, 117291. DOI: 10.1016/j.molliq.2021.117291

Sambhav, K., Kumar, A., & Choudhury, S. K. (2011). Mechanistic force modeling of single point cutting tool in terms of grinding angles. *International Journal of Machine Tools & Manufacture, 51*(10-11), 775–786. DOI: 10.1016/j.ijmachtools.2011.06.007

Santiuste, C., Miguélez, H., & Soldani, X. (2011). Out-of-plane failure mechanisms in LFRP composite cutting. *Composite Structures, 93*(11), 2706–2713. DOI: 10.1016/j.compstruct.2011.05.034

Sarkar, N., & Chaudhuri, B. B. (1992). An efficient approach to estimate fractal dimension of textural images. *Pattern Recognition, 25*(9), 1035–1041. DOI: 10.1016/0031-3203(92)90066-R

Schön, J. (2000). Coefficient of friction of composite delamination surfaces. *Wear, 237*(1), 77–89. DOI: 10.1016/S0043-1648(99)00315-4

Scida, D., Assarar, M., Poilâne, C., & Ayad, R. (2013). Influence of hygrothermal ageing on the damage mechanisms of flax-fibre reinforced epoxy composite. *Composites. Part B, Engineering, 48*, 51–58. DOI: 10.1016/j.compositesb.2012.12.010

Senthilkumar, M., Prabukarthi, A., & Krishnaraj, V. (2018). Machining of CFRP/ Ti6Al4V stacks under minimal quantity lubricating condition. *Journal of Mechanical Science and Technology, 32*(8), 3787–3796. DOI: 10.1007/s12206-018-0731-6

Seward, O., Cepero-Mejías, F., Fairclough, J. P. A., & Kerrigan, K. (2022). Development of a novel friction model for machining simulations in unidirectional composite materials. *Polymers, 14*(5), 847. DOI: 10.3390/polym14050847 PMID: 35267670

Sharma, A. K., Singh, R. K., Dixit, A. R., & Tiwari, A. K. (2017). Novel uses of alumina-MoS2 hybrid nanoparticle enriched cutting fluid in hard turning of AISI 304 steel. *Journal of Manufacturing Processes, 30*, 467–482. DOI: 10.1016/j.jmapro.2017.10.016

Shaw, M. C., & Cookson, J. O. (2005). Metal cutting principles: Vol. 2. *No. 3.* Oxford university press.

Sheikh-Ahmad, J. Y., Almaskari, F., Hafeez, F., & Meng, F. (2019). Evaluation of heat partition in machining CFRP using inverse method. *Machining Science and Technology, 23*(4), 530–546. DOI: 10.1080/10910344.2019.1575401

Sheikh-Ahmad, J., & Mohammed, J. (2014). Optimization of process parameters in diamond abrasive machining of carbon fiber-reinforced epoxy. *Materials and Manufacturing Processes*, *29*(11-12), 1361–1366. DOI: 10.1080/10426914.2014.941869

Sheikh-Ahmad, J., Urban, N., & Cheraghi, H. (2012). Machining damage in edge trimming of CFRP. *Materials and Manufacturing Processes*, *27*(7), 802–808. DOI: 10.1080/10426914.2011.648253

Shen, B., & Shih, A. J. (2009). Minimum quantity lubrication (MQL) grinding using vitrified CBN wheels. *Transactions of NAMRI/SME, 37*, 129-136.

Shen, X., & Zhang, J. (2013). Studies on friction and wear properties of surface produced by ultrasonic vibration-assisted milling. *International Journal of Advanced Manufacturing Technology*, *67*(1), 349–356. DOI: 10.1007/s00170-012-4488-2

Shi, H., Yuan, S., Li, Z., Song, H., & Qian, J. (2019). Evaluation of surface roughness based on sampling array for rotary ultrasonic machining of carbon fiber reinforced polymer composites. *Measurement*, *138*, 175–181. DOI: 10.1016/j.measurement.2019.02.002

Shi, H., Yuan, S., Zhang, C., Chen, B., Li, Q., Li, Z., Zhu, G., & Qian, J. (2020). A cutting force prediction model for rotary ultrasonic side grinding of CFRP composites considering coexistence of brittleness and ductility. *International Journal of Advanced Manufacturing Technology*, *106*(5), 2403–2414. DOI: 10.1007/s00170-019-04730-x

Shi, W., Ma, T., Li, J., Li, J., Zhang, X., Guo, B., Wen, Z., Li, J., Wang, L., & Cao, S. (2025). A review of research on hole machining technology for FRP/LA laminated materials: From mechanical drilling to non-traditional machining. *International Journal of Advanced Manufacturing Technology*, *137*(5), 2109–2156. DOI: 10.1007/s00170-025-15181-y

Shu, J., Liao, W., Zheng, K., & Hussein Youssef, A. M. F. (2021). Surface morphology on carbon fiber composites by rotary ultrasonic milling. *Machining Science and Technology*, *25*(5), 721–737. DOI: 10.1080/10910344.2021.1971705

Silva, L. R., Bianchi, E. C., Fusse, R. Y., Catai, R. E., Franca, T. V., & Aguiar, P. R. (2007). Analysis of surface integrity for minimum quantity lubricant—MQL in grinding. *International Journal of Machine Tools & Manufacture*, *47*(2), 412–418. DOI: 10.1016/j.ijmachtools.2006.03.015

Song, Y. X., Li, C. H., Zhou, Z. M., Liu, B., Sharma, S., Dambatta, Y. S., Zhang, Y., Yang, M., Gao, T., Liu, M., Cui, X., Wang, X., Xu, W., Li, R., & Wang, D. Z. (2025). Nanobiolubricant grinding: A comprehensive review. *Advances in Manufacturing*, *13*(1), 1–42. DOI: 10.1007/s40436-023-00477-7

Soo, S. L., Shyha, I. S., Barnett, T., Aspinwall, D. K., & Sim, W. M. (2012). Grinding performance and workpiece integrity when superabrasive edge routing carbon fibre reinforced plastic (CFRP) composites. *CIRP Annals*, *61*(1), 295–298. DOI: 10.1016/j.cirp.2012.03.042

Sun, X., Wu, Z. Q., & Tian, Y. (2003). *Fractal Principle and Application*. University of Science and Technology of China Press. 53-88.

Sun, J., Qin, F., Chen, P., & An, T. (2016). A predictive model of grinding force in silicon wafer self-rotating grinding. *International Journal of Machine Tools & Manufacture*, *109*, 74–86. DOI: 10.1016/j.ijmachtools.2016.07.009

Sun, Y. J., Gong, H., Wang, Y., & Ni, H. (2021). Tribological viewpoint of the cutting force in rotary ultrasonic vibration-assisted side milling (RUVSM). *International Journal of Advanced Manufacturing Technology*, *114*(9), 2839–2848. DOI: 10.1007/s00170-021-06966-y

Su, Y. (2019). Effect of the cutting speed on the cutting mechanism in machining CFRP. *Composite Structures*, *220*, 662–676. DOI: 10.1016/j.compstruct.2019.04.052

Tawakoli, T., Azarhoushang, B., & Rabiey, M. (2009). Ultrasonic assisted dry grinding of 42CrMo4. *International Journal of Advanced Manufacturing Technology*, *42*(9), 883–891. DOI: 10.1007/s00170-008-1646-7

Tian, J., Bao, Y., Chen, K., Dong, Z., Kang, R., Guo, D., & Sun, J. (2025). Prediction and analysis of grinding burr of CFRP circular tube with a rounded corner grinding wheel. *Journal of Manufacturing Processes*, *133*, 249–259. DOI: 10.1016/j.jmapro.2024.11.059

Tian, J., Kang, R., Dong, Z., Liu, Z., Guo, D., & Bao, Y. (2023). Multi-scale machining damages of CFRP circular cell honeycomb during end face machining. *Journal of Manufacturing Processes*, *86*, 282–293. DOI: 10.1016/j.jmapro.2023.01.006

Verma, G. C., & Pandey, P. M. (2019). Machining forces in ultrasonic-vibration assisted end milling. *Ultrasonics*, *94*, 350–363. DOI: 10.1016/j.ultras.2018.07.004 PMID: 30029793

Virdi, R. L., Chatha, S. S., & Singh, H. (2021). Experimental investigations on the tribological and lubrication behaviour of minimum quantity lubrication technique in grinding of Inconel 718 alloy. *Tribology International*, *153*, 106581. DOI: 10.1016/j.triboint.2020.106581

Voss, R., Henerichs, M., & Kuster, F. (2016). Comparison of conventional drilling and orbital drilling in machining carbon fibre reinforced plastics (CFRP). *CIRP Annals*, *65*(1), 137–140. DOI: 10.1016/j.cirp.2016.04.001

Voss, R., Seeholzer, L., Kuster, F., & Wegener, K. (2019). Analytical force model for orthogonal machining of unidirectional carbon fibre reinforced polymers (CFRP) as a function of the fibre orientation. *Journal of Materials Processing Technology*, *263*, 440–469. DOI: 10.1016/j.jmatprotec.2018.08.001

Wang, C., Li, P., Li, S., Qiu, X., Niu, Q., Li, C., & Ko, T. J. (2023). Study on the mechanism and performance of longitudinal-torsional ultrasonic vibration assisted drilling CFRP/Ti stack. *Journal of Manufacturing Processes*, *92*, 453–465. DOI: 10.1016/j.jmapro.2023.03.005

Wang, D., Jiao, F., & Mao, X. (2020). Mechanics of thrust force on chisel edge in carbon fiber reinforced polymer (CFRP) drilling based on bending failure theory. *International Journal of Mechanical Sciences*, *169*, 105336. DOI: 10.1016/j.ijmecsci.2019.105336

Wang, D., Lu, S., Xu, D., & Zhang, Y. (2020d). Research on material removal mechanism of C/SiC composites in ultrasound vibration-assisted grinding. *Materials (Basel)*, *13*(8), 1918. DOI: 10.3390/ma13081918 PMID: 32325807

Wang, F. J., Zhang, B. Y., Ma, J. W., Bi, G. J., & Hu, H. B. (2019a). Computation of the distribution of the fiber-matrix interface cracks in the edge trimming of CFRP. *Applied Composite Materials*, *26*(1), 159–186. DOI: 10.1007/s10443-018-9687-z

Wang, F., Jin, X., Wang, X., Jia, Z., & Zhang, B. (2022). Numerical investigation on the influences of processing conditions on damage in the CFRTP cutting using a novel Elastic–plastic damage model. *Applied Composite Materials*, *29*(5), 2063–2094. DOI: 10.1007/s10443-022-10041-4

Wang, F., Xuan, S., Chang, Z., Jin, K., Gao, Y., Wang, H., & Song, Q. (2024). Effect of grinding parameters on industrial robot grinding of CFRP and defect formation mechanism. *International Journal of Precision Engineering and Manufacturing-Green Technology*, *11*(2), 427–438. DOI: 10.1007/s40684-023-00561-0

Wang, H., Cong, W., Ning, F., & Hu, Y. (2018a). A study on the effects of machining variables in surface grinding of CFRP composites using rotary ultrasonic machining. *International Journal of Advanced Manufacturing Technology*, *95*(9), 3651–3663. DOI: 10.1007/s00170-017-1468-6

Wang, H., Hu, Y., Cong, W., & Hu, Z. (2019c). A mechanistic model on feeding-directional cutting force in surface grinding of CFRP composites using rotary ultrasonic machining with horizontal ultrasonic vibration. *International Journal of Mechanical Sciences*, *155*, 450–460. DOI: 10.1016/j.ijmecsci.2019.03.009

Wang, H., Hu, Y., Cong, W., Hu, Z., & Wang, Y. (2020c). A novel investigation on horizontal and 3D elliptical ultrasonic vibrations in rotary ultrasonic surface machining of carbon fiber reinforced plastic composites. *Journal of Manufacturing Processes*, *52*, 12–25. DOI: 10.1016/j.jmapro.2020.01.027

Wang, H., Ning, F., Hu, Y., & Cong, W. (2018b). Surface grinding of CFRP composites using rotary ultrasonic machining: A comparison of workpiece machining orientations. *International Journal of Advanced Manufacturing Technology*, *95*(5), 2917–2930. DOI: 10.1007/s00170-017-1401-z

Wang, H., Ning, F., Hu, Y., Fernando, P. K. S. C., Pei, Z. J., & Cong, W. (2016a). Surface grinding of carbon fiber–reinforced plastic composites using rotary ultrasonic machining: Effects of tool variables. *Advances in Mechanical Engineering*, *8*(9), 1687814016670284. DOI: 10.1177/1687814016670284

Wang, H., Ning, F., Hu, Y., Li, Y., Wang, X., & Cong, W. (2018). Edge trimming of carbon fiber-reinforced plastic composites using rotary ultrasonic machining: Effects of tool orientations. *International Journal of Advanced Manufacturing Technology*, *98*(5), 1641–1653. DOI: 10.1007/s00170-018-2355-5

Wang, H., Pei, Z. J., & Cong, W. (2020a). A mechanistic cutting force model based on ductile and brittle fracture material removal modes for edge surface grinding of CFRP composites using rotary ultrasonic machining. *International Journal of Mechanical Sciences*, *176*, 105551. DOI: 10.1016/j.ijmecsci.2020.105551

Wang, H., Pei, Z. J., & Cong, W. (2020b). A feeding-directional cutting force model for end surface grinding of CFRP composites using rotary ultrasonic machining with elliptical ultrasonic vibration. *International Journal of Machine Tools & Manufacture*, *152*, 103540. DOI: 10.1016/j.ijmachtools.2020.103540

Wang, H., Zhang, D., Li, Y., & Cong, W. (2020e). The effects of elliptical ultrasonic vibration in surface machining of CFRP composites using rotary ultrasonic machining. *International Journal of Advanced Manufacturing Technology*, *106*(11), 5527–5538. DOI: 10.1007/s00170-020-04976-w

Wang, J., Zhang, J., & Feng, P. (2017). Effects of tool vibration on fiber fracture in rotary ultrasonic machining of C/SiC ceramic matrix composites. *Composites. Part B, Engineering*, *129*, 233–242. DOI: 10.1016/j.compositesb.2017.07.081

Wang, W., Zhou, Z., Wang, Q., Gao, B., & Mao, C. (2025b). Muti-energy field-assisted grinding of hard and brittle materials: Tools, equipment and mechanisms. *Intelligent and Sustainable Manufacturing*, *2*(2), 10022. DOI: 10.70322/ism.2025.10022

Wang, X. M., & Zhang, L. C. (2003). An experimental investigation into the orthogonal cutting of unidirectional fibre reinforced plastics. *International Journal of Machine Tools & Manufacture*, *43*(10), 1015–1022. DOI: 10.1016/S0890-6955(03)00090-7

Wang, X., Li, C., Zhang, Y., Ali, H. M., Sharma, S., Li, R., Yang, M., Said, Z., & Liu, X. (2022b). Tribology of enhanced turning using biolubricants: A comparative assessment. *Tribology International*, *174*, 107766. DOI: 10.1016/j.triboint.2022.107766

Wang, X., Li, C., Zhang, Y., Ding, W., Yang, M., Gao, T., Cao, H., Xu, X., Wang, D., Said, Z., Debnath, S., Jamil, M., & Ali, H. M. (2020f). Vegetable oil-based nanofluid minimum quantity lubrication turning: Academic review and perspectives. *Journal of Manufacturing Processes*, *59*, 76–97. DOI: 10.1016/j.jmapro.2020.09.044

Wang, X., Song, Y., Li, C., Zhang, Y., Ali, H. M., Sharma, S., Li, R. Z., Yang, M., Gao, T., Liu, M. Z., Cui, X., Said, Z., & Zhou, Z. (2024). Nanofluids application in machining: A comprehensive review. *International Journal of Advanced Manufacturing Technology*, *131*(5), 3113–3164. DOI: 10.1007/s00170-022-10767-2

Wang, X., Wang, C., Shen, X., & Sun, F. (2019). Tribological behaviors of the diamond films sliding against the T800/X850 CFRP laminates. *Wear*, *418*, 191–200. DOI: 10.1016/j.wear.2018.12.007

Wang, X., Wang, F., Gu, T., Jia, Z., & Shi, Y. (2021). Computational simulation of the damage response for machining long fibre reinforced plastic (LFRP) composite parts: A review. *Composites. Part A, Applied Science and Manufacturing*, *143*, 106296. DOI: 10.1016/j.compositesa.2021.106296

Wang, X., Xu, Q., Zheng, Q., Shao, Y., & Shen, J. (2025). Reviews of fiber-reinforced phenolic resin-based thermal protection materials for aircraft. *Energies*, *18*(4), 819. DOI: 10.3390/en18040819

Wang, X., Yang, M., Gao, T., Dong, L., Dambatta, Y. S., Liu, X., Yang, Y. Y., An, Q. L., Zhang, Y. B., & Li, C. (2025a). Lubricant transport mechanism and dynamics model for nepenthes-shaped biomimetic microtexture. *Chinese Journal of Mechanical Engineering*, *38*(1), 36. DOI: 10.1186/s10033-025-01197-8

Wang, Y., Fu, Z., Dong, Y., Liu, S., & Lu, S. (2018). Research on surface generating model in ultrasonic vibration-assisted grinding. *International Journal of Advanced Manufacturing Technology*, *96*(9-12), 3429–3436. DOI: 10.1007/s00170-018-1744-0

Wang, Y., Li, C., Zhang, Y., Yang, M., Li, B., Dong, L., & Wang, J. (2018c). Processing characteristics of vegetable oil-based nanofluid MQL for grinding different workpiece materials. *International Journal of Precision Engineering and Manufacturing-Green Technology*, *5*(2), 327–339. DOI: 10.1007/s40684-018-0035-4

Wang, Y., Li, C., Zhang, Y., Yang, M., Li, B., Jia, D., Hou, Y., & Mao, C. (2016). Experimental evaluation of the lubrication properties of the wheel/workpiece interface in minimum quantity lubrication (MQL) grinding using different types of vegetable oils. *Journal of Cleaner Production*, *127*, 487–499. DOI: 10.1016/j.jclepro.2016.03.121

Wang, Y., Sarin, V. K., Lin, B., Li, H., & Gillard, S. (2016b). Feasibility study of the ultrasonic vibration filing of carbon fiber reinforced silicon carbide composites. *International Journal of Machine Tools & Manufacture*, *101*, 10–17. DOI: 10.1016/j.ijmachtools.2015.11.003

Wang, Y., Wang, H., Wei, J., Lin, B., Xu, J., & Fang, S. (2019b). Finite element analysis of grinding process of long fiber reinforced ceramic matrix woven composites: Modeling, experimental verification and material removal mechanism. *Ceramics International*, *45*(13), 15920–15927. DOI: 10.1016/j.ceramint.2019.05.100

Wang, Z., Liang, Y., Li, H., & Yu, T. (2022a). Milling force and tool wear mechanisms on milling TC21 titanium alloy under different lubrication conditions. *International Journal of Advanced Manufacturing Technology*, *123*(1), 169–185. DOI: 10.1007/s00170-022-10108-3

Wei, C., He, C., Chen, G., Sun, Y., & Ren, C. (2023). Material removal mechanism and corresponding models in the grinding process: A critical review. *Journal of Manufacturing Processes*, *103*, 354–392. DOI: 10.1016/j.jmapro.2023.08.045

Wei, J., Wang, H., Lin, B., Sui, T., Zhao, F., & Fang, S. (2019a). Acoustic emission signal of fiber-reinforced composite grinding: Frequency components and damage pattern recognition. *International Journal of Advanced Manufacturing Technology*, *103*(1), 1391–1401. DOI: 10.1007/s00170-019-03645-x

Wei, J., Wang, H., Lin, B., Sui, T., Zhao, F., & Fang, S. (2019b). A force model in single grain grinding of long fiber reinforced woven composite. *International Journal of Advanced Manufacturing Technology*, *100*(1), 541–552. DOI: 10.1007/s00170-018-2719-x

Wen, Z., Zhao, H., Yan, X., Li, P., & Ding, H. (2024). Grinding performance investigation of multidirectional CF/PEEK composites—A comparative with unidirectional CF/PEEK. *Polymer Composites*, *45*(10), 8889–8904. DOI: 10.1002/pc.28383

Wen, Z., Zhao, H., Yan, X., Peng, Z., & Ding, H. (2023). Grinding performance study of CF/epoxy and CF/PEEK composites. *Polymer Composites*, *44*(8), 4565–4578. DOI: 10.1002/pc.27420

Wickramasinghe, K. C., Sasahara, H., Abd Rahim, E., & Perera, G. I. P. (2020). Green Metalworking Fluids for sustainable machining applications: A review. *Journal of Cleaner Production*, *257*, 120552. DOI: 10.1016/j.jclepro.2020.120552

Wu, W., Li, C., Yang, M., Zhang, Y., Jia, D., Hou, Y., Li, R., Cao, H., & Han, Z. (2019). Specific energy and G ratio of grinding cemented carbide under different cooling and lubrication conditions. *International Journal of Advanced Manufacturing Technology*, *105*(1), 67–82. DOI: 10.1007/s00170-019-04156-5

Wu, X., Gong, J., Gao, T., Wang, Y., Huang, R., & Hou, J. (2024). Experimental study on dynamic wetting behavior of liquid droplets on the aluminum surfaces excited by ultrasonic waves. *Applied Thermal Engineering*, *248*, 123209. DOI: 10.1016/j.applthermaleng.2024.123209

Wu, X., Li, C., Zhou, Z., Nie, X., Chen, Y., Zhang, Y., Cao, H., Liu, B., Zhang, N., Said, Z., Debnath, S., Jamil, M., Ali, H. M., & Sharma, S. (2021). Circulating purification of cutting fluid: An overview. *International Journal of Advanced Manufacturing Technology*, *117*(9), 2565–2600. DOI: 10.1007/s00170-021-07854-1 PMID: 34465936

Wu, Z., & Zhang, L. (2023). Analytical grinding force prediction with random abrasive grains of grinding wheels. *International Journal of Mechanical Sciences*, *250*, 108310. DOI: 10.1016/j.ijmecsci.2023.108310

Xiao, J., Wang, G., Su, H., Huang, P., & Chen, Z. (2022). A study of surface integrity in carbon fiber-reinforced polymer composites cutting with the coupling effect of the multiple machining parameters. *Journal of Composite Materials*, *56*(15), 2385–2397. DOI: 10.1177/00219983221099081

Xie, H., Yu, W., Li, Y., & Chen, L. (2011). Discussion on the thermal conductivity enhancement of nanofluids. *Nanoscale Research Letters*, *6*(1), 124. DOI: 10.1186/1556-276X-6-124 PMID: 21711638

Xie, Z., Liu, Z., Wang, B., Xin, M., Song, Q., & Jiang, L. (2021). Longitudinal amplitude effect on material removal mechanism of ultrasonic vibration-assisted milling 2.5 DC/SiC composites. *Ceramics International*, *47*(22), 32144–32152. DOI: 10.1016/j.ceramint.2021.08.106

Xue, F., Zheng, K., Liao, W., Shu, J., & Dong, S. (2021b). Investigation on fiber fracture mechanism of c/sic composites by rotary ultrasonic milling. *International Journal of Mechanical Sciences*, *191*, 106054. DOI: 10.1016/j.ijmecsci.2020.106054

Xue, F., Zheng, K., Liao, W., Shu, J., & Miao, D. (2021a). Experimental investigation on fatigue property at room temperature of C/SiC composites machined by rotary ultrasonic milling. *Journal of the European Ceramic Society*, *41*(6), 3341–3356. DOI: 10.1016/j.jeurceramsoc.2021.01.046

Xu, J., Feng, P., Feng, F., Zha, H., & Liang, G. (2021). Subsurface damage and burr improvements of aramid fiber reinforced plastics by using longitudinal–torsional ultrasonic vibration milling. *Journal of Materials Processing Technology*, *297*, 117265. DOI: 10.1016/j.jmatprotec.2021.117265

Xu, J., Huang, X., Chen, M., & Paulo Davim, J. (2020). Drilling characteristics of carbon/epoxy and carbon/polyimide composites. *Materials and Manufacturing Processes*, *35*(15), 1732–1740. DOI: 10.1080/10426914.2020.1784935

Xu, J., Lin, T., Li, L., Ji, M., Davim, J. P., Geier, N., & Chen, M. (2022). Numerical study of interface damage formation mechanisms in machining CFRP/Ti6Al4V stacks under different cutting sequence strategies. *Composite Structures*, *285*, 115236. DOI: 10.1016/j.compstruct.2022.115236

Xu, W., Gao, T., Liu, D., Xu, P., An, Q., Wang, Y., Ding, W., & Li, C. (2024). Progress in grinding mechanical modeling of fiber reinforced SiC ceramic matrix composites. *Acta Mater Compos Sin*, *41*(9), 4628–4653.

Xu, W., Li, C., Cui, X., Zhang, Y., Yang, M., Gao, T., Liu, M., Wang, X., Zhou, Z., Sharma, S., & Dambatta, Y. S. (2023). Atomization mechanism and machinability evaluation with electrically charged nanolubricant grinding of GH4169. *Journal of Manufacturing Processes*, *106*, 480–493. DOI: 10.1016/j.jmapro.2023.10.037

Xu, W., Li, C., Zhang, Y., Ali, H. M., Sharma, S., Li, R., Yang, M., Gao, T., Liu, M., Wang, X., Said, Z., Liu, X., & Zhou, Z. (2022). Electrostatic atomization minimum quantity lubrication machining: From mechanism to application. *International Journal of Extreme Manufacturing*, *4*(4), 042003. DOI: 10.1088/2631-7990/ac9652

Xu, W., Zhang, L. C., & Wu, Y. (2014). Elliptic vibration-assisted cutting of fibre-reinforced polymer composites: Understanding the material removal mechanisms. *Composites Science and Technology*, *92*, 103–111. DOI: 10.1016/j.compscitech.2013.12.011

Yamamoto, T., Noda, S., & Kato, M. (2011). A simple and fast method to disperse long single-walled carbon nanotubes introducing few defects. *Carbon*, *49*(10), 3179–3183. DOI: 10.1016/j.carbon.2011.03.040

Yang, M., & Li, C. (Eds.). (2025). *Heat Transfer Mechanism and Temperature Field Model of MQL Micro-Grinding Biomedical Materials*. IGI Global, 2-97.

Yang, H., Zhao, G., Xia, H., Xin, L., Nian, Z., & Li, L. (2024). Effect of fiber orientation on material removal mechanisms in the machining of 3D Cf/SiC composite with a brazed diamond grinding rod: Experiment and simulation. *Journal of Manufacturing Processes*, *114*, 18–38. DOI: 10.1016/j.jmapro.2024.01.066

Yang, M., Hao, J., Wu, W., Li, Z., Ma, Y., Zhou, Z., Gao, T., Liu, M. Z., Cui, X., Zhang, Y. B., Li, B. K., Ma, X., Dambatta, Y. S., & Li, C. (2024a). Critical cutting thickness model considering subsurface damage of zirconia grinding and friction–wear performance evaluation applied in simulated oral environment. *Tribology International*, *198*, 109881. DOI: 10.1016/j.triboint.2024.109881

Yang, M., Kong, M., Li, C., Long, Y., Zhang, Y., Sharma, S., Li, R., Gao, T., Liu, M., Cui, X., Wang, X., Ma, X., & Yang, Y. (2023a). Temperature field model in surface grinding: A comparative assessment. *International Journal of Extreme Manufacturing*, *5*(4), 042011. DOI: 10.1088/2631-7990/acf4d4

Yang, M., Li, C., Luo, L., Li, R., & Long, Y. (2021c). Predictive model of convective heat transfer coefficient in bone micro-grinding using nanofluid aerosol cooling. *International Communications in Heat and Mass Transfer*, *125*, 105317. DOI: 10.1016/j.icheatmasstransfer.2021.105317

Yang, M., Li, C., Said, Z., Zhang, Y., Li, R., Debnath, S., Ali, H. M., Gao, T., & Long, Y. (2021b). Semiempirical heat flux model of hard-brittle bone material in ductile microgrinding. *Journal of Manufacturing Processes*, *71*, 501–514. DOI: 10.1016/j.jmapro.2021.09.053

Yang, M., Li, C., Zhang, Y., Jia, D., Li, R., Hou, Y., & Cao, H. (2019a). Effect of friction coefficient on chip thickness models in ductile-regime grinding of zirconia ceramics. *International Journal of Advanced Manufacturing Technology*, *102*(5), 2617–2632. DOI: 10.1007/s00170-019-03367-0

Yang, M., Li, C., Zhang, Y., Jia, D., Li, R., Hou, Y., Cao, H., & Wang, J. (2019b). Predictive model for minimum chip thickness and size effect in single diamond grain grinding of zirconia ceramics under different lubricating conditions. *Ceramics International*, *45*(12), 14908–14920. DOI: 10.1016/j.ceramint.2019.04.226

Yang, M., Li, C., Zhang, Y., Wang, Y., Li, B., Jia, D., Hou, Y., & Li, R. (2017). Research on microscale skull grinding temperature field under different cooling conditions. *Applied Thermal Engineering*, *126*, 525–537. DOI: 10.1016/j.applthermaleng.2017.07.183

Yang, M., Ma, H., Hao, J., Li, Z., Li, R., Zhou, Z., Gao, T., Liu, M., Cui, X., Wang, X., Zhang, Y., Dambatta, Y., Long, Y., & Li, C. (2024). Droplet size distribution model of needle electrode electrostatic atomization and milling nickel-based alloy performance evaluation. *Journal of Manufacturing Processes*, *119*, 682–698. DOI: 10.1016/j.jmapro.2024.04.011

Yang, M., Ma, H., Li, Z., Hao, J., Liu, M., Cui, X., Zhang, Y., Zhou, Z., Long, Y., & Li, C. (2024b). Force model in electrostatic atomization minimum quantity lubrication milling GH4169 and performance evaluation. *Frontiers of Mechanical Engineering*, *19*(4), 28. DOI: 10.1007/s11465-024-0800-8

Yang, Y., Qu, S., & Gong, Y. (2021a). Investigating the grinding performance of unidirectional and 2.5 DC/SiCs. *Ceramics International*, *47*(4), 5123–5132. DOI: 10.1016/j.ceramint.2020.10.090

Yang, Y., Yang, M., Li, C., Li, R., Said, Z., Ali, H. M., & Sharma, S. (2023b). Machinability of ultrasonic vibration-assisted micro-grinding in biological bone using nanolubricant. *Frontiers of Mechanical Engineering*, *18*(1), 1. DOI: 10.1007/s11465-022-0717-z

Yang, Z., Zhu, L., Zhang, G., Ni, C., & Lin, B. (2020). Review of ultrasonic vibration-assisted machining in advanced materials. *International Journal of Machine Tools & Manufacture*, *156*, 103594. DOI: 10.1016/j.ijmachtools.2020.103594

Yan, L., Zhang, Q., & Yu, J. (2018). Effects of continuous minimum quantity lubrication with ultrasonic vibration in turning of titanium alloy. *International Journal of Advanced Manufacturing Technology*, *98*(1-4), 827–837. DOI: 10.1007/s00170-018-2323-0

Yan, X., Zhao, H., Wen, Z., Wang, Y., Li, P., & Ding, H. (2024). Investigation on grinding temperature characteristics of CF/PEEK and material removal mechanism under temperature treatments. *Composites Science and Technology*, *252*, 110621. DOI: 10.1016/j.compscitech.2024.110621

Yan, Y., Zhao, B., & Liu, J. (2009). Ultraprecision surface finishing of nano-ZrO2 ceramics using two-dimensional ultrasonic assisted grinding. *International Journal of Advanced Manufacturing Technology*, *43*(5-6), 462–467. DOI: 10.1007/s00170-008-1732-x

Yao, L., Liu, Z., Song, Q., Wang, B., & Cai, Y. (2023b). Prediction modelling of cutting force in rotary ultrasonic end grinding 2.5 D woven SiO2f/SiO2 ceramic matrix composite. *Composite Structures*, *304*, 116448. DOI: 10.1016/j.compstruct.2022.116448

Yao, L., Liu, Z., Song, Q., Wang, B., Cai, Y., & Zhao, J. (2023a). Numerical prediction of surface morphology and roughness in rotary ultrasonic face grinding SiO2f/SiO2 composite. *Journal of Materials Research and Technology*, *25*, 5917–5937. DOI: 10.1016/j.jmrt.2023.07.044

Yashiro, T., Ogawa, T., & Sasahara, H. (2013). Temperature measurement of cutting tool and machined surface layer in milling of CFRP. *International Journal of Machine Tools & Manufacture*, *70*, 63–69. DOI: 10.1016/j.ijmachtools.2013.03.009

Yin, J., Li, M., Xu, J., Ding, W., & Su, H. (2022). Edge chipping characteristics in grinding SiCf/SiC composite. *Ceramics International*, *48*(5), 7126–7135. DOI: 10.1016/j.ceramint.2021.11.272

Yin, J., Xu, J., Ding, W., & Su, H. (2021). Effects of grinding speed on the material removal mechanism in single grain grinding of SiCf/SiC ceramic matrix composite. *Ceramics International*, *47*(9), 12795–12802. DOI: 10.1016/j.ceramint.2021.01.140

Yuan, J., Lyu, B., Hang, W., & Deng, Q. (2017a). Review on the progress of ultra-precision machining technologies. *Frontiers of Mechanical Engineering*, *12*(2), 158–180. DOI: 10.1007/s11465-017-0455-9

Yuan, Q., Huang, X., & Zhao, Y. P. (2014). Dynamic spreading on pillar-arrayed surfaces: Viscous resistance versus molecular friction. *Physics of Fluids*, *26*(9), 092104. DOI: 10.1063/1.4895497

Yuan, Q., & Zhao, Y. P. (2013). Wetting on flexible hydrophilic pillar-arrays. *Scientific Reports*, *3*(1), 1944. DOI: 10.1038/srep01944 PMID: 23736041

Yuan, S., Fan, H., Amin, M., Zhang, C., & Guo, M. (2016). A cutting force prediction dynamic model for side milling of ceramic matrix composites C/SiC based on rotary ultrasonic machining. *International Journal of Advanced Manufacturing Technology*, *86*(1), 37–48. DOI: 10.1007/s00170-015-8099-6

Yuan, S., Zhu, G., & Zhang, C. (2017b). Modeling of tool blockage condition in cutting tool design for rotary ultrasonic machining of composites. *International Journal of Advanced Manufacturing Technology*, *91*(5), 2645–2654. DOI: 10.1007/s00170-016-9911-7

Yu, H., Xu, Y., Shi, P., Wang, H., Wei, M., Zhao, K., & Xu, B. (2013). Microstructure, mechanical properties and tribological behavior of tribofilm generated from natural serpentine mineral powders as lubricant additive. *Wear*, *297*(1-2), 802–810. DOI: 10.1016/j.wear.2012.10.013

Zan, Z., Guo, K., Sun, J., Lu, Y., Yang, B., Jia, X., Sivalingam, V., & Xi, J. (2022). Investigation on scratching force and material removal mechanism of 3D SiCf/C–SiC composites during single grain scratching. *Journal of the European Ceramic Society*, *42*(13), 5366–5379. DOI: 10.1016/j.jeurceramsoc.2022.05.079

Zhang B, LU S X, Rabiey M, Axinte, D., & Bleicher, F. (2023). Grinding of composite materials. *CIRP Annals-Manufacturing Technology, 72*(2), 645-671.

Zhang, Y., & Li, C. (2025). *Hybrid-Energy Sustainable Machining: Mechanism and Processability*. Springer Nature, 458-459.

Zhang, Y., & Li, C. (2025). *Hybrid-Energy Sustainable Machining: Mechanism and Processability*. Springer Nature, 596-609.

Zhang, C., Liu, K., Cepero-Mejias, F., Curiel-Sosa, J. L., & Mao, C. (2024a). Numerical investigation on orthogonal cutting and damage response of CFRP/Ti6Al4V stacks. *Journal of Thermoplastic Composite Materials*, *37*(11), 3492–3515. DOI: 10.1177/08927057241234877

Zhang, C., Yuan, S., Amin, M., Fan, H., & Liu, Q. (2016b). Development of a cutting force prediction model based on brittle fracture for C/SiC in rotary ultrasonic facing milling. *International Journal of Advanced Manufacturing Technology*, *85*(1), 573–583. DOI: 10.1007/s00170-015-7894-4

Zhang, H., Zhang, X., Yi, X., He, F., Niu, F., & Hao, P. (2020). Dynamic behaviors of droplets impacting on ultrasonically vibrating surfaces. *Experimental Thermal and Fluid Science*, *112*, 110019. DOI: 10.1016/j.expthermflusci.2019.110019

Zhang, H., Zhao, Z., Li, J., Ye, L., & Liu, Y. (2024). Review on abrasive machining technology of SiC ceramic composites. *Micromachines*, *15*(1), 106. DOI: 10.3390/mi15010106 PMID: 38258225

Zhang, J., Zhao, Y., Tian, F., Zhang, S., & Guo, L. (2015a). Kinematics and experimental study on ultrasonic vibration-assisted micro end grinding of silica glass. *International Journal of Advanced Manufacturing Technology*, *78*(9-12), 1893–1904. DOI: 10.1007/s00170-014-6761-z

Zhang, L. C., Zhang, H. J., & Wang, X. M. (2001). A force prediction model for cutting unidirectional fibre–reinforced plastics. *Machining Science and Technology*, *5*(3), 293–305. DOI: 10.1081/MST-100108616

Zhang, L., Ren, C., Ji, C., Wang, Z., & Chen, G. (2016a). Effect of fiber orientations on surface grinding process of unidirectional C/SiC composites. *Applied Surface Science*, *366*, 424–431. DOI: 10.1016/j.apsusc.2016.01.142

Zhang, L., Wang, S., Li, Z., Qiao, W., Wang, Y., & Wang, T. (2019). Influence factors on grinding force in surface grinding of unidirectional C/SiC composites. *Applied Composite Materials*, *26*(3), 1073–1085. DOI: 10.1007/s10443-019-09767-5

Zhang, L., Wu, Z., Wu, C., & Wu, Q. (2022a). On the numerical modelling of composite machining. *Composites. Part B, Engineering*, *241*, 110023. DOI: 10.1016/j.compositesb.2022.110023

Zhang, M., Shan, C., Xia, Z., Jia, F., & Luo, M. (2024b). Dynamic mechanical model in grinding C/SiC composites. *International Journal of Mechanical Sciences*, *268*, 109042. DOI: 10.1016/j.ijmecsci.2024.109042

Zhang, M., Xia, Z., Shan, C., & Luo, M. (2023). Analytical model of grinding force for ultrasonic-assisted grinding of Cf/SiC composites. *International Journal of Advanced Manufacturing Technology*, *126*(5), 2037–2052. DOI: 10.1007/s00170-023-11257-9

Zhang, S., Li, J. F., & Wang, Y. W. (2012). Tool life and cutting forces in end milling Inconel 718 under dry and minimum quantity cooling lubrication cutting conditions. *Journal of Cleaner Production*, *32*, 81–87. DOI: 10.1016/j.jclepro.2012.03.014

Zhang, X., Li, C., Zhang, Y., Wang, Y., Li, B., Yang, M., Guo, S., Liu, G., & Zhang, N. (2017b). Lubricating property of MQL grinding of Al2O3/SiC mixed nanofluid with different particle sizes and microtopography analysis by cross-correlation. *Precision Engineering*, *47*, 532–545. DOI: 10.1016/j.precisioneng.2016.09.016

Zhang, X., Li, C., Zhou, Z., Liu, B., Zhang, Y., Yang, M., Gao, T., Liu, M., Zhang, N., Said, Z., Sharma, S., & Ali, H. M. (2023). Vegetable oil-based nanolubricants in machining: From physicochemical properties to application. *Chinese Journal of Mechanical Engineering*, *36*(1), 76. DOI: 10.1186/s10033-023-00895-5

Zhang, X., Sun, G., Wang, C., Li, H., & Zhou, S. (2025). A review of structural topology optimization for fiber-reinforced composites. *Composites. Part B, Engineering*, *299*, 112393. DOI: 10.1016/j.compositesb.2025.112393

Zhang, X., Yu, T., & Zhao, J. (2020). An analytical approach on stochastic model for cutting force prediction in milling ceramic matrix composites. *International Journal of Mechanical Sciences*, *168*, 105314. DOI: 10.1016/j.ijmecsci.2019.105314

Zhang, Y. B., Li, L. Y., Cui, X., An, Q. L., Xu, P. M., Wang, W., Jia, D. Z., Liu, M. Z., Dambatta, Y. S., & Li, C. H. (2025). Lubricant activity enhanced technologies for sustainable machining: Mechanisms and processability. *Chinese Journal of Aeronautics*, *38*(6), 103203. DOI: 10.1016/j.cja.2024.08.034

Zhang, Y., Hu, Z., Chen, Y., Yu, Y., Jin, J., Peng, Q., & Xu, X. (2024). Insights into scratching force in axial ultrasonic vibration-assisted single grain scratching. *Journal of Manufacturing Processes, 112*, 150–160. DOI: 10.1016/j.jmapro.2024.01.005

Zhang, Y., Li, C., Jia, D., Li, B., Wang, Y., Yang, M., Hou, Y., & Zhang, X. (2016c). Experimental study on the effect of nanoparticle concentration on the lubricating property of nanofluids for MQL grinding of Ni-based alloy. *Journal of Materials Processing Technology, 232*, 100–115. DOI: 10.1016/j.jmatprotec.2016.01.031

Zhang, Y., Li, C., Jia, D., Zhang, D., & Zhang, X. (2015). Experimental evaluation of the lubrication performance of MoS2/CNT nanofluid for minimal quantity lubrication in Ni-based alloy grinding. *International Journal of Machine Tools & Manufacture, 99*, 19–33. DOI: 10.1016/j.ijmachtools.2015.09.003

Zhang, Y., Li, C., Jia, D., Zhang, D., & Zhang, X. (2015b). Experimental evaluation of MoS_2 nanoparticles in jet MQL grinding with different types of vegetable oil as base oil. *Journal of Cleaner Production, 87*, 930–940. DOI: 10.1016/j.jclepro.2014.10.027

Zhang, Y., Li, C., Ji, H., Yang, X., Yang, M., Jia, D., Zhang, X., Li, R., & Wang, J. (2017). Analysis of grinding mechanics and improved predictive force model based on material-removal and plastic-stacking mechanisms. *International Journal of Machine Tools & Manufacture, 122*, 81–97. DOI: 10.1016/j.ijmachtools.2017.06.002

Zhang, Y., Li, H. N., Li, C., Huang, C., Ali, H. M., Xu, X., Mao, C., Ding, W., Cui, X., Yang, M., Yu, T., Jamil, M., Gupta, M. K., Jia, D., & Said, Z. (2022d). Nano-enhanced biolubricant in sustainable manufacturing: From processability to mechanisms. *Friction, 10*(6), 803–841. DOI: 10.1007/s40544-021-0536-y

Zhang, Z. Q., Jiao, F., Li, Y. X., Wang, X., Niu, Y., & Tong, J. L. (2024). Experimental study on rotary longitudinal-torsional ultrasonic machining of unidirectional CFRP. *Chinese Journal of Aeronautics, 37*(11), 517–534. DOI: 10.1016/j.cja.2024.02.012

Zhang, Z., Sui, M., Li, C., Zhou, Z., Liu, B., Chen, Y., Said, Z., Debnath, S., & Sharma, S. (2022e). Residual stress of grinding cemented carbide using MoS2 nano-lubricant. *International Journal of Advanced Manufacturing Technology, 119*(9), 5671–5685. DOI: 10.1007/s00170-022-08660-z

Zhang, Z., Yuan, S., Gao, X., Xu, W., Zhang, J., & An, W. (2022b). Analytical modelling of side grinding of orthogonal laminated SiCf/SiC composites based on effective elastic properties. *International Journal of Advanced Manufacturing Technology, 120*(9), 6419–6434. DOI: 10.1007/s00170-022-09170-8

Zhang, Z., Yuan, S., Xu, W., Ouyang, X., Zhang, J., & An, W. (2022c). Damage behavior and removal mechanism of different yarn orientations 2.5 D SiCf/SiC composites under single-abrasive scratch test. *Ceramics International*, *48*(20), 30868–30883. DOI: 10.1016/j.ceramint.2022.07.041

Zhao, B., Huang, Q., Cao, Y., Ding, W., Xu, J., & Liu, G. (2023). Thermal analysis of ultrasonic vibration-assisted grinding with moment-triangle heat sources. *International Journal of Heat and Mass Transfer*, *216*, 124552. DOI: 10.1016/j.ijheatmasstransfer.2023.124552

Zhao, B., Wu, B., Yue, Y., Ding, W., Xu, J., & Guo, G. (2023). Developing a novel radial ultrasonic vibration-assisted grinding device and evaluating its performance in machining PTMCs. *Chinese Journal of Aeronautics*, *36*(7), 244–256. DOI: 10.1016/j.cja.2023.01.008

Zhou, Y., Tian, C., Jia, S., Ma, L., Yin, G., & Gong, Y. (2023). Study on grinding force of two-dimensional ultrasonic vibration grinding 2.5 DC/SiC composite material. *Crystals*, *13*(1), 151. DOI: 10.3390/cryst13010151

Zhu, F., Zhang, H., Hu, K., Duan, L., Zhang, K., & Xu, X. (2025). A comparative investigation on surface quality of CFRP laminates between rotary drilling and rotary ultrasonic-assisted drilling. *Composite Structures*, *358*, 118964. DOI: 10.1016/j.compstruct.2025.118964

Zhu, G., Yuan, S., & Chen, B. (2019). Numerical and experimental optimizations of nozzle distance in minimum quantity lubrication (MQL) milling process. *International Journal of Advanced Manufacturing Technology*, *101*(1), 565–578. DOI: 10.1007/s00170-018-2928-3

Zou, F., Dang, J., Wang, X., Zhang, H., Sun, X., An, Q., & Chen, M. (2021). Performance and mechanism evaluation during milling of CFRP laminates under cryogenic-based conditions. *Composite Structures*, *277*, 114578. DOI: 10.1016/j.compstruct.2021.114578

About the Authors

Teng Gao, born in 1993, holds a PhD in Engineering and is a postdoctoral fellow of Qingdao University of Technology (key funding category). He is a postdoctoral innovation talent in Shandong Province, a high-level talent in Qingdao City, a top 2% scientist in the world (for three consecutive years), and the Vice President of Science and Technology in Qingdao City. He is engaged in research in the field of intelligent and clean precision manufacturing. Hosted 6 scientific research/talent program projects, including the National Postdoctoral Researcher Program B (with only 19 students in the Mechanical Engineering discipline in China), Shandong Province Postdoctoral Innovation Talent Program (ranked second in school history), Shandong Province Natural Science Youth Fund, Shandong Province Science and Technology based Small and Medium sized Enterprise Innovation Capability Enhancement Project, etc. As an academic backbone, participated in 7 scientific research projects including the National Key R&D Program, Shandong Province Major Science and Technology Innovation Project, and National Natural Science Foundation. Published 64 high impact papers, including 36 SCI JCR Q1 papers and 25 hot/highly cited papers. Published 13 SCI/EI papers as the first author, including 7 papers from top journals in the 1st district of the Chinese Academy of Sciences, 2 papers from domestic T1 level journals such as the Journal of Mechanical Engineering (1 of which was selected as a high impact paper), 6 ESI highly cited papers, 4 hot papers, 1 SCI journal cover paper, and 2 papers from the 1st district of the Chinese Academy of Sciences that ranked first in citations throughout the year. Obtained 17 intellectual property rights as the first and second inventors, including 9 authorized invention patents from China, the United States, and other domestic and foreign countries, and formulated 3 standards. Received 10 scientific and technological awards, including the Second Prize of Natural Science in Qingdao and the Outstanding Doctoral Dissertation Award in Shandong Province as the first author, and the China Patent Excellence Award as the main author. Served as a column editor for Frontiers of Mechanical Engineering journal (JCR-1 zone, CAS 2 zone) and guest editor for the album "China Surface Engineering". serve as Journal of Manufacturing Processes, Tribology International, Composites Communications, Composites Part A, Materials & Design Expert reviewers for well-known SCI journals in the field of engineering.

Li Changhe is an academician of the Russian Academy of Engineering, a highly cited scientist in the world, a highly cited scholar in China, a top 2% global scientist with lifelong scientific influence of Stanford University, a distinguished expert of Mount Taishan scholar in Shandong Province, a member of the Abrasive Processing Committee of the Chinese Mechanical Engineering Society, a member of the Finishing Committee of the Chinese Mechanical Engineering Society, the founding editor of Intelligent and Sustainable Manufacturing, and the deputy editor of the Chinese Journal of Mechanical Engineering. Outstanding Inventor, Outstanding Science and Technology Worker, Teaching Master, and Outstanding Graduate Guidance Teacher in Shandong Province. Published 257 SCI/EI indexed papers, including 81 SCI/JCR (1-2) papers, 63 ESI highly cited papers, and 20 hot papers, with a total of more than 14000 citations (Web of Science). Ranked 25th on the 2022 China Scholar Academic Influence Ranking (only 70 in the field of mechanical engineering nationwide), and 10th on the 2023 China Scholar Academic Influence Ranking (only 43 in the field of mechanical engineering nationwide). Obtained 69 invention patents in the United States, South Korea, Australia, and other countries, 57 PCT international patents, 137 national invention patents, and 59 software copyrights. Professor Li Changhe is an academician of the Russian Academy of Engineering, a highly cited scientist in the world, a highly cited scholar in China, a top 2% global scientist with lifelong scientific influence of Stanford University, a distinguished expert of Mount Taishan scholar in Shandong Province, a member of the Abrasive Processing Committee of the Chinese Mechanical Engineering Society, a member of the Finishing Committee of the Chinese Mechanical Engineering Society, the founding editor of Intelligent and Sustainable Manufacturing, and the deputy editor of the Chinese Journal of Mechanical Engineering. Outstanding Inventor, Outstanding Science and Technology Worker, Teaching Master, and Outstanding Graduate Guidance Teacher in Shandong Province. Published 257 SCI/EI indexed papers, including 81 SCI/JCR (1-2) papers, 63 ESI highly cited papers, and 20 hot papers, with a total of more than 14000 citations (Web of Science). Ranked 25th on the 2022 China Scholar Academic Influence Ranking (only 70 in the field of mechanical engineering nationwide), and 10th on the 2023 China Scholar Academic Influence Ranking (only 43 in the field of mechanical engineering nationwide). Obtained 69 invention patents in the United States, South Korea, Australia, and other countries, 57 PCT international patents, 137 national invention patents, and 59 software copyrights.

Xuesong Chang is a Postdoctoral fellow of Hisense Group and Shandong University, born in March 1988, graduated with a PhD in Chemical Engineering from East China University of Science and Technology, and currently holds the title of Professorate Senior Engineer. Has been recognized as a Taishan Industry Leading Talent of Shandong Province, Industrial Leading Talent of Qingdao, and Young Expert in Standardization of Shandong Province. Served as an expert committee member or committee member in several national standardization organizations related to plastics(TC15), environmental standardization of electrical and electronic products and systems(TC297), and antibacterial surface performance(TC621). Mainly engaged in research in green materials, green manufacturing, and health-related materials for home appliances. Hosted four scientific research projects supported by the Ministry of Ecology and Environment of China, the Ministry of Industry and Information Technology of China, and the Key Research and Development Program of Shandong Province. Currently, has published more than ten SCI/EI-indexed papers. Obtained 50 intellectual property rights, including 30 invention patents. As the first person to complete the project, has received several provincial and industry-level science and technology awards, including the Second Prize of the Guangdong Provincial Science and Technology Progress Award, the Second Prize of the Science and Technology Progress Award of the China National Light Industry Council, and the Second Prize in the Electrical and Electronics Special Competition of the Green Industry Track at the BRICS Industrial Innovation Contest.

Index

A

B

C

D

E

F

G

N

O

P

R

S

T

U

V

W

www.ingramcontent.com/pod-product-compliance
Lightning Source LLC
LaVergne TN
LVHW080910300926
849162LV00003B/633
* 9 7 9 8 3 3 7 3 9 0 1 3 0 *